NEW PERSPECTIVES IN THE STUDY OF MESOAMERICAN PRIMATES

Distribution, Ecology, Behavior, and Conservation

DEVELOPMENTS IN PRIMATOLOGY: PROGRESS AND PROSPECTS

Series Editor:
Russell H. Tuttle
University of Chicago, Chicago, Illinois

This peer-reviewed book series melds the facts of organic diversity with the continuity of the evolutionary process. The volumes in this series exemplify the diversity of theoretical perspectives and methodological approaches currently employed by primatologists and physical anthropologists. Specific coverage includes: primate behavior in natural habitats and captive settings; primate ecology and conservation; functional morphology and developmental biology of primates; primate systematics; genetic and phenotypic differences among living primates; and paleoprimatology.

ALL APES GREAT AND SMALL

Volume I: African Apes
Edited by Birute' M.F. Galdikas, Nancy Erickson Briggs, Lori K. Sheeran, Gary L. Shapiro and Jane Goodall

THE GUENONS: DIVERSITY AND ADAPTATION IN AFRICAN MONKEYS

Edited by mary E. Glenn and Marina Cords

ANIMAL MINDS, HUMAN BODIES

By W. A. Hillix and Duane Rumbaugh

COMPARATIVE VERTEBRATE COGNITION

Edited by Lesley J. Rogers and Gisela Kaplan

ANTHROPOID ORIGINS: NEW VISIONS

Edited by Callum F. Ross and Richard F. Kay

MODERN MORPHOMETRICS IN PHYSICAL ANTHROPOLOGY

Edited by Dennis E. Slice

BEHAVIORAL FLEXIBILITY IN PRIMATES: CAUSES AND CONSEQUENCES

By Clara B. Jones

NURSERY REARING OF NONHUMAN PRIMATES IN THE 21ST CENTURY

Edited by Gene P. Sackett, Gerald C. Ruppenthal and Kate Elias

NEW PERSPECTIVES IN THE STUDY OF MESOAMERICAN PRIMATES: DISTRIBUTION, ECOLOGY, BEHAVIOR, AND CONSERVATION

Edited by Alejandro Estrada, Paul A. Garber, Mary S. M. Pavelka, and LeAndra Luecke

NEW PERSPECTIVES IN THE STUDY OF MESOAMERICAN PRIMATES

Distribution, Ecology, Behavior, and Conservation

Edited by

Alejandro Estrada
Universidad Nacional Autónoma de México
Mexico City, Mexico

Paul A. Garber
University of Illinois
Urbana, IL, USA

Mary S. M. Pavelka
University of Calgary
Calgary, Alberta, Canada

LeAndra Luecke
Washington University
St. Louis, MO USA

Alejandro Estrada
Estación de Biología Los Tuxtlas, Instituto de Biología
UNAM Apdo. Pos. 176
San Andrés Tuxtla
Veracruz, 95700
Mexico
aestrada@primatesmx.com

Paul A. Garber
Department of Anthropology
University of Illinois
Urbana, IL 61801
USA
p-garber@uiuc.edu

Mary S. M. Pavelka
Department of Anthropology
University of Calgary
Calgary, AB T2N 1N4
Canada
pavelka@ucalgary.ca

LeAndra Luecke
Department of Anthropology
Washington University
St. Louis, MO 63130
USA
lgluecke@artsci.wustl.edu

Library of Congress Control Number: 2005925190

ISBN-10: 0-387-25854-X e-ISBN 0-387-25872-8
ISBN-13: 978-0387-25854-6

Printed on acid-free paper.

Printed in the United States of America. (TB/MVY)

9 8 7 6 5 4 3 2 1

springeronline.com

For Alex, Erika, and Ximena.
Alejandro Estrada

For Sara, Jenni, Seymour, and Sylvia.
Paul A. Garber

For Drew and Cameron, and my mother, Sari Lynn.
LeAndra Luecke

ACKNOWLEDGMENTS

We are grateful to each contributing author for their efforts and enthusiasm in translating ideas, data, theories and perspectives into a single volume. We also are grateful to the American Society of Primatologists. The ASP played an important role in supporting the symposium from which the book project originated, at their annual meetings at the University of Wisconsin-Madison in June of 2004. We also acknowledge the scholarship of participants at the ASP symposium on Mesoamerican primates, some of whom, although not contributors to the volume, nonetheless provided important intellectual insights. We acknowledge the efforts of several graduate students who participated as coauthors in some of the chapters in this book. The quality of their contributions attests to their high level of scientific investigation and to their enthusiasm in conducting field research in the spirit of great primatologists such as C. R. Carpenter and Ch. Southwick. As always, PAG thanks Sara and Jenni for their love, and for just being Sara and Jenni. AE is grateful to Dr. Karen Strier of the Department of Anthropology, University of Wisconsin-Madison, for hosting the sabbatical year during which this book was developed. AE would also like to acknowledge the economic support of the National Autonomous University of Mexico and of the Scott Neotropic Fund in the development of the ASP symposium and book project. Karla Harmon provided help in printing out the final drafts of each chapter. We are grateful to Melanie Luinstra for assistance in formatting the final drafts of all manuscripts.

CONTENTS

LIST OF CONTRIBUTORS

Shelley M. Alexander, Department of Geography, University of Calgary, Alberta, Canada T2N 1N4. Email: smalexan@ucalgary.ca

Victor Arroyo, División de Posgrado, Instituto de Ecología A. C., km 2.5, Carretera Antigua a Coatepec Nro. 351, Congregación del Haya, Xalapa, Veracruz 91070, México. Email: victorarroyo_rodriguez@hotmail.com

Michelle Bezanson, Department of Anthropology, Northern Arizona University, Campus Box 15200, Flagstaff, AZ 86011-5200, USA. Email: michelle.bezanson@nau.edu

Ellen Brown, Yale University School of Forstry and Environmental Studies, M.E.M., New Haven, CT 06511, USA. Email: ellen.l.brown@yale.edu

Nicola H. Bywater, Department of Geography, University of Calgary, Alberta, Canada T2N 1N4. Email: nhbywater@yahoo.com

Sophie Calmé, Colegio de la Frontera Sur, Avenida Centenario, km 5.5, Chetumal, Quintana Roo 77900, México. Email: scalme@ecosur-qroo.mx

Sarah D. Carnegie, Department of Anthropology, University of Calgary, 2500 University Drive NW, Calgary, Alberta, Canada T2N 1N4. Email: sdcarneg@ucalgary.ca

Colin A. Chapman, Anthropology Department and McGill School of Environment, 855 Sherbrooke St. West, McGill University, Montreal, Canada H3A 2T7. Email: colin.chapman@mcgill.ca

Liliana Cortés-Ortiz, Centro de Investigaciones Tropicales, Universidad Veracruzana, Xalapa, Veracruz 91070, México. Email: citro@uv.mx

Holly Noelle DeGama-Blanchet, Department of Anthropology, University of Calgary, Calgary, AB, Canada T2N 1N4. Email: Holly.DeGammaBlanchet@natureconservancy.ca

Luis A. Escobedo-Morales, División de Posgrado, Instituto de Ecología A. C., km 2.5, Carretera Antigua a Coatepec Nro. 351, Congregación del Haya, Xalapa, Veracruz 91070, México. Email: escobedo@ecologia.edu.mx

Alejandro Estrada, Estacion de Biología Los Tuxtlas, Instituto de Biología, Universidad Nacional Autónoma de México, Apartado Postal 176, San Andrés Tuxtla, Veracruz, México. Email: aestrada@primatesmx.com

Linda Marie Fedigan, Department of Anthropology, University of Calgary, Calgary, Alberta, Canada T2N 1N4. Email: fedigan@ucalgary.ca

Susan M. Ford, Department of Anthropology and Center for Systematic Biology, Southern Illinois University, Carbondale, IL 62901-4502, USA. Email: sford@siu.edu

Paul A. Garber, Department of Anthropology, University of Illinois at Urbana-Champaign, 109 Davenport Hall, Urbana, IL 61801, USA. Email: p-garber@uiuc.edu

Kenneth E. Glander, Biological Anthropology and Anatomy, Duke University, Durham, NC 27708, USA. Email: glander@duke.edu

Ana Marie González Di Pierro, Centro de Investigaciones en Ecosistemas, Universidad Nacional Autónoma de México, Apartado Postal 27-3, Morelia, Michoacán 58089, México. Email: agdipierro@yahoo.com.mx

Colin P. Groves, School of Archaeology and Anthropology, Australian National University, Canberra, A.C.T. 0200, Australia. Email:colin.groves@anu.edu.au

Celia Harvey, Departmento de Agricultura y Agroforestería, Centro Agronómico Tropical de Investigación y Enseñanza (CATIE), Turrialba 7170, Costa Rica. Email: charvey@catie.ac.cr

Justin J. H. Hines, School of Archaeology and Anthropology, Australian National University, Canberra A.C.T. 0200, Australia. Email:justin.hines@anu.edu.au

Mariah E. Hopkins, Department of Environmental Science, Policy & Management, Division Insect Biology, University of California, Berkeley, CA 94720, USA. Email: mhopkins@nature.berkeley.edu

Katharine M. Jack, Department of Anthropology, Tulane University, 1021 Audubon Street, New Orleans, LA 70118. Email: kjack@tulane.edu

Petra E. Jelinek, Department of Anthropology, University of Illinois at Urbana-Champaign, 109 Davenport Hall, Urbana, IL 61801, USA. Email: jelinek@uiuc.edu

Clara B. Jones, Department of Psychology, Fayetteville State University, Fayetteville, NC 28301-4298, USA. Emails: cjones@uncfsu.edu; theoretical-primatology@hotmail.com

LeAndra Luecke, Department of Anthropology, Washington University, St. Louis, MO 63130, USA. Email: lgluecke@artsci.wustl.edu

Katherine C. MacKinnon, Department of Sociology and Criminal Justice and Center For International Studies, Saint Louis University, 210 Fitzgerald Hall, 3500 Lindell Blvd., St. Louis, MO, USA. Email: mackinn@slu.edu

Salvador Mandujano, Departamento de Biodiversidad y Ecología Animal, Instituto de Ecología A. C., km 2.5, Carretera Antigua a Coatepec Nro. 351, Congregación del Haya, Xalapa, Veracruz 91070, México. Email: mandujan@ecologia.edu.mx

Katharine Milton, Department of Environmental Science, Policy & Management, Division Insect Biology, University of California, Berkeley, CA 94720, USA. Email: kmilton@berkeley.edu

Russell A. Mittermeier, Conservation International, 1919 M Street NW Suite 600, Washington, DC 20036, USA. Email: r.mittermieir@conservation.org

David Muñoz, Colegio de la Frontera Sur, San Cristóbal de las Casas, Chiapas, México. Email: aullador@primatesmx.com

Eduardo Naranjo, Colegio de la Frontera Sur, San Cristóbal de las Casas, Chiapas, México. Email: enaranjo@sclc.ecosur.mx

Rodolfo Palacios-Silva, División de Posgrado, Instituto de Ecología A. C., km 2.5, Carretera Antigua a Coatepec Nro. 351, Congregación del Haya, Xalapa, Veracruz 91070, México. Email: palacior@ecologia.edu.mx

Mary S. M. Pavelka, Department of Anthropology, University of Calgary, 2500 University Drive NW, Calgary, Alberta, Canada T2N 1N4. Email: pavelka@ucalgary.ca

Gabriel Ramos-Fernández, Pronatura Península de Yucatán A. C., Calle 32 Nro. 269, Colonia Pinzón II, Mérida, Yucatán 97207, México. Email: ramosfer@sas.upenn.edu

Víctor Rico-Gray, Departamento de Ecología Funcional, Instituto de Ecología, A. C., Apartado Postal 63, Xalapa, Veracruz 91070, México. Email: ricogray@ecologia.edu.mx

Andrómeda Rivera, Colegio de la Frontera Sur, Avenida Centenario km 5.5, Chetumal, Quintana Roo 77900, México. Email: andromeda@primatesmx.com

Erika M. Rodríguez-Toledo, División de Posgrado, Instituto de Ecología A. C., km 2.5, Carretera Antigua a Coatepec, No. 351, Congregacion del Haya, Xalapa 91070, Veracruz, México. Email: rodrigt@ecologia.edu.mx

Marleny Rosales-Meda, Escuela de Biología, Edificio T-10, Ciudad Universitaria, Zona 12, Ciudad de Guatemala, Guatemala. Email: marleny_rm@yahoo.com.mx

Anthony B. Rylands, Center for Applied Biodiversity Science, Conservation International, 1919 M Street NW, Suite 600, Washington, DC 20036, USA. Email: a.rylands@conservation.org

Joel Saenz, Programa Regional de Manejo de Vida Silvestre, Universidad Nacional Autónoma (UNA), Heredia, Costa Rica. Email: jsaenz@una.ac.cr

Juan Carlos Serio-Silva, Departamento de Biodiversidad y Ecología Animal, Instituto de Ecología, A. C., Apartado Postal 63, Xalapa, Veracruz 91070, México. Email: serioju@ecologia.edu.mx

Kathryn E. Stoner, Centro de Investigaciones en Ecosistemas, Universidad Nacional Autónoma de México, Apartado Postal 27-3, Morelia, Michoacán 58089, México. Email: kstoner@oikos.unam.mx

Sarie Van Belle, Department of Zoology, University of Wisconsin-Madison, 250 N Mills Street, Madison, WI 53706, USA. Email: sarievanbelle@primatesmx.com

CHAPTER ONE

Overview of the Mesoamerican Primate Fauna, Primate Studies, and Conservation Concerns

Alejandro Estrada, Paul A. Garber, Mary S. M. Pavelka, and LeAndra Luecke

INTRODUCTION

Mesoamerica's unique biological heritage comprises five southern states of Mexico (Chiapas, Tabasco, Yucatan, Campeche, and Quintana Roo) and the Central American countries of Guatemala, Belize, El Salvador, Honduras, Nicaragua, Costa Rica, and Panama. These countries encompass an area of approximately 750,000 km^2 (Figure 1). Mesoamerica's natural ecosystems range

Alejandro Estrada • Field Station Los Tuxtlas, Institute of Biology, National Autonomous University of Mexico, Apdo 176, San Andres Tuxtla, Veracruz, Mexico. **Paul A. Garber** • Department of Anthropology, University of Illinois, Urbana, Illinois, USA. **Mary S. M. Pavelka** • Department of Anthropology, University of Calgary, Calgary, Alberta, Canada. **LeAndra Luecke** • Department of Anthropology, Washington University, St Louis, Missouri, USA.

New Perspectives in the Study of Mesoamerican Primates: Distribution, Ecology, Behavior, and Conservation, edited by Alejandro Estrada, Paul A. Garber, Mary S. M. Pavelka, and LeAndra Luecke. Springer, New York, 2005.

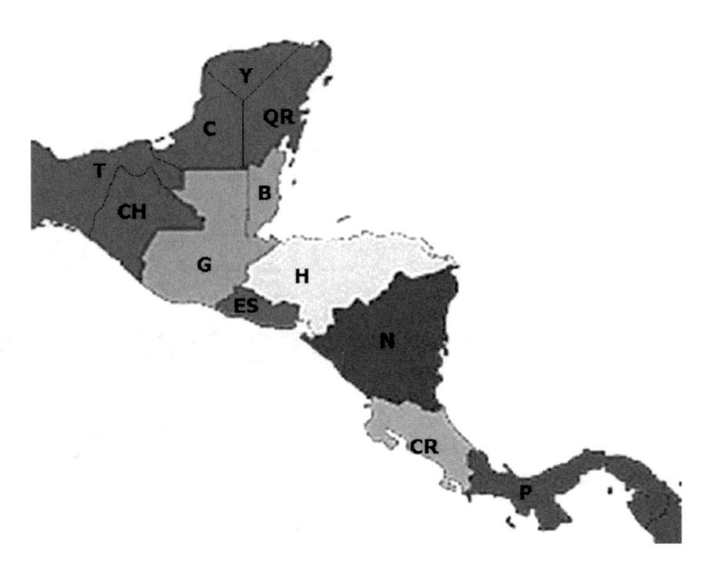

Figure 1. Mesoamerican countries: T, CH, C, Y, and QR are the southern states of Mexico that form part of the Mesoamerican region. G, Guatemala; B, Belize; ES, El Salvador; H, Honduras; N, Nicaragua; CR, Costa Rica; and P, Panama.

from coral reefs and lowland rainforests to pine savannas, semi-arid woodlands, grasslands, and high-mountain forests, constituting about 33 distinct "ecoregions" according to biogeographers (UNDP, 1999; UNEP, 2004). Because of its geographical location, it is a critical migration corridor for many wintering bird species and for some insect groups such as the monarch butterflies (Moore and Simons, 1992; Terborgh, 1989). The region contributes to less than 1.0% of the world's land surface, but it is home to a disproportionate share—about 7–10%—of the planet's biological diversity (Mittermeier *et al.*, 1998).

Mesoamerica is considered to be one of the world's most important centers of origin of genetic diversity and of domestication of important agricultural crops. Its indigenous peoples bred maize, squash, various beans, and chili peppers from wild species endemic to the region (FAO, 2001). The region holds the world's interest as well as a result of having been the cradle of one of the most ancient and sophisticated civilizations of mankind—the Maya (Gómez-Pompa *et al.*, 2003). Indigenous people still inhabit the region and they represent about 16% of the total population (*ca.* 45 million). Although the concentration of the population of indigenous groups is not uniform across Mesoamerican nations, they constitute an important demographic, social, and cultural component of each country in the region (Figure 2). Most of the indigenous people are found

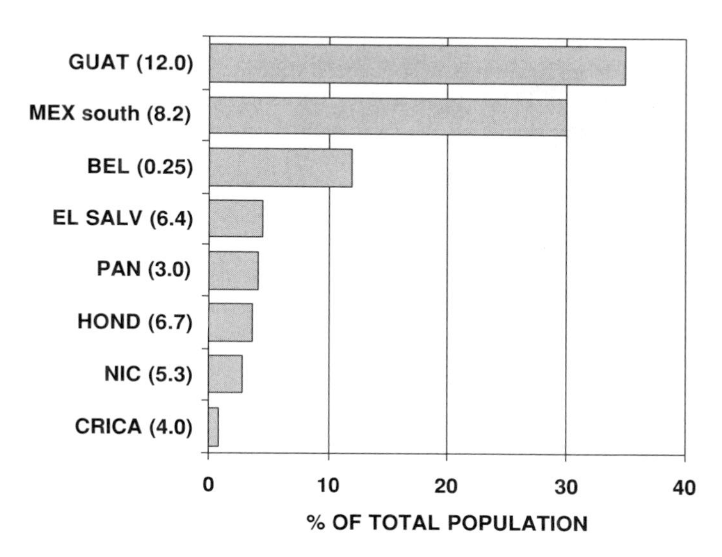

Figure 2. Indigenous population of Mesoamerica as percent of the total population in each country. Total population for each country as millions of people between parenthesis.

in rural and remote areas where tropical forests are still present. It is ironic that amidst such biological wealth in the region, poverty and marginalization are a predominant feature of its human inhabitants (UNEP, 2004). Conservation of areas of native vegetation in the region will need to consider sustainable use and social and economic equity as key elements of any conservation equation (Aguilar-Støen and Dhillon, 2003).

PRIMATE FAUNA

Mesoamerica harbors the northernmost representatives of the Primate Order in the American continent, and these, as the rest of the region's biological richness, are part of the natural and cultural patrimony of its nations. Primate diversity in Mesoamerica is represented by 6 genera (*Saguinus*, *Aotus*, *Alouatta*, *Ateles*, *Saimiri*, and *Cebus*) and some 21 species or subspecies (see Table 1 in Rylands *et al.*, chapter 2). Sixteen of these taxa are endemic to the region (Rodriguez-Luna *et al.*, 1996; Rowe, 1996; Nowak, 1999).

Mesoamerican primates contribute importantly to the richness of arboreal mammals in the tropical forests of the region accounting for approximately 61% of the 55 recognized taxa (15 rodent taxa, 13 marsupial taxa, 4 carnivore taxa, and 2 sloth taxa) (Reid, 1997; Emmons, 1990). Mesoamerican primates occur in various habitats from wet to dry forests to cloud forests to

Table 1. PrimatLit hits regarding documentation on Mesoamerican primates for the 1940–2004 period. Note low number concentration for countries such as Nicaragua, Guatemala, Honduras, and El Salvador, and for taxa such as *Saimiri*, *Saguinus*, and *Aotus*

	Alouatta	*Ateles*	*Cebus*	*Saimiri*	*Saguinus*	*Aotus*	Total
Mexico	302	166					468
Belize	113	31					144
Guatemala	24	25					49
El Salvador	12	16	11				39
Honduras	16	16	16				48
Nicaragua	46	20	15				81
Costa Rica	186	90	190	82			548
Panama	111	65	65	46	51	42	380
Total	810	429	297	128	51	42	1757

mangroves and occupy a wide variety of feeding niches ranging from part-time folivory to frugivory, omnivory, insectivory, gumnivory, or a combination of several of these categories (Rowe, 1996). All taxa are arboreal and, except for one (*Aotus*), diurnal. Social-group living is the rule, with *Aotus* living in pair-bonded family groups, *Ateles* forming fission–fusion communities (Rowe, 1996), *Saguinus* adopting a polyandrous/polygynous mating system associated with cooperative infant care (Garber, 1997), *Saimiri* living in large sex-segregated social groups characterized by males undergoing a fatting stage and female birth synchrony (Boinski, 1987, 1988), *Alouatta* living in small bisexual groups generally containing one or two adult males (*Alouatta pigra*) (see Van Belle and Estrada, chapter 4) or larger multimale–multifemale groups (*Alouatta palliata*) (Carpenter, 1934), and *Cebus capucinus* with a slow life history, living in multimale–multifemale groups with a single dominant male as the primary breeder (see Jack and Fedigan, chapter 13). Mesoamerican primates participate in important ways, as primary consumers, in the recycling of matter, nutrients, and energy in the tropical ecosystem (Estrada and Coates-Estrada, 1993) and play an important role in the natural process of forest regeneration via the dispersal of seeds of a broad spectrum of plants (Chapman, 1988; Estrada and Coates-Estrada, 1984; Garber, 1986).

PRIMATE STUDIES IN MESOAMERICA

Primate studies have a long and important history in Mesoamerica. They began with the investigation by Clarence Ray Carpenter of the behavior of the howler monkeys of Barro Colorado Island (BCI) in Panama. This was the first scientific study of primates in the wild, representing the birth of this phase of our discipline (Carpenter, 1934). A series of consecutive studies in the same site quickly added information on howler monkey behavior and ecology and later expanded to include the other primates (*Cebus*, *Ateles*, and *Aotus*) found or introduced on BCI (Altmann, 1959; Bernstein, 1964; Chivers, 1969; Collias and Southwick, 1952; Hladik and Hladik, 1969; Milton, 1980; Mittermeier, 1973; Moyniham, 1964, 1970; Oppenheimer, 1969; Richard 1970; Thorington *et al.*, 1976; Wagner, 1956).

This development paved the way for the expansion of primate studies in mainland Panama (Baldwin and Baldwin, 1972; Garber, 1980, 1981), Costa Rica (Boinski, 1985, 1986; Fedigan and Baxter, 1984; Glander 1975, 1978), new localities in Panama (Milton and Mittermeier, 1977), Tikal, Guatemala (Cant, 1978; Coelho *et al.*, 1976; Schlichte, 1978) and Los Tuxtlas, Mexico (Estrada, 1982). These efforts provided comparative studies on intraspecific variability in the behavior and ecology of *A. palliata* at different locations in Mesoamerica, and added new data on species such as *Saguinus geoffroyi, Saimiri oerstedii, Cebus capucinus, Ateles geoffroyi,* and *Alouatta*.

Since the 1970s, we witnessed the development of long-term monitoring of populations of *A. palliata* on BCI (Milton, 1977, 1980), in La Pacifica and Santa Rosa (Clarke, 1982, 2002; Fedigan, 1984, 2003; Glander, 1975, 1978, 1992, this volume), and in Los Tuxtlas (Estrada, 1982; Estrada and Coates-Estrada, 1996), and of *A. pigra* in Bermudian Landing, Belize (Bolin, 1981; Horwich, 1983). More recently, field projects have unfolded in the island of Ometepe, Nicaragua (Garber *et al.*, 1999), in Pico Bonito National Park, Honduras (Hines, pers. comm.), in El Salvador (Horwich, pers. comm.), in Belize (Horwich *et al.*, 2001; Ostro *et al.*, 2001; Pavelka *et al.*, 2003), in various sites in southern Mexico, and at Tikal and Lachúa in Guatemala (Estrada *et al.*, 2004). Notwithstanding these past and recent efforts, there are still large areas in Mesoamerica for which we lack essential information regarding the distribution, demography, and ecology of primate species, and the conservation status of the habitats and populations.

PUBLISHED INFORMATION ON MESOAMERICAN PRIMATES

How rich is our database on Mesoamerican primates? A search of the PrimateLit database for the period 1940–2004 resulted in 1757 citations encompassing reports on taxonomy, general biology, ecology, behavior, and conservation

Table 2. Forest cover changes in the Mesoamerican region. See Appendix 1 for sources of raw data. The southern states of Mexico under consideration are Tabasco, Chiapas, Campeche, Yucatan, and Quintana Roo. Countries ranked by current forest cover

	Total land (km^2)	Original forest cover (km^2)	Original Forest as % land area[a]	Forest in 2000 as % land area	Current Forest cover (km^2)	Forest cover change % 1990–2000
Mexico (South)	226,712	204,041	90	28	63,479	−1.10
Honduras	112,520	112,090	100	48	54,009	−1.03
Nicaragua	131,847	130,000	100	25	32,961	−3.01
Panama	75,536	73,254	97	38	28,703	−1.65
Guatemala	108,917	107,801	99	26	28,318	−1.71
Costa Rica	51,113	50,078	98	39	19,934	−0.77
Belize	22,965	21,123	92	59	13,549	−2.32
El Salvador	21,046	20,829	99	6	1,262	−4.60
Total (km^2)	750,656	743,018			242,219	
%		99			32	

[a] Original forest refers to estimated forest cover about 8000 years ago, assuming current climatic conditions (see http://earthtrends.wri.org/pdf_library/country_profiles/).

(Table 1). The frequency distribution of these citations is skewed (80%) toward reports of research conducted in Costa Rica, Mexico, and Panama, reflecting the greater longevity and intensity of primate research in these countries than in the rest of Mesoamerica (Table 1). When the PrimateLit hits are examined for each major taxa, it is evident that almost 87% of the reports provide information on *Alouatta*, *Ateles*, and *Cebus*, while documentation for the other three genera is particularly small (Table 1). This cursory review of the available scientific literature suggests that in spite of almost eight decades of primate research in Mesoamerica, there are critical gaps in the available information for several Mesoamerican countries and taxa that deserve immediate attention. This situation is particularly pressing considering the extensive and intensive loss of primate habitat in the region (see Table 2).

CONSERVATION PROBLEMS

The following paragraphs tackle aspects of rates of tropical rain forest loss, land-use patterns, human population growth trends, as well as regional conservation initiatives by Mesoamerican countries. Statistics reported here were taken from sources listed in Appendix 1. These were used as raw data or as transformed indices and variables to illustrate states, trends, and patterns of forest cover changes over time, land-use patterns, human population density and growth, indigenous population figures, human development indices, and the current system of natural protected areas in Mesoamerica.

It has been suggested that in Mesoamerica, habitat loss and fragmentation are major causes for the loss of plant and animal biodiversity and of local extinction of primates (Estrada and Coates-Estrada, 1996; UNEP, 2004). Further, such processes of land-use undermine the viability of remnant primate populations (Marsh, 2003). General deforestation rates in the region are exceedingly high, estimated at 44 ha/hr or 440,000 ha/yr (FAO, 2000; Sader *et al.*, 1999). Original forest cover (8000 years ago assuming current climatic conditions) for the countries in the region ranged from 90% to 100% (World Resource Institute, 2004). Using FAO statistics and those from the World Resources Institute, our estimates indicate that currently ~70% of the original forest cover present in the region has been lost as a result of human activity (Table 2). Deforestation rates for the period 1990–2000 are highest in El Salvador followed by Nicaragua, Belize, Guatemala, and Panama. Lower forest cover change rates for the period are found in Mexico, Honduras, and Costa Rica (Table 2) (FAO, 2004; World Resources Institute, 2004) (Figure 3).

Major proximate threats to forests in the region are agricultural activities. These are aimed at building up pasture lands for raising cattle and expanding agricultural land to raise food crops. Other threats are brought about by timber extraction, mining, colonization, and hydropower development

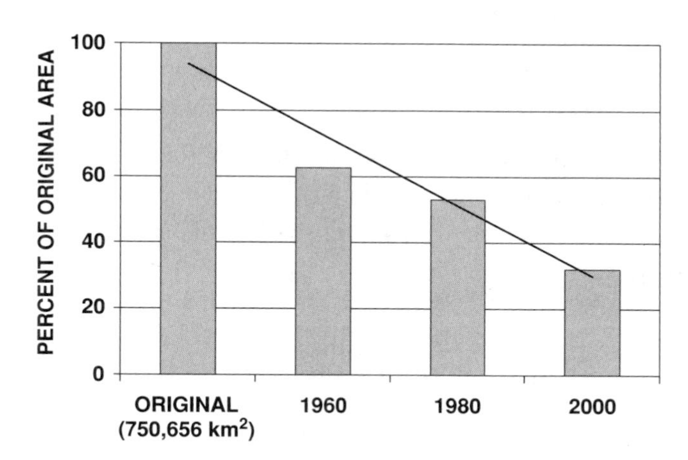

Figure 3. Percent of original forest remaining in Mesoamerica in three time periods. Original forest cover estimated as of 8000 years ago, assuming no climatic changes in the region (see World Resources Institute, 2004).

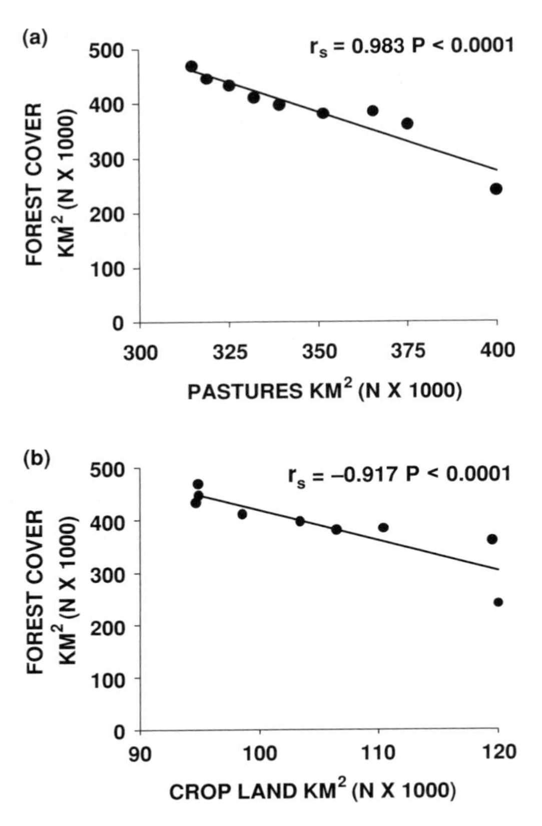

Figure 4. Forest cover as a function of expansion of pastures (a) and of cropland (b) in 5-year intervals for the period 1960–2000. The trend line is shown for illustrative purposes only.

(UNDP, 1999). FAO statistics for land-use patterns for the period 1960–2000, arranged in 5-year intervals, show a clear tendency for significant decreases in forest cover associated with increases in pasture and crop-lands (Figure 4). The changes in forest cover translate into extensive loss of primate habitats throughout Mesoamerica.

HUMAN POPULATION PRESSURES

Current human population in Mesoamerica is estimated to be about 45 million, with a growth rate of 3% since the 1950s. It is expected that the population will double in 20–35 years (FAO, 2000; UNDP, 1999; WWF, 2002; World Resource Institute (WRI), 2004). Demand for land and water and increases in food production to feed growing urban and rural populations are corollaries of population growth in Mesoamerica. Such pressures are enhanced by global market—demands placed upon Mesoamerican countries to produce meat and agricultural products. While population density is 64 people/km^2 (world's average 41.5 people/km^2; WRI, 2004) and is expected to be 128 people/km^2 by 2030, it varies across the countries in the region. El Salvador has the highest population density (296 people/km^2) followed by Guatemala (119 people/km^2) and Costa Rica (74 people/km^2). In the rest of the countries, except Belize where population density is the lowest (11 people/km^2), population density varies from 31 people/km^2 (Mexico) to 57 people/km^2 (Honduras). The percent of forest cover remaining in each country appears to be negatively related to human population density. For example, at one extreme, El Salvador has the lowest percent of forest cover in the region coupled with the highest population density (Figure 5). At the other extreme, Belize retains the highest percent of forest cover and the lowest human population density. Intermediate positions are occupied by Guatemala, Costa Rica, and Panama (Figure 5).

Population growth projections from FAO (2004) for the period 2004–2030 show that Guatemala, Honduras, and Nicaragua have the steepest population growth rates, followed by El Salvador (Figure 6). Ironically, the countries with the greatest population growth are also countries with large expanses of forest cover, high human poverty, and low human development indices (Table 3). These patterns serve to predict where the major pressures upon natural resources are likely to be found in the near future.

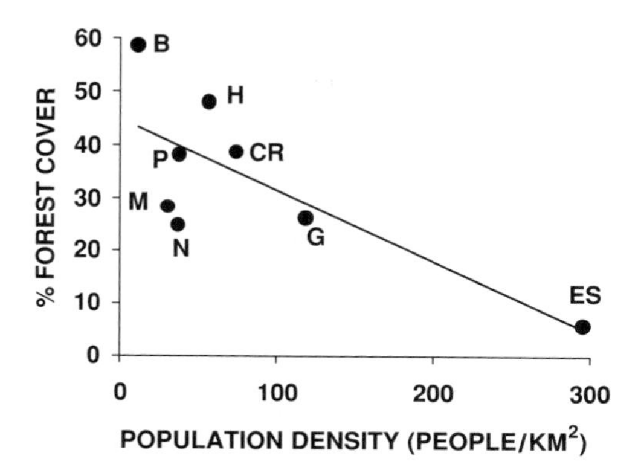

Figure 5. Current percent of forest cover as a function of total land and population density in each Mesoamerican country. The trend line is shown for illustrative purposes only. Codes for countries as in Figure 1, except for M = Mexico (south).

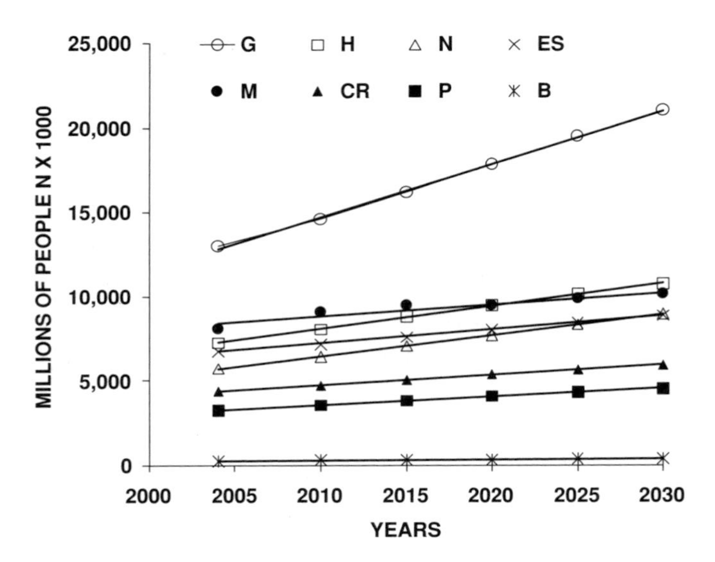

Figure 6. Population growth projection for each Mesoamerican country. Country codes as in Figure 5. Note the steep slopes for Guatemala ($y = 312.8$), Honduras ($y = 134.7$), Nicaragua ($y = 124.7$), El Salvador ($y = 91.6$). Less steep slopes in Mexico ($y = 69.1$), Costa Rica ($y = 59.9$), Panama ($y = 49.7$), and Belize ($y = 4.1$). Graph built with data from FAO (2004; http://faostat.fao.org/).

Table 3. Human population in Mesoamerica, including the indigenous component. Also shown are the Human Poverty Index and the Human Development Index for each country. Countries are ranked according to their HDI

Country	Total land (km^2)	Current forest cover (km^2)	Population	Annual population growth rate (%)	HPI[a]	HDI[b]	Rank HDI
Nicaragua	131,847	32,961	5,335,000	2.6	24.3	0.643	121
Guatemala	108,917	28,318	12,036,000	2.6	22.9	0.652	119
Honduras	112,520	54,009	6,781,000	2.5	19.9	0.667	115
El Salvador	21,046	1,262	6,415,000	1.7	17.2	0.719	105
Belize	22,965	13,549	251,000	2.4	8.8	0.776	67
Panama	75,536	28,703	3,064,000	1.5	7.8	0.788	59
Mexico (South)	226,712	63,479	8,000,000	1.4	8.8	0.800	55
Costa Rica	51,113	19,934	4,094,000	1.8	4.4	0.832	42
Total (km^2)	750,656		45,976,000				

[a] Human Poverty Index (HPI-1). A composite index measuring deprivations in the three basic dimensions captured in the human development index—a long and healthy life, being educated and a decent standard of living.

[b] United Nations Development Program. Introduced in 1990 a new way of measuring development—by combining indicators of life expectancy, educational attainment, and income into a composite human development index, HDI. This index sets a minimum and a maximum for each dimension and then shows where each country stands in relation to these scales—expressed as a value between 0 and 1. HDI involves 175 countries. The following are the 10 top ranking: 1: Norway; 2: Iceland; 3: Sweden; 4: Australia; 5: Netherlands; 6: Belgium; 7: United States; 8: Canada; 9: Japan; 10: Switzerland. For details on how the indices are calculated, see United Nations Development Program. http://hdr.undp.org/reports/global/2003/indicator/index_indicators.html.

CONSERVATION INITIATIVES BY MESOAMERICAN COUNTRIES

Concerned with the conservation of biodiversity and with the aim of curbing the loss of natural habitat, between 1993 and 1994, the governments of Mesoamerica ratified the international convention on biological diversity. This led to the consolidation of existing, and to the creation of new, natural protected areas in each country. It also resulted in the establishment of the Mesoamerican System of Natural Protected Areas (UNDP, 1999; http://www.biomeso.net/). Currently, Mesoamerica has a total of 420 protected areas, encompassing about 15 million ha or *ca.* 20% of the territory under consideration (Figure 7, Table 4) (CCAD, 2003).

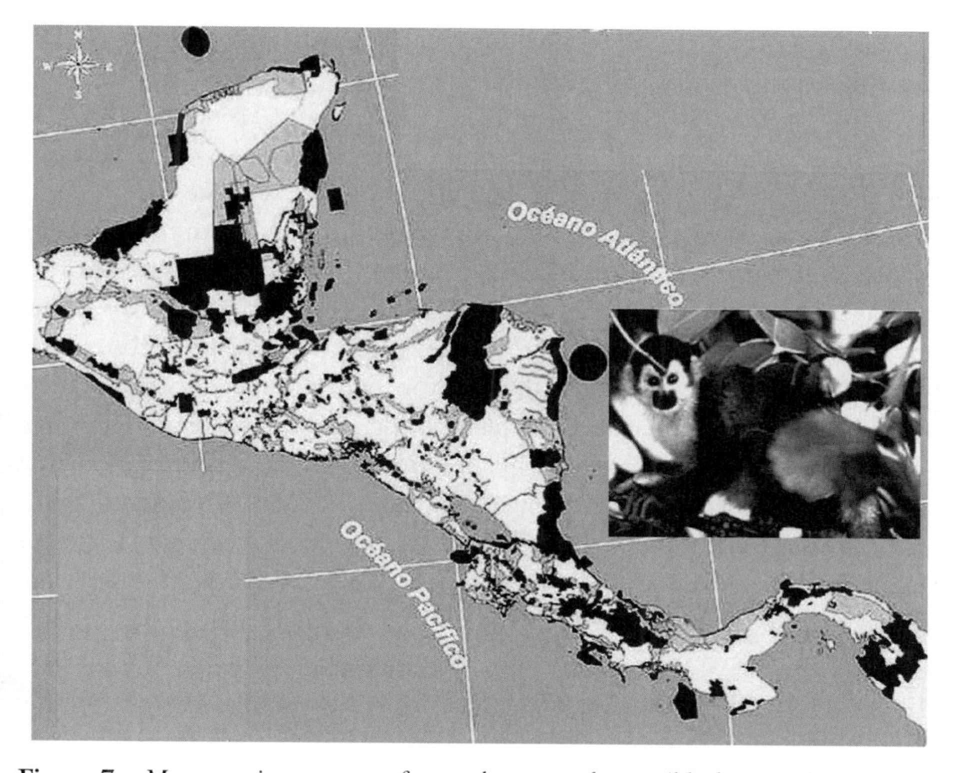

Figure 7. Mesoamerican system of natural protected areas (black areas; $N = 420$; *ca.* 15 million ha or *ca.* 20% of territory) and projected Mesoamerican Biological Corridor interconnecting these and other reserves throughout the region (light gray areas) (map adapted from UNDP, 1999; CCAD, 2003; http://www.biomeso.net/). Reduced effects of isolation in natural protected areas and sustainable use and protection of biodiversity in intermediate areas are key features of corridors in each country.

Table 4. Contribution by each country to the system of natural protected areas in Mesoamerica. Countries ranked by percent of territorial land under protection. For sources of data see Appendix 1

Country	Total land (km^2)	Number of protected areas	Protected area (km^2)	% of territory
Belize	22,965	54	19,670	35
Costa Rica	51,113	126	15,556	31
Panama	75536	42	19,664	26
Mexico (South)	226,712	29	38,902	20
Guatemala	108,917	48	20,614	19
Nicaragua	131,847	75	21,605	18
Honduras	112,520	42	10,703	10
El Salvador	21,046	4	91	0.40
Total	750,656	420	146,805	
%			20	

Major contributors (in km^2) to the system of protected areas are Mexico (38,902), Nicaragua (21,605), and Guatemala (20,614), followed by Belize (19,670), Panama (19,664), Costa Rica (15,556), and Honduras (10,703), with El Salvador having the lowest contribution (91) (Table 4). The countries with the largest areas protected in proportion to their size are Belize, Costa Rica, Panama, Mexico, Guatemala, Nicaragua, Honduras, and El Salvador (Table 4). In spite of these efforts, according to the United Nations Environmental Program, the Mesoamerican system of natural protected areas suffers several problems: few of the areas are actually protected, others remain as paper parks, about 70% are less than 10,000 ha in size, half are not even staffed, only 12% have specific management plans, most are poorly delimited, research projects are only being carried out in a few dozen of them, deforestation rates in surrounding areas are particularly high at approximately 44 ha/hr, and protected areas are virtual islands of vegetation surrounded by altered landscapes (UNEP, 2004; CCAD, 2003). According to UNEP, the lack of steady economic growth in the region makes it difficult or impossible to significantly reduce rural poverty, which in turn will continue to exert enormous pressure on natural habitats and weakly protected reserves.

THE MESOAMERICAN BIOLOGICAL CORRIDOR

Notwithstanding such problems, Mesoamerican countries continue to make efforts to find ways to expand the ecological, social, and economic benefits of conserving their biodiversity, and to mitigate the effects of isolation on existing protected areas. Using remote sensing technology and the results of ground surveys involving sociological, economic, demographic, and biological assessments, each nation has, in a coordinated fashion, proposed a system of biological corridors that will connect protected areas in each country (Figure 7). Management of intermediate areas would involve sustainable use of the land. This proposal gave rise to the Mesoamerican Biological Corridor (MBC) initiative (UNDP, 1999; World Bank, 2004).

The rationale behind the concept of the MBC is that by providing and sustaining connectivity, viability of species and populations in natural protected areas will be enhanced. The cardinal principle of this strategy must be to avoid fragmentation of natural areas and the resulting isolation of vulnerable "islands" of native vegetation surrounded by landscapes altered by human activity. A major focus of the MBC concept is to integrate conservation and sustainable use of biodiversity within the framework of sustainable economic development (NASA, 2004; UNEP, 2004). The idea was proposed in 1992 at the CACB (Central American Convention on Biodiversity), and preparatory phases were completed between 1994 and 1996. The agreement was signed at a presidential summit the following year and diagnostic activities began in 1999 with a 5-year plan that will end in 2004 (UNDP, 1999).

The significance of the MBC for conservation of primate populations and species cannot be underestimated and it may be reduced to four key aspects: (1) increased area of habitat available, (2) increased connectivity, (3) possible increased effective population size, and (4) increased probability of persistence. Important initiatives by primatologists working in the region may be to contact the MBC division in the country where research is being conducted (headquarters for project is found in Nicaragua; see Appendix 2), to explore possibilities of articulating efforts and build interest in protection of primate habitats and populations, to participate in capacity-building by training local biologists, to conduct more diagnostic field projects in undocumented areas and on more primate taxa, and to possibly join forces via the American Society of Primatologists and the International Primatological Society to link conservation efforts with the MBC project.

FINAL COMMENTS

The primates of Mesoamerica represent an important and successful radiation of nonhuman primates that have been under investigation since the 1930s.

Despite many decades of research, documentation is still needed concerning the ecology, diet, social behavior, reproductive biology, and ecological impact of the Mesoamerican primate taxa, and how factors such as deforestation, human disturbance, and habitat change have affected the current distribution, demography, genetic variability, and conservation status of populations in the region. Mesoamerican primates are an integral cultural component of the natural patrimony of all countries in the region. The investigation of their biology, behavior, ecology, and conservation is a fundamental aspect of research providing information for sustaining the biodiversity of the region, the natural history of neotropical primates, their role in forest ecology as seed dispersers, seed predators, pollinators, and agents of forest regeneration, and on the evolution of the Order Primates.

The goal of this volume is to present a comprehensive overview of the most recent advances in primate field research, ecology, and conservation biology in Mesoamerica. This includes information on taxonomy and the historical biogeography of primate origins in Mesoamerica, demographic and population trends from new and long-term field studies, data on feeding ecology, ranging behavior, cognition, and behavioral plasticity, and the effects of habitat disturbance (natural and human induced) on population viability. Chapters are designed to integrate newly collected field data with theoretical perspectives drawn from evolutionary biology, socioecology, biological anthropology, cognitive ecology, and conservation. Several chapters employ innovative methodological techniques such as remote sensing and geographic information systems, experimental field studies, landscape ecology, and reproductive endocrinology to address critical research questions. Data presented in other chapters provide a framework for developing action plans for future research, identifying geographical regions and species for which we continue to lack sufficient information, and highlighting areas for immediate conservation action.

Despite many decades of primate research in Mesoamerica, there continue to remain many unanswered questions. This is highlighted in our volume by the limited behavioral, ecological, and demographic information presented on *S. geoffroyi*, *Aotus zonalis*, and *S. oerstedii*. All three species are restricted in their distribution to the southern region of Mesoamerica. Both *S. geoffroyi* and *S. oerstedii* have not been the focus of long-term field research since mid-1980s. *Aotus zonalis* has never been the focus of a long-term study. We hope this volume will stimulate the development and continuation of both basic field research and applied field studies that will open new lines of inquiry and education focusing on all Mesoamerican primate taxa. We envision a new perspective in which primate research and forest conservation are strengthened by integrating theoretical perspectives and methodological tools from the fields of population genetics, landscape ecology, agroforestry, and behavioral ecology, along multidisciplinary lines. Within this framework, we must consider the needs of the human populations in the region and the progress being made, in spite of overpopulation, poverty, and underdevelopment, by Mesoamerican countries to preserve their natural biodiversity.

REFERENCES

Aguilar-Støen, M. and Dhillon, S. 2003, Implementation of the convention on biological diversity in Mesoamerica: Environmental and developmental perspectives. *Environ. Conserv.* 30:131–138.

Altmann, S. A. 1959, Field observations on a howling monkey society. *J. Mammal.* 40:317–330.

Baldwin, J. D. and Baldwin, J. 1972, The ecology and behavior of squirrel monkeys (*Saimiri oerstedi*) in a natural forest in western Panama. *Folia Primatol.* 18:161–184.

Bernstein. I. S. 1964, A field study of the activities of howler monkeys. *Anim. Behav.* 12:92–97.

Boinski, S. 1985, Status of the squirrel monkey *Saimiri oerstedii* in Costa Rica. *Primate Conserv.* 6:15–16.

Boinski, S. 1986, The ecology of squirrel monkeys in Costa Rica. *Diss. Abstr. Int.* 5:1893.

Boinski, S. 1987, Birth synchrony in squirrel monkeys (*Saimiri oerstedii*): A strategy to reduce neonatal predation. *Behav. Ecol. Sociobiol.* 21:393–400.

Boinski, S. 1988, Sex differences in the foraging behavior of squirrel monkeys in a seasonal habitat. *Behav. Ecol. Sociobiol.* 23:177–186.

Bolin, I. 1981, Male parental behavior in black howler monkeys (*Alouatta palliata pigra*) in Belize and Guatemala. *Primates* 22:349–360.

Cant, J. G. H. 1978, Population survey of the spider monkey *Ateles geoffroyi* at Tikal, Guatemala. *Primates* 19:525–535.

Carpenter, C. R. 1934, A field study of the behavior and social relations of howling monkeys (*Alouatta palliata*). *Comp. Psychol. Monogr.* 10:1–168.

CCAD 2003, Memorias. Primer Congreso Mesoamericano de Areas Protegidas. CCAD (Consejo Centro Americano de Areas Protegidas). Managua, Nicaragua. Document available through http://www.biomeso.net/

Chapman, C. A. 1988, Foraging strategies, patch use, and constraints on group size in three species of Costa Rican primates. *Diss. Abstr. Int.* 9:2573.

Chivers, D. J. 1969, On the daily behaviour and spacing of howling monkey groups. *Folia Primatol.* 10:48–102.

Clarke, M. R. 1982, Socialization, infant mortality, and infant–nonmother interactions in howling monkeys (*Alouatta palliata*) in Costa Rica. *Diss. Abstr. Int.* 4:1217.

Clarke, M. R., Crockett, C. M., Zucker, E. L., and Zaldivar, M. 2002, Mantled howler population of Hacienda La Pacifica, Costa Rica, between 1991 and 1998: Effects of deforestation. *Am. J. Primatol.* 56:155–163.

Coelho, A. M. Jr., Coelho, L. S., Bramblett, C. A., Bramblett, S. S., and Quick, L. B. 1976, Ecology, population characteristics and sympatric association in primates: A sociobioenergetic analysis of howler and spider monkeys in Tikal, Guatemala. *Yrbk Phys. Anthrop.* 20:96–135.

Collias, N. and Southwick, C. 1952, A field study of population density and social organization in howling monkeys. *Proc. Am. Philos. Soc.* 96:143–156.

Emmons, L. 1990, *Neotropical Rainforest Mammals. A Field Guide.* The University of Chicago Press, Chicago.

Estrada A. 1982, Survey and census of howler monkeys (*Alouatta palliata*) in the rain forest of "Los Tuxtlas," Veracruz, Mexico. *Am. J. Primatol.* 2:363–372.

Estrada, A. and Coates-Estrada, R. 1984, Fruit-eating and seed dispersal by howling monkeys in the tropical rain forest of Los Tuxtlas, Mexico. *Am. J. Primatol.* 6:77–91.

Estrada, A. and Coates-Estrada, R. 1993, Aspects of ecological impact of howling monkeys (*Alouatta palliata*) on their habitat: A review, in: A. Estrada, E. Rodriguez Luna, R. Lopez-Wilchis, and R. Coates-Estrada, eds., *Avances en: Estudios Primatologicos en Mexico* I, Asociacion Mexicana de Primatologia, A.C. y Patronatto Pro-Universidad Veracruzana, A. C. Xalapa, Veracruz, Mexico, pp. 87–117.

Estrada, A. and Coates-Estrada, R. 1996, Tropical rain forest fragmentation and wild populations of primates at Los Tuxtlas. *Int. J. Primatol.* 5:759–783.

Estrada, A., Luecke, L., Van Belle, S., Barrueta, E., and Rosales, M. 2004, Survey of black howler (*Alouatta pigra*) and spider (*Ateles geoffroyi*) monkeys in the Maya sites of Calakmul and Yaxchilán, Mexico and Tikal, Guatemala. *Primates* 45:33–39.

FAO 2000, State of Forestry in the Region-2000. Latin American and Caribbean Forestry Commission FAO. Forestry series no 15. FAO http://www.rlc.fao.org

FAO 2001, Food security document. http:/www.faor.org/biodiversity

FAO 2004, FAO Statistical Data Bases. http://faostat.fao.org/

Fedigan, L. M. 2003, Impact of male takeovers on infant deaths, births and conceptions in *Cebus capucinus* at Santa Rosa, Costa Rica. *Int. J. Primatol.* 24:723–741.

Fedigan, L. M. and Baxter, M. J. 1984, Sex differences and social organization in free-ranging spider monkeys (*Ateles geoffroyi*). *Primates* 25:279–294.

Garber, P. A. 1980, Locomotor behavior and feeding ecology of the Panamanian tamarin (*Saguinus oedipus geoffroyi*, Callitrichidae, Primates). *Int. J. Primatol.* 1:185–201.

Garber, P. A. 1981, Locomotor behavior and feeding ecology of the Panamanian tamarin (*Saguinus oedipus geoffroyi*, Callitrichidae, Primates). *Diss. Abstr. Int.* A41:3649.

Garber, P. A. 1986, The ecology of seed dispersal in two species of callitrichid primates (*Saguinus mystax* and *Saguinus fuscicollis*). *Am. J. Primatol.* 10:155–170.

Garber, P. A. 1997, One for all and breeding for one: Cooperation and competition as a tamarin reproductive strategy. *Evol. Anthropol.* 5:135–147.

Garber, P. A., Pruetz, J. D., Lavallee, A. C., and Lavallee, S. G. 1999, A preliminary study of mantled howling monkey (*Alouatta palliata*) ecology and conservation on Isla de Ometepe, Nicaragua. *Neotrop. Primates* 7:13–117.

Glander, K. E. 1975, Habitat description and resource utilization: A preliminary report on mantled howling monkey ecology, in: R. H. Tuttle, ed., *Socioecology and Psychology of Primates*, The Hague, Mouton, pp. 37–57.

Glander, K. E. 1978, Howling monkey feeding behavior and plant secondary compounds: A study of strategies, in: G. G. Montgomery, ed., *The Ecology of Arboreal Folivores,* Smithsonian Institution Press, Washington, D.C., pp. 561–574.

Glander, K. E. 1992, Dispersal patterns in Costa Rican mantled howling monkeys. *Int. J. Primatol.* 13:415–436.

Gómez-Pompa, A., Allen, M. F., Fedick, S. L., and Jiménez-Osornio, J. J. 2003, *The Lowland Maya Area. Three Millenia at the Human–Wildland Interface.* Food Products Press, New York.

Hladik, A. and Hladik, C. M. 1969, Trophic relationships between vegetation and primates in the forest of Barro Colorado (Panama). *Terre et La Vie* 23:25–117.

Horwich, R. H. 1983, Breeding behaviors in the black howler monkey (*Alouatta pigra*) of Belize. *Primates* 24:222–230.

Horwich, R. H., Brockett, R. C., James, R. A., and Jones, C. B. 2001, Population growth in the Belizean black howling monkey (*Alouatta pigra*). *Neotrop. Primates* 9:1–7.

Marsh, L. 2003, *Primates in Fragments: Ecology and Conservation.* Kluwer Academic Press, New York.

Milton, K. 1977, The foraging strategy of the howler monkey in the tropical forest of Barro Colorado Island, Panama. *Diss. Abstr. Int.* B38:1539.

Milton, K. 1980, *The Foraging Strategy of Howler Monkeys: A Study in Primate Economics.* Columbia Press, New York.

Milton, K. and Mittermeier, R. A. 1977, A brief survey of the primates of Coiba Island, Panama. *Primates* 18:931–936.

Mittermeier, R. A. 1973, Group activity and population on dynamics of the howler monkey on Barro Colorado Island. *Primates* 14:1–19.

Mittermeier, R. A., Myers, D., and Thomsen, J. B. 1998, Biodiversity hotspots and major tropical wilderness areas: Approaches to setting conservation priorities. *Conserv. Biol.* 12:516–520.

Moore, F. R. and Simons, T. R. 1992, Habitat suitability and stopover ecology of neotropical landbird migrants, in: J. M. Hagan and D. W. Johnson, eds., *Ecology and Conservation of Neotropical Migratory Birds*, Smithsonian Institution Press, Washington, D.C., pp. 345–355.

Moynihan M. 1964, Some behavior patterns of platyrrhine monkeys. I. The night monkey (*Aotus trivirgatus*). *Smith Miscl. Collec.* 146:1–84.

Moynihan M. 1970, Some behavior patterns of platyrrhine monkeys. II. *Saguinus geoffroyi* and some other tamarins. *Smith Contr. Zool.* 28:1–77.

NASA 2004, Mesoamerican biological corridor project. http://www.ghcc.msfc.nasa.gov/corredor/

Nowak, R. M. 1999, *Walker's Primates of the World.* The John Hopkins University Press, Baltimore.

Oppenheimer, J. R. 1969, Behavior and ecology of the white-faced monkey, *Cebus capucinus*, on Barro Colorado Island, C. Z. *Diss. Abstr. Int.* B30:442–443.

Ostro, L. E. T., Silver, S. C., Koontz, F. W., Horwich, R. H., and Brockett, R. 2001, Shifts in social structure of black howler (*Alouatta pigra*) groups associated with natural and experimental variation in population density. *Int. J. Primatol.* 22:733–748.

Pavelka, M. S. M., Brusselers, O. T., Nowak, D., and Behie, A. M. 2003, Population reduction and social disorganization in *Alouatta pigra* following a hurricane. *Int. J. Primatol.* 24:1037–1055.

Reid, F. 1997, *A Field Guide to the Mammals of Central America and southeast Mexico.* Oxford University Press, London.

Richard, A. 1970, A comparative study of the activity patterns and behavior of *Alouatta villosa* and *Ateles geoffroyi. Folia Primatol.* 12:241–263.

Rodriguez-Luna, E., Cortez-Ortiz, L., Mittermeier, R. A., Rylands, A. B., Wong-Reyes, G., Carrillo, E., Matamoros, Y., Núñez, F., and Motta-Gill, J. 1996, Hacia un plan de accion para los primates Mesoamericanos. *Neotrop. Primates* 4(Suppl.):112–133.

Rowe, N. 1996, *A Pictorial Guide to the Living Primates.* Pogonias Press, Charsletown, RI.

Sader, S. S., Hayes, D. J., Irwin, D. E., and Saatchi, S. S. 1999, Preliminary forest cover change estimates for Central America (1990's), with reference to the proposed Mesoamerican Biological Corridor. NASA Jet Propulsion Lab. http://www.ghcc.msfc.nasa.gov/corredor/

Schlichte H. J. 1978, A preliminary report on the habitat utilization of a group of howler monkeys (*Alouatta villosa pigra*) in the National Park of Tikal, Guatemala, in: G. G. Montgomery, ed., *The Ecology of Arboreal Folivores*, Smithsonian Institution Press, Washington, D.C., pp. 551–559.

Terborgh, J. 1989, *Where Have all the Birds Gone?* Princeton University Press, Princeton, NJ.

Thorington, R. W. Jr., Muckenhirn, N. A., and Montgomery, G. G. 1976, Movements of a wild night monkey (*Aotus trivirgatus*), in: R. W. Thorington, Jr. and P. G. Heltne, eds., *Neotropical Primates: Field Studies and Conservation*. National Academy of Sciences, Washington, D.C., pp. 32–34.

UNDP 1999, Establishment of a Program for the Consolidation of the Mesoamerican Biological Corridor Project document (RLA/97/G31). United Nations Development Programme. Global Environment Facility. Project of the Governments of Belize, Costa Rica, El Salvador, Guatemala, Honduras, Mexico, Nicaragua, Panama. L:\bd\regional\mesoamerica\MBCprodoc.

UNEP 2004, United Nations Environmental Program. www.rolac.unep.mx/recnat/esp/CBM/

Wagner, H. O. 1956, Field observations on spider monkeys. *Zeits Tierpsych* 13:302–313.

World Bank 2004, http://www.worldbank.org/data/

World Resources Institute 2004, http://earthtrends.wri.org/

WWF 2002, Living Planet Report 2002. WWF, UNDP, WCMC. World Wide Fund For Nature. Gland, Switzerland. http://www.panda.org/

APPENDIX 1

Statistics reported here were taken in raw form the sources listed below. These were used raw or as transformed indices and variables to illustrate states, trends and patterns.

Indigenous Populations

http://www.globalgeografia.com/north_america/north_america1.htm
http://www.ciesas.edu.mx/bibdf/ini/webciesas/perfil_nacional.html
http://www.e-mexico.gob.mx/wb2/eMex/eMex_Perfil_Indigena_de_Mexico
UNDP http://geocompendium.grid.unep.ch/data_sets/index_nat_dataset.htm

Population Growth Trends

http://faostat.fao.org/
http://hdr.undp.org/reports/global/2003/indicator/indic_15_1_1.html
http://www.nationmaster.com/index.php

http://www.globalgeografia.com/north_america/north_america1.htm
http://www.worldbank.org/data/

Deforestation Rates

http://faostat.fao.org/default.jsp?language=EN
http://www.panda.org/news_facts/publications/general/livingplanet/index.cf
http://www.panda.org/downloads/general/LPR_2002.pdf
http://geocompendium.grid.unep.ch/data_sets/index_nat_dataset.htm
World Resources Institute http://earthtrends.wri.org/
Global Environmental Facility http://www.gefweb.org/
UNDP http://geocompendium.grid.unep.ch/data_sets/index_nat_dataset.htm;
http://www.worldbank.org/data/

Forest Cover

http://faostat.fao.org/default.jsp?language=EN
http://www.panda.org/news_facts/publications/general/livingplanet/index.cf
http://www.panda.org/downloads/general/LPR_2002.pdf
http://earthtrends.wri.org/gsearch.cfm?kw=indigenous+people&action=results
UNDP http://geocompendium.grid.unep.ch/data_sets/index_nat_dataset.htm
http://geocompendium.grid.unep.ch/geo3_report/index_report.htm
http://www.worldbank.org/data/

Land-Use Patterns

http://faostat.fao.org/default.jsp?language=EN
http://www.panda.org/news_facts/publications/general/livingplanet/index.cf
http://www.panda.org/downloads/general/LPR_2002.pdf
http://earthtrends.wri.org/

UNDP http://geocompendium.grid.unep.ch/data_sets/index_nat_dataset .htm

Social and Economic Development

United Nations Development Programme
http://www.unep.org
http://hdr.undp.org/reports/global/2003/indicator/index_indicators.html
http://hdr.undp.org/statistics/default.cfm
http://hdr.undp.org/reports/global/2003/indicator/indic_15_1_1.html

System of Natural Protected areas and the Mesoamerican Biological Corridor

United Nations Environment Programme, Global Enviornmental Facility
http://www.gefweb.org/
UNEP. 2004. United Nations Environmental Program
http://www.rolac.unep.mx/recnat/esp/CBM/cbm.htm
NASA Mesoamerican biological corridor project. http://www.ghcc.msfc.nasa.gov/corredor/
World Bank http://www.worldbank.org/data/
Mesoamerican Biological Corridor/Corredor Biológico Mesoamericano. http://www.biomeso.net/

APPENDIX 2

Directory of the Mesoamerican Biological Corridor Officials

Regional office, Managua, Nicaragua, Tel: (505) 233-1848, Fax: (505) 233-4455; E-mail: cbm@biomeso.net. Website: www.biomeso.net
Ana Elisa Martínez, Administradora regional, (505) 233-4455, Fax: (505) 233-1848, Nicaragua
Carmen Guevara, carmen.guevara@biomeso.net, Panamá, (507) 232-6717, Panamá
Danilo Saravia danilo.saravia@biomeso.net Esp. Biodiversidad, (505) 233-4455, Fax: (505) 233-1848 Nicaragua

Hector Ruiz Barranco hector.ruiz@biomeso.net, Enlace México, (525) 544-96300 ext. 17130, México
Jorge Mejía Peralta, jorge.mejia@biomeso.net, Asistente Técnico y Comunicaciones, Nicaragua
Lorenzo Cardenal. lorenzo.cardenal@biomeso.net, Coordinador, (505) 233-4455, Fax: (505) 233-1848 Nicaragua
Luis Rojas, luis.rojas@biomeso.net, Enlace Costa Rica, (506) 283-8975, Costa Rica
María Victoria Urquijo, victoria.urquijo@biomeso.net, Esp. Política Ambiental, (505) 233-4455, Fax: (505) 233-1848, Nicaragua
Martin Schneichel, martin.schneichel@biomeso.net, ATP GTZ, (505) 233-4455, Fax: (505) 233-1848, Nicaragua
Rado Barzev, rado.barzev@biomeso.net, Esp. Economía, (505) 233-4455, Fax: (505) 233-1848, Nicaragua

PART ONE

Taxonomy and Biogeography

Introduction: Taxonomy and Biogeography

A. Estrada, P. A. Garber, and M. S. M. Pavelka

Mesoamerica harbors the northernmost representation of Amazonian primates on the American continent, including many taxa that are endemic to particular areas of the region. In spite of the number of decades (at least seven) that Mesoamerican primates have been under investigation, several taxonomic issues remain unresolved and require further investigation. The situation we face is critical given the current rate at which primate habitats and populations are disappearing in the Mesoamerican region as a result of human activity. An improved knowledge of the taxonomy of Mesoamerican primates is critical not only for addressing questions of adaptation and evolutionary history, but also for providing a productive framework to design region-specific and species-specific measures of conservation. The chapters in this section constitute two outstanding contributions to our knowledge of the taxonomy and the historical biogeography of primates in Mesoamerica.

A. Estrada • Estación Biológica Los Tuxtlas, Instituto de Biología, Universidad Nacional Autónoma de México, Apartado Postal 176, San Andrés Tuxtla, Veracruz, México. **P. A. Garber** • Department of Anthropology, University of Illinois at Urbana-Champaign, 109 Davenport Hall, Urbana, IL 61801, USA. **M. S. M. Pavelka** • Department of Anthropology, University of Calgary, 2500 University Drive NW, Calgary, Alberta, Canada T2N 1N4.

New Perspectives in the Study of Mesoamerican Primates: Distribution, Ecology, Behavior, and Conservation, edited by Alejandro Estrada, Paul A. Garber, Mary S. M. Pavelka, and LeAndra Luecke. Springer, New York, 2005.

In their contribution, Rylands and coauthors detail the current taxonomic status of primates in Mesoamerica. They identify 22 primate taxa present in Central America and southern Mexico. Primate taxa are not distributed uniformly across Mesoamerica. Panama (with eight species) has the richest primate community; and Costa Rica has four species (five if night monkeys, *Aotus*, are included, but their existence in Costa Rica remains questionable). Capuchin monkeys (*Cebus capucinus*) extend north as far as Nicaragua and Honduras, and only spider monkeys (*Ateles geoffroyi*) and howling monkeys (*Alouatta palliata* and *A. pigra*) exist in Belize, Guatemala, and Mexico. The authors report that only spider monkeys are currently recorded from El Salvador and that Geoffroy's tamarin (*Saguinus geoffroyi*) and the night monkey (*Aotus*) are both regionally restricted to Panama. There are two presently accepted subspecies of squirrel monkey, *Saimiri oerstedii* and *S. o. citrinellus.* The former occurs across a small area of the Pacific lowlands of Panama and Costa Rica, and the latter is restricted to small, highly fragmented populations along the Pacific coast of Costa Rica. These authors indicate that the white-throated capuchin, (*Cebus capucinus*), extending from Panama to northern Honduras, may comprise three subspecies, although their validity is uncertain.

When referring to the genus *Alouatta*, Rylands and coauthors point out that there are two distinct howling monkey species, the mantled howler (*Alouatta palliata*) and the black howler (*A. pigra*). They note that howling monkeys of Coiba Island and the Azuero Peninsula exhibit certain distinct morphological features that argue for their classification as a third species, *A. coibensis*, but that recent studies based on molecular genetics have failed to distinguish it from *A. palliata*. Finally, Rylands and coauthors list three possible subspecies of *A. palliata*.

With respect to spider monkeys (*Ateles geoffroyi*), there exists considerable morphologically variability. Seven subspecies are listed, and there is the possibility of an eighth yet undescribed subspecies in northern Honduras. Variability in body mass, coat color, and population genetics in *Ateles* are poorly understood, however, and the possibility remains that a number of taxa are not valid. The Colombian black spider monkey (*Ateles fusciceps rufiventris*) extends into eastern Panama and appears to be a very recent migrant into Mesoamerica (Ford, this volume).

An important complement to the paper by Rylands and coauthors is the contribution by Susan Ford on the biogeographic history of Mesoamerican primates. Ford proposes that Mesoamerican primates derive from distinct source populations that were likely isolated in northwestern Colombia approximately 8 mya with the rise of the northern Andes. She points out that this community of monkeys must have included squirrel monkeys in addition to relatives of the other Mesoamerican taxa, although squirrel monkeys are now extinct or absent from this region of Colombia. She further explains that with the complete emergence and establishment of a land connection across the Darién region around 3.5 mya, primates quickly moved widely into Mesoamerica. Part of this land corridor may have been submerged periodically sometime between 2–3 mya, because there appears to have been a second major cycle of emergence/dispersal around 2 mya. Further evidence suggests a possible third dispersal/emergence cycle at 1 mya, with a filter present today. Ford clearly documents that the modern distributions suggest that primates entered Mesoamerica in at least three and likely four waves. The first wave included ancestors of *Alouatta pigra* and *Saimiri oerstedii*, with the initial major emergence of the isthmus. Today, these taxa exist only as relic and endangered populations. The second wave was likely characterized by an explosive entry and rapid dispersal of ancestral populations of *Alouatta palliata*, *Ateles geoffroyi*, and *Cebus capucinus* into the isthmus. Ford proposes that as gene flow between populations was interrupted by highlands, grasslands, and periodic rises in sea-level, groups differentiated, including the distinctive howlers of the Azuero Peninsula and Isla de Coiba. The third and fairly recent wave brought tamarins (*Saguinus geoffroyi*) and owl monkeys (*Aotus zonalis*) into the Mesoamerican region. The final invader was *Ateles fusciceps*. This species may have entered through a filter that permitted back migrations of tamarins, capuchins, howlers, and owl monkeys into northwestern Colombia (although these may be part of the ancestral population that remained in this region). Ford also proposes that three recent primate immigrants into northwestern Colombia (*Alouatta seniculus*, *Cebus albifrons*, and *Cebus apella*) may eventually invade the isthmus, placing pressure on the unique primate fauna of the Mesoamerican region.

In summary, the two chapters bring up-to-date the taxonomy, history, and biogeography of primate species in Mesoamerica and open up new lines of investigation regarding the evolutionary history of the primate fauna of the region. Moreover, the information in these chapters stands as an important frame of reference for continuing and future research on Mesoamerican primates.

CHAPTER TWO

Taxonomy and Distributions of Mesoamerican Primates

Anthony B. Rylands, Colin P. Groves, Russell A. Mittermeier, Liliana Cortés-Ortiz, and Justin J. H. Hines

INTRODUCTION

Geoffroy's tamarin, a squirrel monkey, a night monkey, the white-throated capuchin, two or three species of howling monkey, and one or two spider monkeys comprise the primate fauna of Middle America, historically throughout the subtropical and tropical forests from about 24°N in Tamaulipas, Mexico, extending south along the coast of the Gulf of Mexico, through Central America to the border of Colombia and Panama. This is the simple description, and hides a remarkable, and still poorly understood, diversity of 7–9 species and up to 22 taxa,

Anthony B. Rylands • Center for Applied Biodiversity Science, Conservation International, 1919 M Street NW, Suite 600, Washington, DC 20036. **Colin P. Groves** • School of Archaeology and Anthropology, Australian National University, Canberra, Australia. **Russell A. Mittermeier** • Conservation International, Washington DC. **Liliana Cortés-Ortiz** • Centro de Investigaciones Tropicales, Universidad Veracruzana, Xalapa, Veracruz, México. **Justin J. H. Hines** • School of Archaeology and Anthropology, Australian National University, Canberra.

New Perspectives in the Study of Mesoamerican Primates: Distribution, Ecology, Behavior, and Conservation, edited by Alejandro Estrada, Paul A. Garber, Mary S. M. Pavelka, and LeAndra Luecke. Springer, New York, 2005.

all with ranges restricted to Middle America and west of the Andes, through Colombia and Ecuador to the Tumbes region of extreme northern Peru.

In this review, we will follow, as far as the evidence permits, the Phylogenetic Species Concept (PSC), as outlined by Groves (2001). This means that homogeneous taxa, diagnosable by unique, apparently consistent (fixed) heritable features, are ranked as species. Subspecies are geographic segments within a species, characterized by high frequency but not fixed differences from other such segments.

While Geoffroy's tamarin, *Saguinus geoffroyi*, and the night monkey, *Aotus*, are considered distinct and monotypic, some have considered them to be subspecies, others species. In recent years, the night monkey has been assigned four different names as a result of differing opinions concerning its affinities with, and the taxonomy of, the diverse forms in Colombia. There are two broadly accepted subspecies of squirrel monkey, *Saimiri oerstedii*, restricted to a small area of the Pacific lowlands of Panama and Costa Rica. They are separated from all other squirrel monkeys, their nearest relatives being east of the Río Magdalena in Colombia, and in the past were considered to have arisen from human introduction in pre-Columbian times. A genetic study by Cropp and Boinski (2000) indicated that this is unlikely, however, and their isolated presence in Central America in this case can only be explained by prehistoric geographic and climatic changes and the extinction of the intervening populations through vegetation changes. The white-throated capuchin, *Cebus capucinus*, extending from Panama to northern Honduras, may comprise three subspecies—Hershkovitz (1949) listed them, but neither Hernández-Camacho and Cooper (1976) nor Groves (2001) accepted their validity.

Lawrence (1933), Smith (1970), Horwich (1983), and Cortés-Ortiz *et al.* (2003) have consolidated evidence for the existence of two howling monkeys in Central America, the black howler, *Alouatta pigra*, from the Yucatán Peninsula, and the mantled howling monkey, *A. palliata*, from southeastern Mexico into Colombia and Ecuador. Froehlich and Froehlich (1986, 1987) argued that the diminutive Coiba Island howler was also a distinct species, *A. coibensis*. Groves (2001) accepted their arguments, but Cortés-Ortiz *et al.* (2003) were unable to confirm this in their study of the molecular genetics of the genus. Groves (2001) otherwise found the evidence insufficient to distinguish a further three mantled howlers listed here: *aequatorialis*, *mexicana*, and *trabeata*.

Perhaps, the most confusion surrounds the spider monkeys. For many years, the taxonomy was based on Kellogg and Goldman's (1944) careful revision of cranial morphology and pelage. They recognized two species for Middle America, *A. geoffroyi* and *A. fusciceps* (Froehlich *et al.* [1991] and Collins and Dubach [2000] have suggested that *A. fusciceps* is a synonym of *A. geoffroyi*), and all but two of the forms they described are still recognized today. The strong indications are that they gave the wrong name to the Colombian black spider monkey (Heltne and Kunkel, 1975); Napier (1976) argued that *A. geoffroyi panamensis* is a synonym of *A. g. ornatus*; and Silva-López *et al.* (1995, 1996) argued that *A. g. pan* is a synonym of *A. g. vellerosus*. It is only recently that there has been a tendency to further lump the subspecies of *A. geoffroyi*. Collins (1999) and Collins and Dubach (2000) divided them into two: northern *geoffroyi* and southern Central American *geoffroyi*. Groves (2001) provisionally recognized only five of Kellogg and Goldman's (1944) nine subspecies of *A. geoffroyi*.

Biogeographical considerations are of course fundamental for our understanding of the diversity of these primates. For this reason, we also present here a review of the current information regarding the historical ranges of the various species and subspecies. Historical is the key word. Our understanding of where these animals occurred naturally is increasingly dependent on the relatively scarce collections in museums. They are hunted and their forests are now severely reduced, degraded, and fragmented throughout Middle America (Estrada and Coates-Estrada, 1984; Horwich and Johnson, 1984, 1986; Luecke, 2004; Silva-López *et al.*, 1995; Estrada *et al.*, this volume). This diminution of the geographic extent of their occurrence makes it extremely difficult to achieve an understanding of the full diversity of the species in terms of pelage variation, morphology and genetics, and as such to make confident decisions concerning their taxonomy. This is particularly critical for the Central American spider monkeys, a group that is evidently still very poorly known in many regions (Konstant *et al.*, 1985; Hines, 2004).

The destruction of the Middle American forests has severely reduced population diversity over the majority of the ranges of all the Middle American primates. The distribution maps provided here are hypotheses of the historical ranges—they overestimate by far the actual area of occupation. The reality today is that each of the taxa is restricted to few and isolated forest fragments. Now, the real distributions are scattered and isolated localities—remnant forest patches—and there is an urgent need for regionwide and detailed surveys to identify and map them, to determine the status of the populations remaining. The GIS is a powerful tool for mapping these forests and populations, with an accuracy and scale never achieved before (for example, Luecke, 2004; Estrada *et al.*, 2004; Pavelka *et al.*, this volume). The conservation of these remnant forests is vital: in the future, the ranges of these primates (Table 1) will undoubtedly be described as lists of protected areas.

SPECIES AND SUBSPECIES OF MESOAMERICAN PRIMATES

Saguinus geoffroyi (Pucheran, 1845)

Geoffroy's tamarin, red-crested bare-face tamarin (Hershkovitz, 1977; Reid, 1997), rufous-naped tamarin (Moynihan, 1970), tití or bichichi in Colombia and Panama (Hernández-Camacho and Cooper, 1976; Reid, 1997). Panama, Colombia (Figure 1).

Type: The type specimen is a mounted skin and a (separate) skull of a female in the Muséum National d'Histoire Naturelle, Paris. Skin No. 112, Skull No. 621. Originally donated to the Jardin des Plantes, and died there on 25 August, 1845 (Hill, 1957; Hershkovitz, 1977).
Type locality: Panama. Restricted by Hershkovitz (1949) to the Canal Zone.

There are a considerable number of synonyms, listed and discussed by Hill (1957), Hershkovitz (1949, 1977), and Groves (2001). *Oedipomidas spixi* (Reichenbach, 1862) was the name used by Hill (1957), following the recommendation of Cabrera (1940), who argued that the specific name *geoffroyi* was preoccupied by *Simia geoffroyi* (Humboldt, 1812), the white-face marmoset of the Brazilian Atlantic forest. Hershkovitz (1949) disagreed with Cabrera (1940) (as pointed out by Hill [1957, p. 260] himself) and listed the tamarin as *Marikina geoffroyi*. Hershkovitz (1949) argued at length that the two species had never been placed in the same genus and as such "a real state of homonymy never existed" (1977; p. 759). Cabrera (1958) evidently later accepted Hershkovitz's argument, listing the Panamanian tamarin as *Leontocebus geoffroyi* Pucheran. Eisenberg (1989) confused the authorship, attributing *geoffroyi* to Reichenbach (1862), author in fact of the junior synonym *spixi*.

Oedipomidas salaquiensis was the name given by Elliot (1912b) to a specimen from the Chocó, Río Salaquí (a tributary of the Río Atrato), a skin and skull in the American Museum of Natural History, New York. Elliot distinguished *salaquiensis* by its larger and differently proportioned skull, chestnut rather than burnt umber crown and nape, and buffy yellow (instead of pure white)

Table 1. The primates of Mesoamerica.

Taxon	Common name	Distribution
Callitrichidae		
Saguinus geoffroyi (Pucheran, 1845)	Geoffroy's tamarin, rufous-naped tamarin	Colombia, Panama
Cebidae		
Saimiri oerstedii oerstedii (Reinhardt, 1872)	Black-crowned Central American squirrel monkey	Costa Rica, Panama
Saimiri oerstedii citrinellus[a] (Thomas, 1904)	Grey-crowned Central American squirrel monkey	Costa Rica
Cebus capucinus capucinus (Linnaeus, 1758)	White-throated capuchin	Colombia, Panama
Cebus capucinus imitator[b] Thomas, 1903	Panamanian white-throated capuchin	Costa Rica, Nicaragua, Panama
Cebus capucinus limitaneus[b] Hollister, 1914	Honduran white-throated capuchin	Honduras, Nicaragua
Aotidae		
Aotus zonalis Goldman, 1914	Panamanian night monkey	Colombia, Costa Rica (?), Panama
Atelidae		
Alouatta palliata palliata (Gray, 1849)	Golden-mantled howling monkey	Costa Rica, Honduras, Nicaragua, Panama
Alouatta palliata mexicana[b] Merriam, 1902	Mexican howling monkey	Mexico, Guatemala
Alouatta palliata aequatorialis[b] Festa, 1903	Ecuadorean mantled howling monkey	Colombia, Ecuador, Panama
Alouatta coibensis coibensis[a] Thomas, 1902	Coiba Island mantled howling monkey	Panama
Alouatta coibensis trabeata[b] Lawrence, 1933	Azuero mantled howling monkey	Panama
A. pigra (Lawrence, 1933)	Black howling monkey	Belize, Guatemala, Mexico
Ateles geoffroyi geoffroyi Kuhl, 1820	Geoffroy's or Nicaraguan spider monkey	Costa Rica, Nicaragua
Ateles geoffroyi azuerensis Bole, 1937	Azuero spider monkey	Panama
Ateles geoffroyi frontatus (Gray, 1842)	Black-browed spider monkey	Costa Rica, Nicaragua
Ateles geoffroyi grisescens Gray, 1866	Hooded spider monkey	Panama, Colombia (?)
Ateles geoffroyi ornatus Gray, 1870	Ornate spider monkey	Costa Rica, Nicaragua
Ateles geoffroyi vellerosus Gray, 1866	Mexican spider monkey	El Salvador, Honduras, Guatemala,Mexico
Ateles geoffroyi yucatanensis[b] Kellogg and Goldman, 1944	Yucatán spider monkey	Belize, Guatemala, Mexico
Ateles fusciceps rufiventris[a] Sclater, 1871	Colombian black spider monkey	Colombia, Panama

[a] Subspecific versus specific status needs further examination.
[b] Validity dubious.

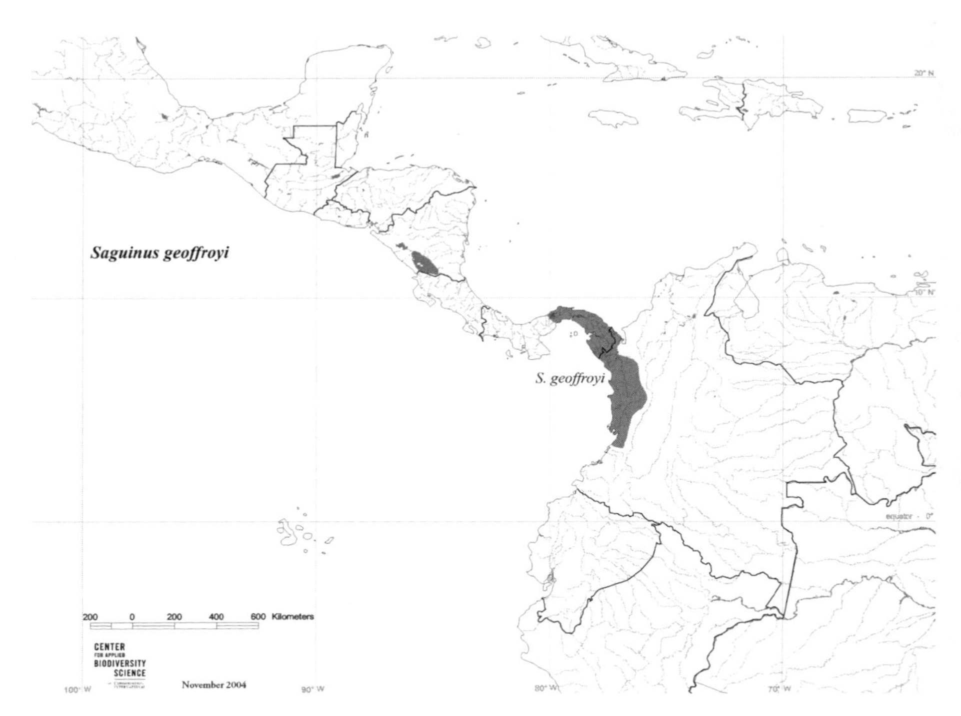

Figure 1. The distribution of *Saguinus geoffroyi*. Based on Eisenberg (1989), Emmons and Feer (1997), Hernández-Camacho and Cooper (1976), Hershkovitz (1977), Matamoros and Seal (2001), Mast *et al.* (1993), Reid (1997), Rodríguez-Luna *et al.* (1996), Rylands *et al.* (1993), and Skinner (1985). Map drawn by Mark Denil and Kimberly Meek (Center for Applied Biodiversity Science, Conservation International, Washington, DC.

underparts. It was, however, not listed in Elliot's (1913) "*A Review of the Primates*", and Elliot (1914) reported that the investigation of further material from Colombia had indicated that the yellowish underparts were due to staining (Hershkovitz [1977] argued that it is in fact natural) and that the skull size, although large, was within the natural variation of that found for *O. geoffroyi*. Hernández-Camacho and Cooper (1976, p. 41) recorded that "of the few museum specimens of *S. geoffroyi* known for Colombia as well as those seen in captivity (largely from the region of Acandí), a large percentage have distinct sulfurous yellowish underparts, including lightly pigmented areas of the limbs." Hershkovitz (1977) found that *S. geoffroyi* does in fact get paler from south to north. The most saturate series he examined was from Sandó (locality 28) in the Chocó. Anthony (1916) remarked on the yellowish underparts of animals he observed in Panama (along the Canal Zone, the Maxon Ranch [Río Trinidad], and localities on the Río Tuyra, Darién—Boca de Cupe, Chepigana, Cituro, Tacarcuna, and Tapalisa), concluding that it was variable and had no diagnostic value.

Hershkovitz (1977) classified the Panamanian tamarin as a subspecies of *S. oedipus* (the cotton-top tamarin from northern Colombia) based on pelage patterns and color, cranial and mandible morphology, and pinna size. Mittermeier and Coimbra-Filho (1981; see also Mittermeier *et al.*, 1988; Rylands, 1993) regarded the forms *oedipus* and *geoffroyi* to be distinct species, arguing that there is no evidence of intergradation between them and that "*S. oedipus* and *S. geoffroyi* are at least as differentiated from one another as are the members of the *Callithrix jacchus* group" (which they also argued to be valid species). Also influential was the suggestion of Thorington (1976) that the cotton-top tamarin was more closely related to *S. leucopus* (the silvery-brown bare-face tamarin of northern Colombia) than to *S. geoffroyi*. Hanihara and Natori (1987) carried out a multivariate comparative analysis of the dental morphology of a number of species of *Saguinus*, and confirmed Thorington's (1976) view. Skinner (1991) examined body weight and a number of morphological characters and found that *S. geoffroyi* was significantly larger than *S. oedipus*, and morphologically more similar to *S. leucopus* than to *S. oedipus* in 16 of the 17 morphological characters studied. Skinner also discussed the pelage color and patterns of the three forms (emphasizing differences rather than the similarities demonstrated by Hershkovitz, 1977), along with aspects concerning hybridization and intergradation in *Saguinus* in general. Moore and Cheverud (1992, p. 73) concluded that "... A variety of multivariate statistical analyses including discriminant function and cluster analysis suggest that *S. oedipus* and *S. geoffroyi* differ morphologically at a level consistent with species-level distinctions. The extent of differences between these taxa is large ... " and later " ... a comparison of collecting localities revealed that the variation we observed among *S. oedipus* and *S. geoffroyi* was not clinal but presented a large morphological discontinuity at the boundary between taxa ... ". Like Skinner (1991), they found that *S. leucopus* was more similar to *S. oedipus*.

Elliot (1913), Hershkovitz (1977), Eisenberg (1989), Rylands *et al.* (1995), and Emmons and Feer (1997) all indicated that *S. geoffroyi* occurs from northwest Colombia, through Panama to the border with Costa Rica, entering its southeasternmost tip on the Pacific coast. Reid (1997) pointed out that this is based on a sight record by Carpenter (1935), who noted that the tamarins were very scarce in the Cotó region. Hershkovitz (1977, p. 924) listed two localities in his gazetteer, which would evidently be mistakes in this case: "Puntarenas, Cotó Region, 8°35′N, 83°05′W, C. R. Carpenter, June 1932, February–March, 1933", and an unspecified locality in Chiriquí, Panama, "arbitrarily indicated on the map, fig. XIII.3". Baldwin and Baldwin (1976) reported on a survey of 71 forest areas in Chiriquí between August 1968 and December 1970, and made no mention at all of *S. geoffroyi*. Reid (1997) cited a Panamanian zoogeographer (F. Delgado, in litt. to D. Engleman) as saying that the record is questionable and certainly does not reflect the species' current distribution, which is limited to central and eastern Panama (and Colombia). The exact western limit is not clearly defined, but marked by Reid (1997) at just a little west of the Canal Zone. Their range is restricted to the east of the Azuero peninsula.

In discussing habitat preference in Panama, Moynihan (1970) stated that "Rufous-naped tamarins are abundant in some parts of the Pacific coastal region, and also occur in some central areas approximately equidistant from both coasts. To our knowledge however, they are completely absent from the whole of the Atlantic coast of the isthmus, except for one small, highly modified or "unnatural" area." (p. 2). The exception he mentioned is around the Canal Zone, the city of Colón, and Lake Gatún where the original forest has been almost entirely destroyed, and Moynihan (1970, 1976) argued that their occurrence there is the result of a recent range extension. The map of localities provided by Hershkovitz (1977, p. 915) confirms Moynihan's observation, with only two records on the Atlantic side of the isthmus except in the vicinity of the Canal Zone. The two outlying Atlantic coast records listed in the gazetteer (p. 925) are: Locality 6c, San Blas, Mandinga, 9°27′N, 79°04′W, C. O. Handley, Jr., May 1957, a series of six specimens in New York; and locality 6d, San Blas, Armila, Quebrada Venado, 8°40′N, 77°28′W, C. O. Handley, Jr., February–March 1963, a series of 12 specimens, also in the US National Museum. Moynihan (1970, 1976) suggested that their absence from the Atlantic coast was related to a preference for drier forests ("of moderate humidity") typical of the Pacific coast. Skinner (1985) confirmed their occurrence in San Blas and reported the presence of *S. geoffroyi* in 21 sites all in moist tropical forest from the western Río Chagres basin to the Darién, from the Atlantic to the Pacific coasts.

In Colombia, it occurs along the Pacific coast, south as far as the Río San Juan. The Río Atrato was believed to be the eastern limit to its range (Hernández-Camacho and Cooper, 1976; Hershkovitz, 1977), but Vargas (1994, cited in

Defler, 2003) found the species occurring around the National Natural Park of Las Orquídeas in the vicinity of the village of Mandé, Antioquia, at elevations as high as 1000 m, extending its range to the west of the upper Río Cauca. Barbosa *et al.* (1988, in Mast *et al.*, 1993) also recorded the species at Quibdo, a town just east of the upper Río Atrato.

Saimiri oerstedii (Reinhardt, 1872)

Central American squirrel monkey, mono ardilla, mono tití.
Costa Rica, Panama (Figure 2).

The two subspecies listed here are recognized by Hill (1960) and Hershkovitz (1984). Cabrera (1958) and Thorington (1985), on the other hand, regarded the Central American squirrel monkey to be a subspecies of

Figure 2. The distribution of *Saimiri oerstedii*. Based on Boinski (1985, 1987), Boinski *et al.* (1998), Reid (1997), Sierra *et al.* (2003), Matamoros and Seal (2001), Rodríguez-Luna *et al.* (1996), and Wong (1990). Map drawn by Mark Denil and Kimberly Meek (Center for Applied Biodiversity Science, Conservation International, Washington, DC).

S. sciureus (Linnaeus, 1758); both sharing the Gothic-arch superciliary pattern. Thorington (1985) wrote, however, that his classification resulted from him also placing the form *boliviensis* (d'Orbigny, 1834) as a subspecies of *sciureus*—"In coat color and pattern and in craniometric analyses, *oerstedii* seems no more different from *S. sciureus sciureus* in Colombia than does *boliviensis*. Because I am treating *boliviensis* as a subspecies of *S. sciureus*, I should treat *oerstedii* as a subspecies of *sciureus* as well. A demonstration that *boliviensis* is a valid species would change the way I treat *oerstedii*." (p. 22). Hershkovitz (1984) and Groves (2001) recognized *sciureus*, *boliviensis*, and *oerstedii* as distinct species, and *citrinellus* as a subspecies of *oerstedii*. Costello *et al.* (1993) and Silva *et al.* (1993) recognized the distinctiveness of *S. oerstedii* compared to all other squirrel monkeys, which they lump as *S. sciureus*. Studies by Boinski and Cropp (1999) using mtDNA, behavioral and morphological data, and Cropp and Boinski (2000) using two nuclear genes (IRBP and ZFX) and one mitochondrial (D-Loop) also confirmed that *sciureus*, *bolivensis*, and *oerstedii* should be considered distinct species. Their DNA study included specimens of both putative subspecies of *S. oerstedii*, and these formed homogeneous clades, raising the question of whether they might better be ranked as two distinct species (*S. oerstedii* and *S. citrinellus*); there is a need for further study to determine whether the described phenotypic differences are also consistent.

Hershkovitz (1969, 1984) presented a number of circumstantial arguments that these squirrel monkeys were introduced into Central America by humans, probably by sea from the Pacific coast of Ecuador or Peru. They included: their tameness; the "beach-head sized range"; their discrete distributions (well separated from *S. sciureus* to the east of the Río Magdalena), which cannot be explained by natural dispersal; and the extremely derived pelage color patterns of the two forms. Furthermore, Hershkovitz (1984) argued that *oerstedii* is the more derived of the two subspecies, and yet *citrinellus* is geographically the most peripheral. A study of nuclear and mtDNA by Cropp and Boinski (2000), however, provided divergence dates (3–4.4 mya—mtDNA; 420,000–260,000 years ago—nuclear DNA) that clearly negate Hershkovitz's (1969) introduction hypothesis.

Hill (1960), Hershkovitz (1984), and Groves (2001) provide descriptions of pelage color patterns. Both *S. o. oerstedii* and *S. o. citrinellus* are predominantly orange to reddish-orange with a characteristic dark cap. In *S. o. oerstedii*, both males and females have a black cap. Compared to *S. o. citrinellus*, the limbs are more yellowish, and there is a stronger yellowish tinge in the underparts (abdomen, groin, and medial aspects of thigh) (Hill, 1960). The outer side of the leg is orange like the arms (Groves, 2001). Thomas (1904) distinguished *citrinellus* by its less black head and less yellow limbs. Elliot (1913) regarded these features as variable (the head color with age) and considered *citrinellus* a synonym of *oerstedii*.

Saimiri oerstedii oerstedii (Reinhardt, 1872)

Black-crowned Central American squirrel monkey, red-backed squirrel monkey, or Panamanian red-backed squirrel monkey (Hill, 1960), tití, mono ardilla, mono tití.
Costa Rica, Panama (Figure 2).

Type: Skin and skull in the Zoological Museum, Copenhagen. Collected by A. S. Örsted.
Type locality: Vicinity of David, Chiriquí, Panama.

S. o. oerstedii occurs along the Pacific coast of Costa Rica, from the left bank of the Río Grande de Térraba to the Osa Pensinsula, along the coast of the Golfo Dulce and the Burica Peninsula to the western part of the Chiriquí Province, mouth of the Río Fonseca, including the Archipelago of the Golfo de Chiriquí, in Panama (Hershkovitz, 1984; Boinski *et al.*, 1998). Surveys by Baldwin and Baldwin (1972, 1976) recorded its presence on the Burica Peninsula, but indicated that it is now restricted to a narrow strip of scattered lowland coastal forest fragments, not extending to the type locality David, although it possibly occurred as far east as Remedios (well to the east of David) prior to the 1950s. Altitudinal range is 0–500 m asl (Hershkovitz, 1984). Rodríguez-Vargas (2003) mapped the remaining populations in Panama.

Saimiri oerstedii citrinellus Thomas, 1904

Grey-crowned Central American squirrel monkey or Costa Rican red-backed squirrel monkey (Hill, 1960), tití, mono ardilla, mono tití.
Costa Rica (Figure 2).

Type: Male, skin and skull in British Museum (Natural History), No. 1904.2.72, collected 31 May, 1902, by C. F. Underwood (Napier, 1976; Hershkovitz, 1984).
Type locality: Pozo Azul, Río Pirris or Parrita, San José, Costa Rica. According to Carriker (1910, p. 349, see Hershkovitz, 1984, p. 197), Pozo Azul is a

locality on the Río Grande de Pirris about 10 miles from the Pacific Ocean, reached by cart-road from San José.

The historic range of *S. o. citrinellus* is along the Pacific coast of Costa Rica, to altitudes of up to 500 m asl. The northeastern limit is marked by the Río Tulín in the north Herradura Mountains (9° 40′N, 84° 35′W) and Dota Mountains (9° 37′N, 84° 35′W), and the southern limit is the north bank of the Río Grande de Térraba (8° 25′N, 84° 25′W) (Arauz, 1993; Sierra *et al.*, 2003). Its occurrence is sporadic, and the surviving populations are entirely fragmented (Alfaro, 1987; Wong, 1990; Sierra *et al.*, 2003). As mentioned above, we recommend further study of the differences between this subspecies and nominotypical *S. o. oerstedii* to determine if they are correctly ranked taxonomically.

Cebus capucinus (Linnaeus, 1758)

White-throated capuchin, mono carablanca, cariblanco, mono capuchino. Colombia, Costa Rica, Ecuador, Honduras, Nicaragua, Panama (Figure 3).

Cebus capucinus is the only capuchin monkey in Central America, ranging from Honduras in the north, through Nicaragua, Costa Rica, and Panama and through the Chocó-Darién into Colombia (Hernández-Camacho and Cooper 1976, Rodríguez-Luna *et al.*, 1996, Reid, 1997, Marinero and Gallegos, 1998). It is easily distinguished from other members of the genus in the black pelage on the crown, nape, back, flanks, limbs, and tail. The face, forehead, sides of head, throat, sides and front of neck, shoulders, and chest are white (off-white or slightly yellowish). The Colombian *C. c. nigripectus* was distinguished by Elliot (1913) for its black chest (but see below). Hershkovitz (1949) listed five subspecies of *Cebus capucinus*: *C. c. nigripectus* (from the upper Río Cauca in Colombia), *C. c. capucinus* (Colombia and eastern Panama), *C. c. imitator* (western Panama, Coiba Island [Panama], and Costa Rica), *C. c. limitaneus* (Honduras and Nicaragua), and *C. c. curtus* (Gorgona Island, Colombia, possibly introduced in the 16th or 17th century); yet he went on to say that "None of the distinguishing characters attributed to the described races of *C. capucinus* appears to be valid" (pp. 346–347), while considering it desirable to retain the named subdivisions pending a thorough study. Hernández-Camacho and Cooper (1976) argued that variability in populations on the upper Río Cauca did not support the validity of *C. c. nigripectus*, and considered that the Central American populations were subject to the same limitation. Hernández-Camacho and Defler (1991) and Defler (1994) listed just two subspecies of

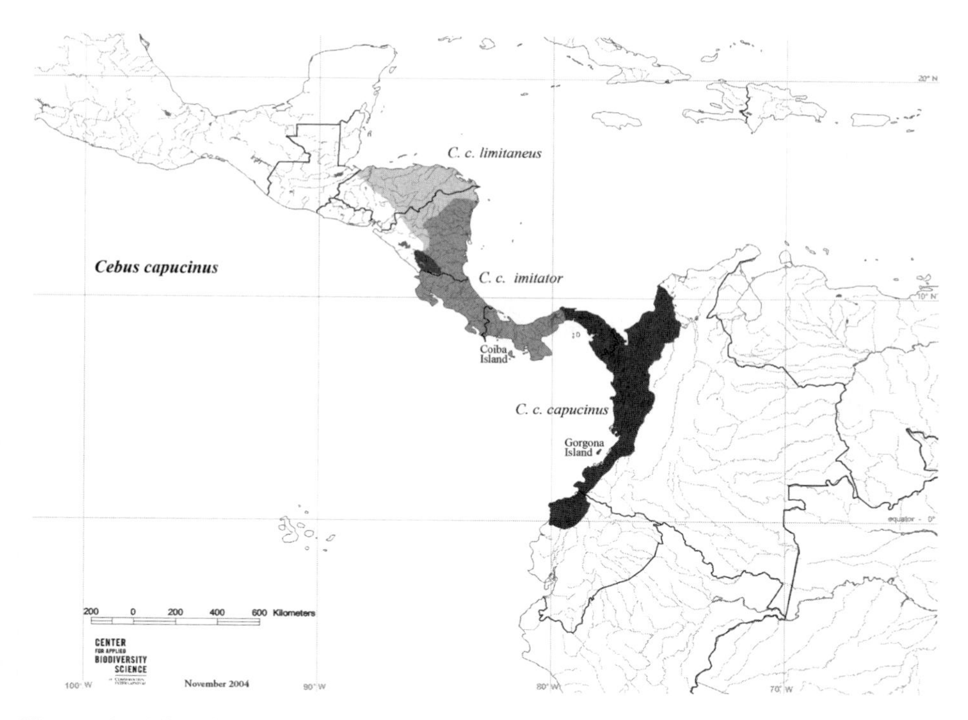

Figure 3. The distribution of *Cebus capucinus.* Based on Defler (2003), Eisenberg (1989), Hill (1960), Hall (1981), Hernández-Camacho and Cooper (1976), Marineros and Gallegos (1998), Matamoros and Seal (2001), Reid (1997), and Rodríguez-Luna *et al.* (1996). The numbers indicate two localities where capuchin monkeys have been reported but their presence has yet to be confirmed; 1: Mayan Mountains of western Belize (the Chiquebul forest and in the region of the Trio and Bladen branches of the Monkey River); 2: Sierra del Espíritu Santo near the Guatemala–Honduras border (see text). Map drawn by Mark Denil and Kimberly Meek (Center for Applied Biodiversity Science, Conservation International, Washington, DC).

C. capucinus for Colombia: *C. c. capucinus* and *C. c. curtus.* Groves (1993, 2001), having reviewed specimens in the US National Museum, the American Museum of Natural History, and the Museum of Comparative Zoology at Harvard University, also regarded all the subspecies listed by Hershkovitz (1949) and Hill (1960) to be junior synonyms of a monotypic *C. capucinus.* Here, we list the three Mesoamerican forms separately to draw attention to each one, but this should not be taken to imply that we endorse their status as separate subspecies. There are no records of this species in El Salvador or Mexico. There are unconfirmed reports of its occurrence in southern Belize and eastern Guatemala.

Cebus capucinus capucinus (Linnaeus, 1758)

White-throated capuchin, mono carablanca.

Type: None exists.
Type locality: Not known, but Hershkovitz (1949) indicated "Northern Colombia".
Colombia, Ecuador, Panama (Figure 3).

According to Hill (1960, p. 425), *C. c. capucinus* differs from other subspecies in the "general whiteness of the pallid areas of head, neck, shoulders, arms, and underparts, including the chest". It is larger than *C. c. limitaneus*, and the females do not have the frontal tufts of *C. c. imitator*. In Colombia, the white-throated capuchin occurs south from the Panamanian border along the Pacific Coast, west of the Andes into northwestern Ecuador. It is apparently restricted to the west bank of the Río Cauca and extends north across the Río Sinu into Cordoba, Sucre, and Atlantico to the town of Barranquilla on the northern coast of Colombia (Hernández-Camacho and Cooper, 1976; Defler, 2003). In Central America, *C. c. capucinus* extends west as far the Panama Canal (Hall, 1981).

Cebus capucinus imitator Thomas, 1903

Panamanian white-throated capuchin, mono carablanca.
Costa Rica, Nicaragua, Panama (Figure 3).

Type: Adult female (skin and skull) in British Museum (Natural History), No. 1903.3.3.13. Collected 15 October, 1902, by H. J. Watson (Napier, 1976).
Type locality: Chiriquí, Boquete, western Panama, altitude 4000 ft.

Much resembling typical *C. c. capucinus*, but females have elongated frontal tufts, entirely altering the facial appearance. Hall (1981) places this subspecies in western Panama, west from the Canal, and in adjacent areas of Costa Rica. Populations also occur on the islands of Coiba and nearby Jicarón. Baldwin and Baldwin (1976, 1977) documented the occurrence of *C. capucinus* in a number of localities in the Province of Chiriquí, southwestern Panama. Crockett *et al.* (1997) listed localities in Nicaragua, and Allen (1908, 1910) recorded specimens of *C. capucinus* (referred to as *C. hypoleucus* in Allen, 1908), from Ocotal (northern highlands, 4500 ft), and localities on the east slope of the highlands, Savala (800 ft), Chontales (lowlands east of Lake Nicaragua, altitudes 500–1500 ft), and the Río Tuma (500 ft) and Muy Muy (Matagalpa Province, 1500–2000 ft).

Cebus capucinus limitaneus Hollister, 1914

Honduran white-throated capuchin monkey, white-faced capuchin monkey.
Honduras, Nicaragua (Figure 3).

Type: Adult male, skin and skull in United States National Museum, collected by C. H. Townsend in 1887 (Hill, 1960).
Type locality: Segovia River, eastern Honduras. Restricted by Hershkovitz (1949) to Cabo Gracias a Dios at the mouth of the river, eastern border between Honduras and Nicaragua.

Described by Hollister (1914) as similar to *C. c. imitator* Thomas of western Panama and Costa Rica, but slightly smaller with a "decidedly smaller skull". This is the most northerly population of the species and genus. Besides the type locality, Hollister (1914) recorded specimens from Patuca, Honduras, and the Río Escondido, Nicaragua. In Honduras, Marineros and Gallegos (1998) recorded it from throughout the north (Departments of Gracias a Dios, Colón, Atlantida, and Cortés) besides Santa Bárbara in the northwest, and Olancho and El Paraíso in the east.

Hollister (1914) also listed a skin from British Honduras (Belize). There have been unauthenticated reports of capuchins in the Mayan Mountains of western Belize (the Chiquebul forest and in the region of the Trio and Bladen branches of the Monkey River) and in Sarstoon National Park on the southern border. Its occurrence in Belize has never been confirmed (McCarthy, 1982; Dahl, 1984, 1987; Hubrecht, 1986). Silva-López *et al.* (1995; Silva-López, 1998) also reported on the possible occurrence of *C. c. limitaneus* in Guatemala, in the Sierra del Espíritu Santo near the Guatemala–Honduras border. This also remains to be substantiated.

Aotus zonalis Goldman, 1914

Panamanian night monkey, owl monkey, mono de noche, marteja, jujuná (Reid, 1997).
Colombia, Costa Rica (?), Panama (Figure 4).

Type: In the National Museum of Natural History, Washington, DC, Accession No. USNM 171231, collected 29 April, 1922, by E. A. Goldman.
Type locality: Lake Gatún, Canal Zone (Panama), altitude 100 ft.

Hershkovitz (1949) recognized two night monkeys in northern Colombia and Central America, both as subspecies of *A. trivirgatus* (Humboldt, 1812):

Figure 4. The distribution of *Aotus zonalis*. Based on Defler (2003), Hall (1981), Hernández-Camacho and Cooper (1976), Hershkovitz (1983), Matamoros and Seal (2001), Reid (1997), Rodríguez-Luna *et al.* (1996), and Timm (1988). The numbers indicate two unconfirmed localities in Costa Rica: 1: La Selva Biological Reserve, a field station of the OTS, 1 km south of Puerto Viejo de Sarapiquí, Heredia (10°26′N, 83°59′W, altitude 35–150 m) (three sightings; Timm, 1988); 2: Near Bribri, Limón Province, about 70 km north northwest of Isla Bastimentos, Panama, the northernmost documented population of night monkeys (reported; Vaughan, 1983). Map drawn by Mark Denil and Kimberly Meek (Center for Applied Biodiversity Science, Conservation International, Washington, DC).

A. t. lemurinus (I. Geoffroy, 1843) and *A. t. griseimembra* Elliot, 1912a. He proposed that *griseimembra* in the northern lowlands and far northwestern Colombia (*Type locality*: Hacienda Cincinnati, northeast of Santa Marta, northwestern slope of the Sierra Nevada de Santa Marta, Magdalena, Colombia, altitude 1480 m) was the form extending into Panama, and included *zonalis* Goldman, 1914 as a synonym. Hill (1960) provided a very similar appraisal, one perhaps significant difference being that he placed the form *bipunctatus* Bole, 1937 from the Azuero Peninsula as a synonym of *griseimembra*, whereas Hershkovitz (1949, p. 404) had stated that "The night monkey of the Azuero

Peninsula, Panama, described as *A. bipunctatus*, is certainly a member of the common species but requires further comparison with additional material to determine its exact relationship to *griseimembra*. Most characters of *bipunctatus* described as distinctive, appear to be, rather, individual variables". It is not clear on what basis Hill (1960) synonymized *bipunctatus* with *griseimembra*, however, when Hershkovitz (1949) was reluctant to do so.

Hill (1960) listed the form *rufipes* Sclater, 1872 as a subspecies of *Aotes [sic] trivirgatus*. This was a live animal received by the Zoological Society of London from San Juan del Norte, Nicaragua. Night monkeys have never otherwise been recorded from Nicaragua, and Hershkovitz (1949) argued that the original description and color plate identify the animal as having come from Brazil. Allen (1910) simply said that the locality was unquestionably erroneous, a sentiment repeated by Elliot (1913).

Hernández-Camacho and Cooper (1976) restricted both *lemurinus* (Colombian Andes, elevations from 1000 to 1500 m up to 3000 to 3200 m) and *griseimembra* (northern lowlands, Santa Marta mountains, west to Río Sinú, Río San Jorge, lower Río Cauca, and lowlands of middle and upper Río Magdalena) to Colombia, while recognizing the form *zonalis* as the night monkey of northwestern Colombia (Chocó) and Panama. Hershkovitz (1983) continued to recognize *lemurinus* and *griseimembra* as distinct, but considered them to be subspecies of a single species; he made no mention of the name *zonalis*, but as he ascribed Central American night monkeys to *A. lemurinus lemurinus*, by implication he was regarding it as a synonym of this latter form. Unfortunately, a full explanation of his research and views regarding *Aotus* taxonomy was never published, but this switch from his 1949 arrangement was probably due to interpretation of the variable diploid numbers in the genus (*A. l. lemurinus* $2n = 55/56$; *A. l. griseimembra* $2n = 52/53/54$).

Reviewing the entire taxonomy and distributions of the night monkeys, Ford (1994) carried out multivariate analyses of craniodental measures and pelage patterns and color, and also took into consideration chromosomal data and blood protein variations. She concluded that there was "good support" for just two species north of the Río Amazonas: *A. trivirgatus* (Humboldt, 1812) to the east and north of the Rio Negro, and the polymorphic *A. vociferans* to the west of the Rio Negro. *A. vociferans*, as such, would include all the forms north of the Río Amazonas/Solimões in Brazil (west of the Rio Negro), Peru, Colombia, and Ecuador, and in the Chocó, northern Colombia and Colombian Andes, and Panama: *brumbacki*, *lemurinus*, *griseimembra*, and *zonalis*. Torres *et al.* (1998) identified six karyomorphs in Colombia, but concluded that a larger sample is required (both in numbers and geographic spread) in order to elucidate whether they represent different species.

Groves (2001) followed Hernández-Camacho and Cooper (1976) in recognizing *zonalis* as the form in Panama, and listed it as a subspecies of *lemurinus* along with *griseimembra* and *brumbacki* Hershkovitz, 1983. Defler *et al.* (2001) concluded that the karyotype of *A. hershkovitzi* Ramirez-Cerquera, 1983; (from the upper Río Cusiana, Boyacá, Colombia; $2n = 58$) was in fact that of true *lemurinus*, and that the karyotypes that Hershkovitz (1983) had considered to be those of *lemurinus* were in fact of *zonalis*. Defler *et al.* (2001; Defler, 2003; Defler and Bueno, 2003) concluded that *A. lemurinus* of Hershkovitz (1983) is in fact three karyotypically well-defined species, and that the night monkeys of the lowlands of Panama and the Chocó region of Colombia belong to the species *A. zonalis*, and those of the Magdalena valley to *A. griseimembra*, while those above altitudes of 1500 m should correctly be referred to as *A. lemurinus*. *A. zonalis* is distinguished from *griseimembra* by the darker upper surfaces of the hands and feet; blackish in Panama, but brownish in Colombia (Hershkovitz, 1949; Hernández-Camacho and Cooper, 1976). In the Canal Zone, they are brownish in overall body color, but grade into paler, grayer forms along the upper Río Tuira (Hershkovitz, 1949). In Colombia, they again have a brownish tinge to the pelage.

Spix's night monkey, *A. vociferans* (Spix, 1823), is recognized by Hershkovitz (1983), Groves (2001), and Defler (2003) as the form occupying a large part of the Colombian Amazon, north of the Río Amazonas, north to the Río Tomo (Hershkovitz, 1984) or Río Guaviare (Defler, 2003). It extends into Venezuela, Brazil, Peru, and Ecuador.

The distribution in Panama was mapped by Hall (1981; see also Hershkovitz, 1949; Reid, 1997). It would appear that it occurs west as far as the Río San Pedro in Veraguas along the Pacific coast, and from there is restricted to the Atlantic side of Panama through the province of Bocas del Toro, west as far as the Río Changuinola. It is absent from Chiriquí (Baldwin and Baldwin, 1976). Anthony (1916) recorded specimens from the Río Tuyra (Tuira), and Darién (Boca de Cupe and Tapalisa).

There have been a number of unconfirmed reports of night monkeys in Costa Rica (Reid, 1997). Timm (1988) examined the curious history and confusion of a specimen collected by Dr. van Patten in the highlands of Costa Rica that was recorded by Sclater (1872) and found by him to be the same as night monkeys from Bogotá. There was, however, confusion about the locality, and the specimen has been lost. Timm (1988; Timm *et al.*, 1989) argued that night monkeys should still be widely distributed in the eastern Caribbean lowlands of Costa Rica. This is based on three sightings in La Selva Biological Reserve, a field station of the Organization for Tropical Studies (OTS), 1 km south of Puerto Viejo de Sarapiquí, Heredia (10°26′N, 83°59′W, altitude 35–150 m). Lowland evergreen forest is predominant there. Timm (1988) also cited Vaughan (1983) who obtained information indicating the presence of night monkeys in Limón Province, around Bribri, near the Panamanian border, and only about 70 km north northwest of Isla Bastimentos, Panama, the northernmost documented population of night monkeys. In Colombia, *A. zonalis* occurs in the Pacific lowlands, south at least to the Río Raposo, south of Buenaventura, the region of Urabá and east to the Sinú valley, possibly through the San Jorge valley to the region of Puerto Valdivia in northern Antioquia (Hernández-Camacho and Cooper, 1976; Defler *et al.*, 2001).

Alouatta palliata (Gray, 1849)

Mantled howling monkey.
Colombia, Costa Rica, Ecuador, Honduras, Guatemala, Mexico, Nicaragua, Panama (Figure 5).

The current taxonomy of the mantled howling monkey is based on a thorough and detailed study of the pelage, crania, and taxonomic history of the Mesoamerican howlers by Lawrence (1933). It was inspired by the arrival at Harvard University's Museum of Comparative Zoology of several specimens from Herrara Province on the Azuero Peninsula, Panama, which Lawrence was unable to identify. Her findings resulted in her describing the form *trabeata* for the newly arrived specimens, besides *pigra* from Guatemala (replacing the name *villosa* in use previously) and *luctuosa* from Belize. She recognized seven subspecies in all.

Hill (1962) followed the taxonomy of Lawrence (1933) but continued to list the monotypic Guatemalan howling monkey, *A. villosa* (Gray, 1845), as a separate species, following Elliot (1913), but noting that *pigra* might turn out to be a synonym of *A. villosa*. In this, he was followed by Hall and Kelson (1959) and Napier (1976). The type of *A. villosa* is a skull (skin untraceable) of an adult female in the British Museum (Natural History), Accession No. 1843.9.14.3, unsexed (Napier, 1976). Smith (1970), however, regarded the name *villosa* as

Figure 5. The distributions of *Alouatta palliata* and *Alouatta coibensis*. Based on Aquino and Encarnación (1994), Curdts (1993), Defler (2003), Eisenberg (1989), Estrada and Coates-Estrada (1984), Froehlich and Froehlich (1987), García-Orduña and Canales-Espinosa (1995), García-Orduña *et al.* (1999), Hall (1981), Hernández-Camacho and Cooper (1976), Horwich and Johnson (1986), Reid (1997), Rodríguez-Luna *et al.* (1996, 2001), Silva-López *et al.* (1995), Smith (1970), Tirira (2001), Watts *et al.* (1986), and Watts and Rico-Gray (1987). Map drawn by Mark Denil and Kimberly Meek (Center for Applied Biodiversity Science, Conservation International, Washington, DC).

indeterminable, as had Lawrence (1933), and used *A. pigra* for the Guatemalan black howler. The type locality was given as "Brazils" by Gray (1845), but Sclater (1872) was convinced that it was attributable to a skin of a black howler from Vera Paz, Guatemala: a specimen collected by a Mr. Salvin, who also provided an account of howler monkeys in Guatemala. The account was published by Sclater (1872), and parts were reproduced *verbatim* in Elliot (1913) and Lawrence (1933), listing a number of localities in the Petén region of Guatemala. Apart from being all black, Sclater (1872) indicated that the direction of the hairs on the head was diagnostic, but this was ruled out by both Elliot (1913) and

Lawrence (1933) as too variable to be of use. The tenuous connection between Gray's *villosa* and Salvin's black howler from Guatemala, and the lack of a skin for the holotype (the skull is not diagnostic), resulted in Lawrence (1933, p. 336) concluding that it was "advisable to reidentify the howler monkeys of this region [Guatemala] and to regard *M[ycetes]. villosus* as indeterminable due to the absence of a type locality and the imperfect condition of the type." Hence, she described the form *A. palliata pigra* for the Guatemalan howler; the name used today, and recognized as a species distinct from *A. palliata* following the analyses of Smith (1970), Horwich (1983), and Horwich and Johnson (1984).

Lawrence's (1933) taxonomy included another all-black howler she described as *A. p. luctuosa* from a single specimen, adult male (skin and skull) from Mountain Cow, Cayo District, British Honduras (Belize), collected by O. L. Austin Jr., 12 April, 1928, and kept in the Museum of Comparative Zoology at Harvard University (Accession No. 20459). Smith (1970, p. 375) examined it, however, and found that it fell well within the range of individual variation observed in *pigra*, and considered it a junior synonym as a result. The resulting taxonomy of a monotypic *A. pigra* and five subspecies comprising *A. palliata* (*A. p. palliata*, *A. p. aequatorialis*, *A. p. mexicana*, *A. p. trabeata*, and *A. p. coibensis*) was accepted by Hall (1981), and modified only slightly by Froehlich and Froehlich (1986, 1987; in litt. to RAM 17 March 1987) whose study of dermatoglyphs convinced them that the forms *coibensis* and *trabeata* were quite distinct from the rest, and placed them as subspecies of *A. coibensis* Thomas, 1902.

Rylands *et al.* (2000) listed three species of howling monkeys in Mesoamerica: *A. palliata* (with three subspecies), *A. coibensis* (two subspecies), and *A. pigra*. Groves (2001) likewise recognized these three howling monkey species, but none of their subspecies. The forms *mexicana* Merriam, 1902 and *aequatorialis* Festa, 1903 were considered by him to be synonyms of *A. palliata*, and he listed *trabeata* Lawrence, 1933 as a synonym of *coibensis*. Smith (1970; see also Hall, 1981) found that the cranial characteristics (size and shape), dental cusp pattern and stylar development of the upper molars, and the color of the pelage (all features which he demonstrated clearly distinguished *A. p. mexicana* from sympatric *A. pigra*) were very similar in the forms *palliata*, *mexicana*, *trabeata*, and *coibensis*, indicating not just that they are closely allied but only weakly definable as subspecies. He, like others who followed, however, was reticent about actually subsuming all as junior synonyms because of the lack of an extensive analysis of geographic variation (Hall, 1981; Rylands *et al.*, 2000; Groves, 2001; Cortes-Ortiz *et al.*, 2003). To date, nobody has superseded the detailed, considerate, and thorough study of Lawrence (1933), but we recommend a reexamination of the status of *coibensis* and *trabeata*.

Alouatta palliata palliata (Gray, 1849)

Golden-mantled howling monkey or Nicaraguan mantled howling monkey (Hill, 1962), mono congo, mono aullador.
Costa Rica, Guatemala, Honduras, Nicaragua, Panama (?) (Figure 5).

Type: Syntypes, an adult female (skin and skull) and adult male (skin) in the British Museum (Natural History), collected by A. Sallé. Accession Nos. 1848.10.26.1 and 1848.10.26.2, respectively (Napier, 1976).
Type locality: Shores of Lake Nicaragua (*fide* Sclater, 1872; given in original description by Gray [1849] as Caracas, Venezuela).

Lawrence (1933) remarked that, excepting the small *coibensis* and the black howlers from Guatemala, she found it very difficult to distinguish the subspecies of *A. palliata* due to individual differences almost conciding with variation over the total range of the species. She described *A. p. palliata* as a large race of generally black pelage relieved by light yellowish flank hairs, but showing much individual variation. Especially difficult she found was the separation of *palliata* from *aequatorialis* in the region of Panama where they intergrade. *A. p. palliata* differs from *aequatorialis* in being generally blacker and with more rufous than yellowish golden hairs forming the mantle. It differs from *A. p. mexicana* mainly in some aspects of skull morphology (Lawrence, 1933).

The range limits separating *A. p. aequatorialis* from *A. p. palliata* are not clear. Lawrence (1933) cited a specimen of *A. p. palliata* from Cotó, extreme western Panama, and Hill (1962, p. 106) mentioned that specimens from Sevilla Island, western Panama, collected by J. H. Batty were "manifestly" *A. p. palliata*. Hall (1981), on the other hand, lists Sevilla Island, and Puerto Cortez, Costa Rica, as marginal records for *A. p. aequatorialis*. Many individuals from Panama are intermediate (Lawrence, 1933). From eastern Costa Rica, at least, *A. p. palliata* extends through Nicaragua to northern Honduras and, according to Curdts (1993), it just extends into Guatemala to the Río Motagua and possibly along the coast a short distance to the Cabo de Tres Puntas, where it meets *A. pigra*. It is not known to occur in El Salvador to the south (Burt and Stirton, 1961).

Alouatta palliata mexicana Merriam, 1902

Mexican howling monkey, mono aullador, mono aullador pardo, saraguato, mono zambo.
Mexico, Guatemala (Figure 5).

Type: Adult male, skin and skull, US National Museum, Accession No. 79398, collected by E. W. Nelson and E. A. Goldman, 23 April, 1906. Biological Survey Collection.
Type locality: Minatitlán, Vera Cruz, Mexico (Hill, 1962; Groves, 2001).

According to Merriam (1902), *A. p. mexicana* is similar to but much smaller than *palliata* (Gray, 1849), and he also provided a number of distinguishing (qualitative) cranial features. Lawrence (1933) was unable to establish any consistent difference in size from *A. p. palliata*, but recognized the subspecies due to some differences in pelage and certain cranial traits. The main feature is a more diffuse distribution of light-banded hairs over the back, and the paler more silvery bases of the hairs on the flank and on parts of the dorsum. The head, shoulders, limbs, tail, and (occasionally) spinal region are black.

The range of *A. p. palliata* extends eastward from southeastern Mexico, provinces of Vera Cruz, Tabasco, and northern Chiapas and Oaxaca. As discussed by Smith (1970), in Tabasco *A. p. mexicana* meets, and is sympatric with, *A. pigra* in a region 5 miles southeast of Macuspana. García-Orduña *et al.* (1999) found mixed populations of the two species in small habitat fragments in Tabasco (see also Rodríguez-Luna *et al.*, 2001). From there it extends east in a swathe through central Guatemala, skirting the southern limits of the range of *A. pigra*, but not extending south into El Salvador. Historically, at least, it would meet *A. p. palliata* only on the border with Honduras. Whether the two subspecies, *mexicana* and *palliata*, are still in contact is not known.

Alouatta palliata aequatorialis Festa, 1903

South Pacific blackish howling monkey, Ecuadorian mantled howling monkey. Peru: Mono coto de Tumbes, coto mono or coto mono de Tumbes (Aquino and Encarnación, 1994; Encarnación and Cook, 1998). Colombia: Aullador negro (Defler, 2003). Ecuador: Aullador de la costa, coto negro, mono mongón (Tirira, 2001). Panama: mono negro.
Colombia, Costa Rica (?), Ecuador, Panama, Peru (Figure 5).

Type: Four cotypes, two adult males, one female, and one young (skin numbers 101, 102, 103, and 104, and skull numbers 4688, 4886, 4692, and 4693), Museum of Zoology and Comparative Anatomy, University of Turin. Collected in September, year uncertain but between 1895 and 1898 by Enrico Festa.
Type locality: Vinces, Guayas Province, west coast of Ecuador.

Described by Elliot (1913)—who listed it as a species—as "similar to *A. palliata* but general color chocolate-brown instead of black". Lawrence (1933) found the mantle hairs to be golden-ochraceous, slightly shorter than in *palliata*, and most numerous posteriorly, hardly extending as far forward as the axillary region (note difference in this aspect to *trabeata*). She noted the original account of Festa who said that the general color was chocolate-brown with the bases of the hairs yellowish fulvous, the tips yellow, and the flanks golden yellow. The females, according to Festa, are browner with less golden than the males. Lawrence (1933) pointed out that the overall color is actually quite variable and can range from the bright-colored individuals to "quite black". The general color of the paler forms, however, is very different from that of *palliata*—paler and more golden brown than the orange-rufous of *palliata*, and the bright coloring extends farther down the hind limbs than in *palliata*. Even where the bright mantle is almost totally absent, the back is still broadly and more evenly sprinkled with paler hairs than is found in *palliata*. Lawrence (1933) failed to find any cranial characters to distinguish *aequatorialis* from *palliata*. Smith (1970) found that *aequatorialis* resembles typical *palliata* in most respects, and the few specimens he examined from Panama seem to indicate the presence of a well-defined zone of intergradation. In some ways, *aequatorialis* resembles *A. pigra*—both are large (Smith [1970] reported that the dental arcade does not exhibit such a marked trend in size reduction as seen in typical *palliata*), and the typical mantle coloration of *palliata* is frequently reduced or lost completely in *aequatorialis*. Cortés-Ortiz *et al.* (2003), who analyzed mtDNA of 19 *A. palliata* (from Panama to southern México), found a maximum level of sequence divergence of 0.5% but "a minor phylogeographic break separating northern and southern *A. palliata* [. . .] near Panama's Sona Pensinsula" (p. 75).

A. p. aequatorialis occurs in Panama, from the southern limit to the range of *A. p. palliata* (either in western Panama or extreme eastern Costa Rica), through the Serranía del Darién (Anthony, 1916; Lawrence, 1933) into western Colombia, north through the basins of the Ríos Sinú and Atrato to the Caribbean coast, and south through the Serranía del Baudó (Defler, 2003) and the foothills, lowlands, and lower montane areas west of the Andes to the Pacific coast, through Colombia and Ecuador, just into the Tumbes region of northern Peru (Aquino and Encarnación, 1994; Encarnación and Cook, 1998; Tirira, 2001).

Alouatta coibensis Thomas, 1902

Coiba Island howling monkey.
Panama (Figure 5).

Froehlich and Froehlich (1987) concluded that the howlers on Coiba Island and the Azuero Peninsula (*trabeata*) are close to, but quite distinct from, *A. palliata* in Panama, Costa Rica, and Nicaragua. Their argument was based on an analysis of fingerprint data, which they used as a surrogate to indicate genetic distance. Citing Bartlett and Barghoorn (1973), Froehlich and Froehlich (1987) indicated that the islands of Coiba and Jicarón were last connected to the mainland about 24,000 to 15,000 years ago, and they argued that *coibensis* should be considered a distinct species, with two subspecies—*coibensis* and *trabeata*. *A. coibensis coibensis* is smaller and has a less distinctive (duller) color than *trabeata*.

Cortés-Ortiz *et al.* (2003) found, however, that both *trabeata* and *coibensis* shared mtDNA haplotypes with *A. palliata* and were unable to substantiate the classification of *coibensis* (or *trabeata*) as a distinct species. The mitochondrial DNA divergence between *A. palliata* and *A. coibensis* was very low, showing only 0.1% sequence divergence—more than an order of magnitude fewer nucleotide substitutions than were observed between any other pair of *Alouatta* species. Divergence between *A. palliata* and *A. coibensis* was found to be similar to mitochondrial DNA distances observed between geographically separated populations within each of these two species. This, of course, does not by itself mean that the species *A. coibensis* should be sunk, but it does suggest that the morphological characters should be reassessed for their consistency. A morphometric study by Guadalupe Méndez is indicating that the howler monkeys from Azuero and Coiba are well differentiated from other Central American forms, and that they are certainly distinct subspecies (A. Cuarón, in litt. 21 May, 2003). Rylands *et al.* (2000) and Groves (2001) followed Froehlich and Froehlich (1987) in recognizing *coibensis* as a full species. We continue to recognize both *trabeata* and *coibensis* as distinct, but fully accept the possibility that they should be considered subspecies of *A. palliata*.

Alouatta coibensis coibensis Thomas, 1902

Coiba Island howling monkey.
Panama (Figure 5).

Type: An old male (skin and skull, Accession No. 1902.3.5.9) in the British Museum (Natural History), collected 18 May, 1901, by J. H. Batty (Napier, 1976).
Type locality: Coiba Island, Pacific coast of Panama.

This howling monkey is known only from Coiba Island and neighboring Jicarón, off the Pacific coast of Panama. It is smaller than other Central American howling monkeys, and has a duller pelage than the closely related form from the Azuero Peninsula, *A. c. trabeata*. Compared to *A. c. trabeata*, the mantle is more confined to the flanks. Hill (1962) described the head and fore part of the back as "seal brown, appearing almost black in most lights". The lower back is paler and the rump and proximal parts of the hind limbs are walnut. The flank hairs are elongated—orange-rufous to cinnamon-rufous according to Hill (1962), and golden as described by Groves (2001). The type specimen has a large pedunculated and unpigmented scrotum (Hill, 1962). The females are similar in color to the males, but smaller.

Alouatta coibensis trabeata Lawrence, 1933

Azuero howling monkey, golden howling monkey (Froehlich and Froehlich, 1987).
Panama (Figure 5).

Type: Adult male (skin and skull, Accession No. 29545) in the Museum of Comparative Zoology, Harvard, collected by Thomas Barbour, in March 1933.
Type locality: Capina, Herrera Province, Panama.

According to Hill (1962), this howling monkey is distinguished principally by its golden flanks and loins (golden-ochraceous tips to hairs), together with a browner appearance of the rest of the body. Lawrence (1933) described it as having a walnut-colored back and very long silky golden flank hairs extending from the axilla to the groin. Besides this, she noted a greater degree of sexual dimorphism in skull measurements than in other populations of Central American howlers. Froehlich and Froehlich (1987) found it to be more closely related to *coibensis* Thomas, 1902 than to other Central American forms and, recognizing *coibensis* as a full species, placed it as a subspecies. Although listing it as a synonym of *coibensis*, Groves (2001, p. 180) recognized that "the mainland

and insular populations of this species [*coibensis*] differ considerably and are presumably (at least?) subspecifically distinct." Froehlich and Froehlich (1987) provide an interesting discussion regarding the zoogeography of the region and how and why the Azuero peninsula may have been relatively isolated in the past, resulting in the differentiation of its howlers (and spider monkeys) and a relatively depauperate mammal fauna. Rylands *et al.* (2000) and Groves (2001) followed Froehlich and Froehlich (1987) in recognizing *trabeata* as a subspecies of *coibensis.* It is endemic to the Azuero Peninsula, Panama (Froehlich and Froehlich, 1987; Rowe, 2000).

Alouatta pigra Lawrence, 1933

Black howling monkey, Lawrence's howler monkey (Hall, 1981), Yucatán black howler (Reid, 1997), saraguato negro (Mexico).
Belize, Guatemala, Mexico (Figure 6).

Figure 6. The distribution of *Alouatta pigra.* Based on Curdts (1993), Eisenberg (1989), Hall (1981), Horwich and Johnson (1986), Jones *et al.* (1974), Reid (1997), Rodríguez-Luna *et al.* (1996), Silva-López *et al.* (1995), Smith (1970), and Watts and Rico-Gray (1987). Map drawn by Mark Denil and Kimberly Meek (Center for Applied Biodiversity Science, Conservation International, Washington, DC).

Type: Adult male, skin and skull, in the Museum of Zoology, University of Michigan, collected 4 May, 1931 by A. Murie. One of a series of 12 specimens (five adult males, five adult females, and a young female; Lawrence, 1933).
Type locality: Uaxactún, Petén, Guatemala.

Although placed as a subspecies of *A. palliata* by Lawrence (1933) and Hill (1962), Smith (1970) (see also Jones *et al.*, 1974; Horwich, 1983; Horwich and Johnson, 1984) demonstrated that the black howling monkey from the Yucatán peninsula (Mexico), Belize, and northern Guatemala is a valid species. *A. pigra* is larger than typical *palliata* and distinguished from other Central American howlers by absence in both sexes of light areas along flanks. Smith (1970) found a zone where the two species are sympatric in Tabasco, Mexico (5 miles SE of Macuspana; see also García-Orduña *et al.*, 1999; Rodríguez-Luna *et al.*, 2001; Serio-Silva and Rico-Gray, 2004), and compared the pelage, dental and cranial morphology, and the articulation of the mandible in the two species. Hall (1981) described the cranial differences between *A. pigra* and *A. palliata*. Silva-López (1998) recorded that it occurs in sympatry with *A. palliata* in the Biotopos Chocón Machacas and the Mario Dary Rivera Biosphere Reserve, and there is a need to study the mechanics of their coexistence in these areas.

The westernmost locality given by Hall (1981) is at Frontera, in the Mexican state of Tabasco; *A. palliata* has been recorded just west of there along the coast, 6 miles south of Cárdenas. Further localities that define the western and southern limits of its range include 5 miles southeast of Macuspana, Tabasco, and San Mateo Ixtatán (*ca.* 11,000 ft) in Guatemala. Hall (1981) identified the southern limits of its range in the east with three localities along the Río Motagua basin in Guatemala, including Quirigua and Zacapa (right bank of the river). Curdts (1993), on the other hand, found that the southern and southwestern limits of the range of *A. pigra* in Guatemala were defined by the Lago de Izabel, El Golfete, and the Río Dulce. He noted large numbers of *A. pigra* in the Río Polochic delta, entering the west end of the Lago de Izabel. These are just to the north of the Río Motagua, where Curdts (1993) identified *A. palliata*.

Ateles geoffroyi Kuhl, 1820

Geoffroy's spider monkey, mono araña.
Belize, Colombia, Costa Rica, El Salvador, Honduras, Guatemala, Mexico, Nicaragua, Panama (Figure 7).

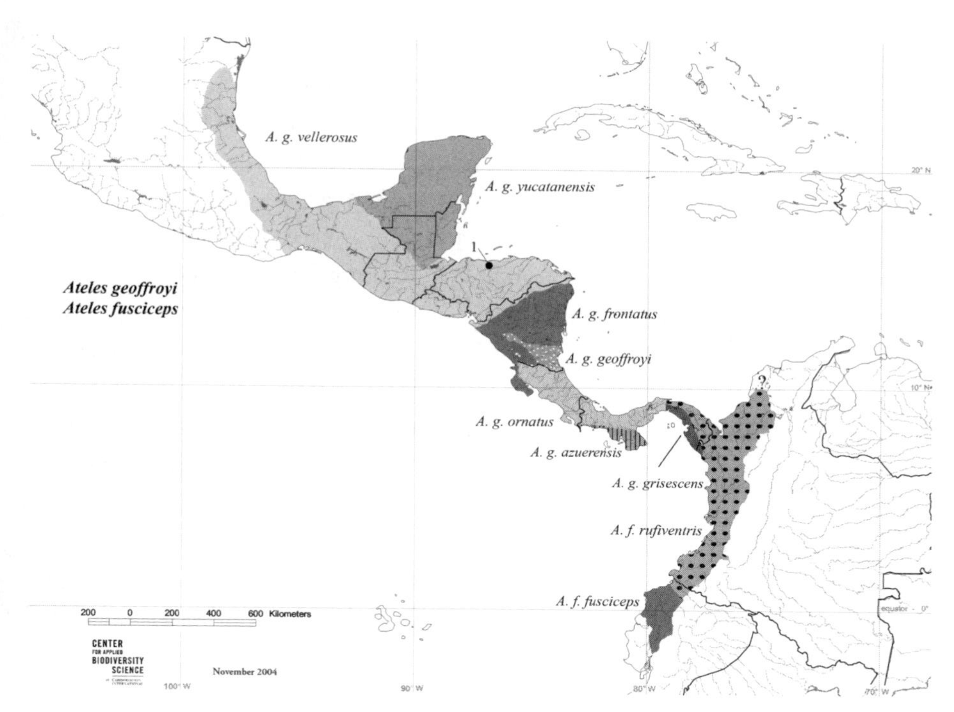

Figure 7. The distributions of *Ateles geoffroyi* and *Ateles fusciceps.* Based on Defler (2003), Eisenberg (1989), Estrada and Coates-Estrada (1984), Hall (1981), Heltne and Kunkel (1975), Hernández-Camacho and Cooper (1976), Hershkovitz (1949), Horwich and Johnson (1986), Kellogg and Goldman (1944), Konstant *et al.* (1985), Marineros and Gallegos (1998), Reid (1997), Rodríguez-Luna *et al.* (1996), Silva-López *et al.* (1995), Tirira (2001), Watts *et al.* (1986), and Watts and Rico-Gray (1987). 1: Pico Bonito National Park, Honduras. Map drawn by Mark Denil and Kimberly Meek (Center for Applied Biodiversity Science, Conservation International, Washington, DC).

The Mesoamerican spider monkeys are variable in their pelage color and are difficult to resolve taxonomically. The classic study of Kellogg and Goldman (1944) resulted in the recognition of 16 taxa of spider monkeys, 10 of them in Mesoamerica: 9 taxa of *A. geoffroyi* and, extending into Panama from northwestern Colombia, *A. fusciceps robustus.* Over the last 60 years, little has been done that has shaken the foundation laid down by Kellogg and Goldman (1944): Hill (1962), Hall (1981), and Konstant *et al.* (1985) maintained their taxonomy of the Central American spider monkeys and, till recently, all but one have stood the test of time. Schultz (1960) who studied geographic variation in the crania of 203 adult *A. geoffroyi* concurred with the taxonomic arrangement of Kellogg and Goldman (1944). Silva-López *et al.* (1996) concluded that *A. geoffroyi pan* Schlegel, 1876, a very dark form from Guatemala, was a variant of the Mexican spider monkey, *A. g. vellerosus*, while Napier (1976) showed that *A. g. panamensis* is a synonym of *A. g. ornatus*.

Having disqualified *A. g. pan* and *panamensis*, however, the remaining eight Mesoamerican spider monkeys identified by Kellogg and Goldman (1944) are still poorly defined. Their taxonomy was based on cranial morphology, body size, and pelage color patterns although the cranial differences were minimal. Silva-López *et al.* (1996) suspected that, like *pan*, the form *A. geoffroyi yucatanensis* is merely a color variant of *A. g. vellerosus*. The validity of the remaining forms requires a good understanding of the geographic patterns of natural variation; something which is increasingly difficult to attain due to the widespread loss and fragmentation of their forests and populations.

Collins (1999) and Collins and Dubach (2000) divided the subspecies into two groups: northern (*vellerosus* and *yucatanensis*) and southern (*frontatus*, *ornatus*, *geoffroyi*, *panamensis*, and *grisescens*). Groves (2001) recognized only five subspecies: *A. g. yucatanensis*, *A. g. vellerosus* (synonym *A. g. pan*), *A. g. geoffroyi* (synonym *A. g. frontatus*), *A. g. ornatus* (synonyms *azuerensis* and *panamensis*), and *A. g. griscescens*. The taxonomy we follow here is essentially that of Kellogg and Goldman (1944).

Ateles geoffroyi geoffroyi Kuhl, 1820.

Nicaraguan spider monkey (Kellogg and Goldman, 1944).
Costa Rica, Nicaragua (Figure 7).

Type: Adult female, Muséum National d'Histoire Naturelle, Paris (menagerie specimen acquired in 1819; registered in I. Geoffroy, Catalog méthodique del la collection des mammifères, pt 1 (Catalog des Primates), p. 49, 1851) (Kellogg and Goldman, 1944).
Type locality: Unknown, but restricted to San Juan del Norte (Greytown), Nicaragua by Kellogg and Goldman (1944) who refer to a specimen of Sclater (1862) from the Río Rana, Gorgon Bay, near San Juan del Norte, which was listed by Gray (1870) as "*Ateles hybridus*" from St. Juan, Nicaragua.

The nominotypical subspecies of Geoffroy's spider monkey is silvery to brownish gray on the back, upper arms, and thighs (Konstant *et al.*, 1985). The black on the elbows, knees, and upper and lower arms and legs is variable, but the hands and feet are always black. Chest is similar to the back but the lower abdomen can be quite golden. The face is black with flesh-colored eye rings. According to Kellogg and Goldman (1944), it most closely resembles *frontatus* from northwestern Costa Rica, and the light buff (silvery to brownish gray) color of the back contrasts with that of *ornatus* of eastern Costa Rica, which they described as rich rufescent. The skull is very similar to *ornatus* but apparently smaller (Kellogg and Goldman, 1944). Groves (2001) considered the form *frontatus*, also from Costa Rica and Nicaragua, to be a junior synonym of *A. g. geoffroyi*, but this needs further study (see below).

Its distribution is given by Kellogg and Goldman (1944) as the coastal region around San Juan del Norte or Martina Bay, southeastern Nicaragua; probably ranging across the lowlands to the vicinity of Lake Managua and Lake Nicaragua on the Pacific coast. It possibly extends into northern Costa Rica. Specimens examined by Kellogg and Goldman (1944) were from Monagua, Nicaragua.

Ateles geoffroyi azuerensis Bole, 1937

Azuero spider monkey.
Panama (Figure 7).

Type: Adult female, Cleveland Museum of Natural History, Accession No. 1235.
Type locality: Altos Negritos, 10 miles east of Montijo Bay (part of the spur forming south drainage divide of Río Negro), Mariato Suay Lands, Azuero Peninsula, Veraguas Province, Panama; altitude 1500 ft.

Distinguished from neighboring forms by a general color of light tawny or ochraceous-tawny. Konstant *et al.* (1985) summarized the description of *azuerensis* (two skins from the type locality) by Kellogg and Goldman (1944), as follows: the back is grayish brown, and a little darker than the underside; outer surfaces of limbs black, but top of head, black or blackish brown. Believed by Groves (2001) to be a probable junior synonym of *A. g. ornatus* of Nicaragua and Costa Rica.

Definitely known only from the western (Veraguas) side of the forested mountains of the Azuero peninsula in the vicinity of Ponuga, where it appears to be isolated. Kellogg and Goldman (1944) indicated that it may occur to the west along the Pacific coast to the Burica Peninsula, near the Panama–Costa Rica border. Kellogg and Goldman (1944) tentatively attributed a series of 25 skulls from the collection of Adolph H. Schultz (no skins, but reported to have been light-colored) from Río La Vaca, near Puerto Armmuelles, Burica Peninsula to *A. g. azuerensis*. Baldwin and Baldwin (1976) found no evidence that spider monkeys ever occurred in the Province of Chiriquí. Konstant *et al.*

(1985) reported that the Azuero Pensinula was widely deforested and it is likely to be surviving only in western parts. It occurs in the Cerro Hoya National Park (Matamoros and Seal, 2001).

Ateles geoffroyi frontatus (Gray, 1842)

Black-browed spider monkey, black-foreheaded miriki (Kellogg and Goldman, 1944).
Costa Rica, Nicaragua (Figure 7).

Type: Adult female with young (paratype) shot by Capt. Sir Edward Belcher (skin and skull), British Museum (Natural History). Accession No. 1842.10.30.4 (Napier, 1976).
Type locality: South America (= harbor of Culebra, León = Culebra, Bay of Culebra, Guanacaste, northwestern Costa Rica) according to Gray (1843 in Kellogg and Goldman, 1944).

According to Kellogg and Goldman (1944), *frontatus* is similar in color pattern (restriction of black areas to top of head and, irregularly, to outer surfaces of limbs) to *geoffroyi* of southeastern Nicaragua, but the body is darker, with the upperparts brown and underparts honey yellow to tawny, rather than light buff. It differs from *panamensis* in having a brownish instead of deep ferruginous general body color, and from *vellerosus* of Veracruz in the restriction of black areas to the anterior part of the back and more yellowish tone of the lumbar region. Apart from the type in the British Museum from northwestern Costa Rica, little is definitely known of its characters or distribution. The genetic analyses of Collins and Dubach (2000) included a sample from a Nicaraguan spider monkey, which, by its pelage, they tentatively identified as *A. g. frontatus*. It was quite distinct from *panamensis* and *vellerosus/yucatanensis*, and their findings suggested that it was a sister clade to northern, or even all, *A. geoffroyi*. Groves (2001) did not recognize this form as a valid subspecies, considering it a synonym of *A. g. geoffroyi*.

A. g. frontatus is believed to range through northwestern Costa Rica and extreme western and northern Nicaragua (Kellogg and Goldman, 1944). Specimens from Nicaragua examined by Kellogg and Goldman (1944) were from the following localities: Lavala; Peña Blanca; Río Siquia; Río Yoya, a tributary of the Río Princapolca; Tuma; and Uluce. Allen (1908, 1910) recorded *A. geoffroyi* from the east slope of the Nicaraguan highlands, Savala (800 ft), Tuma (1000 ft), Peña Blanca (high point in low Atlantic coast forests, 1500 ft), and

Uluce (about 1000 ft), and in the highlands of northern Nicaragua at Matagalpa (2000 ft).

Ateles geoffroyi grisescens Gray, 1866

Hooded spider monkey.
Panama, Colombia (?) (Figure 7).

Type: Skin of an adult, sex unknown, in the British Museum (Natural History), Accession No. 1865.4.20.2 (Napier, 1976).
Type locality: Unknown, but restricted by Kellogg and Goldman (1944) to the Río Tuyra, southeastern Panama.

According to Kellogg and Goldman (1944), the adults have long, lax pelage and a peculiar dusky coloration, with a general admixture of yellowish gray or golden hairs, the hairs on the upperparts are golden at the base. The skull, they concluded, indicates a close relationship to *panamensis*, despite the latter's contrasting deep reddish color. Specimens examined by Kellogg and Goldman (1944) were from Chepigana, Darién. Konstant *et al.* (1985) examined specimens (no locality given) that were much paler than the descriptions of Kellogg and Goldman (1944) and Hernández-Camacho and Cooper (1976) would indicate.

Kellogg and Goldman (1944) presumed that it occurred in the valley of the Río Tuyra and probably southeastward through the Serranía del Sapo of extreme southeastern Panama and the Cordillera de Baudó of northwestern Colombia. Matamoros and Seal (2001) indicate its occurrence in the basin of the lower Río Tuira in Panama and the frontier zone with Colombia. Heltne and Kunkel (1975) indicated Cerro Pirre or Río Tucutí as marking the limits of its range with *A. f. rufiventris* to the north. Hernández-Camacho and Cooper (1976) indicated that *grisescens* occurs in Colombia: "... [it] is known only from the vicinity of Juradó very near the Panamanian border on the Pacific coast. It is undoubtedly restricted by the Baudó Mountains to a narrow coastal strip that may extend as far south as Cabo Corrientes." (p. 66). Defler *et al.* (2003) recorded that there is no recent information regarding its presence or otherwise along the Panamanian border, but that colonists near the northern parts of the Serranía de Baudó region talk of two "types" of *Ateles*, one in the lowlands (definitely *A. fusciceps*) and another form above 500–600 m altitude (J. V. Rodríguez-M., unpubl. data): the only real suggestion is that this taxon might actually be present in Colombia. *A. fusciceps* in the central part of the

Sierra de Baudó would indicate that the occurrence of *grisescens* there would be limited to the portion immediately abutting Panama, and not the entire mountain range (Defler *et al.*, 2003).

Ateles geoffroyi ornatus Gray, 1870

Ornate spider monkey (Kellogg and Goldman, 1944), Mono Colorado.
Costa Rica, Nicaragua (Figure 7).

Type: Juvenile (skin and skull) of unknown sex, British Museum (Natural History), Accession No. 1850.1.26.2 (Napier, 1976).
Type locality: Unknown, but restricted by Kellogg and Goldman (1944) to Cuabre, Talamanca region, southeastern Costa Rica.

Type of *panamensis*: Adult female, skin and skull, U.S. National Museum, Accession No. 171489 (Biological Surveys collection); collected by E. A. Goldman, 8 June, 1911. Original No. 21165.
Type locality: Cerro Brujo, about 15 miles southeast of Portobello, Province of Colón, Panama; altitude 2000 feet (Kellogg and Goldman, 1944).

This most intensely red of the Central American spider monkeys was described by Kellogg and Goldman (1944), under the name *panamensis*, as a rather large, deeply rufescent race, similar to *ornatus* of the Caribbean slope of Costa Rica, but with a more intense reddish tone (back of shoulders to base of tail, backs of thighs, and sides of body), the back less obscured by overlying dusky hairs; inner side of upper arms pinkish cinnamon to ferruginous. It differs from *azuerensis* in its deep reddish instead of cinnamon or tawny general coloration. A black (sometimes freckled with a pale skin) face.

Kellogg and Goldman (1944) described *ornatus* as being "a dark golden yellowish subspecies, the upper parts in strong light having a glossy, golden yellow sheen, owing to the yellowish subterminal bands of hairs". Napier (1976: 88) found that the type specimen falls well within the range of variation of Panamanian specimens, and implied that Kellogg and Goldman (1944) would not have described *panamensis* had they not been prevented by wartime constraints from traveling to London to examine the type of *ornatus*, which they knew only from the somewhat misleading type description.

The ornate spider monkey is found in forested regions of Panama, east of the Canal Zone (Cordillera San Blas), and west through Chiriquí to central western Costa Rica. Heltne and Kunkel (1975) give the following localities as marking the eastern limit of its range: San Juan, Cerro Brujo, Cerro Azul, and Río Pequeñi—all on or within the boundary line of the Madden Lake watershed, and nowhere more than 30 miles east of the Panama Canal. The Río Bayano basin just to the east is occupied by *A. fusciceps rufiventris* (see Handley, 1966; Heltne and Kunkel, 1975). This is the spider monkey of the Osa Peninsula, Corcovado National Park, and Carara Biological Reserve in Costa Rica (Matamoros and Seal, 2001). The population on the Island of Barro Colorado is introduced (Carpenter, 1935; J. F. Eisenberg, pers. comm. in Konstant *et al.*, 1985). Crockett *et al.* (1997; see also Cody, 1994; Querol *et al.*, 1996) observed spider monkeys in the Refugio Bartola/ Reserva Indio-Maíz (300,000 ha), along the Río Bartola, north of the Río San Juan along the frontier with Costa Rica. They were unable to identify the subspecies but said that, unlike *A. g. geoffroyi*, they were "distinctly reddish on the back and on the top of the tail; the ends of the limbs were dark" (p. 73).

Ateles geoffroyi vellerosus Gray, 1866

Mexican spider monkey (Kellogg and Goldman, 1944). Mono araña.
El Salvador, Honduras, Guatemala, Mexico (Figure 7).

Type: Skin (Accession No. 1845.11.2.2) and skull (Accession No. 1845.12 .8.16) in the British Museum (Natural History). Napier (1976) inferred that it was a female. Figured in Sclater (1872).
Type locality: Originally assigned by Gray to Brazil, but restricted by Kellogg and Goldman (1944) to Mirador, about 15 miles northeast of Huatasco, Veracruz, Mexico; altitude 2000 feet (the type locality of the junior synonym *A. neglectus*, Reinhardt, 1873).

A. g. vellerosus occurs in the forests of Veracruz and eastern San Luis Potosí and southeastward through Tabasco, across the Isthmus of Tehuantepec in eastern Oaxaca, including the highlands of Guatemala (thought by Kellogg and Goldman to have been occupied by *A. g. pan*, here considered a synonym) through El Salvador and Honduras, including the north coast to the lowlands of the Mosquitia in the Department of Gracias a Dios.

Kellogg and Goldman (1944, p. 33) described *A. g. vellerosus* as "a subspecies distinguished by a combination of black or brownish-black top of head, neck, and shoulders, in contrast with buffy lumbar region, and pinkish-buff to cinnamon-buff underparts. Differs from *yucatanensis* of Quintana Roo in deeper buff underparts (underparts in *yucatanensis* are silvery-white or light buff)". According to Konstant *et al.* (1985), dorsal surfaces range from black to dark brown, except for a light band across the lumbar region, and contrast strongly with its lighter abdomen and inner limbs. Exposed flesh-colored skin is often present about the eyes. Silva-López *et al.* (1996) reported that this description is compatible with *A. g. vellerosus* at Sierra de Santa Marta, Veracruz, Mexico, although there is also considerable variation, for example, in lighter dorsal surfaces, a less distinct band across the lumbar region, and lack of contrast between the color and tones of the dorsal surfaces and the inner limbs. Konstant *et al.* (1985) also indicated the absence, or marked reduction, of the white triangular forehead patch and sideburns (present in *A. belzebuth* and the darker *A. hybridus*). Some spider monkeys at the Sierra Santa Marta have distinct white forehead triangles, and Silva-López *et al.* (1996) found that *vellerosus* there is quite variable, with the pelage ranging from very dark to very pale. In Tikal, Guatemala, they observed whitish *vellerosus*, with a darker distal third of the tail. In El Salvador, Burt and Stirton (1961, p. 21) described *vellerosus* as follows "Top of head, arms, legs and tip of tail nearly black; from shoulders to rump golden slightly washed with dark brown; cheeks, throat, belly, and undersides of limbs whitish (washed with pale yellow on breast)". According to Marineros and Gallegos (1998) in Honduras, it has a black pelage, paler on the back (grizzled coffee color) and underparts, with pale circles of naked skin around their eyes.

The very dark *A. g. pan* from Cobán, Alta Vera Paz, Guatemala (co-types: an adult male and two adult females in Leiden) was listed by Kellogg and Goldman (1944) as the species of the central highlands of Guatemala. Konstant *et al.* (1985) noted its similarity to the darker *vellerosus*, differing only in the relative lack of contrast between dorsal and ventral color and lack of a lighter-colored saddle on its lumbar region, and doubted its validity. Its supposed range is broadly covered by pine forest, dominated by *Pinus*, *Quercus*, and *Liquidambar* with some remnants of tropical forest in the lowlands of Alta Verapaz and Quiché (including the locality of Barillas), to the north; near Chilascó and in the Biotopo Mario Dary Rivera, in the east; and in Escuintla and Retalhuleu, in the south (Silva-López *et al.*, 1996). Only howling monkeys have been found in the Sierra de Chamá (Alta Verapaz, Quiché), 300–1500 m asl; the Sierra de Chuacús (Baja Verapaz, Quiché), 600–2100 m asl; and the Sierra de los Cuchumatanes (Huehuetenango, Quiché), 1500–2700 m asl (Silva-López *et al.*, 1996). Kellogg and Goldman (1944) believed that *A. g. pan* intergraded with *vellerosus*, Konstant *et al.* (1985) were doubtful of its validity, and Silva-López *et al.* (1996) concluded that it was not a valid taxon.

The spider monkeys of Honduras have been very poorly documented. They are based on samples from the Tegucigalpa area (Cantoral and Guaymaca), Olancho (Catacamas), and Octopeque (El Chorro), all from central and southern Honduras, south of the Cordillera Nombre de Dios. Recent studies in the Pico Bonito National Park in northern Honduras (Hines, 2004) have indicated that the spider monkeys there are neither *A. g. vellerosus* nor *A. g. yucatanensis.* Unlike *vellerosus*, the North Honduran *Ateles* have a bright-reddish-orange back, similar to *panamensis. A. g. yucatanensis* is a much darker auburn-brown. The underparts in the northern Honduran *Ateles* are closer to the silver-white of *A. g. yucatanensis*, although the lower stomach tends towards a darker buff color. The white on the inside of the arms and legs of the northern Honduran specimens extend to the ankles and wrists, as in *A. g. yucatanensis*, whereas in *A. g. vellerosus*, the light color generally extends only as far as the elbows and knees. The Honduran specimens examined by Kellogg and Goldman (1944) were from the central and southern parts of of the country where the climate is much drier than along the northern coast. The climate is markedly drier on the southern side of Pico Bonito, particularly in the Ahuan Valley, which has desert-like conditions and a flora that contrast with the more humid coastal side of the park. It is quite common to encounter agave and cacti in the Ahuan Valley and throughout the areas along the southern side of the park. Specimens from southern Honduras are less intense in the red-orange color on their back, but retain the similar bright silver-white upper chest, and a darker buff coloration on the lower chest and stomach.

Ateles geoffroyi yucatanensis Kellogg and Goldman, 1944

Yucatán spider monkey (Kellogg and Goldman, 1944).
Belize, Guatemala, Mexico (Figure 7).

Type: Adult male, skin an skull, U. S. National Museum, Accession No. 108531 (Biological surveys collection), collected 2 April, 1901, by E. W. Nelson and E. A. Goldman; original number 14652.
Type locality: Puerto Morelos, northeast coast of Quintana Roo, Mexico; altitude 100 ft.

Kellogg and Goldman (1944, p. 35) wrote that *A. g. yucatanensis* is a "rather small, slender, light-colored race with underparts silvery whitish or very pale buff, pelage short and thin. Size about as in *vellerosus* of Veracruz but decidedly paler, especially on the underparts where in typical specimens a whitish silvery tone extends to neck and inner sides of limbs; underside of tail cream-buff to near callosity; frontal outline of skull more prominent". Konstant *et al.* (1985) described it as having a brownish-black head, neck, and shoulders, lighter brown on the lower back and contrasting with silvery-white underside, inner limbs, and sideburns. In the south of its range (Campeche and Guatemala), Kellogg and Goldman (1944) noted that specimens from Apazote, Campeche, and Uaxactúm, Guatemala, are referable to *yucatanensis* but with slightly darker and more buffy underparts, indicating gradation towards neighboring *vellerosus.*

Jones *et al.* (1974) studied the crania of spider monkeys from the Yucatán peninsula, from Veracruz and Oaxaca (*vellerosus*) and from Nicaragua (*frontatus*) and found that they differ mainly in breadth dimensions. They also examined pelage color, and concluded that whereas *frontalis* from Nicaragua was quite distinct (almost entirely yellowish except for a blackish area on the head and neck), specimens from the Yucatán did not differ from adjacent *vellerosus*, and therefore considered *yucatanensis* a synonym, while Konstant *et al.* (1985) noted that *yucatanensis* can be confused with lighter individuals of *vellerosus.* Silva-López and Rumiz (1995) reported that spider monkeys in the Río Bravo Conservation and Management Area in Belize resembled the descriptions of *vellerosus* more than *yucatanensis*, and noted that inter-individual variation in the color made it difficult to assign individuals to a particular subspecies. The genetic studies of Collins and Dubach (2000) indicated that *vellerosus* and *yucatanensis* were inseparable in mtDNA (based on three individuals: one from Belize, second from Yucatán, and the third from the Guatemala). Further morphological and genetic studies and most importantly field observations and a modern review of pelage variation are needed to clarify the validity or otherwise of this taxon (Silva-López *et al.*, 1995), but the evidence that *yucatanensis* is separable appears poor.

A. g. yucatanensis occurs in the forests of the Yucatán peninsula, northeastern Guatemala, and adjoining parts of Belize, intergrading to the south in Mexico (Campeche) and Guatemala with *vellerosus.* Parra Lara and Jorgenson (1998) reported on a survey of 36 localities in the state of Quintana Roo. They confirmed the presence of spider monkeys in 11 of them, and received reports of their occurrence in a further 19, extending from the Ejido Tres Garantias in the south to locations way in the north, near Cancún, at Cenote Notnozot. Ramos-Fernández and Ayala-Orozco (2003) have studied the population size and habitat use of *A. g. yucatanensis* around the Punta Laguna, Quintana Roo.

Ateles fusciceps rufiventris Sclater, 1871

Colombian black spider monkey (Kellogg and Goldman, 1944). Panama: mono araña, mono negro.
Panama, Colombia (Figure 7).

Type: Juvenile skin (date and collector unknown), BM 1876.1.31.24 (Napier, 1976: 95).
Type locality: Río Atrato, Darién, Colombia.

The Colombian black spider monkey was described by Kellogg and Goldman (1944) as nearly all black, except for a brownish tinge on the forehead of one individual they examined, and a few inconspicuous whitish hairs on the chin and around the mouth. Heltne and Kunkel (1975) examined pelage color and patterns in detail, and added that the specimens they examined from eastern Panama had white or golden hairs on the cheeks and reddish or golden-banded hairs on the ventral surface of the trunk and limbs to a varying extent. Only 6 of the 24 specimens they examined were completely black on the frontal region. A series from the region of Tacarcuna showed all possible combinations of the distribution of white hairs on the facial and frontal areas and all black or brick-red tinged hairs in the ventral (genital) region, extending to the inner thigh.

There is still some confusion as to the taxonomy of the Colombian black spider monkey, despite the fact that a careful reading of Heltne and Kunkel (1975) leaves no doubt regarding the validity of the name *rufiventris* Sclater, 1871 as opposed to *robustus* Allen, 1914. Mittermeier and Coimbra-Filho (1981), Groves (2001), and Defler (2003) listed *A. fusciceps rufiventris* (= *robustus*), whereas Konstant *et al.* (1985) and Mittermeier *et al.* (1988) listed *Ateles f. robustus* (= *rufiventris*). Hernández-Camacho and Cooper (1976) also used the name *robustus*. Rylands *et al.* (2000) misidentified the author—listing *rufiventris* but ascribing it to Allen (1914) rather than Sclater (1871). Basing themselves only on the description of Sclater, an illustration, and a more detailed description of the type by Elliot (1913), Kellogg and Goldman (1944) argued that Sclater's *rufiventris* was probably a young female *A. p. aequatorialis*. Hershkovitz (1949) concluded, without saying why, that *rufiventris* was a color variant of *A. g. grisescens*; in this, he was followed by Napier (1976). Hill (1962) studied the type of *rufiventris* and decided it was a valid species. While not comparing it with *robustus*, his notes on pelage variation showed it to be similar, and especially similar to *Ateles dariensis* Goldman, 1915, from "near head of Río Limón, Mount Pirre, eastern Panama; altitude 5200 feet", which was considered later by Goldman himself to be a synonym of Allen's *robustus* (Kellogg and Goldman, 1944). It is of interest that, contra Hershkovitz (1949), Hill (1960: 502) found that "of all the races of *A. geoffroyi*, *A. g. grisescens* shows the least resemblance to *A. rufiventris*".

Another question, yet to be resolved, is whether *rufiventris* is a subspecies of the brown-headed spider monkey, *A. fusciceps* Gray, 1866, of Ecuador, or should be aligned with *geoffroyi*, or should be regarded as a distinct species. Having decided that the pelage of the type of *A. rufiventris* "merely represents a pattern variant certainly within the spectrum of variation implied by the USNM [US National Museum] series of *A. f. robustus*" (p. 98), Heltne and Kunkel (1975) pointed out that none of the USNM specimens they examined, and only one reported by Kellogg and Goldman (1944), showed the slightly brownish tinge on the forehead—the character (along with some cranial details) that Kellogg and Goldman (1944) used to align it with *fusciceps*. Kellogg and Goldman (1944, pp. 3–4) indicated that "perhaps the most clearly defined line of demarcation between the species, as we understand them, is in eastern Panama, where the range of the deep reddish *panamensis*, a member of the *geoffroyi* group, meets or closely approaches the range of the nearly all black *robustus* [*rufiventris*]". Color it would seem is the basis for them separating *fusciceps* from *geoffroyi*, but Kellogg and Goldman (1944) commented later (p. 30) that "Despite the marked contrast in color between this black form [*robustus*, here a synonym of *rufiventris*] and the red monkey of eastern Panama [*panamensis*], the agreement in nearly all cranial details suggests close relationship". Hernández-Camacho and Cooper (1976) recorded that the southernmost specimens in Colombia (Barabacoas, Department of Nariño) show nothing of the brownish color typical of Ecuadorian *A. f. fusciceps*. Cranial and dental morphometric analysis led Froehlich *et al.* (1991) to lump all northwestern South American spider monkeys (*fusciceps* and *hybridus*) as subspecies of *geoffroyi*. Rossan and Baerg (1977) bred a hybrid between *rufiventris* and *panamensis*, and recorded two specimens from the wild that resembled this animal, although they were careful to add that the two taxa are evidently quite homogeneous, and the (anecdotally reported) putative hybrid zone must be very narrow. Medeiros *et al.* (1997) concluded that *A. f. rufiventris* may be genetically isolated from both *hybridus* and *geoffroyi* subspecies (differs in two chromosome pairs, 5 and 6, according to Kunkel *et al.*, 1980), and argued that the mere occurrence of hybrids is inconclusive unless the degree of fertility is established. In their mtDNA analysis, Collins and Dubach (2000), like Froehlich *et al.* (1991), found that *A. f. robustus* (as they called it) formed a clade with the subspecies of *A. geoffroyi*. As a result, they too recommended that it be regarded as another subspecies of *A. geoffroyi*. Within this clade, however, all the "*robustus*" specimens (from three different localities) formed one subclade, and all *A. geoffroyi* subspecies formed another; so the two taxa are consistently different in this character. Collins and Dubach (2000) were unable to sample *A. f. fusciceps*, so we do not know whether "*robustus*" (i.e., *rufiventris*) is distinct from this taxon or not.

Ateles f. rufiventris ranges from the western cordillera of the Andes from southwestern Colombia, northward on the west side of the Río Cauca to eastern Panama (Cerro Pirre and the basin of the Río Bayano of the Pacific coast). The Cerro Pirre or the Río Tucutí mark the border with *A. g. grisescens*. In Colombia, *A. f. rufiventris* occurs throughout the Pacific lowlands except for Juradó, northwestern part of the Department of Chocó, supposedly the domain of *A. g. grisescens* (Hernández-Camacho and Cooper, 1976; Defler, 2003). It occurs in the Urabá region in northwestern Antioquia, Córdoba, Sucre, and northern Bolívar east to the lower Río Cauca along the western bank to south-central Antioquia. The most southerly record in Colombia is Barabacoas, Department of Nariño, and the most northerly is southern bank of the Canal del Dique, Cartagena. Hernández-Camacho and Cooper (1976) believed that it formerly occurred as far north as Pendales.

SUMMARY

In this chapter, we review the taxonomy and distributions of the 21 primate taxa occurring in Central America and southern Mexico, from about 24°N in Tamaulipas, Mexico, extending south along the coast of the Gulf of Mexico, through Central America to the border of Colombia and Panama. In our appraisal, we follow the PSC, as outlined by Groves (2001). Panama (with eight species) has the richest primate community; Costa Rica has four species (five if night monkeys, *Aotus* are included). Capuchin monkeys, *C. capucinus*, extend north as far as Nicaragua and Honduras, and only spider monkeys (*A. geoffroyi*) and howling monkeys (*A. palliata* and *A. pigra*) occur in Belize, Guatemala, and Mexico. Only spider monkeys have been recorded from El Salvador.

Geoffroy's tamarin, *S. geoffroyi*, and the night monkey, *Aotus*, both regionally restricted to Panama, are considered distinct and monotypic. There are two broadly accepted subspecies of squirrel monkey, *S. oerstedii*, occurring in a small area of the Pacific lowlands of Panama and Costa Rica. The white-throated capuchin, *C. capucinus*, extending from Panama to northern Honduras, may comprise three subspecies, although their validity is doubtful. There are two distinct howling monkey species, the mantled howler (*A. palliata*) and the black howler (*A. pigra*). The howling monkeys of Coiba Island and the Azuero Peninsula have some distinct morphological features that argue for their classification as a third species, *A. coibensis*, but a recent molecular genetics' study failed to distinguish them from *A. palliata*. We list three subspecies of *A. palliata* but they are of doubtful validity. The spider monkeys, *A. geoffroyi*, are highly variable. Seven subspecies are listed, and there is the possibility of an eighth undescribed subspecies in northern Honduras. The variability is still poorly understood, however, and the possibility remains that a number of taxa are not valid. The Colombian black spider monkey, *A. fusciceps rufiventris*, extends a short way into Panama.

A notable finding while researching this review was the lack of modern published revisions of the taxonomy and distributions of the region's primates; the major references are still those of Kellogg and Goldman (1944), Hershkovitz (1949), and Hall (1981, based on Hall and Kelson, 1959). The spider monkeys, howler monkeys, and capuchin monkeys are in urgent need of major taxonomic revision, while it is probable that the establishment of the precise historic distributions of all of the Mesoamerican primates is now an impossible task due to introductions, hunting, and forest loss and fragmentation. The widespread loss of population diversity makes taxonomic and biogeographic research on the Mesoamerican primates an increasingly difficult task. All are now restricted to few, diminishing, and isolated forest fragments, and there is an urgent need for regionwide and detailed surveys to identify and map them, to determine the status of the populations remaining.

REFERENCES

Alfaro, A. 1897, *Mamíferos de Costa Rica*. Tipografía Nacional, San José, Costa Rica, 51 pp.

Allen, J. A. 1908, Mammals from Nicaragua. *Bull. Am. Mus. Nat. Hist.* 24:647–670.

Allen, J. A. 1910, Additional mammals from Nicaragua. *Bull. Am. Mus. Nat. Hist.* 28:87–115.

Allen, J. A. 1914, New South American monkeys. *Bull. Am. Mus. Nat. Hist.* 33:647–655.

Anthony, H. E. 1916, Panama mammals collected in 1914–1915. *Bull. Am. Mus. Nat. Hist.* 35:357–375.

Aquino, R. and Encarnación, F. 1994, Primates of Peru/Los Primates del Perú. *Primates Rep.* (35):1–127.

Arauz, J. 1993, Estado de conservación del mono tití (*Saimiri oerstedi citrinellus*) en su área de distribución original, Manuel Antonio, Costa Rica. MSc Thesis, Universidad Nacional Autónoma (UNA), Heredia, Costa Rica.

Baldwin, J. D. and Baldwin, J. I. 1972, The ecology and behavior of squirrel monkeys (*Saimiri oerstedi*) in a natural forest in western Panama. *Folia Primatol.* 18: 161–184.

Baldwin, J. D. and Baldwin, J. I. 1976, Primate populations in Chiriquí, Panama, in R. W. Thorington Jr. and P. G. Heltne, eds., *Neotropical Primates: Field Studies and Conservation*, National Academy of Sciences, Washington, DC, pp. 20–31.

Baldwin, J. D. and Baldwin, J. I. 1977, Observations on *Cebus capucinus* in southwestern Panama. *Primates* 18:937–941.

Barbosa, C., Fajardo-P., A., Giraldo, H., and Rodríguez-M., J. V. 1988, Evaluación del hábitat y status del mono tití de cabeza blanca, *Saguinus oedipus* Linnaeus, 1758, en Colombia. Unpublished Final Report of Status of Cotton-top Tamarin in Colombia Project, INDERENA, Bogotá.

Bartlett, A. S. and Barghoorn, E. S. 1973, Phytogeographic history of the Isthmus of Panama during the past 12,000 years, in A. Graham, ed., *Vegetation and Vegetational History of Northern Latin America*, Elsevier, New York, pp. 203–299.

Boinski, S. 1985, Status of the squirrel monkey (*Saimiri oerstedi*) in Costa Rica. *Primate Conserv.* (6):15–16.

Boinski, S. 1987, The status of *Saimiri oerstedi citrinellus* in Costa Rica. *Primate Conserv.* (8):67–72.

Boinski, S. and Cropp, S. 1999, Disparate data sets resolve squirrel monkey (*Saimiri*) taxonomy: Implications for behavioral ecology and biochemical usage. *Int. J. Primatol.* 20:237–256.

Boinski, S., Jack, K., Lamarsh, C., and Coltrane, J. A. 1998, Squirrel monkeys in Costa Rica: Drifting to extinction. *Oryx* 32:45–58.

Burt, W. H. and Stirton, R. A. 1961, The mammals of El Salvador. *Misc. Publ. Mus. Zool. Univ. Mich.* (117):1–69.

Cabrera. A. 1940, Los nombres científicos de algunos monos americanos. *Ciencia México* 1(9):402–405.

Cabrera, A. 1958, Catalogo de los mamíferos de América del Sur. *Rev. Mus. Argentino de Cienc. Nat. "Bernardino Rivadavia"* 4(1):1–307.

Carpenter, C. R. 1935, Behavior of red spider monkeys in Panama. *J. Mamm.* 16:171–180.

Carriker Jr., M. A. 1909–1910, An annotated list of the birds of Costa Rica including Cocos Island. *Ann. Carnegie Mus.* 6:314–915.

Cody, M. L. 1994, Introduction to Gran Reserva Biológica "Río Indio-Maíz": Site, Flora and Fauna, in M. L. Cody, ed., *Refugio Bartola and the Gran Reserva Biológica "Río Indio-Maíz": Field Ecology Reports, Lowland Tropical Rainforest, Southeastern Nicaragua*, Department of Biology, University of California, Los Angeles, pp. 4–31.

Collins, A. C. 1999, Species status of the Colombian spider monkey, *Ateles belzebuth hybridus. Neotrop. Primates* 7(2):39–41.

Collins, A. C. and Dubach, J. 2000, Phylogenetic relationships of spider monkeys (*Ateles*) based on mitochondrial DNA variation. *Int. J. Primatol.* 21(3):381–420.

Cortés-Ortiz, L., Bermingham, E., Rico, C., Rodríguez-Luna, E., Sampaio, I., and Ruiz-Garcia, M. 2003, Molecular systematics and biogeography of the Neotropical monkey genus, *Alouatta. Mol. Phylogenet. Evol.* 26:64–81.

Costello, R. K., Dickinson, C., Rosenberger, A. L., Boinski, S., and Szalay, F. S. 1993, Squirrel monkey (genus *Saimiri*) taxonomy: A multidisciplinary study of the biology of the species, in W. H. Kimbel and L. B. Martin, eds., *Species, Species Concepts, and Primate Evolution,* Plenum Press, New York, pp. 177–210.

Crockett, C. M., Brooks, R. D., Crockett Meacham, R., Crockett Meacham, S., and Mills, M. 1997, Recent observations of Nicaraguan primates and a preliminary conservation assessment. *Neotrop. Primates* 5(3):71–75.

Cropp, S. J. and Boinski, S. 2000, The Central American squirrel monkey (*Saimiri oerstedii*): Introduced hybrid or endemic species? *Mol. Phylogenet. Evol.* 16:350–365.

Curdts, T. 1993, Distribución geográfica de las dos especies de mono zaraguate que habitan en Guatemala, *Alouatta palliata* y *Alouatta pigra*, in A. Estrada, E. Rodríguez-Luna, R. Lopes-Wilchis, and R. Coates-Estrada, eds., *Estudios Primatológicos en México*, Vol. 1, Universidad Veracruzana, Veracruz, pp. 317–329.

Dahl, J. F. 1984, Primate survey in proposed reserve area in Belize. *IUCN/SSC Primate Specialist Group Newsl.* (4):28–29.

Dahl, J. F. 1987, Conservation of primates in Belize, Central America. *Primate Conserv.* (8):119–121.

Defler, T. R. 1994, La conservación de primates en Colombia. *Trianea (Act. Cien. INDERENA)* 5:255–287.

Defler, T. R. 2003, *Primates de Colombia.* Conservación Internacional Serie de Guías Tropicales de Campo, Conservación Internacional Colombia, Bogotá.

Defler, T. R. and Bueno, M. L. 2003. Karyological guidelines for *Aotus* taxonomy. *Am. J. Primatol.* 60(Suppl. 1):134–135.

Defler, T. R., Bueno, M. L., and Hernández-Camacho, J. I. 2001, Taxonomic status of *Aotus hershkovitzi*: Its relationship to *Aotus lemurinus lemurinus. Neotrop. Primates* 9:37–52.

Defler, T. R., Rodríguez-M., J. V., and Hernández-Camacho, J. I. 2003, Conservation priorities for Colombian primates. *Primate Conserv.* (19):10–18.

Eisenberg, J. F. 1989, *Mammals of the Neotropics: The Northern Neotropics Vol: 1, Panama, Colombia, Venezuela, Guyana, Suriname, French Guiana*. The Chicago University Press, Chicago.

Elliot, D. G. 1912a, New species of monkeys of the genera *Seniocebus*, *Alouatta*, and *Aotus*. *Bull. Am. Mus. Nat. Hist.* 31:31–33.

Elliot, D. G. 1912b, Description of a new species of *Oedipomidas*. *Bull. Am. Mus. Nat. Hist.* 31:137.

Elliot, D. G. 1913, *A Review of Primates*. Monograph Series, American Museum of Natural History, New York.

Elliot, D. G. 1914, The genera *Oedipomidas* and *Seniocebus*. *Bull. Am. Mus. Nat. Hist.* 33:643–645.

Emmons, L. H. and Feer, F. 1997, *Neotropical Rainforest Mammals: A Field Guide*. 2nd Edn. The University of Chicago Press, Chicago.

Encarnación, F. and Cook, A. G. 1998, Primates of the tropical forest of the Pacific coast of Peru: The Tumbes Reserved Zone. *Primate Conserv.* (18):15–20.

Estrada, A. and Coates-Estrada, R. 1984, Some observations on the present distribution and conservation of *Alouatta* and *Ateles* in southern Mexico. *Am. J. Primatol.* 7:133–137.

Estrada, A., Van Belle, S., Luecke, L., and Rosales, M. 2004, Primate populations in the protected forests of Mayan archaeological sites in southern Mexico. *Am. J. Primatol.* 62(Suppl. 1):77.

Ford, S. M. 1994, Taxonomy and distribution of the owl monkey, in J. F. Baer, R. E. Weller, and I. Kakoma, eds., *Aotus: The Owl Monkey,* Alan R. Liss, New York, pp. 1–57.

Froehlich, J. W. and Froehlich, P. H. 1986, Dermatoglyphics and subspecific systematics of mantled howler monkeys (*Alouatta palliata*), in D. M. Taub and F. A. King, eds., *Current Perspectives in Primate Biology*. Van Nostrand Reinhold, New York, pp. 107–121.

Froehlich, J. W. and Froehlich, P. H. 1987, The status of Panama's endemic howling monkeys. *Primate Conserv.* (8):58–62.

Froehlich, J. W., Supriatna, J., and Froehlich, P. H. 1991, Morphometric analyses of *Ateles*: Systematic and biogeographic implications. *Am. J. Primatol.* 25:1–22.

García-Orduña, F. and Canales-Espinosa, D. 1995, Situación de poblaciones de *Alouatta palliata* (mono aullador) en dos localidades del Estado de Veracruz, México. *Neotrop. Primates* 3(2):37–38.

García-Orduña, F., Canales-Espinosa, D., Vea, J., and Rodríguez-Luna, E. 1999, Current distribution of *Alouatta palliata mexicana* and *Alouatta pigra* in Tabasco, Mexico: Preliminary Report. *VII Simposio Nacional de Primatologia, Catemaco, Veracruz, 6–8 Septiembre, 1999*, Asociación Mexicana de Primatología, Catemaco, Veracruz, México, In Spanish, p. 3.

Goldman, E. A. 1914, Descriptions of five new mammals from Panama. *Smithson. Misc. Coll.* 63(5):1–7.

Goldman, E. A. 1915, A new spider monkey from Panama. *Proc. Biol. Soc. Wash.* 28:101–102.

Gray, J. E. 1843, *List of the Specimens of Mammalia in the Collection of the British Museum*. British Museum, London.

Gray, J. E. 1845, On the howling monkeys (*Mycetes*, Illiger). *Ann. Mag. Nat. Hist.*, ser. 1, 16:217–221.

Gray, J. E. 1849, Description of two species of Mammalia from Caraccas. *Proc. Zool. Soc. Lond.* 1849:7–10.

Gray, J. E. 1870, *Catalogue of Monkeys, Lemurs and Fruit-eating Bats in the Collection of the British Museum*. Trustees of the British Museum, London.

Groves, C. P. 1993, Order Primates, in D. E. Wilson and D. M. Reeder, eds., *Mammal Species of the World: A Taxonomic and Geographic Reference*, 2nd Edn., Smithsonian Institution Press, Washington, DC, pp. 243–277.

Groves, C. P. 2001, *Primate Taxonomy*. Smithsonian Institution Press, Washington, DC.

Hall, E. R. 1981, *The Mammals of North America*. Vol. 1. John Wiley and Sons, New York.

Hall, E. R. and Kelson, K. R. 1959, *The Mammals of North America*. Vol. 1. The Ronald Press Co., New York.

Handley, Jr., C. O. 1966, Checklist of the mammals of Panama, in R. L. Wetzel and V. J. Tipton, eds., *Ectoparasites of Panama*, Field Museum of Natural History, Chicago, pp. 753–795.

Hanihara, T. and Natori, M. 1987, Preliminary analysis of numerical taxonomy of the genus *Saguinus* based on dental measurements. *Primates* 28(4):517–523.

Heltne, P. G. and Kunkel, L. M. 1975, Taxonomic notes on the pelage of *Ateles paniscus paniscus*, *A. p. chamek* (*sensu* Kellogg and Goldman, 1944) and *A. fusciceps rufiventris* (= *A. f. robustus*, Kellogg and Goldman, 1944). *J. Med. Primatol.* 4:83–102.

Hernández-Camacho, J. and Cooper, R. W. 1976, The non-human primates of Colombia, in R. W. Thorington, Jr. and P. G. Heltne, eds., *Neotropical Primates: Field Studies and Conservation*, National Academy of Sciences, Washington, DC, pp. 35–69.

Hernández-Camacho, J. and Defler, T. R. 1991, Algunos aspectos de la conservación de primates no-humanos en Colombia, in C. J. Saavedra, R. A. Mittermeier, and I. B. Santos, eds., *La Primatología en Latinoamérica*, World Wildlife Fund, Washington, DC, pp. 67–100.

Hershkovitz, P. 1949, Mammals of northern Colombia. Preliminary report No. 4: Monkeys (Primates), with taxonomic revisions of some forms. *Proc. U.S. Nat. Mus.* 98:323–427.

Hershkovitz, P. 1969, The evolution of mammals on southern continents. VI. The recent mammals of the neotropical region: A zoogeographic and ecological review. *Quart. Rev. Biol.* 44(1):1–70.

Hershkovitz, P. 1977, *Living New World Monkeys (Platyrrhini) with an Introduction to Primates,* Vol. 1. Chicago University Press, Chicago.

Hershkovitz, P. 1983, Two new species of night monkeys, genus *Aotus* (Cebidae, Platyrrhini): A preliminary report on *Aotus* taxonomy. *Am. J. Primatol.* 4:209–243.

Hershkovitz, P. 1984, Taxonomy of squirrel monkeys, genus *Saimiri* (Cebidae, Platyrrhini): A preliminary report with description of a hitherto unnamed form. *Am. J. Primatol.* 7:155–210.

Hill, W. C. O. 1957, *Primates. Comparative Anatomy and Taxonomy III. Hapalidae.* Edinburgh University Press, Edinburgh.

Hill, W. C. O. 1960, *Primates. Comparative Anatomy and Taxonomy IV. Cebidae Part A.* Edinburgh University Press, Edinburgh.

Hill, W. C. O. 1962, *Primates. Comparative Anatomy and Taxonomy V. Cebidae Part B.* Edinburgh University Press, Edinburgh.

Hines, J. J. 2004, Taxonomic status of *Ateles geoffroyi* in northern Honduras. *Am. J. Primatol.* 62(Suppl. 1):79–80.

Hollister, N. 1914, Four new mammals from tropical America. *Proc. Biol. Soc. Wash.* 27:103–106.

Horwich, R. H. 1983, Species status of the black howler monkey, *Alouatta pigra*, of Belize. *Primates* 24:288–289.

Horwich, R. H. and Johnson, E. D. 1984, Geographic distribution and status of the black howler monkey. *IUCN/SSC Primate Specialist Group Newsl.* (4):25–27.

Horwich, R. H. and Johnson, E. D. 1986, Geographical distribution of the black howler (*Alouatta pigra*) in Central America. *Primates* 27:53–62.

Hubrecht, R. C. 1986, Operation Raleigh primate census in the Maya Mountains, Belize. *Primate Conserv.* (7):15–17.

Jones Jr., J. K., Genoways, H. H., and Smith, J. D. 1974, Annotated checklist of mammals of the Yucatán Peninsula. III. Marsupialia, Insectivora, Primates, Edentata, Lagomorpha. *Occ. Pap. Texas Tech University* (23):1–12.

Kellogg, R. and Goldman, E. A. 1944, Review of the spider monkeys. *Proc. U. S. Nat. Mus.* 96:1–45.

Konstant, W., Mittermeier, R. A., and Nash, S. D. 1985, Spider monkeys in captivity and in the wild. *Primate Conserv.* (5):82–109.

Kunkel, L. M., Heltne, P. G., and Borgaonkar, D. S. 1980, Chromosomal variation and zoogeography in *Ateles. Int. J. Primatol.* 1(3):223–232.

Lawrence, B. 1933, Howler monkeys of the *palliata* group. *Bull. Mus. Comp. Zool.* (Harvard University) 75:314–354.

Luecke, L. 2004, Distribution of the black howler monkey (*Alouatta pigra*) in Mesoamerica: A GIS analysis. *Am. J. Primatol.* 62(Suppl. 1):75–76.

Marineros, L. and Gallegos, F. M. 1998, *Guía de Campo de los Mamíferos de Honduras.* Instituto Nacional de Ambiente y Desarrollo, Tegucigalpa, Honduras.

Mast, R. B., Rodríguez, J. V., and Mittermeier, R. A. 1993, The Colombian cotton-top tamarin in the wild, in N. K. Clapp, ed., *A Primate Model for the Study of Colitis and Colonic Carcinoma: The Cotton-Top Tamarin (Saguinus oedipus)*, CRC Press, Inc., Oak Ridge, pp. 3–43.

Matamoros, Y. and Seal, U. S. eds. 2001, *Informe Final. Taller para la Conservación Asesoramiento y Manejo Planificado para los Primates Mesoamericanos, Parque Zoológico Simón Bolívar, San José, Costa Rica, 23–25 Junio 1997.* IUCN/SSC Conservation Breeding Specialist Group (CBSG), Apple Valley, Minnesota, 314 pp.

McCarthy, T. J. 1982, *Chironectes, Cyclopes, Cabassous* and probably *Cebus* in southern Belize. *Mammalia* 46:397–400.

Medeiros, M. A. A., Barros, R. M. S., Pieczarka, J. C., Nagamachi, C. Y., Ponsa, M., Garcia, M., Garcia, F., and Egozcue, J. 1997, Radiation and speciation of spider monkeys, genus *Ateles*, from the cytogenetic viewpoint. *Am. J. Primatol.* 42: 167–178.

Merriam, C. H. 1902, Five new mammals from Mexico. *Proc. Biol. Soc. Wash.* 15:67–69.

Mittermeier, R. A. and Coimbra-Filho, A. F. 1981, Systematics: Species and subspecies, in A. F. Coimbra-Filho and R. A. Mittermeier, eds., *Ecology and Behavior of Neotropical Primates*, Vol. 1, Academia Brasileira de Ciências, Rio de Janeiro, pp. 29–110.

Mittermeier, R. A., Rylands, A. B., and Coimbra-Filho, A. F. 1988, Systematics: Species and subpecies—an update, in R. A. Mittermeier, A. B. Rylands, A. F. Coimbra-Filho, and G. A. B. da Fonseca, eds., *Ecology and Behavior of Neotropical Primates,* Vol. 2, World Wildlife Fund, Washington, DC, pp. 13–75.

Moore, A. J. and Cheverud, J. M. 1992, Systematics of the *Saguinus oedipus* group of the bare-faced tamarins: Evidence from facial morphology. *Am. J. Phys. Anthropol.* 89:73–84.

Moynihan, M. 1970, Some behavior patterns of platyrrhine monkeys II. *Saguinus geoffroyi* and some other tamarins. *Smithson. Contrib. Zool.* (28):1–77.

Moynihan, M. 1976, *The New World Primates: Adaptive Radiation and the Evolution of Social Behavior, Languages and Intelligence.* Princeton University, Princeton, NJ.

Napier, P. H. 1976, *Catalogue of the Primates in the British Museum (Natural History). Part I. Families Callitrichidae and Cebidae.* British Museum (Natural History), London.

Parra Lara, A. del C. and Jorgenson, J. P. 1998, Notes on the distribution and conservation status of spider and howler monkeys in the state of Quintana Roo, Mexico. *Primate Conserv.* (19):25–29.

Querol, D. *et al.* 1996, *Especies útiles de un bosque húmedo tropical: Güises Montaña Experimental, Río San Juan, Nicaragua.* F. Campodónico F., Industria Gráfica S. A., Lima, Peru.

Ramos-Fernández, G. and Ayala-Orozco, B. 2003, Population size and habitat use of spider monkeys at Punta Laguna, Mexico, in L. K. Marsh, ed., *Primates in Fragments: Ecology and Conservation*, Kluwer Academic/Plenum Publishers, New York, pp. 191–209.

Reid, F. A. 1997, *A Field Guide to the Mammals of Central America and Southeast Mexico.* Oxford University Press, New York.

Rodríguez-Luna, E., Cortés-Ortiz, L., Mittermeier, R. A., and Rylands, A. B. 1996, *Plan de Acción para los Primates Mesoamericanos.* IUCN/SSC Primate Specialist Group, Xalapa, Veracruz, Mexico, 121 pp.

Rodríguez-Luna, E., Vea, J., García-Orduña, F., Canales-Espinosa D., and Cortés-Ortiz, L. 2001, Zone of sympatry of *Alouatta palliata* and *Alouatta pigra* in Tabasco (Mexico): Mixed populations in fragmented habitat. *I Congresso Mexicano de Primatología, Merida, Yucatán, 2–5 Septiembre, 2001: Programa y Resumenes*, Asociación Mexicana de Primatología, Merida, Yucatán, México, In Spanish, p. 27.

Rodríguez-Vargas, A. R. 2003, Analysis of the hypothetical population structure of the squirrel monkey (*Saimiri oerstedii*) in Panamá, in L. K. Marsh, ed., *Primates in Fragments: Ecology and Conservation*, Kluwer Academic/Plenum Publishers, New York, pp. 53–62.

Rossan, R. N. and Baerg, D. C. 1977, Laboratory and feral hybridization of *Ateles geoffroyi panamensis* Kellogg and Goldman, 1944 and *A. fusciceps robustus* Allen, 1914 in Panama. *Primates* 18(1):235–237.

Rowe, N. 2000, Records of howlers (*Alouatta*) on the Azuero Peninsula and Canal Zone of Panama. *Neotrop. Primates* 8:154–156.

Rylands, A. B. 1993, The bare-face tamarins *Saguinus oedipus oedipus* and *Saguinus oedipus geoffroyi*: Subspecies or species? *Neotrop. Primates* 1(2):4–5.

Rylands, A. B., Coimbra-Filho, A. F., and Mittermeier, R. A. 1993, Systematics, distributions, and some notes on the conservation status of the Callitrichidae, in A. B. Rylands, ed., *Marmosets and Tamarins: Systematics, Behaviour, and Ecology*, Oxford University Press, Oxford, pp. 11–77.

Rylands, A. B., Mittermeier, R. A., and Rodríguez-Luna, E. 1995, A species list for the New World primates (Platyrrhini): Distribution by country, endemism, and conservation status according to the Mace-Lande system. *Neotrop. Primates* 3(Suppl.):113–160.

Rylands, A. B., Schneider, H., Langguth, A., Mittermeier, R. A., Groves, C. P., and Rodríguez-Luna, E. 2000, An assessment of the diversity of New World primates. *Neotrop. Primates* 8(2):61–93.

Schultz, A. H. 1960, Age changes and variability in the skulls of and teeth of Central American monkeys, *Alouatta*, *Cebus* and *Ateles*. *Proc. Zool. Soc. Lond.* 133: 337–390.

Sclater, P. L. 1871, Additions to the Society's menagerie in April, 1871. *Proc. Zool. Soc. Lond.* 1871:478.

Sclater, P. L. 1872, On the Quadrumana found in America north of Panama. *Proc. Zool. Soc. Lond.* 1872:2–9.

Serio-Silva, J. C. and Rico-Gray, V. 2004, Mexican primates in the Yucatán Peninsula: Priority area for conservation in Mesoamerica. *Am. J. Primatol.* 62(Suppl. 1):79.

Sierra, C., Jiménez, I., Altrichter, M., Fernández, M., Gómez, G., González, J., Hernández, C., Herrera, H., Jiménez, B., López-Arévalo, H., Millán, J., Mora, G., and E. Tabilo, E. 2003, New data on the distribution and abundance of Saimiri oerstedtii citrinellus. *Primate Conserv.* (19):5–9.

Silva, B. T. F., Sampaio, M. I. C., Schneider, H., Schneider, M. P. C., Montoya, E., Encarnación, F., Callegari-Jacques, S. M., and Salzano, F. M. 1993, Protein electrophoretic variability in *Saimiri* and the question of its species status. *Am. J. Primatol.* 29:183–193.

Silva-López, G. 1998, Distribution and status of the primates of Guatemala. *Primate Conserv.* (18):30–41.

Silva-López, G., Motta-Gill, J., and Sánchez-Hernández, A. I. 1995, The Primates of Guatemala: Distribution and Status. Unpublished Report. NYZS, The Wildlife Conservation Society, New York.

Silva-López, G., Motta-Gill, J., and Sánchez-Hernández, A. I. 1996, Taxonomic notes on *Ateles geoffroyi*. *Neotrop. Primates* 4(2):41–44.

Silva-López, G. and Rumiz, D. 1995, Los primates de la Reserva Río Bravo, Belice. *Ciencia y el Hombre* (20):49–64.

Skinner, C. 1985, Report on a field study of Geoffroy's tamarin in Panama. *Primate Conserv.* (5):22–24.

Skinner, C. 1991, Justification for reclassifying Geoffroy's tamarin from *Saguinus oedipus geoffroyi* to *Saguinus geoffroyi*. *Prim. Rep.* 31:77–83.

Smith, J. D. 1970, The systematic status of the black howler monkey, *Alouatta pigra* Lawrence. *J. Mammal.* 51:358–369.

Thomas, O. 1904, New forms of *Saimiri*, *Saccopteryx*, *Balantiopteryx*, and *Thrichomys* from the Neotropical region. *Ann. Mag. Nat. Hist.* 13(7):188–196.

Thorington Jr., R. W. 1976, The systematics of New World monkeys. *First Interamerican Conference on Conservation and Utilization of American Nonhuman Primates in Biomedical Research*. Pan American Health Organization (PAHO), Washington, DC, pp. 8–18.

Thorington Jr., R. W. 1985, The taxonomy and distribution of squirrel monkeys (*Saimiri*). In: L. A. Rosenblum and C. L. Coe, eds., *Handbook of Squirrel Monkey Research*. Plenum Press, New York, pp. 1–33.

Timm, R. M. 1988, A review and appraisal of the night monkey, *Aotus lemurinus* (Primates, Cebidae) in Costa Rica. *Rev. Biol. Trop.* 36(2B):537–540.

Timm, R. M., Wilson, D. E., Clauson, B. L., LaVal, R. K., and Vaughan, C. S. 1989, Mammals of La Selva—Braulio Carrillo complex, Costa Rica. *North American Fauna, US Fish Wildl. Serv. Publ.* 75:1–162.

Tirira, S. D. ed. 2001, *Libro Rojo de Los Mamíferos de Ecuador*. Sociedad para la Invetigación y Monitoreo de la Biodiversidad Ecuatorania (SIMBIOE)/Ecociencias/ Ministerio del Ambiente/UICN. Serie Libros Rojos del Ecuador, Tomo 1. Publicación Especial sobre los Mamíferos del Ecuador.

Torres, O. M., Enciso, S., Ruiz, F., Silva, E., and Yunis, I. 1998, Chromosome diversity in the genus *Aotus* from Colombia. *Am. J. Primatol.* 44:255–275.

Vargas, T. N. 1994, Evaluación de las poblaciones de primates en dos sectores del Parque Nacional Natural "Las Orquídeas", Departamento de Antioquia. Unpublished Manuscript.

Vaughan, C. 1983, A report on dense forest habitat for endangered wildlife species in Costa Rica. Universidad Nacional, Heredia, Costa Rica, 66 pp.

Watts, E. S. and Rico-Gray, V. 1987, Los Primates de la Península de Yucatán, México: Estudio preliminar sobre su situación actual y estado de conservación. *Biótica* 12(1):57–66.

Watts, E. S., Rico-Gray, V., and Chan, C. 1986, Monkeys of the Yucatán Peninsula, Mexico: Preliminary survey of their distribution and status. *Primate Conserv.* (7):17–22.

Wong, G. 1990, Uso del hábitat, estimación de la composición y densidad poblacional del mono tití (*Saimiri oerstedii citrinellus*) en la zona de Manuel Antonio, Quepos, Costa Rica. MSc Thesis, Universidad Nacional Autónoma (UNA), Heredia, Costa Rica.

CHAPTER THREE

The Biogeographic History of Mesoamerican Primates

Susan M. Ford

INTRODUCTION

New World monkeys ranged into Mesoamerica along with the mass migration of South American fauna (and flora) northward during the Great American Interchange (Marshall *et al.*, 1982; Marshall, 1988; Stehli and Webb, 1985; Webb, 1991, 1999), as a result of the emergence of the Panamanian isthmus around 3.5 mya (Coates *et al.*, 2003; Cronin and Dowsett, 1996). This interchange involved a major influx of previously unrepresented southern taxa into Mesoamerica, and an even larger movement of northern (North American) groups into South America. However, uncertainty remains about the number of independent invasions of Mesoamerica by New World monkeys and other fauna, the timing of these invasions, and the speed and direction of movement into various Mesoamerican regions. In addition, the degree of isolation and eventual genetic separation of various groups into distinctive subspecies or even species remains controversial; this last question is addressed in other

---

Susan M. Ford • Department of Anthropology, Center for Systematic Biology, Southern Illinois University, Carbondale, IL 62901-4502.

New Perspectives in the Study of Mesoamerican Primates: Distribution, Ecology, Behavior, and Conservation, edited by Alejandro Estrada, Paul A. Garber, Mary S. M. Pavelka, and LeAndra Luecke. Springer, New York, 2005.

contributions to this volume, particularly Rylands *et al.* (this volume; see also Groves, 2001).

Here, the modern distribution of primates in Mesoamerica is interpreted against the backdrop of the geographic landscape across the region (including mountain ranges, lowlands, and habitats), the geologic history of the formation of the isthmian connection, and the phylogenetic ties of Mesoamerican primates to their South American relatives.

GEOGRAPHIC AND GEOLOGIC BACKGROUND

Key to this analysis is an understanding of the basic geography of the Central or Mesoamerican isthmus. It is a single, very long, narrow strip of land marked by Pacific and Atlantic coastal lowlands separated by high mountain ranges down the middle for much of its length (see Figure 1). In several places, these highlands reach nearly to the coast (especially in northwestern Costa Rica and across Honduras and El Salvador). These central mountains

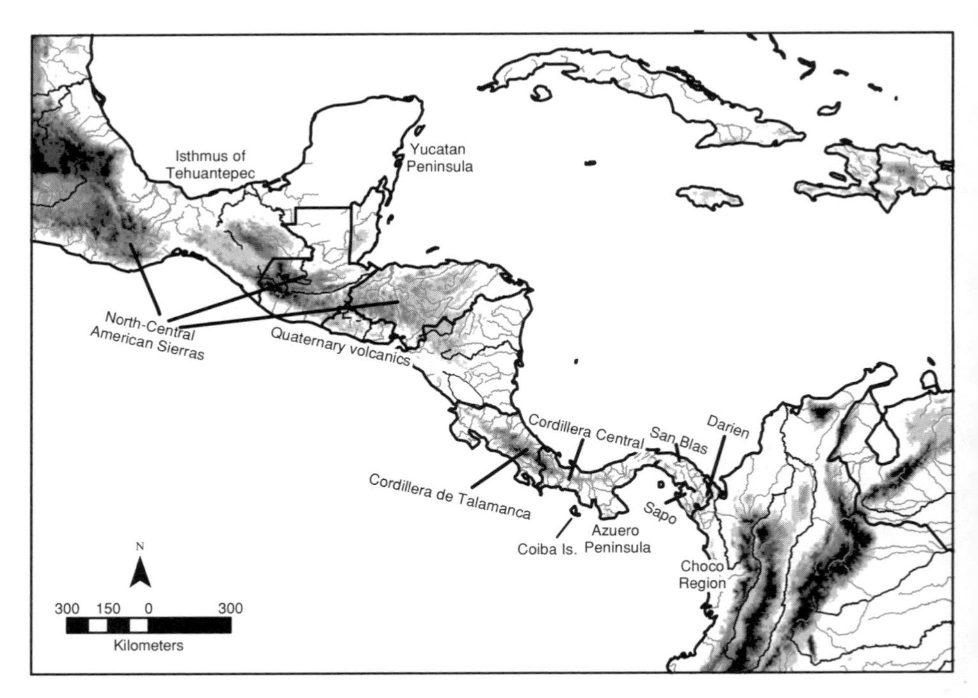

Figure 1. Map of Mesoamerican landscape, showing major topographic features.

Table 1. Approximate maximum elevation limits for Mesoamerican primates (from Reid, 1997)

Taxon	Max. elevation
Alouatta palliata	2500 m
Alouatta pigra	<500 m
Aotus zonalis	650+ m
Ateles geoffroyi	1800 m
Ateles fusciceps	>2000 m
Cebus capucinus	2000 m
Saguinus geoffroyi	900 m
Saimiri oerstedii	<500 m

are of varying age (de Cserna, 1989; Savage, 2002; Weyl, 1980): the North Central American Sierras, from southern Mexico through Guatemala and Honduras into northern Nicaragua, originated pre-Cenozoic but have experienced some additional Pliocene uplift. West of these, the Central American Tertiary Volcanics (Guatemala through Nicaragua) resulted from uplift and volcanism from the Miocene through Pliocene. This time frame also saw the uplift of the central mountains of Costa Rica (the Cordillera de Talamanca) and western Panama (the Cordillera Central). In the Quaternary and continuing today, active volcanic ridges have developed along the Pacific from southernmost Mexico through central Costa Rica. Most of these ranges include areas of high elevation that preclude habitation by modern monkeys except along forest river valleys. Mesoamerican primates are lowland fauna, with maximum elevations reported for *Alouatta palliata* at 2500 m and the others significantly lower (see Table 1 and reviews in Reid, 1997; Rylands *et al.*, this volume).

The Pacific coast tends to be drier than the Atlantic, as in South America, but there often remain areas of forest with continuous canopy (Savage, 2002). The presence of subhumid–semiarid forested corridors along both the coasts was likely the case through much of the Pleistocene (Colinvaux, 1993, 1996), despite arguments for periodic more arid conditions by others (e.g., Webb and Rancy, 1996; Whitmore and Prance, 1987). However, these have been divided by the central uplands throughout the late Cenozoic, and there may have been intermittent breaks in the corridors during particular cold–dry cycles (Savage, 2002) or during rises in sea-level flooding the coastal regions (Nores, 1999; Eberhard and Bermingham, 2004). Currently, major grass and shrub areas exist in the Azuero Peninsula, the Pacific coast of northwest Costa Rica and Nicaragua, and the northeastern corner of the Yucatan Peninsula.

There are three major areas of coast-to-coast lowlands: the Isthmus of Tehuantepec in southern Mexico; lowlands angling across from southwestern Nicaragua to northeastern Costa Rica (and now partly filled with Lake Nicaragua); and the Gatún region of Panama, the location of the modern Panama Canal. These represent areas for easy exchange of lowland fauna from the east and west coasts (including monkeys). A fourth, extra-isthmus lowland connection between the Atlantic and Pacific coasts occurs across extreme northwestern Colombia, in the Chocó/Atrato River region.

The exception to this pattern (of coastal lowlands separated by central highlands) is in eastern Panama, at the southeastern terminus of the isthmus. Here, the Serranía Darién now ranges across the entire isthmian terminus, at the border with Colombia, rising up to 2000 m in elevation and forming a formidable barrier to faunal exchange of lowland taxa. In eastern Panama, the highlands separate to form Atlantic (Serranía San Blas) and Pacific (Serranía Sapo-Baudo) coastal ranges separated by a central lowland region, with the coastal ranges diminishing as they reach the central Gatún lowland region. This narrowest part of the isthmus, which was an important corridor for transit of lowland fauna in the past, is now disrupted by the Panama Canal and various large lakes formed early in the 20th century.

Just west of this lowland region, the central ranges begin and the Pacific coast has a large peninsula, the Azuero Peninsula, one of the most arid regions of Mesoamerica and covered by grassland and shrub forest. Offshore, Isla de Coiba shares a similar habitat. It is separated by about 50 km, and was likely last connected to the mainland from about 24,000 to 15,000 yBP (Bartlett and Barghoorn, 1973; Froehlich and Froehlich, 1987). The peninsula was likely separated from the mainland as well, as the northern end is much lower than the southern and currently covered by arid grasslands (Bennett, 1968; Froehlich and Froehlich, 1987; see below). Both the Azuero Peninsula and especially Isla de Coiba have a markedly depauperate mammalian fauna, suggesting filtered migration and isolation (*Ibid.*), probably from both periodic flooding of the lowland regions and extensive grasslands when emergent.

None of the rivers of this narrow landform attain great size or length. Therefore, none of the isthmian rivers appears to represent a significant barrier to the movement of primates. The modern exception to this would be the recent addition of the Panama Canal to the landscape.

Platyrrhine primates have been able to move from South America to Mesoamerica only since the Pliocene connection of the two land areas. The geologic history of the southern isthmus has been recently reviewed in Gregory-Wodzicki (2000) and Ford (in prep.) and of the entire isthmus in Savage (2002); key aspects are presented here. Although the lower Central American region was largely under water for much of the Cenozoic (Pindell and Barrett, 1990), by the Mid Miocene (around 10 mya), the southern portion corresponding to modern Costa Rica and western Panama became an "extensively emergent archipelago" of volcanic islands associated with the uplift of the central mountains, while eastern Panama in the region of the Darién and the Chocó remained under deep water (Coates *et al.*, 2003: 271; also Coates, 1999; Collins *et al.*, 1996a). On the South American mainland, ongoing uplift of the Andean chain from the Miocene resulted in the increasing isolation of northwestern Colombia and western Ecuador, particularly of the modern Magdalena, Cauca, and Chocó/Atrato River basins and the Maracaibo Basin in northwestern Venezuela from the Amazon/Orinoco basins, although the individual northwestern basins continued to be forested tropical lowlands (Rull, 1998). By 9 mya, the freshwater fish of the Atrato Basin of far northwestern Colombia (which directly borders the Mesoamerican isthmus) became isolated from those in the other Colombian basins (Martin and Bermingham, 2000), suggesting that at least parts of the Chocó remained emergent (with freshwater) from then on. Late Miocene (8–5 mya) saw a general subsidence of the lower isthmus region under deepening water (Aubry and Berggren, 1999; Coates *et al.*, 2003; Collins *et al.*, 1996a). Most of the lower isthmus/contact region, including Costa Rica (Collins *et al.*, 1995; McNeill *et al.*, 1999), the Darién of southern Panama (Collins *et al.*, 1998), and the Atrato Basin of northwestern Colombia (Duque-Caro, 1990; Coates *et al.*, 2003), remained under deep water through this period. Although land migrations into Mesoamerica are unlikely this early, Collins *et al.* (1996b) suggest there may have been emergence sufficient to disrupt gene flow between Atlantic and Pacific marine foraminiferans around 8 mya. Based on molecular dating estimates, Salazar-Bravo *et al.* (2001) suggest possible migration south of northern field mice of the genus *Calomys* and Engel *et al.* (1998) for sigmodontine rodents in general into South America in an early pulse of exchange, and Perdices *et al.* (2002) use molecular dates to suggest a northern dispersal of *Rhamdia* freshwater fish around 6 mya.

The Early Pliocene (5–3 mya) saw significant shallowing and ocean regression in Costa Rica (McNeill *et al.*, 1999) and reef formation in Panama (Coates and Obando, 1996; Coates *et al.*, 2003; Collins and Coates, 1999). By the Late Pliocene, *ca.* 3.5–2.0 mya, there was complete emergence of the isthmus

(Coates *et al.*, 2003), with sea level 100 m lower than today around 3.4 mya (Vail and Hardenbol, 1979). This allowed the Great American Interchange to proceed in both northward and southward directions, involving terrestrial fauna (Eisenberg, 1989; Marshall *et al.*, 1982; Marshall, 1988; Savage, 2002; Simpson, 1980; Stehli and Webb, 1985; Webb, 1999), aerial forms (Eberhard and Bermingham, 2004; Hoffman and Baker, 2003), and freshwater fish (Martin and Bermingham, 2000; Perdices *et al.*, 2002). However, there is increasing evidence that this initial connection and exchange was transitory in nature. It is now apparent that between 2.8 and 2.5 mya there was a major exchange of Atlantic and Pacific marine fauna (Cronin and Dowsett, 1996) with subsidence of the lower isthmus, followed by a second wave of terrestrial dispersals around 2.0 mya (Savage, 2002). Mounting evidence of at least two separate periods of freshwater fish dispersals since 3.5 mya exists as well, but Martin and Bermingham (2000) suggest that the later dispersal was as late as 1.0 mya. Taken together, these strongly indicate at least two and perhaps as many as four separate and distinct periods of faunal migrations across the Darién region (filtered dispersal 8–6 mya, major exchange 3.5–3.0 mya, later exchanges 2.0 mya and perhaps 1.0 mya). In addition, there appears to be filtered exchange across the Darién today, along with probable recent (Pleistocene-Recent) introductions of Amazonian-based primates into the northwestern Colombian basins (Ford, in prep., for *Ateles fusciceps* and *Alouatta seniculus*, at least; see also Hoffman and Baker, 2003, on complex history of short-tailed bats in this region).

MODERN DISTRIBUTIONS OF MESOAMERICAN PRIMATES

I follow the species and subspecies usage of Rylands *et al.* (this volume). Figures 2–7 show the distribution of primate genera in Mesoamerica, superimposed on the topography of the region. These distributions are based on two major sources. They are drawn primarily from the literature as reviewed and presented in Rylands *et al.* (this volume; see also Henderson and Adams, 2002; Reid, 1997). Where my maps differ, I offer range expansions based on museum catalog records. The locality datapoints represent collecting localities (in Colombia, Panama, Costa Rica, and Nicaragua) associated with all primate specimens from four of the world's major museum collections of Neotropical primates: the Smithsonian Institution, Washington, D.C. (USNM), the American Museum of Natural History in New York (AMNH), the Field Museum of

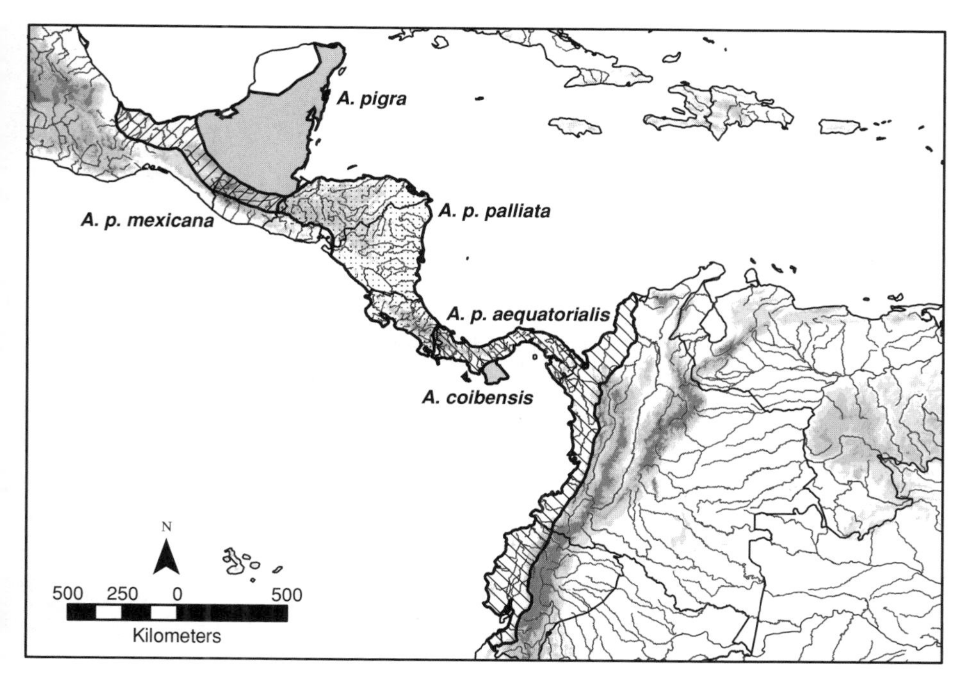

Figure 2. Map of distribution of *Alouatta coibensis*, *Alouatta palliata*, and *Alouatta pigra*. Changes in elevation shadings correspond to 500, 1000, 2000, and >2500 m.

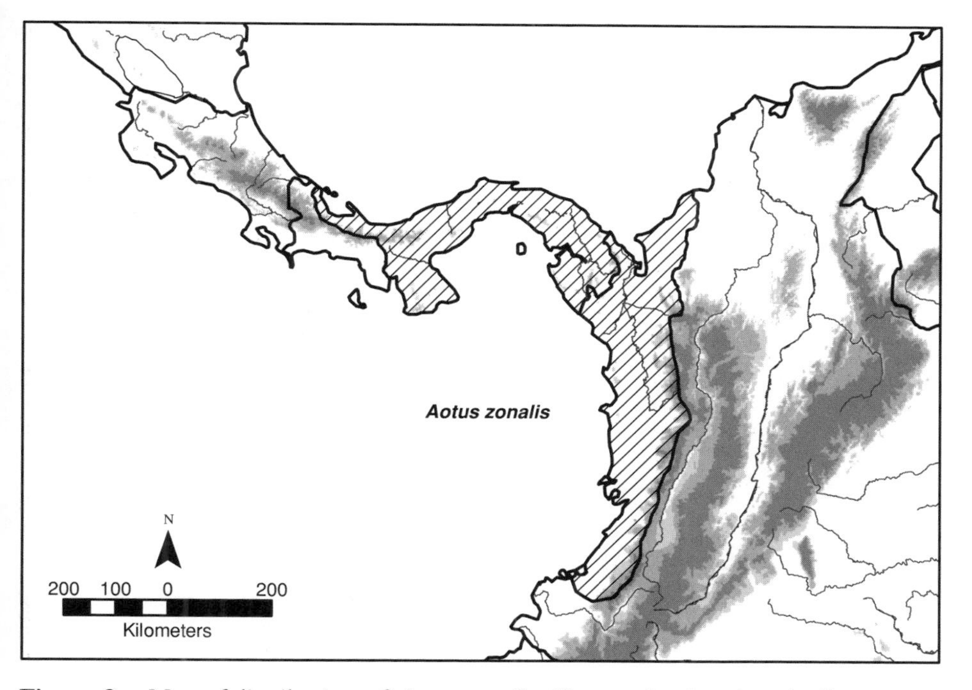

Figure 3. Map of distribution of *Aotus zonalis*. Changes in elevation shadings correspond to 650, 1000, and >2000 m.

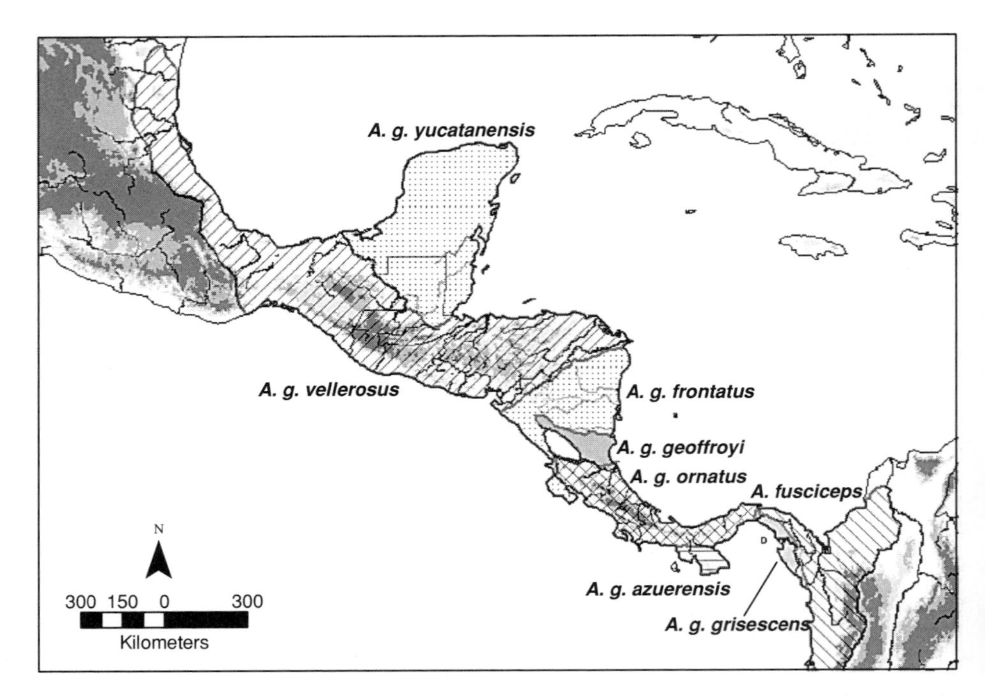

Figure 4. Map of distribution of *Ateles geoffroyi* and *Ateles fusciceps*. Changes in elevation shadings correspond to 500, 1000, and >1800 m.

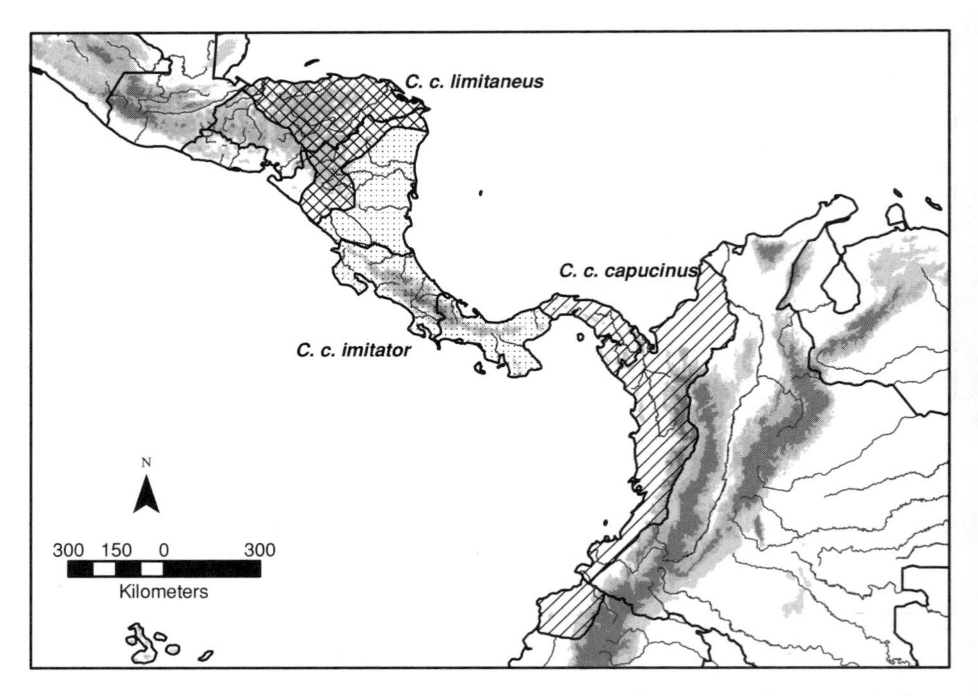

Figure 5. Map of distribution of *Cebus capucinus*. Changes in elevation shadings correspond to 500, 1000, and >2000 m.

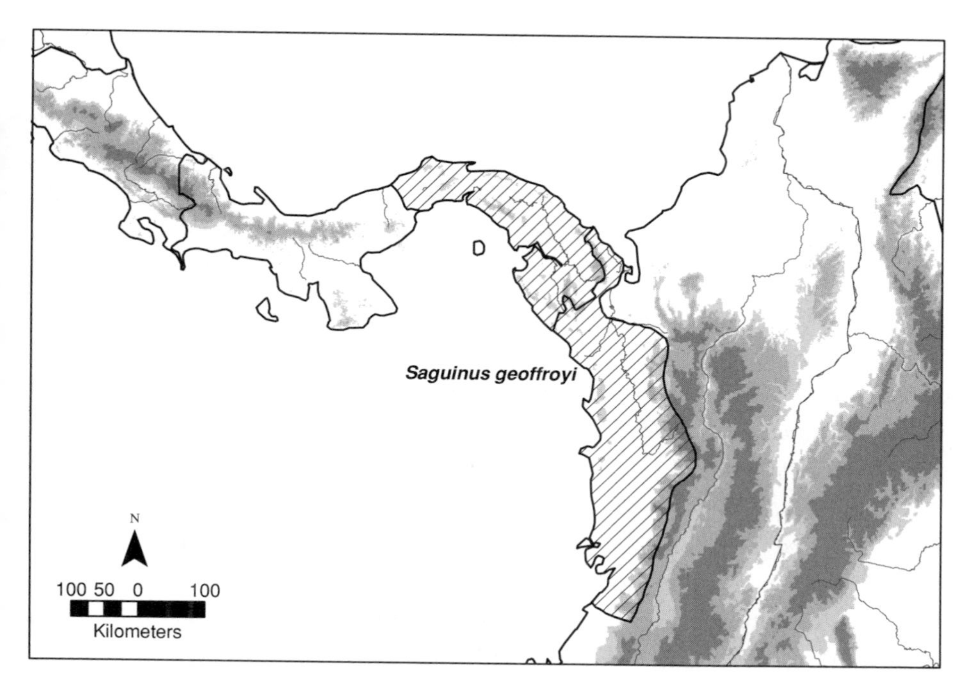

Figure 6. Map of distribution of *Saguinus geoffroyi*. Changes in elevation shadings correspond to 500, 1000, and >2000 m.

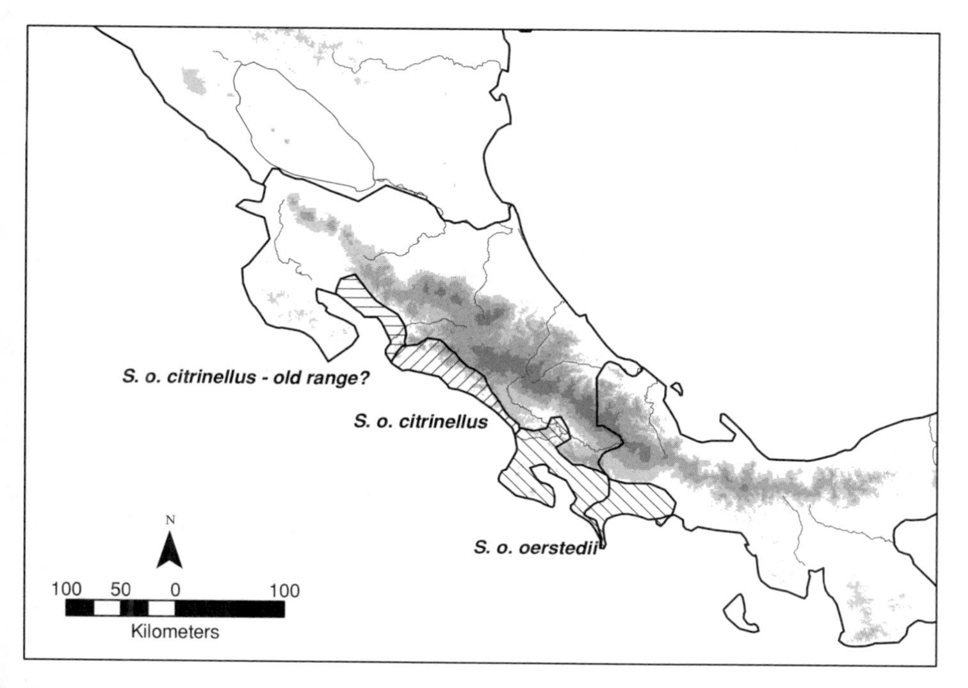

Figure 7. Map of distribution of *Saimiri oerstedii*. Changes in elevation shadings correspond to 500, 1000, and >2000 m.

Natural History in Chicago (FMNH), and the British Museum of Natural History in London (BM(NH)). These data were used to supplement the published range information that forms the basis of the maps in Rylands *et al.* (this volume). As part of a study of the biogeographic patterns of primates in northern South America, I have determined latitude and longitude values for these localities using a variety of gazetteers (Burt and Stirton, 1961; Goodwin, 1942, 1946; Hershkovitz, 1977; Paynter, 1982, 1993, 1997; United States, Geographic Names Division, gazetteers for each country, 1957–1985) and maps (*Ecuador—Atlas Histórico-Geográfico*, 1942; *General Map of Nicaragua Canal Region*, 1899; *Nicaragua*, 1979; *Panama*, 1981; *República de Panamá—Mapa Físico y Mapa Político*, 1993; *Republic of Panama*, 1967; *South America North West 4th Ed.*, 2000; *Travel Map of Ecuador*). All localities for which a latitude and longitude were determined (or the nearest landmark/community) were located on one or more maps of the area. In many cases, there is more than one place with the same name; use of this database covering multiple museum collections which could be sorted by collector and of the Harvard bird gazetteers (Paynter, 1982, 1993, 1997) allowed quite detailed information on the travels and locations of individual collectors, so that information on collector and date could aid in identifying localities. Also, maps from different time periods were used when possible, since some localities given by early collectors no longer exist.

These museum records and differing interpretations of published information did result in several differences from Rylands *et al.*'s (this volume) distributions. I expand the range of *Cebus capucinus* westward in Nicaragua and the range of *Saguinus geoffroyi* slightly farther westward in Panama, based on actual collecting records. *Saimiri oerstedii citrinellus* was collected north of its current range in 1902; I indicate this likely historical extension of its range separately in Figure 7. In the Brooks Parsimony Analysis described below, I also allow for the possibility that *Aotus zonalis* ranged into southeastern Costa Rica, following Timm (1994), but this possible range expansion is not indicated on the map in Figure 3. However, as will be seen, should this extension of the owl monkey's range be incorrect, the deletion has little effect on the overall scenario of primate biogeography in Mesoamerica. Last, the range maps of Rylands *et al.* cover broad areas that almost certainly include regions from which the monkeys are absent due to habitat or elevation restrictions. My map boundaries are slightly altered to accommodate presumed elevation limits; microhabitat effects will require more detailed data.

BROOKS PARSIMONY ANALYSIS

Methods

Using ArcView 8.2 GIS software (ESRI, 2002), distributions of primates are mapped and overlain on both topographic (Figure 1) and ecozone/habitat maps (not figured). The topographic map is provided by ESRI within ArcView, as is the ecozone map, which is based on data from the World Wildlife Fund.

As one means of exploring various historical biogeographic reconstructions, possible scenarios were tested using Brooks Parsimony Analysis (BPA) (Brooks and McLennan, 2001, 2002; Brooks *et al.*, 2001). In BPA, individual geographic areas become the operational taxa, and the presence or absence of individual species and their "ancestors" (from a known phylogenetic tree) become the characters for each area. Each species and each ancestral node for the phylogenetic trees are numbered; if a species is found in an area, then each of its ancestral nodes is coded as present as well. Repeated parsimony analyses are performed, duplicating areas to indicate independent occupations, until little or no homoplasy remains (ideally). These duplicate occupations of areas indicate separate vicariance events, dispersals, extinctions, or other modes of speciation or biogeographic processes beyond a simple unfolding of vicariance events associated with the original area cladogram.

Here, BPA was compromised by the fact that there are many species and subspecies of *Alouatta, Cebus*, and *Ateles* identified across the region, with no good phylogenies within each genus. For the subspecies of *Alouatta palliata* and *C. capucinus*, which are well dispersed ranging up the isthmus, assumptions were made that those farther from Colombia were more recently connected phylogenetically than those closest to Colombia. However, for the many subspecies of *Ateles geoffroyi*, scattered all over the isthmus, no such assumptions could be reasonably made (Collins, 2004; Collins and Dubach, 2000a,b, 2001; Madeiros *et al.*, 1997; Silva-Lópes *et al.*, 1996), leading to a large multichotomy. As a result, the trees produced in the BPA also never could be fully resolved, leading to ambiguity and limiting its usefulness. The phylogenetic tree for Mesoamerican platyrrhines used as a base for BPA is shown in Figure 8 and is derived in part from Collins (2004; Collins and Dubach, 2000a,b, 2001) and Cortés-Ortiz *et al.* (2003).

Sixteen distinct biogeographic zones within the isthmus, as well as two external but neighboring zones, were identified and used as the basis of discussion and analysis. While defining zones is a critical part of any BPA study, the

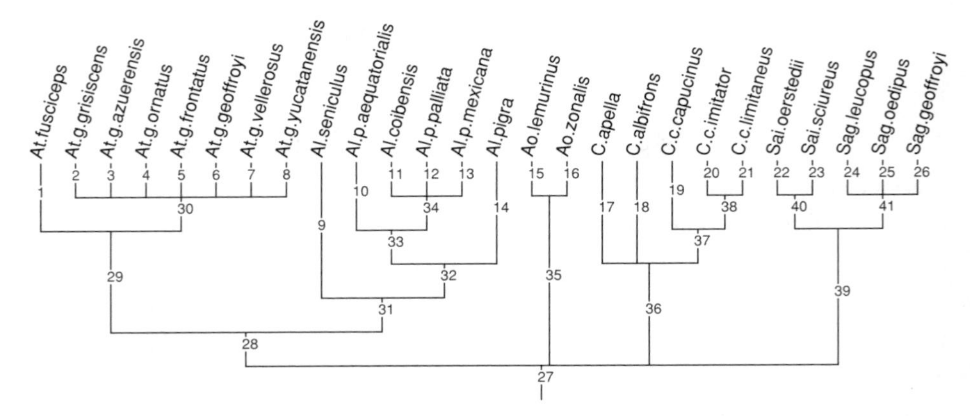

Figure 8. Cladogram of Mesoamerican primates, derived in part from Collins (2004; Collins and Dubach, 2000a,b, 2001) and Cortés-Ortiz *et al.* (2003). Numbered nodes indicate terminal taxa and "ancestors" as coded in the Brooks Parsimony Analysis.

geologic, geographic, and biologic history of Mesoamerica remains poorly understood. Therefore, the boundaries of the zones used here should not be interpreted as absolute but as hypotheses of "meaningful" biogeographic areas. The two mainland zones represent the Chocó/Atrato region, corresponding to a well-defined refuge area west of the Andean rise (Haffer, 1969, 1982), and the northern Colombian area between the Cordillera Occidental and Cordillera Oriental of the Andean area. The Andean uplift effectively isolated many trans-Andean faunal and floral elements from the Amazonian/Orinoco region (cis-Andean) to the east and south of this intra-Andean region around 8 mya (Díaz de Gamero, 1996; Haq *et al.*, 1987; Hoorn, 1993; Hoorn *et al.*, 1995; Lovejoy *et al.,* 1998; Martin and Bermingham, 2000; Montoya-Burgos, 2003; Reis, 1997; Rull, 1998; Sivasundar *et al.*, 2001; Van der Hammen, 1989; Vari, 1988).

The zones within the Mesoamerican isthmus were based on several criteria. Primary was a general consideration of the apparent boundaries for the modern distributions of subspecies of Mesoamerican primates. The distributions of the most supspeciose taxa, particularly *Ateles* and *Alouatta*, and the most restricted taxa (*Saguinus* and *Saimiri*) were given high consideration in drawing zone boundaries. In addition, Nores (1999) has suggested a 100 m rise in sea level during periods in the Pleistocene. Figures 9a and 9b show what the Mesoamerican landscape would look like, with all areas below 100 m elevation under water. It is apparent that such an event would entirely isolate the Azuero Peninsula, break the isthmus at the Gatún and Costa Rica/Nicaragua areas, and create

(a)

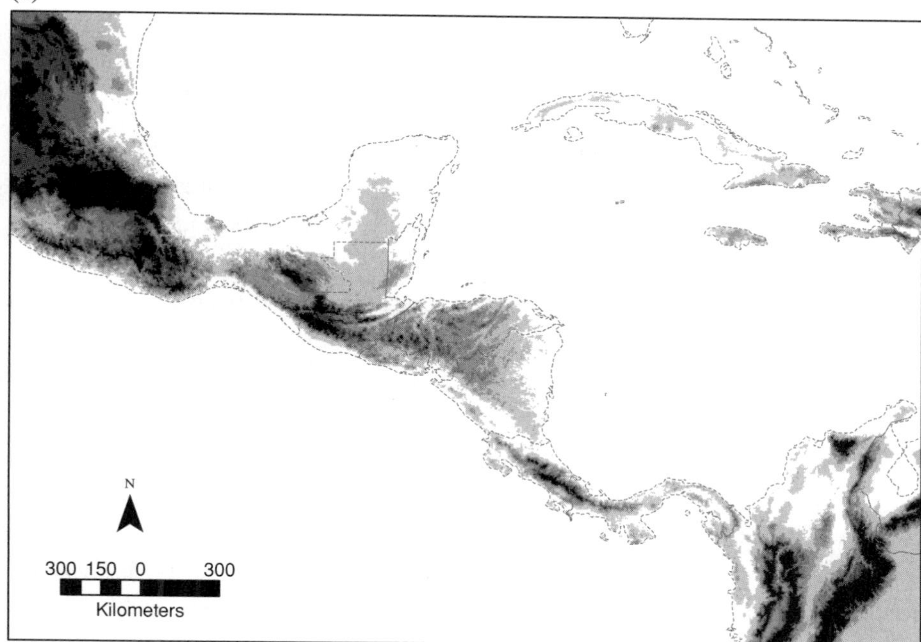

(b)

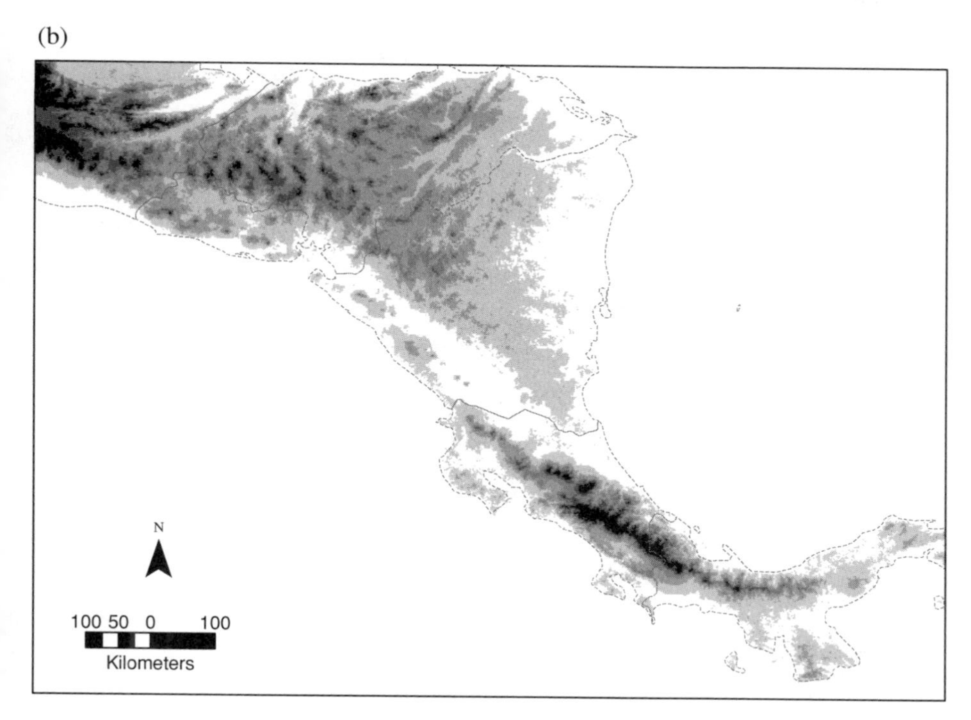

Figure 9. Map of Mesoamerica as it would appear with a 100 m rise in sea level. (a) lower Mesoamerica; (b) close-up of western Panama through Honduras. Only the shaded areas would be emergent.

major barriers to dispersal along the Pacific coast of Nicaragua and the Atlantic coast at the Honduras/Guatemala border. These gaps in land contact were also considered in defining biogeographic zones. Where central mountain ranges exist (much of the isthmus), the boundaries were drawn along the continental divide; this also conformed extremely well to posited boundaries for many individual subspecies in Rylands *et al.* (this volume; see Figures 2–7) and the divide is often of higher elevation than the reported elevational limits of most Mesoamerican primates. Finally, marked contrasts in ecozone which currently restrict primate taxa were also considered, although ecozones are more labile and changing over the last 3 my than geography.

The resulting zones should be interpreted as initial hypotheses of meaningful biogeographic zones in Mesoamerica, to be refined through this analysis and as more data become available on faunal and floral ranges across the region. The 18 zones are shown in Figure 10, and include: OUT—the Colombian northern areas west of the Cordillera Oriental, including the Magdalena and Cauca valleys; A—the Chocó/Atrato region of northwest Colombia and

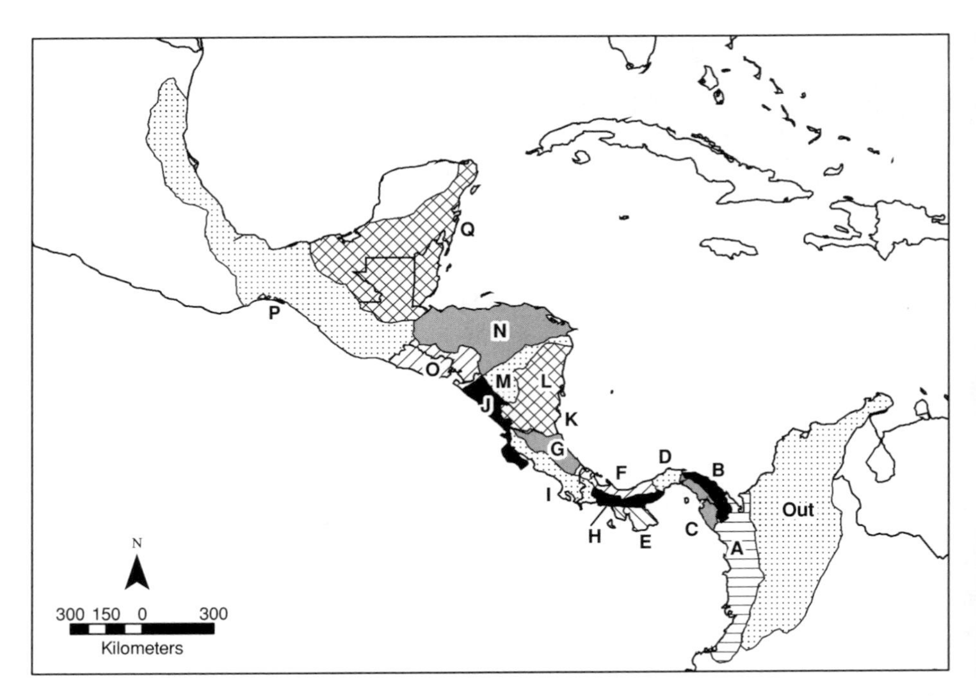

Figure 10. Map of seventeen distinct biogeographic zones plus "out" zone used for Brooks Parsimony Analysis of Mesoamerican platyrrhines. Zones are based on distributions of primate taxa and topographic features, as defined in text and used in Table 2.

western Ecuador; B—Atlantic and central region of the Darién of Panama; C—Pacific coast of the eastern Darién; D—central lowlands of Panama, including current Canal Zone; E—Azuero Peninsula and Isla de Coiba; F—Atlantic coast of Panama just into Costa Rica; G—Atlantic coast of Costa Rica and central lowlands to Lake Nicaragua; H—Pacific coast of western Panama almost to Costa Rica; I—far southwestern Panama and Pacific coast of southern Costa Rica (lusher woodlands, ends where drier forests begin); J—Pacific coast of northern Costa Rica and Nicaragua (more arid wood/shrublands), and including Ometepe Island; K—Atlantic Coast of southern Nicaragua, to Lake Nicaragua; L—highlands of northeastern and northern Nicaragua; M—northeastern and central borderland of Nicaragua; N—Atlantic Honduras and southeastern Guatemala; O—Pacific Honduras and extreme northwestern Nicaragua, El Salvador, and western Guatemala; P—southern Mexico and northeastern Guatemala; Q—prehistorically forested regions of the Yucatan, Belize, and northeastern Guatemala. The distribution of monkeys in these areas is given in Table 2.

The data (consisting of 18 geographic zones as taxa and the presence or absence of taxa and their ancestral nodes in a zone as characters) were entered and final trees produced with MacClade 4.0 (Maddison and Maddison, 2000). All parsimony analyses were done with PAUP* 4.10 (Swofford, 2002), with the goal of reducing homoplasy (Homoplasy Index), raising consistency (Consistency Index), and improving retention (Retention Index). All analyses were heuristic searches, with random addition of taxa, ACCTRAN, and retention of all shortest trees. Bootstrap analyses (600 replicates) were also performed, to produce a consensus tree (generally identical to the Strict Consensus Tree resulting from the heuristic search). Multiple analyses were run, from an initial exploration of a common inhabitation of each zone by all constituents to various separate migrations up the isthmus by different combinations of primate genera, following protocol in Brooks and McLennan (2001, 2002).

Results

As expected, given the poorly resolved phylogenies of Mesoamerican primate species, the Brooks Parsimony analytical runs never achieved zero homoplasy, or anything approaching it. The best run, duplicating nearly every area at least once, resulted in a consistency index of only 0.64, with a homoplasy index of 0.36, similar to the original analysis with only 18 zones, but it raised the

Table 2. Mesoamerican primate distributions by zone

	ZONE																	
	OUT	A	B	C	D	E	F	G	H	I	J	K	L	M	N	O	P	Q
Ateles																		
A. fusciceps	×	×	×	×														
A.g. grisescens		×		×														
A.g. ornatus					×		×	×	×	×								
A.g. azuerensis						×												
A.g. frontatus											×		×	×				
A.g. geoffroyi												×						
A.g. vellerosus															×	×	×	
A.g. yucatanensis																		×
Alouatta																		
A. seniculus	×	×																
A.p. aequatorialis	×	×	×	×	×		×		×									
A.p. palliata								×		×	×	×	×	×	×			
A.p. mexicana																	×	
A. coibensis						×												
A. pigra																		×
Aotus																		
A. lemurinus	×	×																
A. zonalis	×	×	×	×	×	×	×		×									

Cebus																		
C. apella	×																	
C. albifrons	×	×																
C.c. capucinus	×	×	×	×	×													
C.c. imitator						×	×	×	×	×	×	×	×					
C.c. limitaneus											×			×	×			
Saguinus																		
S. leucopus	×																	
S. oedipus	×	×																
S. geoffroyi		×	×	×	×													
Saimiri																		
S. oerstedii										×								

The biogeographic zones are defined in the text and illustrated in Figure 10. They include: OUT—the Colombian northern areas west of the Cordillera Oriental, including the Magdalena and Cauca valleys; A—the Chocó/Atrato region of northwest Colombia and western Ecuador; B—Atlantic and central region of the Darién of Panama; C—Pacific coast of the eastern Darién; D—central lowlands of Panama, including current Canal Zone; E-Azuero Peninsula and Isla de Coiba; F—Atlantic coast of Panama just into Costa Rica; G—Atlantic coast of Costa Rica and central lowlands to Lake Nicaragua; H—Pacific coast of western Panama almost to Costa Rica; I—far southwestern Panama and Pacific coast of southern Costa Rica (lusher woodlands, ends where dry forest begins); J—Pacific coast of northern Costa Rica and Nicaragua (more arid wood/shrublands), and including Ometepe Island; K—Atlantic Coast of southern Nicaragua, to Lake Nicaragua; L—highlands of northeastern and northern Nicaragua; M—northeastern and central borderland of Nicaragua; N—Atlantic Honduras and southeastern Guatemala; O—Pacific Honduras and extreme northwestern Nicaragua, El Salvador, and western Guatemala; P—southern Mexico and northeastern Guatemala; Q—prehistorically forested regions of the Yucatan, Belize, and northeastern Guatemala.

retention index markedly to 0.86, suggesting a stronger hypothesis of the connections between areas. This run included separate codings of the areas inhabited by *Alouatta pigra* and *Saimiri oerstedii* from all others, separate dispersal events by *Alouatta palliata/coibensis*, *Ateles geoffroyi*, plus *Cebus capucinus* from that of *Aotus zonalis* and *Saguinus geoffroyi*, and separate from *Ateles fusciceps*. This suggests a total of four to five (if *A. pigra* and *S. oerstedii* dispersed in separate events) distinct invasions into the isthmus region from the Chocó.

When branches with less than 70% bootstrap support are collapsed in this final run, the separate isthmian invasions remain, but most branchings within these Mesoamerican biogeographic zones disappear. A few but important patterns remain, however. The areas with *Alouatta palliata*, *Alouatta coibensis*, *Ateles geoffroyi*, and *Cebus* form a single, large, almost entirely unresolved bush. Nested within this is an area clade for zones defined by the presence of *Saguinus* and *Aotus*. All analyses, including this preferred one, indicate close ties between zones B and D (especially) and C, identity of zones F and H, and ties between zones G, L, J, and M, with I (not including the presence of *Saimiri*) nearly as close. These suggest that within each major dispersal, there was broad exchange and similarity across eastern Panama through the central lowlands, continued broad exchange between Atlantic and Pacific coastal regions of Panama near the start of the central range, and across northern and central Nicaragua along with Pacific coastal Nicaragua/Costa Rica. While these zones were defined based on individual range boundaries of primate species, they are probably not meaningfully different biogeographic areas.

The largest differences (with most changes in faunal elements) are between Colombian Area A and the clade of all Mesoamerican regions (bootstrap of 94), a break between elsewhere and the zones for *Saimiri oerstedii* (bootstrap of 86) and for *Alouatta pigra* (bootstrap of 70), and between the zones including both *Saguinus* and *Aotus* (zones B, C, and D) versus the rest of Mesoamerica (bootstrap of 87).

Thus, even with the strong limitations on a Brooks Parsimony Analysis due to poor phylogenetic resolution, the results suggest a model of at least four separate introductions to the isthmus: (1) *Alouatta pigra* and *Saimiri oerstedii*; (2) *Alouatta geoffroyi*, *Alouatta palliata* and its offshoot *A. coibensis*, and *Cebus capucinus*; (3) *Aotus zonalis* and *Saguinus geoffroyi*; and (4) *Ateles fusciceps* a last and very recent entrant on the isthmus, based on its highly restricted range near the southern entrance to the isthmus. The lack of *A. pigra* and *S. oerstedii* in other areas along the isthmus is almost certainly due to their extinction in intervening areas (see below). In addition, the BPA strongly suggests that certain zones used here are not separate biogeographic units but largely integrated and inter-connected regions for the later immigrants (B–C–D, F–H, and G–I–J–L–M).

A MODEL FOR THE BIOGEOGRAPHIC HISTORY OF MESOAMERICAN PRIMATES

A dominating aspect of the invasion of Mesoamerica by South American taxa is the continental effects of the Andean orogeny. This uplift long predates the establishment of contact with Mesoamerica and occurred over a 27+ my period in the last half of the Cenozoic. By 8–10 mya, the Cordillera Oriental in Colombia and Venezuela effectively isolated many taxa to either side of the northern Andes (see above). Subsequent to that time, there has been only limited movement of eastern Neotropical taxa (particularly Amazonian) around this barrier into the northwestern regions of Colombia and Ecuador that border the Mesoamerican isthmus.

Therefore, once a land connection was formed around 3.5 mya, the only lowland taxa available for migration were those already present in the northwestern area, occupying particularly the Chocó and also the Cauca and Magdalena river valleys and northern coast of Colombia. For primates, this limited source area had dramatic effects on the populating of Mesoamerica. The Andean barrier has kept many genera of primates from the northwestern source area, including all pitheciines, titi monkeys (*Callicebus*), and Goeldi's monkeys (*Callimico*). The Amazon and other southern barriers have further restricted most marmosets (*Callithrix*, *Mico*, and *Callibella*), golden lion tamarins (*Leontopithecus*), and muriquis (*Brachyteles*). While woolly monkeys (*Lagothrix*) are found in the headwaters region of the Cauca and Magdalena rivers, they appear to be recent migrants to this area from over Andean passes and do not range far enough north to disperse into the isthmus (Ford, in prep.). Pygmy marmosets (*Cebuella*) were collected in southern valleys in the Cordillera Central of Colombia, but there are no clear records that they have ranged into the northwestern basins. Certainly, there are no indications that *Cebuella* ever invaded the isthmus.

As a result, the only primates known to have been geographically available to disperse into the opening Mesoamerican region were howlers (*Alouatta*), spider monkeys (*Ateles*), owl monkeys (*Aotus*), capuchins (*Cebus*), and tamarins (*Saguinus*). All of these did indeed disperse northward—using the emergent isthmus as an open highway from northwestern Colombia. However, there are substantial differences in the degree of genetic isolation of the Mesoamerican populations from those found south of the Darién in Colombia and from one another ranging up the isthmus (see Groves, 2001; Rylands *et al.*, this volume). Howlers and spider monkeys range the farthest, reaching to Mexico, with a unique species of howler in the Yucatan region (*A. pigra*) and another on Azuero Peninsula and Isla de Coiba (*A. coibensis*). Genetic evidence suggests that *A. pigra* has been distinct from *A. palliata* and *A. coibensis* for some time (Cortés-Ortiz *et al.*, 2003; see also Froehlich *et al.*, 1991). Genetic relationships among other Mesoamerican primates are still uncertain.

Given recent evidence for at least two, and possibly more, separate waves of introductions of fauna to the isthmus since the connection was first made 3.5 mya, the potential is there for multiple independent invasions by monkeys, and Brooks Parsimony Analysis supports this model.

Far less resolved are relationships between neighboring Mesoamerican zones, particularly in Costa Rica and Nicaragua. The possibility for east–west migrations in the past through the southern Nicaraguan lowlands, currently largely blocked by Lake Nicaragua, and through valleys in the highlands of northern Nicaragua, in particular, may have allowed complex mixing and separations of populations through this region. Current taxonomy and exact ranges of subspecies in this area remain uncertain (see Rylands *et al.*, this volume; Groves, 2001). Nonetheless, a broad outline of movements into and up the isthmus emerges.

Wave One

Initial invaders, with the earliest development of an emergent pathway at 3.5–3.0 mya, were the precursors of *Alouatta pigra* and *Saimiri oerstedii*. Source populations of howlers were and are in northwestern Colombia (Zones A and OUT). However, no squirrel monkeys are currently present. After ruling out human transport (in agreement with Cropp and Boinski, 2000; Rylands *et al.*, this volume), the only possibility is that squirrel monkeys were present in the late Miocene or Pliocene in northwestern Colombia, isolated from Amazonian populations by the rise of the northern Andes around 8 mya along with other primates. Certainly, ancestral squirrel monkeys were in an area bordering this while the proto-Andes were still quite low, at La Venta in the late Miocene—currently in the Colombian Andes (Kay *et al.*, 1997). In the intervening time, squirrel monkeys in northwestern Colombia have become extinct. Once on the isthmus, squirrel monkeys may have migrated up the Pacific coast of eastern Panama into their current home in Pacific Costa Rica/Panama (Zone I). There is little obvious barrier to their expansion north up to the area where drier shrublands develop, and collecting records indicate they did fill this region at least in the beginning of the 20th century. Large collections made in the mid-20th century may have contributed to their diminished range. Squirrel monkeys are currently the most endangered Mesoamerican primate (Cropp and Boinski, 2000; Reid, 1997), limited to very low elevations and edge and disturbed forests.

Howlers, on the other hand, may have migrated up the Atlantic coast once past the central Gatún area, traveling eventually all the way to the Yucatan (Zone Q). Perhaps, as a result of competition with later invading monkeys (see also Cortés-Ortiz *et al.*, 2003) or other fauna, both howlers of *A. pigra* type and squirrel monkeys were extirpated from areas outside their current range, leaving relict populations in two far-flung pockets. While Cortés-Ortiz *et al.* suggest that *A. pigra* may have been pushed up the isthmus by the later invading *A. palliata*, the remarkable dispersal abilities of howlers (evidenced in fact by *A. palliata* itself) suggest that *A. pigra* may have already spread far north before the advent of *A. palliata*. Its current distribution represents a last stand against competition from *A. palliata*. With its back to the sea, *A. pigra* has nowhere to go should *A. palliata* continue its advance into the peninsula. Reid (1997) indicates far narrower adaptive choices for *A. pigra*, with no populations recorded above 500 m, while *A. palliata* has been found as high as 2500 m.

The differentiation between *S. oerstedii oerstedii* and *S. oerstedii citrinellus* is almost certainly the result of isolation during high water periods. *Saimiri* is restricted to lowland settings (below 500 m, see Reid, 1997), and a finger of the Costa Rican central range extends nearly to the Pacific between the two. Nores (1999) suggested a sea-level rise of approximately 100 m in the Pleistocene. The effects of such rise can be seen in Figures 7 and 9b—the distribution of *S. oerstedii* would be cut into two, precisely at the boundary of the current subspecies. On the other hand, testament to the ability of *Saimiri* to take advantage of low water stands is its presence (at least in 1902) on Sevilla Island and Almijas Island (= Isla Sabaneta) off the Pacific coast of western Panama.

The current limitation of *A. pigra* and *S. oerstedii* to isolated, far-flung areas of Mesoamerica with their presumed extinction elsewhere remains difficult to explain beyond competition with later primate immigrants to the isthmus. However, it is worth noting again that these two monkeys are the most restricted in terms of elevational range to lowland areas below 500 m (Reid, 1997). Pavelka and Chapman (this volume) describe the striking effect of a hurricane on a population of *A. pigra*, resulting in a dramatic decline in population, continuing over more than 2 years after the hurricane. They believe that this population decline may be due to a combination of effects from the hurricane, including loss of food trees, increased parasite loads, and social disruption. Black howlers and squirrel monkeys, due to their restriction to low lying, mostly coastal (in the narrow Mesoamerican isthmus) regions would have been most vulnerable over time to the ravages of storms that are common occurrences in the Mesoamerican region. The long-term effects of this type of random, brief, but dramatic event may be impossible to test for over the history of primates in Mesoamerica, but continued work on the short-term impact will help us understand the potential role of storms and catastrophic occurrences on population size, structure, and survival of Mesoamerican primates.

Wave Two

A second wave of introductions would have occurred with the re-emergence of a terrestrial connection around 2.0 mya. This wave included the ancestors of *Alouatta palliata* (spun from the same source population as the earlier *A. pigra*), *Ateles geoffroyi*, and *Cebus capucinus*. All spread broadly up both coasts of the isthmus, at least through Costa Rica (Zones B–I), and all successfully moved across the filter barrier into the Azuero Peninsula, differentiating in the process. *Alouatta* appears to have differentiated more completely (particularly *A. coibensis*, isolated during the Pleistocene high water levels predicted by Nores, 1999, and perhaps earlier, as seen in Figure 9), but in any event, the biogeographic implications are the same—offshoots of *A. palliata*, *A. geoffroyi*, and *C. capucinus* in the central (Zone D) and/or Pacific coastal (Zone H) area ferreted their way onto the peninsula, becoming isolated for some time.

The near absence of *A. geoffroyi* from Zone B, in the northeastern and central Darién of Panama, is almost certainly quite recent and due perhaps to competitive exclusion from the invading *Ateles fusciceps* (see below). Some interbreeding may be occurring in area C in the central valley (Rossan and Baerg, 1977), suggesting that genetic isolation of these two species is not complete despite perhaps 2 million years of separation. *A. geoffroyi* continues to range widely, up both coasts, to the northernmost extent of primates in the southern states of Mexico. Its northern boundaries appear to include the Sierra Madre mountains and Atlantic coastal grasslands in Mexico. However, it is so successful that it is the only primate to range currently on the west side of the continental divide in El Salvador and Pacific coastal Mexico.

A. palliata appears to have been only slightly less successful. As howlers ranged northward along the Pacific coast, the mountains of northwestern Nicaragua and El Salvador were effective barriers to their continued dispersal up the Pacific coast. Once into northern Nicaragua and eastern Honduras, however, *A. palliata* has successfully moved northward, skirting the range of *A. pigra*, nearly as far north as spider monkeys.

The difference in degree of adaptability of spider and howler monkeys from other areas of their sympatry, most notably western Venezuela, is notable. In Venezuela, howler monkeys are far better able to move into somewhat inhospitable habitats, following gallery forest into the Llanos grasslands. In contrast, spider monkeys range around the highlands of western Guatemala and Honduras to disperse along the forests of the Pacific coast, where howlers are absent. The effectiveness of the high mountains in northwestern Nicaragua and southwestern El Salvador as a barrier to *Alouatta* and *Cebus* suggests that they were also effective against *Ateles*. *Ateles* is the only monkey in Pacific coastal Mexico, and it likely migrated south from this region into El Salvador. Only substantial genetic information on the affinities of these populations will provide an adequate test of these hypotheses.

The range of *C. capucinus* is nearly as broad, but with less apparent differentiation between populations (the degree of differentiation is controversial; see Rylands, this volume) and they never make it beyond Honduras. This could have two possible explanations. The first is that *Cebus* arrived later, with *Saguinus* and *Aotus*, but was able to disperse farther. While this is possible, and its broader elevational range (see Table 1) might support this scenario, at present the Brooks Parsimony Analysis would support the second alternative: *Cebus* arrived in this earlier cycle (Wave Two), and its restriction to areas south of Mexico and Belize are due to some barrier that *A. geoffroyi* and *A. palliata* were able to cross. The form of such barrier is unclear, and unsubstantiated reports of the occurrence of *Cebus* into Belize exist (see Rylands *et al.*, this volume; Reid, 1997). However, it is notable that comparing the range of *Cebus* (Figure 5) with a world flooded by 100 m (Figure 9b) shows a perfect match with a low region on the Honduran/Guatemalan border that would have flooded back into the high mountains of the interior. However, this did not form a barrier for *Alouatta* and *Ateles*.

One other difference is their differential presence on islands, likely related to the greater home range needs of spider monkeys versus howlers (Bernstein *et al.*, 1976; Chapman, 1988; Crockett and Eisenberg, 1987; Kinzey, 1997a,b; Palacios and Rodriguez, 2001; Wallace *et al.*, 1998; Yoneda, 1990). *Ateles* is not reported from any offshore islands along the isthmus. *Alouatta*, on the other hand, has been collected on many islands, both large and small, all likely connected to the mainland during low water cycles in the Pleistocene. These include not only Isla de Coiba, but also Isla Colón on the Atlantic side of Panama, and other small islands dotting the coasts, particularly of Panama. *Cebus* has also been collected on Isla Colón and Isla Bastimentos, as well as the Pacific coastal islands of Panama of C'baco, Coiba, Insoleta, Brava, and Sevilla. Whether or not *Ateles* ever migrated to these areas during periods of connection, the small size of the islands apparently cannot support their ranging needs today.

Several studies of other Mesoamerican immigrants from the south document explosive radiations, with likely rapid spread up the isthmus into Mexico followed by later divergence into separate taxa. These include work on short-tailed bats (Hoffman and Baker, 2003), parrots (Eberhard and Bermingham, 2004), and freshwater fish (Bermingham and Martin, 1998; Perdices *et al.*, 2002), as well as a recent study on howler monkeys (Cortés-Ortiz *et al.*, 2003). In all of these cases, mtDNA analyses fail to clearly indicate branchings between various Mesoamerican populations, supporting models of rapid expansion across the region.

Wave Three

Without well-dated fossil localities, it is conjecture whether the remaining established Mesoamerican primates , *Saguinus geoffroyi* and *Aotus zonalis,* entered the isthmus with Wave Two or as part of a postulated independent, younger invasion, associated with a possible influx of freshwater fish around 1.0 mya (Martin and Bermingham, 2000). Both *Aotus* and *Saguinus* exhibit only limited dispersal northward. *Saguinus* only reaches Zone D (the lowland Canal Zone area of Panama). *Aotus* extends beyond this, both on the Atlantic side of the Panamanian Central Range and into Azuero Peninsula. It does reach, with certainty, nearly to the Costa Rican border. There is little apparent geographic barrier to a spread into Atlantic coastal Costa Rica, and Timm (1994; but see Rylands *et al.*, this volume) has argued for its presence in southeastern Costa Rica. In either event, there is no evidence for its ever having extended farther up the isthmus on either coast. This abbreviated presence coupled with their lack of divergence strongly suggests a late entry to the isthmus, particularly in light of the widespread successful dispersal of both these genera throughout much of tropical South America.

Aotus (collected on Isla Colón and Isla Bastimentos on the Atlantic coast of Panama) has also dispersed to island areas, like *Alouatta*, *Cebus*, and *Saimiri*. These distributions suggest that *Aotus* was present in the region during the low water cycles of the later Pleistocene or that the emergent low water pathways were available in the very recent past.

Wave Four

A last wave would be recent, filtered invasions by *Ateles fusciceps*, particularly into the northeastern (Atlantic) region of Panama. This is almost certainly very recent, given the highly limited presence of *A. fusciceps* in Mesoamerica. The Mesoamerican *A. palliata*, *C. capucinus*, *A. zonalis*, and *S. geoffroyi* all share extremely similar southern range extensions into northwestern Colombia and Pacific coastal Ecuador. These may be remnants of the original source populations for these Mesoamerican monkeys, or they may be the result of recent back migrations into South America through the same filter route being used by *A. fusciceps* to extend north into the isthmus.

Last, in South American we find *Alouatta seniculus*, *Cebus albifrons*, and — farther east—*Cebus apella* making inroads across northern Colombia, although none currently occur on the isthmus. All three are widespread in northern South America east of the Andean range. While they could represent source populations for the isthmian taxa, it is most likely that they are very recent immigrants who have managed to skirt the Andean range. The evidence of the absence in northwestern Colombia of other monkeys found east of the Andean ranges, such as pitheciines, *Saimiri*, *Callicebus*, and *Cebuella*, as well as the genetic and morphologic distinctions between all other trans-Andean versus cis-Andean primates suggest that the cross-Andean distributions of these three are highly unusual. All three are very adaptable primates, found in wide varieties of habitats and elevations, and their absence from the isthmus or the Chocó reinforces the hypothesis that they are recent immigrants to northern Colombia. As these successful monkeys continue to move westward, they may enter the isthmus via the same filter through the Darién Range that has been exploited by *A. fusciceps*, putting further pressure on the native monkeys of Mesoamerica.

SUMMARY

Mesoamerican primates derive from distinct source populations that were likely isolated in northwestern Colombia approximately 8 mya with the rise of the northern Andes. This community of monkeys must have included squirrel monkeys in addition to relatives of the other Mesoamerican taxa, although squirrel monkeys are now extinct/absent in the region. All primates known to be distinct parts of the trans-Andean Colombian fauna migrated into the isthmus.

With the complete emergence and establishment of a land connection across the Darién region around 3.5 mya, primates quickly moved widely into Mesoamerica. Evidence from a variety of sources suggests that the connection subsided again periodically over the last 3 my, resulting in at least a second major cycle of emergence/dispersal around 2 mya. Some evidence suggests a third subsidence/emergence cycle around 1 mya, with a filter present today. Filtered exchange of land fauna may have also occurred pre-emergence, around 6–8 mya. Although primates would have been present in the source Chocó region, there is no current evidence that they utilized any tenuous early connection that may have existed.

Modern distributions suggest that primates entered Mesoamerica in at least three and likely four waves. The first wave included ancestors of *Alouatta pigra* and *Saimiri oerstedii*, with initial major emergence of the isthmus. These now exist only in relict areas where they are endangered, with their ancestors elsewhere on the isthmus, and in the case of squirrel monkeys in northern Colombia, now extinct. The second wave was likely an explosive entry and rapid dispersal up the isthmus of ancestral *Alouatta palliata*, *Ateles geoffroyi*, and *Cebus capucinus.* As gene flow between populations was interrupted by highlands, grasslands, and periodic rises in sea level, groups differentiated, including the distinctive howlers of Azuero Peninsula and Isla de Coiba. The third and fairly recent wave brought tamarins (*Saguinus geoffroyi*) and owl monkeys (*Aotus zonalis*). The final invader has been *Ateles fusciceps*, through a filter that may also have allowed back migrations of tamarins, capuchins, howlers, and owl monkeys into northwestern Colombia, although these may be part of the ancestral population pool that remained in this region. Three recent immigrants into northwestern Colombia (*Alouatta seniculus, Cebus albifrons,* and *Cebus apella*) may eventually invade the isthmus, placing pressure on the unique primate fauna of the Mesoamerican region.

Relationships between the areas inhabited by the various named subspecies of *A. geoffroyi*, *A. palliata*, and *C. capucinus* in Mesoamerica remain obscure.

Current models, derived from mtDNA analyses of howlers and other fauna, suggest explosive dispersal throughout the region followed by differentiation. A test of this model is needed; mtDNA data from populations of howlers, spider monkeys, and capuchins should show equidistant relationships between monkeys in each of the biogeographic zones identified here if this model is correct.

ACKNOWLEDGMENTS

I thank Dr. Richard Thorington, Jr. and Dr. Brian Patterson for making the catalogs of their respective museums available to me in online form. Dr. Patterson and Dr. Thorington consulted with me about issues in geography and collectors, Dr. Colin Groves consulted on issues of taxonomy, and Dr. Harry Davis assisted me with locating appropriate maps. Dr. Anthony Rylands provided his distribution maps and lengthy correspondence on distributions. My growing knowledge of Neotropical fishes owes much to ongoing discussions with Dr. Brooks Burr, Dr. Uriel Buitrago, and Mr. Matthew Thomas. Mr. Kevin Davie of Morris Library, SIUC, has provided invaluable and ongoing assistance in using ArcView, and Ms. Kendra Hicks helped in final editing.

REFERENCES

Aubry, M. -P. and Berggren, W. O. 1999, Appendix 1, new biostratigraphy, in: L. S. Collins and A. G. Coates, eds., *A Paleobiotic Survey of Caribbean Faunas from the Neogene of the Isthmus of Panama*, Bulletins of American Paleontology, no. 357, pp. 38–40.

Bartlett, A. S. and Barghoorn, E. S. 1973, Phytogeographic history of the Isthmus of Panama during the past 12,000 years, in: A. Graham, ed., *Vegetation and Vegetational History of Northern Latin America*, Elsevier, New York, pp. 203–299.

Bennett, C. F. 1968, *Human Influences on the Zoogeography of Panama*. University of California Press, Berkeley.

Bermingham, E. and Martin, A. P. 1998. Comparative mtDNA phylogeography of Neotropical freshwater fishes: Testing shared history to infer the evolutionary landscape of lower Central America. *Molec. Ecol.* 7:499–517.

Bernstein, I. S., Balcaen, P., Dresdale, L., Gouzoules, H., Kavanagh, M., Patterson, T., and Neyman-Warner, P. 1976, Differential effects of forest degradation on primate populations. *Primates* 17:401–411.

Brooks, D. R. and McLennan, D. A. 2001, A comparison of a discovery-based and an event-based method of historical biogeography. *J. Biogeogr.* 28:757–767.

Brooks, D. R. and McLennan, D. A. 2002, *The Nature of Diversity: An Evolutionary Voyage of Discovery.* The University of Chicago Press, Chicago.

Brooks, D. R., Van Veller, M., and McLennan, D. A. 2001, How to do BPA, really. *J. Biogeogr.* 28:345–358.

Burt, W. H. and Stirton, R. A. 1961, *The Mammals of El Salvador.* Misc. Publ. Mus. Zool., Univ. Mich., No. 117, pp. 1–69.

Chapman, C. A. 1988, Patch use and patch depletion by the spider and howling monkeys of Santa Rosa National Park, Costa Rica. *Behaviour* 105:99–116.

Coates, A. G. 1999, Lithostratigraphy of the Neogene strata of the Caribbean coast from Limón, Costa Rica, to Colon, Panama, in: L. S. Collins and A. G. Coates, eds., *A Paleobiotic Survey of Caribbean Faunas From the Neogene of the Isthmus of Panama*, Bulletins of American Paleontology, no. 357, pp. 17–38.

Coates, A. G., Aubry, M. -P., Berggren, W. A., Collins, L. S., and Kunk, M. 2003, Early Neogene history of the Central American arc from Bocas del Toro, western Panama. *GSA Bull.* 115:271–287.

Coates, A. G. and Obando, J. A. 1996, The geologic evolution of the Central American isthmus, in: J. B. C. Jackson, A. F. Budd, and A. G. Coates, eds., *Evolution and Environment in Tropical America*, University of Chicago Press, Chicago, pp. 21–56.

Colinvaux, P. A. 1993, Pleistocene biogeography and diversity in tropical forests of South America, in: P. Goldblatt, ed., *Biological Relationships Between Africa and South America*, Yale University Press, New Haven, pp. 473–499.

Colinvaux, P. A. 1996, Quaternary environmental history and forest diversity in the Neotropics, in: J. B. C. Jackson, A. F. Budd, and A. G. Coates, eds., *Evolution and Environment in Tropical America*, University of Chicago Press, Chicago, pp. 359–405.

Collins, A. C. 2004, Atelinae phylogenetic relationships: The trichotomy revived? *Am. J. Phys. Anthropol.* 124:285–296.

Collins, A. C. and Dubach, J. M. 2000a, Phylogenetic relationships of spider monkeys (*Ateles*) based on mitochondrial DNA variation. *Int. J. Primatol.* 21:381–420.

Collins, A. C. and Dubach, J. M. 2000b, Biogeographic and ecological forces responsible for speciation in *Ateles. Int. J. Primatol.* 21:421–444.

Collins, A. C. and Dubach, J. M. 2001, Nuclear DNA variation in spider monkeys (*Ateles*). *Mol. Phylogenet. Evol.* 19:67–75.

Collins, L. S., Budd, A. F., and Coates, A. G. 1996b, Earliest evolution associated with closure of the tropical American seaway. *Proc. Natl. Acad. Sci. USA* 93:6069–6072.

Collins, L. S. and Coates, A. G., eds., 1999, *A Paleobiosurvey of Caribbean Faunas from the Neogene of the Isthmus of Panama.* Bulletins of American Paleontology, no. 357, pp. 1–351.

Collins, L. S., Coates, A. G., Aubry, M. -P., and Berggren, W. A. 1998, The Neogene depositional history of Darien, Panama. *Geological Soc. Am. Abstr. Programs,* 30(7):A26.

Collins, L. S., Coates, A. G., Berggren, W. A., Aubry, M. -P., and Zhang, J. 1996a, The late Miocene Panama isthmian strait. *Geology* 24:687–690.

Collins, L. S., Coates, A. G., Jackson, J. B. C., and Obango, J. A. 1995, *Timing and Rates of Emergence of the Limón and Bocas del Toro Basins: Caribbean Effects of Coco Ridge Subduction?* Geological Society of America Special Paper No. 295, pp. 263–289.

Cortés-Ortiz, L., Bermingham, E., Rico, C., Rodríguez-Luna, E., Sampaio, I., and Ruiz-García, M. 2003, Molecular systematics and biogeography of the Neotropical monkey genus, *Alouatta. Mol. Phylogenet. Evol.* 26(1):64–81.

Crockett, C. M. and Eisenberg, J. F. 1987, Howlers: Variation in group size and demography, in: B. B. Smuts, D. L. Cheney, R. M. Seyfarth, R. W. Wrangham, and T. T. Struhsaker, eds., *Primate Societies,* University of Chicago Press, Chicago, pp. 54–68.

Cronin, T. M. and Dowsett, H. J. 1996, Biotic and oceanographic response to the Pliocene closing of the Central American isthmus, in: J. B. C. Jackson, A. F. Budd, and A. G. Coates, eds., *Evolution and Environment in Tropical America,* University of Chicago Press, Chicago, pp. 76–104.

Cropp, S. J. and Boinski, S. 2000, The Central American squirrel monkey (*Saimiri oerstedii*): Introduced hybrid or endemic species? *Mol. Phylogenet. Evol.* 16:350–365.

de Cserna, Z. 1989, An outline of the geology of Mexico, in: A. W. Bally and A. R. Palmer, eds., *The Geology of North America, Vol. A, The Geology of North America: An Overview,* Geological Society of America, Boulder, pp. 233–264.

Díaz de Gamero, M. L. 1996, The changing course of the Orinoco river during the Neogene: A review. *Palaeogeogr. Palaeoclimatol. Palaeoecol.* 123:385–402.

Duque-Caro, H. 1990, Neogene stratigraphy, palaeoceanography and palaeobiogeography in northwest South America and the evolution of the Panama Seaway. *Palaeogeogr. Palaeoclimatol. Palaeoecol.* 77:203–234.

Eberhard, J. R. and Bermingham, E. 2004, Phylogeny and biogeography of the *Amazona ochrocephala* (Aves: Psittacidae) complex. *The Auk* 121(2):318–332.

Ecuador—Atlas Histórico-Geográfico 1942, Prof. Juan Morales y Eloy, Ministerio de Relaciones Exteriores, Quito.

Eisenberg, J. F. 1989, *Mammals of the Neotropics: The Northern Neotropics, Vol. 1: Panama, Colombia, Venezuela, Guyana, Suriname, French Guiana (Mammals of Neotropics*). University of Chicago Press, Chicago.

Engel, S. R., Hogan, K. M., Taylor, J. F., and Davis, S. K. 1998, Molecular systematics and paleobiogeography of the South American sigmodontine rodents. *Mol. Biol. Evol.* 15:35–49.

Environmental Systems Research Institute (ESRI), Inc. 2002, *ArcView GIS 8.2*. ESRI, Redlands, CA.

Ford, S. M. (in prep.), Biogeographic patterns of the Atelinae across the northern tier of South America.

Froehlich, J. W. and Froehlich, P. H. 1987, The status of Panama's endemic howling monkeys. *Primate Conserv.* 8:58–62.

Froehlich, J. W., Supriatna, J., and Froehlich, P. H. 1991, Morphometric analysis of *Ateles*: Systematics and biogeographic implications. *Am. J. Primatol.* 25:1–22.

General Map of Nicaragua Canal Region (map—1″ = 8 miles) 1899, Nicaragua Canal Commission.

Goodwin, G. G. 1942, Mammals of Honduras. *Bull. Am. Museum Nat. Hist.* 79(2):107–195.

Goodwin, G. G. 1946, Mammals of Costa Rica. *Bull. Am. Museum Nat. Hist.* 87(5):274–458.

Gregory-Wodzicki, K. M. 2000, Uplift history of the central and northern Andes: A review. *GSA Bull.* 112:1091–1105.

Groves, C. 2001, *Primate Taxonomy*. Smithsonian Institution Press, Washington, D.C.

Haffer, J. 1969, Speciation in Amazon forest birds. *Science* 185:131–137.

Haffer, J. 1982, General aspects of the refuge theory, in: G. T. Prance, ed., *Biological Diversification in the Tropics*, Columbia University Press, New York, pp. 6–24.

Haq, B. U., Hardenbol, J., and Vail, P. R. 1987, Chronology of fluctuating sea levels since the Triassic. *Science* 235:1156–1167.

Henderson, C. L. and Adams, S. 2002, *Field Guide to the Wildlife of Costa Rica*. University of Texas Press, Austin.

Hershkovitz, P. 1977, *Living New World Monkeys*, vol. 1, University of Chicago Press, Chicago.

Hoffman, F. G. and Baker, R. J. 2003, Comparative phylogeography of short-tailed bats (*Carollia*: Phyllostomidae). *Mol. Ecol.* 12:3403–3414.

Hoorn, C. 1993, Marine incursions and the influence of Andean tectonics on the Miocene depositional history of Northwestern Amazonia: Results of a palynostratigraphic study. *Palaeogeogr. Palaeoclimatol. Palaeoecol.* 105:267–309.

Hoorn, C., Guerrero, J., Sarmiento, G. A., and Lorente, M. A. 1995, Andean tectonics as a cause for changing drainage patterns in Miocene northern South America. *Geology* 23:237–240.

Kay, R. F., Madden, R. H., Flynn, J. J., and Cifelli, R., eds., 1997, *Vertebrate Paleontology in the Neotropics: The Miocene Fauna of La Venta, Colombia*, Smithsonian Institution Press, Washington, D.C.

Kinzey, W. G. 1997a, Synopsis of New World primates (16 genera): *Ateles*, in: W. G. Kinzey, ed., *New World Primates: Ecology, Evolution, and Behavior*, Aldine de Gruyter, New York, pp. 192–199.

Kinzey, W. G. 1997b, Synopsis of New World primates (16 genera): *Alouatta*, in: W. G. Kinzey, ed., *New World Primates: Ecology, Evolution, and Behavior*, Aldine de Gruyter, New York, pp. 174–185.

Lovejoy, N. R., Bermingham, E., and Martin, A. P. 1998, Marine incursion into South America. *Nature* 396:421–422.

Maddison, D. and Maddison, W. 2000, *MacClade 4: Analysis of Phylogeny and Character Evolution*. Sinauer Associates, Inc., Sunderland, MA.

Madeiros, M. A., Barros, R. M. S., Pieczarka, J. C., Nagamachi, C. Y., Ponsa, M., Garcia, M., Garcia, F., and Egozcue, J. 1997, Radiation and speciation of spider monkeys, genus *Ateles*, from the cytogenetic viewpoint. *Am. J. Primatol.* 42:167–178.

Marshall, L. G. 1988, Land mammals and the Great American interchange. *Am. Sci.* 76:380–388.

Marshall, L. G., Webb, S. D., Sepkoski, J. J., and Raup, D. M. 1982, Mammalian evolution and the Great American interchange. *Science* 215:1351–1357.

Martin, A. P. and Bermingham, E. 2000, Regional endemism and cryptic species revealed by molecular and morphological analysis of a widespread species of Neotropical catfish. *Proc. R. Soc. Lond.* B 267:1135–1141.

McNeill, D. R., Coates, A. G., Budd, A. F., and Borne, P. F. 1999, Integrated biological and paleomagnetic stratigraphy of the Neogene deposits around Limón, Costa Rica: A coastal emergence record of the Central American isthmus. *GSA Bull.* 112:963–981.

Montoya-Burgos, J. I. 2003, Historical biogeography of the catfish genus *Hypostomus* (Siluriformes: Loricariidae), with implications on the diversification of Neotropical ichthyofauna. *Mol. Ecol.* 12:1855–1867.

Nicaragua (map—1:1,500,000) 1979, Central Intelligence Agency, Washington, D.C.

Nores, M. 1999, An alternative hypothesis for the origin of Amazonian bird diversity. *J. Biogeogr.* 26:475–485.

Palacios, E. and Rodriguez, A. 2001, Ranging pattern and use of space in a group of red howler monkeys (*Alouatta seniculus*) in a southeastern Colombian rainforest. *Am. J. Primatol.* 55:233–251.

Panama (map- 1:1,500,000) 1981, Central Intelligence Agency, Washington, D.C.

Pavelka, M. S. M. and Chapman, C. A. 2005, Population structure of black howlers (*Alouatta pigra*) in southern Belize and responses to Hurricane Iris, in: A. Estrada, P. A. Garber, M. S. M. Pavelka, and L. Luecke, eds., *New Perspectives in the Study of Mesoamerican Primates: Distribution, Ecology, Behavior and Conservation. Developments in Primatology: Progress and Prospects*, R. A. Tuttle (series ed.), Kluwer, New York, pp. xx–xx.

Paynter, R. A. 1982, *Ornithological Gazetteer of Venezuela*, Museum of Comparative Zoology, Harvard University, Cambridge.

Paynter, R. A. 1993, *Ornithological Gazetteer of Ecuador*, 2nd edn., Museum of Comparative Zoology, Harvard University, Cambridge.

Paynter, R. A. 1997, *Ornithological Gazetteer of Colombia*, 2nd edn., Museum of Comparative Zoology, Harvard University, Cambridge.

Perdices, A., Bermingham, E., Montilla, A., and Doadrio, I. 2002, Evolutionary history of the genus *Rhamdia* (Teleostei: Pimelodidae) in Central America. *Mol. Phylogenet. Evol.* 25:172–189.

Pindell, J. L. and Barrett, S. F. 1990, Geological evolution of the Caribbean Region; a plate-tectonic perspective, in: G. Dengo and J. E. Case, eds., *The Caribbean Region: The Geology of North America Volume H*, Geological Society of America, Boulder, pp. 405–432.

Reid, F. A. 1997, *A Field Guide to the Mammals of Central America and Southeast Mexico*. Oxford University Press, Oxford.

Reis, R. E. 1997, Systematics, biogeography, and the fossil record of the Callichthyidae: A review of the available data, in: L. R. Malabarba, R. E. Reis, R. P. Vari, Z. M. A. Lucena, and C. A. S. Lucena, eds., *Phylogeny and Classification of Neotropical Fishes*, Edipucrs, Porto Alegre, Brazil, pp. 351–362.

República de Panamá—Mapa Físico y Mapa Político (map—1:1,000,000) 1993, El Instituto Geografico Nacional "Tommy Guardia", Agosto.

Republic of Panama (map in three sheets—1:500,000) 1967, USARCARIB, Inter-American Geodetic Survey.

Rossan, P. N. and Baerg, D. C. 1977, Laboratory and feral hybridization of *Ateles geoffroyi panamensis* Kellogg and Goldman, 1944, in Panama. *Primates* 18:235–237.

Rull, V. 1998, Biogeographical and evolutionary considerations of *Mauritia* (Arecaceae), based on palynological evidence. *Rev. Paleobot. Palynol.* 100:109–122.

Rylands, A. B., Mittermeier, R. A., Hines, J., Groves, C., and Cortés-Ortiz, L. 2005, Taxonomy and distributions of Mesoamerican primates, in: A. Estrada, P. A. Garber, M. S. M. Pavelka, and L. Luecke, eds., *New Perspectives in the Study of Mesoamerican Primates: Distribution, Ecology, Behavior and Conservation. Developments in Primatology: Progress and Prospects*, R. A. Tuttle (series ed.), Kluwer, New York, pp. xx–xx.

Salazar-Bravo, J., Dragoo, J. W., Tinnin, D. S., and Yates, T. L. 2001, Phylogeny and evolution of the Neotropical rodent genus *Calomys*: Inferences from mitochondrial DNA sequence data. *Mol. Phylogenet. Evol.* 20:173–184.

Savage, J. M. 2002, *The Amphibians and Reptiles of Costa Rica: A Herpetofauna Between Two Continents, Between Two Seas.* University of Chicago Press, Chicago.

Silva-Lópes, G., Motta-Gill, J., and Sánchez Hernández, A. I. 1996, Taxonomic notes on *Ateles geoffroyi. Neotrop. Primates* 4(2):41–44.

Simpson, G. G. 1980, *Splendid Isolation: The Curious History of South American Mammals.* Yale University Press, New Haven.

Sivasundar, A., Bermingham, E., and Ortí, G. 2001, Population structure and biogeography of migratory freshwater fishes (*Prochilodus*: Characiformes) in major South American rivers. *Mol. Ecol.* 10:407–417.

South America North West 4th Ed. (map- 1:4,000,000) 2000, International Travel Maps, Vancouver, B.C., Canada.

Stehli, F. G. and Webb, S. D., eds., 1985, *The Great American Biotic Interchange*, Plenum Press, New York.

Swofford, D. L. 2002, *PAUP**: *Phylogenetic Analysis Using Parsimony (and Other Methods) 4.0 Beta*. Sinauer Associates, Inc., Sunderland, MA.

Timm, R. M. 1994, Mammals, in: L. A. McDade, K. S. Bawa, H. A. Hespenheide, and G. S. Hartshorn, eds., *La Selva: Ecology and Natural History of a Neotropical Rain Forest*, University of Chicago Press, Chicago, pp. 394–397.

Travel Map of Ecuador.

United States Board on Geographic Names 1957, *Ecuador: Official Standard Names Gazetteer*, Office of Geography, Department of Interior, Washington, D.C.

United States Board on Geographic Names 1969, *Panama and the Canal Zone: Official Standard Names Gazetteer*, Geographic Names Division, U.S. Army Topographic Command, Washington, D.C.

United States Board on Geographic Names 1985, *Gazetteer of Nicaragua*, 3rd edn., Defense Mapping Agency, Washington, D.C.

Vail, P. R. and Hardenbol, J. 1979, Sea-level changes during the Tertiary. *Oceanus* 22:71–80.

Van der Hammen, T. 1989, History of the montane forests of the northern Andes. *Plant Syst. Evol.* 162:109–114.

Vari, R. P. 1988, The Curimatidae, a lowland Neotropical fish family (Pisces: Characiformes); distribution, endemism, and phylogenetic biogeography, in: W. R. Heyer and P. E. Vanzolini, eds., *Proceedings of a Workshop on Neotropical Distribution Patterns*, Academia Brasileira de Ciências, Rio de Janeiro, pp. 343–377.

Wallace, R. B., Painter, R. L. E., and Taber, A. B. 1998, Primate diversity, habitat preferences, and population density estimates in Noel Kempff Mercado National Park, Santa Cruz Department, Bolivia. *Am. J. Primatol.* 46:197–211.

Webb, S. D. 1991, Ecogeography and the great American interchange. *Paleobiology* 17:266–280.

Webb, S. D. 1999, The great American faunal interchange, in: A. G. Coates, ed., *Central America: A Natural and Cultural History*, Yale University Press, New Haven, pp. 97–122.

Webb, S. D. and Rancy, A. 1996, Late Cenozoic evolution of the Neotropical mammal fauna, in: J. B. C. Jackson, A. F. Budd, and A. G. Coates, eds., *Evolution and Environment in Tropical America*, University of Chicago Press, Chicago, pp. 335–358.

Weyl, R. 1980, Geology of Central America. *Beiträge zur regionalen Geologie der Erde* 15:1–371.

Whitmore, T. C. and Prance, G. T. 1987, *Biogeography and Quaternary History in Tropical America.* Clarendon Press, Oxford.

Yoneda, M. 1990, Habitat utilization of six species of monkeys in Rio Duda, Colombia. *Field Studies of New World Monkeys*, La Macarena, Colombia, vol. 1, pp. 39–45.

PART TWO

Population Responses to Disturbance

Introduction: Population Responses to Disturbance

A. Estrada, P. A. Garber, and M. S. M. Pavelka

Natural catastrophic events such as hurricanes, floods, earthquakes, volcanic activity, and disease as well as extensive deforestation caused by humans have critical impacts on primate population viability. Natural and human-induced destruction results in habitat loss and fragmentation, population reduction, and in demographic, social, and reproductive disruption. Primate populations can persist in fragmented landscapes if remnant forest fragments are large enough to provide sufficient resources and expand in area through regeneration over time. The chapters in this section examine issues of primate behavior, ecology, diet, and mating strategies from different perspectives and offer important research methods for evaluating behavioral plasticity and the response of primate species to habitat contraction and fragmentation, resulting from natural or artificial causes.

Demographic data on primate populations in continuous forests with minimal human disturbance provide needed baseline information about species'

A. Estrada • Estación Biológica Los Tuxtlas, Instituto de Biología, Universidad Nacional Autónoma de México, Apartado Postal 176, San Andrés Tuxtla, Veracruz, México. **P. A. Garber** • Department of Anthropology, University of Illinois at Urbana-Champaign, 109 Davenport Hall, Urbana, IL 61801, USA. **M. S. M. Pavelka** • Department of Anthropology, University of Calgary, 2500 University Drive NW, Calgary, Alberta, Canada T2N 1N4.

New Perspectives in the Study of Mesoamerican Primates: Distribution, Ecology, Behavior, and Conservation, edited by Alejandro Estrada, Paul A. Garber, Mary S. M. Pavelka, and LeAndra Luecke. Springer, New York, 2005.

natural population parameters. Comparison of demographic traits of populations of the same species living in continuous versus fragmented forests enhances our knowledge of the variability in demographic responses of a given primate species to habitat loss and fragmentation. In this context, the chapter by Van Belle and Estrada reports contrasts in demographic features among populations of *Alouatta pigra* existing in continuous forests and in adjacent fragmented landscapes in Mexico and Guatemala. On the basis of surveys of eight large protected forests, they report that black howler monkey troops in large protected forests are more commonly multimale-multifemale in structure and exhibit population densities that range from 12.7 to 44.1 individuals/km^2. In contrast, in fragmented landscapes, howler monkey troops are more commonly unimale with 1–2 adult females. The authors suggest that adult males living in fragmented forests may respond to the effects of habitat saturation by increased intrasexual intolerance and isolation. The authors also report that population densities are, on average, five times higher in forest fragments than in continuous forests, suggesting that the long-term persistence of isolated groups of black howler monkeys in fragmented forests may place severe pressures on ecological resources and negatively impact the social interactions and mating patterns of the troop.

Pavelka and Chapman investigated the effects of hurricane Iris on habitat change and population demography of *Alouatta pigra* in a coastal site in Belize. Severe habitat damage and a decline of 43% in howler monkey prehurricane population levels were recorded a year after the hurricane. Further declines in population numbers were noted up to three years after the hurricane. The authors indicate that the most apparent factor explaining both the initial drop in population density and the subsequent decline over 29 months was the change in food availability that resulted from hurricane damage. They hypothesize that a combination of nutritional and social stress interacting with increased parasite loads (and possibly increased predation) may have led groups and individuals to leave the area in search of more suitable habitat.

In the third chapter of this section, DeGamma-Blanchet and Fedigan examine the effects of habitat type on monkey densities in the Guanacaste region of Costa Rica. They found that forest fragment age was an important explanatory variable for capuchin and howler monkey density (higher densities were found in older areas of forest) and that forest fragment isolation and size made little contribution to explaining the density of howler, capuchin, and spider monkeys in Guanacaste. The authors conclude that large size of the forest patches studied (up to 95.71 km^2) may not constrain primate population demography in the same way as would a smaller forest fragment (e.g.10-ha fragment). The relatively large Guanacaste forest fragments may be of sufficient size and productivity, such that current isolation and fragment size has little or no effect on primate population densities. However, these variables failed to explain the density of spider monkeys in these same forest fragments, suggesting that variability in the response of different primate species merit further investigation. DeGamma-Blanchet and Fedigan also found that the presence of evergreen forests in the Guanacaste landscape mosaic is important for explaining the absolute density of all three primate species. Independent of fragment size, primate population densities were higher in fragments containing evergreen forest.

Andrómeda Rivera and Sophie Calmé link dietary flexibility to conservation in a study comparing feeding preferences of black howlers (*Alouatta pigra*) in response to forest fragmentation in the Calakmul area of Mexico. The authors argue that the manner in which different primate species respond to seasonal changes in the availability and distribution of resources is critical for developing conservation and management policies. Their comparison is based on black howlers living in protected and continuous forest and in forest fragments managed by local communities. Their study suggests that black howlers exploit fewer fruit and leaf species in nonfragmented forests but display greater dietary diversity in forest fragments. They also point out that in the forest fragments investigated, trees exploited by howlers for feeding have commercial dimensions, and most of these species are commercially logged. As a result, in forest fragments that are logged, howlers have to face the lack of continuity in the canopy and the loss of many vital trees for feeding, affecting their persistence in these habitats.

The final chapter of this section addresses a different kind of disturbance, gastrointestinal parasite load. Primates are particularly vulnerable to parasitic infections because many species live in cohesive social groups that facilitate parasite transmission between individuals. Parasitic infections may increase host susceptibility to predation or decrease the competitive fitness of the individual. The study of parasitic infections in nonhuman primates is important for understanding ecological and evolutionary host-parasite relationships and for recommending conservation strategies. Moreover, parasitic infections may be exacerbated by habitat contraction and isolation resulting from human activity and presence. Addressing these issues, Stoner and González Di Pierro report the result of an investigation on gastrointestinal parasites in a population of black howler monkeys (*A. pigra*) in the Lacandon forest, Mexico. Only a few studies document endoparasitic infections in neotropical primates, and their study represents the first documentation of intestinal parasites in *Alouatta pigra*. Stoner and González Di Pierro evaluated the importance of demographic factors (i.e., density), age-sex class, environmental factors (i.e. seasonality or humidity), and diet in determining parasitic infections. They identified eight species of parasites in the populations investigated. They found a higher incidence and intensity of infection in the largest troop studied, a higher incidence and intensity of infection in females and subadults compared with males and all adults, and that parasitic infections varied by season. They also found that the howler group with the highest parasite incidence was the one living in the vicinity of humans. Lastly, the authors explore the issue of self-medication by howlers, finding a negative relationship between incidence of parasites and the time howlers spent feeding on figs. This study by Stoner and González Di Pierro describes an important research tool for examining the ecology, health, and conservation status of primate populations.

CHAPTER FOUR

Demographic Features of *Alouatta pigra* Populations in Extensive and Fragmented Forests

Sarie Van Belle and Alejandro Estrada

INTRODUCTION

The black howler monkey of Mesoamerica, *Alouatta pigra*, is a primate species endemic to the area shared by Mexico, Belize, and Guatemala (Horwich and Johnson, 1986; Rylands *et al.*, chapter 2). Eighty percent of the geographic range of *A. pigra* is found in Mexico, specifically, the states of Tabasco and Chiapas, and it is the only *Alouatta* species present in the Yucatan peninsula (Smith, 1970; Horwich and Johnson, 1986; Watts and Rico-Gray, 1987; Rylands *et al.*, this volume).

Until now, information on population parameters for *A. pigra* and on its conservation status was only available from three localities in Belize. The Community Baboon Sanctuary (CBS), an area of about 4700 ha, encompasses remnants

Sarie Van Belle • Department of Zoology, University of Wisconsin-Madison, 250 N Mills Street, Madison, WI 53706. **Alejandro Estrada** • Estación de Biología Los Tuxtlas, Instituto de Biología, Universidad Nacional Autónoma de México.

New Perspectives in the Study of Mesoamerican Primates: Distribution, Ecology, Behavior, and Conservation, edited by Alejandro Estrada, Paul A. Garber, Mary S. M. Pavelka, and LeAndra Luecke. Springer, New York, 2005.

of riparian forest along the Belize River and small rainforest fragments amidst agriculture fields and pasture lands (Horwich *et al.*, 2001a; Ostro *et al.*, 2001). Another site, Monkey River, in southern Belize, is a forest fragment, 52 ha in size (Pavelka *et al.*, 2003). The third site, Cockscomb Basin Wildlife Sanctuary, is a large reserve of 34,700 ha, but the howler population was re-introduced after *A. pigra* had gone extinct (Horwich, 1998; Ostro *et al.*, 2001). Data on *A. pigra* in undisturbed extensive forests (>5 km^2) exist only from two sites throughout its geographic distribution. For these sites, only mean troop size and population density were reported. One is the Muchukux forest in the state of Quintana Roo, Mexico (Gonzalez-Kirchner, 1998). The other is Tikal National Park in Guatemala (Coelho *et al.*, 1976; Schlichte, 1978). Basic population parameters, such as population density, population size, and age–sex composition of social groups, are poorly known for *A. pigra* existing in extensive forests. Such information is essential to understand the behavior and life history traits of this primate species (Strier, 2003a).

Throughout much of the geographic distribution of *A. pigra*, tropical rainforests have been degraded, fragmented, and converted to agriculture and pasture lands. Currently, it is estimated that only 28% tropical forest cover remains in Mexico, 59% in Belize, and 26% in Guatemala (Estrada *et al.*, chapter 1). Deforestation is continuing at a rate of −1.10% in Mexico, −2.32% in Belize, and −1.71% in Guatemala (Estrada *et al.*, chapter 1). The drastic conversion of tropical forests to anthropogenic landscapes places populations of this regionally endemic primate species at risk. *A. pigra*'s IUCN conservation status has been revised recently from "least concerned" in 2000 to "endangered" due to habitat loss and better available information (Cuarón *et al.*, 2003). However, the paucity of data on demographic parameters, ecology, and behavior of *A. pigra* in both natural protected habitats and in habitats associated with marked human disturbance makes conservation assessments of this primate species particularly difficult (Estrada and Coates-Estrada, 1988; Rylands *et al.*, 1995).

In this chapter, we present information on population density, troop size, and age–sex composition of eight populations of *A. pigra* existing in extensive forest tracks in Mexico and Guatemala. We assess the variability in these demographic parameters for populations of *A. pigra* existing in fragmented landscapes by contrasting these population characteristics for two of the populations living in extensive forest with those living in the abutting fragmented landscape. Finally, we compare literature reports of troop composition and population density of *A. pigra* populations in fragmented landscapes with the information for the eight sites of extensive forests.

METHODS

Sites and Data Collection

Between 2000 and 2003, population surveys of *A. pigra* were conducted in the eight sites located in extensive forest reserves in southern Mexico [El Tormento (ET), Calakmul (CAL), Palenque (PAL), Montes Azules along Rio Lacantun (LMA), Reforma ejido's reserve near Rio Lacantun (LRE), and Yaxchilán (YAX)] and in northern central Guatemala [Municipal Reserve Salinas Nueve Cerros, Lachuá ecoregion (LAC) and Tikal (TIK)] (see Baumgarten, 2000; Estrada *et al.*, 2002a,b,c; Barrueta, 2003; Barrueta *et al.*, 2003; Rosales Meda, 2003; Estrada *et al.*, 2004). We refer to extensive forest as forests >8.5 km^2 and protected by local governments.

Four of the sites (ET, LAC, LRE, and PAL: Figure 1) were large forest patches surrounded by anthropogenic landscapes dominated by pasture lands and agriculture fields. The other four sites were part of several larger reserves (>300 km^2) connected to each other to form the largest protected forested landmass in Mesoamerica of about 4 million hectares of tropical rainforest (Estrada *et al.*, 2004). All sites, including these sites in this large landmass, are sufficiently far apart (>100 km) to consider the primate populations as independent demographic units (Table 1).

In addition to the primate populations surveyed in extensive forest tracks, we surveyed *A. pigra* populations living in two fragmented landscapes abutting two of the extensive reserves, Palenque in Mexico, and Lachuá in Guatemala (Figure 1). Forest fragments refer to unprotected forests <4 km^2. In Palenque, we surveyed 22 forest fragments occupied by black howler monkeys. These had a mean area of 10.9 ± 9.4 ha (range 1.9–35 ha) (see Estrada *et al.*, 2002b, for details). In Lachuá, we surveyed 26 fragments in which black howler monkeys were present. The fragments had a mean area of 66 ± 109 ha (range 1.0–387 ha) (see Rosales Meda, 2003 for details).

Primate Surveys

At each site, we determined the relative location of all troops, defined as a social unit having at least one adult male and one adult female, living in the study area. This was accomplished using early morning (05:00–07:00 h) triangulation of their morning choruses and subsequent ground surveys (see Estrada *et al.*, 2004; Estrada *et al.*, chapter 19). Each troop was followed and counted repeatedly until a consensus of troop size and age and sex composition was

Figure 1. A map of southern Mexico, Belize, and northern Guatemala with the locations of the eight extensive forests sites studied. In Mexico: **1**: Palenque National Park; **2**: Yaxchilán National Monument; **3**: Montes Azules along Rio Lacantun; **4**: Reforma Community Reserve; **5**: El Tormento; **6**: Calakmul Biosphere Reserve; In Guatemala: **7**: Municipal Reserve Salinas Nueve Cerros; **8**: Tikal National Park. The shaded areas indicate the system of national protected areas in southern Mexico, Belize, and northern Guatemala.

reached. Troops were searched for on subsequent days to confirm their composition and approximate location to reduce the probability of counting a troop more than once.

Individual howler monkeys in troops were classified as infants (clinging ventrally or dorsally to mother), juveniles (independent of mother and 1/4–1/2 the size of adults), and adults (all large and robust individuals) (Izawa *et al.*, 1979). The sex of adults and juveniles could be reliably determined. Solitary individuals and extratroop social units encountered were noted. We cannot be certain that we encountered all solitary individuals or extratroop units since they are more silent and less conspicuous than larger and more vocal units.

Table 1. Features of study sites investigated. Study area refers to the area encompassed by our surveys at each site

Site[a]	Reserve area (ha)	Study area (ha)	Annual rainfall (mm)	Dry season	Mean ± SD temperature (°C)
Calakmul	700,000	400	820	November–April	25.0 ± 2.2
El Tormento	1400	1400	1380	December–May	24.1 ± 2.0
Lacantun-MA	300,000	836	2874	February–April	25.2 ± 1.8
Lacantun-RE	1700	450	2874	February–April	25.2 ± 1.8
Lachuá	850	850	2252	February–May	26.3 ± 0.6
Palenque	1771	600	2200	January–April	26.4 ± 2.2
Tikal	57,600	500	1762	December–April	25.0 ± 8.5
Yaxchilán	2700	100	1951	December–April	25.5 ± 2.2

[a] Calakmul: Archeological Mayan site in center of the Biosphere Reserve Calakmul; El Tormento: Forestry Reserve “El Tormento”; Lacuntun-MA: Montes Azules Biosphere Reserve by river Lacantun; Lacantun-RE: Community reserve of Reforma by river Lacantun; Lachuá: Municipal Reserve Salinas Nueve Cerros; Palenque: Palenque National Park; Tikal: Tikal National Park; Yaxchilán: Natural Monument Yaxchilán.

Data Processing and Statistics

For each site, mean troop size and mean troop composition were calculated. The mean adult sex ratios and mean number of infants per adult female and juveniles per adult female were calculated based on the ratios for each troop. The juvenile sex ratio per site was calculated from the sum of juvenile males and females encountered in all troops to avoid losing information from troops with juveniles of one sex only. Overall mean troop size, troop composition, and the above mentioned ratios were calculated from the population means for the eight extensive forest sites. We expressed population density figures (individuals/km^2, including solitary individuals and extratroop units) in terms of the area sampled. For two sites (LMA and YAX), density calculations include the area sampled from boat surveys along the rivers abutting the study area (see Estrada *et al.*, 2004, for details).

Mean troop composition and density of *A. pigra* populations were compared between extensive forests and forest fragments occurring in the two fragmented landscapes we investigated (Lachuá and Palenque) and in the fragmented landscape of the CBS, Belize, as reported by Ostro *et al.* (2001).

Comparisons among populations in extensive forests and between populations in extensive versus fragmented forests were assessed via the nonparametric Kruskal–Wallis test and Mann–Whitney U test. The Spearman rank

Table 2. Relative percentage of each age–sex class, including solitary individuals, within each *A. pigra* population in extensive forest (AM: adult male; AF: adult female; JM: juvenile male; JF: juvenile female; J: juvenile, sum of JF and JM; I: infant; IMM: immature, sum of J and I)

Site[a]	AM	AF	JM	JF	J	I	IMM
Calakmul	34.4	29.5	13.1	9.8	23.0	13.1	36.1
El Tormento	28.7	36.0	10.7	7.3	18.0	17.4	35.4
Lacantun-MA	36.0	40.0	4.0	8.0	12.0	12.0	24.0
Lacantun-RE	37.5	43.8	6.3	6.3	12.5	6.3	18.8
Lachuá	32.2	32.9	NA[b]	NA[b]	24.5	10.5	35.0
Palenque	33.1	27.2	11.8	14.0	24.3	14.0	38.3
Tikal	27.8	32.2	6.7	13.3	20.0	20.0	40.0
Yaxchilán	42.6	29.6	7.4	5.6	13.0	14.8	27.8
Mean	34.0	33.9	8.6	9.2	18.4	13.5	31.9
SD	4.8	5.7	3.3	3.3	5.4	4.2	7.5

[a] See Table 1 for site information.
[b] Juvenile sex not noted for Lachuá.

correlation was used to assess relations between demographic parameters and density (WinSTAT 3.0, 1994; SPSS 12.0, 2003).

RESULTS

Population surveys at these sites resulted in a total of 801 black howler monkeys encountered in the eight extensive forests studied. Of these, 769 individuals were members of 120 troops, and 32 were solitary or in extratroop social units. Thirty-four percent were adult males and 34% were adult females, 18% were juveniles, and 14% were infants (Table 2).

Troop Composition of Extensive Forest Populations

For the populations investigated, the overall mean troop size was 6.57 ± 1.20 individuals (Table 3). A troop had on average 2.07 ± 0.41 adult males, 2.26 ± 0.33 adult females, 1.28 ± 0.48 juveniles, and 0.96 ± 0.44 infants (Table 3). The overall mean adult sex ratio in a troop was 1.36 ± 0.20 females per male (Table 4). The mean juvenile sex ratio for the populations was 1.18 ± 0.58 juvenile females per juvenile male (Table 4). Troops had on average 1.10 ± 0.39 immatures per adult female or 0.65 ± 0.28 juveniles and 0.46 ± 0.16 infants per adult female (Table 4).

Troop size in the eight populations varied from 2 to 12 individuals, and the number of adult males and females in a troop both ranged from 1 to 5. Unimale

Table 3. The mean age and sex composition of troops and mean troop size of *A. pigra* populations living in extensive forests. (*N*: number of troops; *n*: individuals counted; Sol. Ind.: solitary individuals; AM: adult male; AF: adult female; J: juveniles; I: infants; IMM: immatures including infants and juveniles; bold only for clarity)

Site[a]	*N*	*n*	Sol. Ind.		AM	AF	J	I	IMM	Troop size
Calakmul	8	60	1	**Mean**	**2.50**	**2.25**	**1.75**	**1.00**	**2.75**	**7.50**
				SD	0.93	0.71	1.28	0.53	1.49	1.93
				Range	1–3	2–4	0–3	0–2	0–4	4–9
El Tormento	26	173	5	**Mean**	**1.77**	**2.46**	**1.23**	**1.19**	**2.42**	**6.65**
				SD	0.82	1.14	0.76	0.90	1.24	2.28
				Range	1–4	1–5	1–3	0–3	0–5	3–11
Lacantun-MA	13	72	3	**Mean**	**1.85**	**2.31**	**0.69**	**0.69**	**1.38**	**5.54**
				SD	0.80	0.63	0.63	0.75	0.96	1.51
				Range	1–3	1–3	0–2	0–2	0–3	3–8
Lacantun-RE	12	61	3	**Mean**	**1.75**	**2.33**	**0.67**	**0.33**	**1.00**	**5.08**
				SD	0.75	1.23	0.65	0.49	0.95	2.07
				Range	1–3	1–5	0–2	0–1	0–3	2–10
Lachuá	24	134	9	**Mean**	**1.58**	**1.92**	**1.46**	**0.63**	**2.08**	**5.58**
				SD	0.83	0.65	0.83	0.65	1.06	1.84
				Range	1–4	1–3	0–3	0–2	0–4	3–10
Palenque	19	128	8	**Mean**	**2.05**	**1.91**	**1.74**	**1.00**	**2.71**	**6.74**
				SD	0.91	1.03	1.37	0.94	1.88	2.77
				Range	1–4	1–4	0–4	0–3	0–4	2–12
Tikal	10	88	2	**Mean**	**2.30**	**2.90**	**1.80**	**1.80**	**3.60**	**8.80**
				SD	0.67	0.99	0.79	0.92	1.17	2.15
				Range	1–3	1–4	1–3	1–3	2–5	6–12
Yaxchilán	8	53	1	**Mean**	**2.75**	**2.00**	**0.88**	**1.00**	**1.88**	**6.63**
				SD	1.58	0.53	0.64	0.76	1.13	3.13
				Range	1–5	1–3	0–2	0–2	1–4	4–10
Total	120	769	32	**Mean**	**2.07**	**2.26**	**1.28**	**0.96**	**2.23**	**6.57**
				SE	0.41	0.33	0.48	0.44	0.83	1.20

[a] See Table 1 for site information.

Table 4. Mean adult sex ratio (AF/AM), mean adult female:juvenile ratio (J/AF), mean adult female:infant ratio (I/AF), and mean adult female:immature ratio (IMM/AF) of troops, percentage single male troops (% 1 AM), and density (individuals/km²) of *A. pigra* populations living in extensive forests (bold only for clarity)

Site[a]		AF/AM	JF/FM	J/AF	I/AF	IMM/AF	% 1 AM	Density (individuals/km²)
Calakmul	**Mean**	**1.25**	**0.75**	**0.88**	**0.50**	**1.38**	**25.0**	**15.2**
	SD	1.21		0.64	0.27	0.74		
El Tormento	**Mean**	**1.67**	**0.68**	**0.60**	**0.50**	**1.10**	**38.5**	**12.7**
	SD	1.11		0.38	0.40	0.61		
Lacantun-MA	**Mean**	**1.49**	**2.00**	**0.31**	**0.35**	**0.65**	**38.5**	**44.1**
	SD	0.80		0.30	0.31	0.51		
Lacantun-RE	**Mean**	**1.47**	**1.00**	**0.27**	**0.18**	**0.46**	**41.7**	**25.6**
	SD	0.92		0.31	0.32	0.48		
Lachuá	**Mean**	**1.40**	**NA**[b]	**0.85**	**0.35**	**1.21**	**58.3**	**15.8**
	SD	0.63		0.60	0.38	0.79		
Palenque	**Mean**	**1.14**	**1.06**	**1.07**	**0.54**	**1.61**	**31.6**	**23.0**
	SD	0.87		0.99	0.50	1.33		
Tikal	**Mean**	**1.40**	**2.00**	**0.68**	**0.71**	**1.38**	**10.0**	**17.8**
	SD	0.75		0.30	0.51	0.65		
Yaxchilán	**Mean**	**1.08**	**0.75**	**0.50**	**0.54**	**1.04**	**25.0**	**12.8**
	SD	0.94		0.38	0.42	0.70		
Total	**Mean**	**1.36**	**1.18**	**0.65**	**0.46**	**1.10**		**20.9**
	SE	0.20	0.58	0.28	0.16	0.39		

[a] See Table 1 for site information.
[b] Juvenile sex not noted for Lachuá.

Table 5. Percentage of troops ($N = 120$) according to number of adult males and females for the extensive forests

	1 AM	2 AM	3 AM	4 AM	5 AM	Total
1 AF	12.5	6.7	1.7	–	–	20.8
2 AF	15.0	15.8	13.3	2.5	1.7	48.3
3 AF	5.8	8.3	5.0	–	–	19.2
4 AF	5.0	3.3	1.7	–	–	10.0
5 AF	–	0.8	0.8	–	–	1.7
Total	38.3	35.0	22.5	2.5	1.7	

troops and troops with two adult males accounted for 38.3% and 35.0% of the troops, respectively. Forty-eight percent of the troops had two adult females (Table 5). Troops with a multimale–multifemale structure accounted for 53.3%. Among these, troops with two adult males and two adult females were the most frequent (15.8%; Table 5). The percentage of unimale troops per population ranged from 10.0% (TIK) to 58.3% (LAC) (Table 4).

Comparison of Extensive Forest Populations

Mean troop sizes were significantly different among the eight populations existing in extensive forests (KW $H_7 = 21.77$, $p = 0.003$). There were no differences in the mean number of females in a troop (KW $H_7 = 11.92$, $p = 0.104$) and in the mean adult sex ratio (KW $H_7 = 10.77$, $p = 0.149$) among these populations. The difference in mean troop size was due to differences in mean number of adult males (KW $H_7 = 14.61$, $p = 0.048$), and perhaps most importantly due to differences in the mean number of immatures among the populations (KW $H_7 = 27.60$, $p < 0.0001$). Mean infant:adult female ratios did not differ significantly among populations (KW $H_7 = 12.02$, $p = 0.100$), but mean juvenile:adult female ratios did (KW $H_7 = 18.91$, $p = 0.008$).

While troop composition differed among populations, the proportion of each age–sex class, including solitary individuals, within each population was not different from the expected mean proportions for all populations (χ^2-test, d.f. $= 4$, $p > 0.900$) (Table 2).

For the eight populations living in extensive forests, density ranged from 12.7 to 44.1 individuals/km^2. Across all populations, mean density was 20.9 individuals/km^2 (Table 4). Mean troop size ($r_s = -0.41$, $p = 0.320$) (Figure 2), mean number of adult males ($r_s = -0.29$, $p = 0.493$), mean number of adult females ($r_s = -0.02$, $p = 0.955$), and mean number of immatures ($r_s = -0.36$, $p = 0.385$) were not correlated with population density. The percentage of unimale troops within a population ($r_s = 0.24$, $p = 0.565$), and the

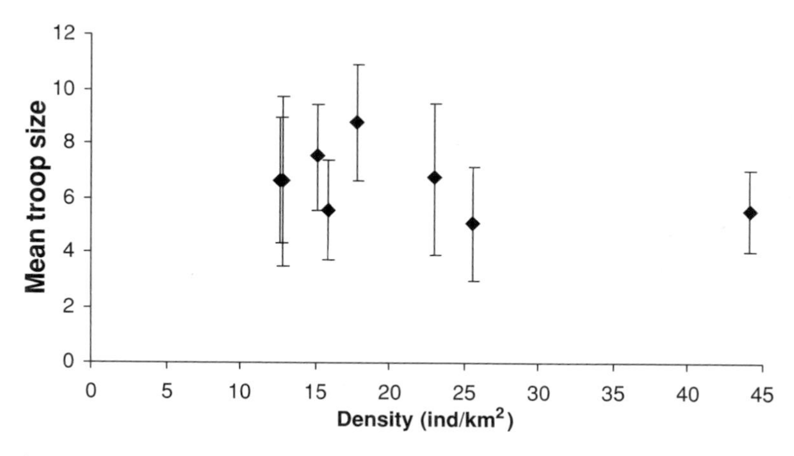

Figure 2. Relationship between mean troop size ± SD and population density for the eight populations existing in the extensive forests investigated (Spearman $r_s = -0.41$, NS).

relative age class distribution in the populations (r_s in all cases $p > 0.050$) also were not correlated with population density.

Troop Composition in Forest Fragments

Palenque, Mexico

In the 22 forest fragments occupied by black howler monkeys at Palenque, we counted a total of 115 individuals. Of these, 107 were members of 18 troops, 5 males formed 2 all male social units, and 3 males were solitary. One forest fragment harbored both a troop and an all male unit. All other fragments contained only one social unit. Troop size ranged from 2 to 15 individuals (mean = 5.94 ± 3.08 individuals, median = 5.0 individuals). A troop had on average 1.39 ± 0.50 adult males (range 1–2), 1.83 ± 0.92 adult females (range 1–4), 0.78 ± 0.81 juvenile males (range 0–3), 1.11 ± 1.08 juvenile females (range 0–4), and 0.83 ± 0.86 infants (range 0–3). Mean adult sex ratio was 1.39 ± 0.78 females per male. There were on average 1.05 ± 0.74 juveniles, 0.43 ± 0.42 infants, or 1.48 ± 0.73 immatures per adult female in a troop. Sixty-one percent of the troops had only one male. Mean ecological density was 119.2 ± 83.0 individuals/km^2, ranging between 11.3 and 315.8 individuals/km^2. While mean troop size and age–sex composition did not significantly differ between the protected forests of Palenque National Park and the fragmented landscape surrounding this reserve, mean number of adult males was significantly smaller in forest fragments than in the extensive forest ($U = 99.5$, $p = 0.018$).

Lachuá, Guatemala

The 26 forest fragments surveyed in the fragmented landscape of Lachuá yielded a total count of 225 howler monkeys, of which 223 were troop members of 43 troops and 2 were solitary males. Twenty-two fragments harbored 1 troop, and 4 fragments (50, 18, 60, and 216 ha) harbored 2, 3, 7, and 9 troops, respectively. Troop size ranged from 2 to 11 individuals (mean = 5.19 ± 1.97 individuals, median = 5.5 individuals). A troop had on average 1.63 ± 0.85 adult males (range 1–4), 1.95 ± 1.00 adult females (range 1–5), 0.30 ± 0.56 juvenile males (range 0–2), 0.51 ± 0.59 juvenile females (range 0–2), and 0.79 ± 0.94 infants (range 0–4). Mean adult sex ratio was 1.50 ± 0.99 females per male. A troop had on average 0.48 ± 0.46 juveniles, 0.41 ± 0.45 infants, or 0.89 ± 0.68 immatures per adult female. Fifty-five percent of the troops had a unimale structure. Mean ecological density was 107.18 ± 178.73 individuals/km^2, ranging between 1.0 and 700.0 individuals/km^2. There was no difference in mean troop size and age–sex composition of troops between Finca Municipal Salinas Nueve Cerros and the fragmented area abutting this reserve, except for the mean number of juveniles and mean number of juveniles per adult female, which was significantly smaller in forest fragments than in the extensive forest (J: $U = 292.5$, $p = 0.002$; J/AF: $U = 324.5$, $p = 0.010$).

Troop Composition and Population Density in Fragmented Landscapes Versus Extensive Forests

To assess the broader consistency of these patterns, we compared data available on mean population density and mean troop composition for *A. pigra* in three fragmented landscapes (CBS, LAC, and PAL) with the eight sites of extensive forest (Table 6). We found no significant difference in mean troop size ($U = 9.0$, $p = 0.540$), mean number of adult females ($U = 8.0$, $p = 0.414$), mean adult sex ratio ($U = 10.0$, $p = 0.682$), mean number of immatures ($U = 12.0$, $p = 1.000$), and mean adult female to immature ratio ($U = 12.0$, $p = 1.000$) (Figure 3). However, the mean number of adult males per troop was significantly smaller in forest fragments than in extensive forests ($U = 1.0$, $p = 0.025$) (Figure 3). Populations in forest remnants lived, on average, at ecological densities 5.4 times higher in comparison with populations in extensive forests ($U = 0.0$, $p = 0.007$).

Based on the 99 troops encountered in the fragmented landscapes of Palenque, Lachuá, and CBS (Table 7), troops with one and two males accounted for 57.6% and 34.3%, respectively. Troops with one and two females accounted for 30.3% and 46.5%, respectively. Troops with one adult male and two adult females were the most common accounting for 28.3%, while multimale–multifemale troops accounted for 31.2% (Table 7).

DISCUSSION

New survey data on population characteristics of *A. pigra* suggest considerable variability in population density and troop size. Further, data suggest important differences between populations of this howler monkey species existing in extensive and fragmented forests. In the fragmented forest, mean population

Table 6. Troop size and composition and population desnsity of *A. pigra* populations in fragmented landscapes: Belize: Community Baboon Sanctuary (CBS); Mexico: Palenque (PAL); Guatemala: Lachuá (LAC) (AM: adult male; AF: adult female; IMM: immature; bold only for clarity)

	Sites	Troops		Density (individuals/km^2)	Troop size	AM	AF	IMM	AF/AM	IMM/AF	Tot cum area (ha)	Avg frag size (ha)
CBS[a]	5	38	**Mean**	**123.2**	**5.66**	**1.40**	**2.42**	**2.21**	**1.63**	**1.08**	**171**	**42.7**
			SD	95.4	0.62	0.68	0.31	0.56	0.37	0.47		12.3
PAL	22	18	**Mean**	**119.2**	**5.90**	**1.39**	**1.83**	**2.72**	**1.39**	**1.48**	**160**	**10.9**
			SD	83.0	3.08	0.50	0.92	2.14	0.78	0.73		9.4
LAC	26	43	**Mean**	**107.2**	**5.19**	**1.63**	**1.95**	**1.60**	**1.51**	**0.98**	**1716**	**66**
			SD	178.7	1.97	0.85	1.00	1.33	1.06	0.78		109.0

[a] Ostro *et al.* (2001).

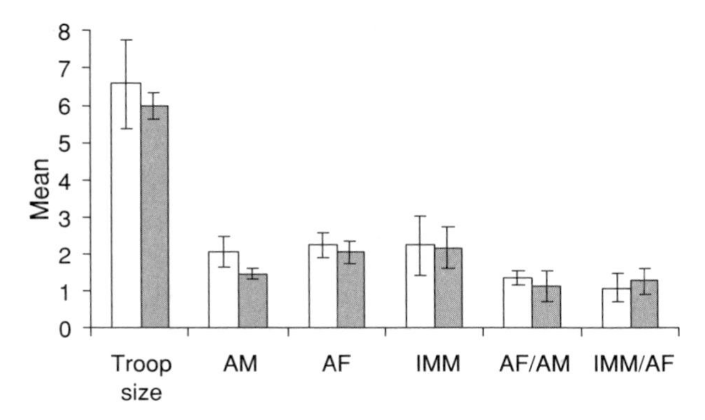

Figure 3. Comparison (Mann–Whitney *U* test) of mean troop size (± SE) and mean troop composition (± SE) between the fragmented landscapes (Community Baboon Sanctuary, Palenque, and Lachuá) (gray bars) and the eight sites of extensive forests (white bars).

Table 7. Percentage of troops ($N = 99$) according to number of adult males and females in the fragmented landscapes of Palenque, Lachuá, and Community Baboon Sanctuary

	1 AM	2 AM	3 AM	4 AM	Total
1 AF	19.2	8.1	1.0	2.0	30.3
2 AF	28.3	14.1	4.0	–	46.5
3 AF	7.1	10.1	1.0	–	18.2
4 AF	3.0	1.0	–	–	4.0
5 AF	–	1.0	–	–	1.0
Total	57.6	34.3	6.1	2.0	

density increases and mean number of adult males in troops decreases compared with the protected forests >5 km^2 in area. In the next paragraphs, we discuss some of these patterns.

Variation in Demographic Features of Populations in Extensive Forests

Black howler mean troop sizes in the extensive forests studied ranged from 5.08 to 8.80 individuals. The earlier figure of mean troop size reported for the *A. pigra* population in Tikal, Guatemala, of 6.25 individuals falls within this range (Coelho *et al.*, 1976). However, mean troop size of 3.16 individuals reported by Gonzalez-Kirchner (1998) for *A. pigra* in the Muchukux forest,

Quintana Roo, Mexico, is considerably lower. In the case of Tikal, mean troop size appears to have increased, during the 30-year interval between censuses, from 6.25 individuals (Coelho *et al.*, 1976) to 8.80 individuals, associated with an increase in number of troops encountered in the study area probably in response to a population growth (Estrada *et al.*, 2004).

Until now, troop composition had only been described for troops living in fragmented forests in Belize, and troops have been depicted as predominantly monogamous (Bolin, 1981) or as larger unimale troops (Horwich *et al.*, 2001a). Data presented here suggest that most *A. pigra* troops living in extensive forests have a multimale–multifemale social structure, with two adult males and two adult females being the most common. The number of males in a troop ranged between one and five, with troops of one, two, and three adult males being the most frequent. The number of adult females in a troop, on the other hand, was less variable. Almost half of the troops encountered had two adult females, but their number also ranged between one and five. This supports the idea that in *A. pigra* female group size is limited due to either female intolerance or a limited number of female breeding opportunities in troops (Pope, 2000a,b). However, Crockett and Janson (2000) have argued that in *A. seniculus* female group size is constrained by the risk of infanticide. They suggest that troops with greater than four adult females are more attractive to extratroop males and that during male takeovers, infanticide may occur. Whether this provides an explanation of female group size in *A. pigra* remains unclear. Alternatively, Garber *et al.* (1993) have proposed that small multimale–multifemale groups, like the *A. pigra* troops in extensive forests, are likely to represent the primitive condition for platyrrhines, and that compared to Old World monkeys, in many New World monkeys female breeding opportunities within groups are more limited.

Mean troop size significantly differed among the *A. pigra* populations studied in extensive forests. This was principally due to differences in mean number of immatures, but no clear pattern in population differences in mean number of infants and juveniles could be discerned. The mean infant:adult female ratio did not significantly differ among the populations, while the mean juvenile:adult female ratio did significantly differ but this was only for the LMA and LRE populations, which had, on average, fewer juveniles per adult female in a troop compared to the other populations. Differences in mean number of immatures between populations could be attributed to differences in fertility rates, mortality rates of the young age classes, dispersal age and rate, birth seasonality (observed in *A. pigra* in CBS, Brockett *et al.*, 2000), and period of population survey (Strier, 2003a,b). Long-term monitoring of each population will be needed to assess whether populations truly differ. Notwithstanding the dissimilarities in mean troop compositions among the eight populations of extensive forest, there were no significant differences in relative number of the age–sex classes among them. Hence, our study indicates that a black howler population in extensive forests almost consistently comprises 34% adult males, 34% adult females, 18% juveniles, and 14% infants, independent of density or mean troop size.

For the extensive forest sites, *A. pigra* densities ranged from 12.7 to 44.1 individuals/km^2, and densities of 25.0 individuals/km^2 reported earlier for Tikal (Coelho *et al.*, 1976), and of 16.5 individuals/km^2 for Muchukux (Gonzalez-Kirchner, 1998) fall within this range. The lack of association between mean troop size and density implies that populations of *A. pigra* living at higher densities did not consistently have larger troops. This contrasts with data from populations of *A. caraya* in Argentina (Rumiz, 1990), of *A. palliata* in Costa Rica (Fedigan *et al.*, 1998; Fedigan and Jack, 2001), of *A. pigra* in CBS (Horwich *et al.*, 2001b), and of *A. seniculus* in Venezuela (Rudran and Fernandez-Duque, 2003), where mean troop size positively correlated to population density. These populations have been reported to have experienced significant population fluctuations, apparently caused by epidemic pathogens, such as yellow fever and botfly infestations, by food shortages, or by natural disasters such as hurricanes (Chapman and Balcomb, 1998; Rudran and Fernandez-Duque, 2003, Pavelka *et al.*, 2003; Pavelka and Chapman, this volume). It is likely that the eight populations we investigated in the extensive forests have also experienced such population fluctuations. For example, in Tikal the number of troops increased from 4 to 10, and also underwent an increase in mean troop size during the 30-year interval between the surveys by Coelho *et al.* (1976) and by Estrada *et al.* (2004).

Population growth in *A. caraya* (Rumiz, 1990), *A. palliata* (Fedigan *et al.*, 1998), *A. pigra* (Horwich *et al.*, 2001a), and *A. seniculus* (Rudran and Fernandez-Duque, 2003) has been associated with the expansion of the established troops until a maximum troop size has been reached, determined by the biotic and aboitic characteristics of the local habitat. The second stage of population increase is primarily due to formation of new and smaller troops by dispersing individuals from the established troops. When the habitat becomes saturated, recently established troops also increase their size until they reach maximum troop size (Rudran and Fernandez-Duque, 2003). It is possible that the populations of *A. pigra* investigated in extensive forests were at different growth stages, which, in combination with possible differences in maximum troop size and carrying capacity among the eight sites studied, could explain why no correlation was found between mean troop size and density. Thus, the results we described here may only indicate the extent of population fluctuations experienced by populations of *A. pigra*. Only long-term studies of *A. pigra* will offer insight into population dynamics and the processes driving these changes in troop size, composition, and density (Rudran and Fernandez-Duque, 2003; Strier, 2003a). For now, we have documented demographic baseline data on *A. pigra* populations existing in extensive forests and data on the variability of several demographic features among these populations. This information is useful in an initial assessment of how black howler populations, social groups, and individuals are influenced by habitat fragmentation.

General Fragmentation Effects

Based on a comparison between populations of extensive and adjacent fragmented forests, it appears that habitat fragmentation had certain important effects on troop composition in *A. pigra*. In Palenque, Mexico, and in Lachuá, Guatemala, mean troop sizes in fragments, although not statistically significant, were, in general, smaller than in the extensive forest. Troops living in the fragmented landscape of Palenque had, on average, fewer males than neighboring troops living in the protected forest. In Lachuá, troops living in forest fragments had, on average, half as many juveniles and juveniles per adult female as the troops dwelling in the extensive forest reserve. While the observed troop composition values in the fragmented landscapes investigated fell within the range documented for troops living in extensive tropical forests, the observed differences might be a result of demographic or environmental stochastic events, since the time of fragmentation (>30 years).

Comparing data available on mean population density and mean troop composition for *A. pigra* in three fragmented landscapes (CBS, LAC, and PAL) with data for the eight sites of extensive forest shows that the populations in forest fragments live, on average, at ecological densities 5.4 times higher in comparison with the populations in extensive forests, suggesting overcrowding (Estrada *et al.*, 2002b). However, troops in forest fragments did not differ in mean troop size and composition from those in extensive forests, except for the mean number of adult males, which was significantly smaller for populations living in forest fragments than in extensive forests. Also, unimale troops were more common (57.6%) in the fragmented landscapes than in the extensive forest reserves (38.3%). This suggests that adult males may be most sensitive to habitat saturation or isolation. For example, in a study about injuries and scars of *A. palliata mexicana* troops in forest fragments, Cristóbal-Azkarate *et al.* (2004) found that 90% of the observed injuries were facial scars on adult males. They suggested that these scars most likely result from fights during attempts of troop take-over. Injuries inflicted during these male-to-male fights can be severe and lead to the death of contestants (Crockett, 2003).

Marsh (1999) suggested that behavioral adjustment to habitat fragmentation occurs before troop composition changes in response to fragmentation. However, she found no significant difference in daily activity patterns and food diversity in the *A. pigra* troops living in different sized forest fragments in CBS, Belize. Similar results have been reported for other howler monkey species (Bicca-Marques, 2003), suggesting that *Alouatta*'s flexible diet and small home ranges make them more able to cope with habitat fragmentation than other large-bodied monkeys (Jones, 1995; Estrada *et al.*, 1999; Bicca-Marques, 2003).

Although fragmentation may have limited effects on the behavioral ecology of *A. pigra*, troops living in forest fragments are increasingly exposed to hunting and/or predation by domestic animals (e.g., alpha male was killed by a dog in CBS, Belize, Kitchen, 2004). Additionally, reduced tree species diversity may lead to nutritional stress and/or to the use of suboptimal resources. This and greater exposure to sun radiation due to continued habitat degradation may result in increased mortality (Estrada *et al.*, 1999). In Lachuá, for example, the mean number of juveniles in troops living in forest fragments was about half its counterpart for troops living in extensive forests, suggesting an important effect of fragmentation on survivorship of immatures.

Furthermore, in the fragmented landscapes dispersal opportunities may be lower. For example, most forest patches we studied harbored only one troop, and dispersal-induced-mortality is thought to increase when howlers travel in open agricultural fields and cattle pasture (e.g., *A. palliata*, Rodriguez-Toledo *et al.*, 2003). Dispersal costs may become so high that individuals may stay in their natal troop, as we suspect in the case for a troop of 15 individuals encountered in an isolated forest fragment in Palenque (Estrada *et al.*, 2002b), therefore increasing the potential of inbreeding depression.

Finally, small isolated troops are more vulnerable to typical demographic and environmental stochasticity, resulting in increased extinction probabilities for remaining isolated populations. Although observations suggest that howlers can survive in forest fragments, they may not do well in the medium and long term (Estrada *et al.*, 2002b).

While our study provides information on the variability of demographic features of the *A. pigra* populations in extensive forests, and on the influence of habitat fragmentation on these demographic features, more long-term data are needed from more *A. pigra* populations in both forest conditions to understand *A. pigra*'s population dynamics and persistence potential in habitat fragments.

SUMMARY

Until now, little information was known about demographic features of the regionally endemic black howler populations living in large tracks of tropical forests. Here, we report results from population surveys conducted in eight extensive forests sites (>8.5 km^2) in Mexico and Guatemala. We also surveyed the black howler population in the fragmented landscape bordering two of the extensive forest reserves. Based on the 120 troops encountered in the extensive forest sites, mean troop size was 6.57 (SE = 1.20) individuals (range 2–12 individuals). Troops had most commonly a multimale–multifemale social structure, with two adult males and two adult females being the most common. Population densities ranged from 12.7 to 44.1 individuals/km^2. Mean troop size and composition were not correlated to population densities. For the two fragmented landscapes bordering the extensive forests, mean densities in forest fragments were, on average, five times higher than in extensive forests. Mean troop composition was not significantly different between forest fragments and extensive forests, but data suggest a tendency of fewer males in troops living in forest fragments. Long-term monitoring in extensive forests is necessary to document variability in population dynamics and to make adequate assessments of the consequences of fragmentation on demography, behavior, and life history of *A. pigra*.

ACKNOWLEDGMENTS

We are grateful to the Cleveland Zoo Scott Neotropical Fund, to the American Society of Primatologists, to Primate Conservation Inc., and Universidad Nacional Autónoma de México for support to sustain the Maya Sites and Primate Conservation project. We thank the Mexican Institute of Anthropology (INAH) for permission to work at all the Maya sites explored in Mexico, and we are grateful to the government of Guatemala for permission to conduct the primate surveys in Tikal and in Lachúa, and for providing invaluable logistical support. SVB acknowledges the support of the Flemish Government of Belgium, of the Belgian American Educational Foundation, of Secretaria de Relaciones Exteriores de México, and of Universidad Nacional Autónoma de México. We thank Lucía Castellanos, Adrián Mendoza, LeAndra Luecke, Marleny Morales, Yasminda García, Amrei Baumgarten, and David Muñoz for assistance in the field, and we are specially grateful to the many local Maya indigenous people who assisted us while doing field work in southeast Mexico and in Guatemala. We want to thank Karen Strier, Paul Garber, and Mary Pavelka for insightful comments on earlier versions of the manuscript.

REFERENCES

Barrueta, T. 2003, Reconocimiento demográfico y dieta de *Alouatta pigra* en El Tormento, Campeche. M. Sc. Thesis. Colegio de la Frontera Sur, Mexico.

Barrueta, T., Estrada, A., Pozo, C., and Calmé, S. 2003, Reconocimiento Demográfico de *Alouatta pigra* y *Ateles geoffroyi* en la Reserva El Tormento, Campeche, México. *Neotrop. Primates* 11:163–167.

Baumgarten, A. 2000, Estructura poblacional y uso del hábitat del mono aullador negro (*Alouatta pigra* Lawrence, Cebidae) en el área central de Guatemala. B. Sc. Thesis Universidad de San Carlos de Guatemala, Guatemala.

Bicca-Marques, J. C. 2003, How do howler monkeys cope with habitat fragmentation?, In: L. K. Marsh, ed., *Primates in Fragments: Ecology and Conservation*, Kluwer Academic/Plenum Publishers, New York, pp. 283–303.

Bolin, I. 1981, Male parental behavior in black howler monkeys (*Alouatta palliata pigra*) in Belize and Guatemala. *Primates* 22:349–360.

Brockett, R. C., Horwich, R. H., and Jones, C. B. 2000, Reproductive seasonality in the Belizean black howling monkey (*Alouatta pigra*). *Neotrop. Primates* 8:136–138.

Chapman, C. A., and Balcomb, S. R. 1998, Population characteristics of howlers: Ecological conditions or group history. *Int. J. Primatol.* 19:385–403.

Coelho, A. M. Jr., Coelho, L., Bramblett, C., Bramblett, S., and Quick, L. 1976, Ecology, population characteristics, and sympatric associations in primates: A socioenergetic analysis of howler and spider monkeys in Tikal, Guatemala. *Yearb. Phys. Anthropol.* 20:96–135.

Cristóbal-Azkarate, J., Dias, P. A. D., and Veà, J. J. 2004, Causes of intraspecific aggression in *Alouatta palliata mexicana*: Evidence from injuries, demography, and habitat. *Int. J. Primatol.* 25:939–953.

Crockett, C. M., and Janson, C. H. 2000, Infanticide in red howlers: Female group size, male membership, and possibly link to folivory, in C. P. van Schaik and C. H. Janson, eds., *Infanticide by Males and its Implications,* Cambridge University Press, Cambridge, pp. 75–98.

Crockett, C. M. (2003). Re-evaluating the sexual selection hypothesis for infanticide by Alouatta males. In: Jones, C. B. (ed.), Sexual Selection and Reproductive Competition in Primates: New Perspectives and Directions. American Society of Primatology: special topics in primatology. Volume 3. The American Society of Primatologists, Norman, Oklahoma, pp. 327–365.

Cuarón, A. D., de Grammont, P. C., Cortés-Ortiz, L., Wong, G., and Silva, J. C. S. 2003, *Alouatta pigra*. In: IUCN 2003. *2003 IUCN Red List of Threatened Species* (May 1, 2004); http://www.redlist.org.

Estrada, A., Castellanos, L., Ibarra, A., Garcia Del Valle, Y., Muñoz, D., Rivera, A., Franco, B., Fuentes, E., and Jimenez., C. 2002a, Survey of the population of the black howler monkey, *Alouatta pigra*, at the Mayan site of Palenque, Chiapas, Mexico. *Primates* 44:51–58.

Estrada, A., and Coates-Estrada, R. 1988, Tropical rain forest conversion and perspectives in the conservation of wild primates (*Alouatta* and *Ateles*) in Mexico. *Am. J. Primatol.* 14:315–327.

Estrada, A., Juan Solano, S., Ortíz Martínez, T., and Coates-Estrada, R. 1999, Feeding and general activity patterns of a howler monkey (*Alouatta palliata*) troop living in a forest fragment at Los Tuxtlas, Mexico. *Am. J. Primatol.* 48:167–183.

Estrada, A., Luecke, L., Van Belle, S., Barrueta, E., and Rosales-Meda, M. 2004, Survey of black howler (*Alouatta pigra*) and spider (*Ateles geoffroyi*) monkeys in the Mayan sites of Calakmul and Yaxchilán, Mexico and Tikal, Guatemala. *Primates* 45: 33–39.

Estrada, A., Luecke, L., Van Belle, S., French, K., Muñoz, D., García, Y., Castellanos, L., and Mendoza, A. 2002c, The black howler monkey (*Alouatta pigra*) and spider monkey (*Ateles geoffroyi*) in the Mayan site of Yaxchilán, Chiapas, Mexico: A preliminary survey. *Neotrop. Primates* 10:89–95.

Estrada, A., Mendoza, A., Castellanos, L., Pacheco, R., Van Belle, S., García, Y., and Muñoz, D. 2002b, Population of the black howler monkey (*Alouatta pigra*) in a fragmented landscape in Palenque, Chiapas, Mexico. *Am. J. Primatol.* 58:45–55.

Fedigan, L. M., and Jack, K. 2001, Neotropical primates in a regenerating Costa Rican dry forest: A comparison of howler and capuchin population patterns. *Int. J. Primatol.* 22:689–713.

Fedigan, L. M., Rose, L. M., and Avila, R. M. 1998, Growth of mantled howler groups in a regenerating Costa Rican dry forest. *Int. J. Primatol.* 19:405–432.

Garber, P. A., Encarnación, F., Moya, L., and Pruetz, J. D. 1993, Demographic and reproductive patterns in moustached tamarin monkeys (*Saguinus mystax*): Implications for reconstructing platyrrhine mating systems. *Am. J. Primatol.* 29:235–254.

Gonzalez-Kirchner, J. P. 1998, Group size and population density of the black howler monkey (*Alouatta pigra*) in Muchukux Forest, Quintana Roo, Mexico. *Folia Primatol.* 69:260–265.

Horwich, R. H. 1998, Effective solutions for howler conservation. *Int. J. Primatol.* 19:579–598.

Horwich, R. H., Brockett, R. C., James, R. A, and Jones, C. B. 2001a, Population growth in the Belizean black howling monkey (*Alouatta pigra*). *Neotrop. Primates* 9:1–7.

Horwich, R. H., Brockett, R. C., James, R. A, and Jones, C. B. 2001b, Population structure and group productivity of the Belizean black howling monkey (*Alouatta pigra*): Implication for female socioecology. *Primate Rep.* 61:47–65.

Horwich, R. H., and Johnson, E. D. 1986, Geographical distribution of the black howler (*Alouatta pigra*) in Central America. *Primates* 27:53–62.

Izawa, K., Kimura, K., and Samper-Nieto, A. 1979, Grouping of the wild spider monkeys. *Primates* 20:503–512.

Jones, C. B. 1995, Howler monkeys appear to be preadapted to cope with habitat fragmentation. *Endangered Species UPDATE* 12:9–10.

Kitchen, D. M. 2004, Alpha male black howler monkey responses to loud calls: Effect of numeric odds, male companion behaviour and reproductive investment. *Anim. Behav.* 67:125–139.

Marsh, L. K. 1999, Ecological effect of the black howler monkey (*Alouatta pigra*) on fragmented forests in the Community Baboon Sanctuary, Belize. Ph. D. thesis, Washington University, St. Louis.

Ostro, L. E., Silver, S. C., Koontz, F. W., Horwich, R. H., and Brockett, R. 2001, Shifts in social structure of black howler (*Alouatta pigra*) groups associated with natural and experimental variation in population density. *Int. J. Primatol.* 22:733–748.

Pavelka, M. S. M., Brusselers, O. T., Nowak, D., and Behie, A. M. 2003, Population reduction and social disorganization in *Alouatta pigra* following a hurricane. *Int. J. Primatol.* 24:1037–1055.

Pope, T. R. 2000a, Reproductive success increases with degree of kinship in cooperative coalitions of female red howler monkeys (*Alouatta seniculus*). *Behav. Ecol. Sociobiol.* 48:253–267.

Pope, T. R. 2000b, The evolution of male philopatry in Neotropical monkeys, in P. M. Kappeler, ed., *Primate Males: Causes and Consequences of Variation in Group Composition*. Cambridge University Press, Cambridge, pp. 219–235.

Rumiz, D. I. 1990, *Alouatta caraya*: Population density and demography in northern Argentina. *Am. J. Primatol.* 21:279–294.

Rodriguez-Toledo, E. M., Mandujano, S., and Garcia-Orduña, F. 2003, Relationships between forest fragments and howler monkeys (*Alouatta palliata mexicana*) in southern Veracruz, Mexico, in L. K. Marsh, ed., *Primates in Fragments: Ecology and Conservation.* Kluwer Academic/Plenum Publishers, New York, pp. 79–97.

Rosales Meda, M. M. 2003, Abundancia, distribución y composición de tropas del mono aullador negro (*Alouatta pigra*) en diferentes remanentes de bosque en la eco región Lachuá. B. Sc. Thesis Universidad de San Carlos de Guatemala, Guatemala.

Rudran, R., and Fernandez-Duque, E. 2003, Demographic changes over thirty years in a red howler population in Venezuela. *Int. J. Primatol.* 24:925–947.

Rylands, A. B., Mittermeier, R. A., and Rodriquez Luna, E. 1995, A species list for New World primates (Platyrrhini): Distribution by country, endemism, and conservation status according to the Mace-Land system. *Neotrop. Primates* 3(suppl.):133–160.

Schlichte, H. 1978, A preliminary report on the habitat utilization of a group of howler monkeys (*Alouatta villosa pigra*) in the National Park of Tikal, Guatemala, in G. G. Montgomery, ed., *The Ecology of Arboreal Folivores.* Smithsonian Institute Press, Washington, D.C., pp. 551–561.

Smith, J. D. 1970, The systematic status of the black howler monkeys, *Alouatta pigra* Lawrence. *J. Mammal.* 51:358–369.

SPSS. 2003, *SPPS for Windows 12.0.* SPSS Inc.

Strier, K. B. 2003a, Primatology comes of age: 2002 AAPA luncheon address. *Yb. Phys. Anthropol.* 46:2–13.

Strier, K. B. 2003b, Demography and the temporal scale of sexual selection, in C. B Jones, ed., *Sexual Selection and Reproductive Competition in Primates: New Perspectives and Directions. American Society of Primatology: Special Topics in Primatology,* Vol. 3, The American Society of Primatologists, Norman, Oklahoma, pp. 45–63.

Watts, E., and Rico-Gray, V. 1987, Los primates de la península de Yucatán, México: estudio preliminar sobre su distribución actual y estado de conservación. *Biótica* 12:57–66.

WinSTAT. 1994, *WinSTAT Statistics for Windows, version 3.0.* Kalmia Co. Inc.

CHAPTER FIVE

Population Structure of Black Howlers (*Alouatta pigra*) in Southern Belize and Responses to Hurricane Iris

Mary S. M. Pavelka and Colin A. Chapman

INTRODUCTION

A fundamental issue in ecology is determining factors regulating the density of animal populations. A variety of potential factors have been proposed including external factors, such as food resources, weather, predation, and disease, and internal conditions, such as territoriality and aggressive behaviors and life history traits (Andrewartha and Birch, 1954; Boutin, 1990; Krebs, 1978; Nicholson, 1933). The importance of understanding determinants of animal abundance

Mary S. M. Pavelka • Department of Anthropology, University of Calgary, 2500 University Drive NW, Calgary, Alberta, T2N 1N4, Canada. **Colin A. Chapman** • Anthropology Department and McGill School of Environment, 855 Sherbrooke St West, McGill University, Montreal, Canada, H3A 2T7 and Wildlife Conservation Society, 2300 Southern Boulevard, Bronx, NY 10460.

New Perspectives in the Study of Mesoamerican Primates: Distribution, Ecology, Behavior, and Conservation, edited by Alejandro Estrada, Paul A. Garber, Mary S. M. Pavelka, and LeAndra Luecke. Springer, New York, 2005.

has increased with the need to develop informed management plans for endangered or threatened species. With respect to primates, these theoretical issues are critical because tropical forests occupied by primates are undergoing rapid anthropogenic transformation and modification (National Research Council, 1992). Cumulatively, countries with primate populations are losing approximately 125,000 km^2 of forest annually; based on global estimates of primate densities, this results in the loss of 32 million primates per year (Chapman and Peres, 2001). Other populations are being affected primarily by a subcategory of anthropogenic disturbance in the form of forest degradation due to logging, fire, and hunting. However, understanding and predicting factors that determine the abundance of particular primate species have proven extremely difficult, and examining the importance of internal conditions has proven to be the most difficult factor to quantify.

Natural disturbances to ecosystems, such as hurricanes, provide a unique opportunity to tease apart the importance of different factors that may determine the density of animal populations. However, few opportunities exist to document the effect of major natural habitat disturbance on primate populations, due to the unpredictable nature of natural disasters, such as hurricanes, and the absence of predisturbance data. From 1871 to 1964, an average of 4.6 hurricanes hit the Caribbean each year (Walker *et al.*, 1991), frequently causing severe damage to forests and animal populations. When Hurricane Iris struck the Belizean coastline on October 8, 2001, a Central American black howler (*Alouatta pigra*) study site was directly in its path. The population of monkeys inhabiting a 52-ha study area had been closely monitored for 3 years preceding the hurricane, and their exact population structure was known (Pavelka, 2003). Hurricane Iris was a small (winds extended for 40 km from the center), but powerful storm (category four on a scale of five) with sustained winds of 233 km/h and gusts of 282 km/h that caused massive structural damage to the forest and complete defoliation of the trees. For a description of the initial impact of the storm and its impact on the diet, activity, and food supply of the howler population, see Pavelka *et al.* (2003), Behie and Pavelka (2005), and Pavelka and Behie (2005). Continued monitoring over the subsequent 29 months has allowed us to document the affect of the hurricane on the population and to begin to consider which factors appear to be most important in determining population recovery and density. In this paper, we document changes to the population during this time period. Based on these documented changes and observations over this period, we speculate on which of several factors, proposed in the literature to determine animal density, has played a primary role in this black howler population. We also outline what is currently being done to quantitatively determine the relative importance of the different factors.

A. pigra, previously one of the least well known of the currently recognized six howler species, is rapidly becoming well studied in Mexico and Belize (van Belle and Estrada, this volume; Brockett and Clark, 2000; Brockett *et al.*, 1999; 2000a,b; Clark and Brockett, 1999; Estrada *et al.*, 2002a,b,c, this volume; Gonzalez-Kirchner, 1998; Horwich *et al.*, 2001a,b; Ostro *et al.*, 1999, 2000, 2001; Silver *et al.*, 1998; Silver and Marsh, 2003; Treves *et al.*, 2001). Significantly smaller group sizes are reported for *A. pigra* (2–10 individuals per group) when compared to those of its geographic neighbor and close relative, *A. palliata* (2–45 individuals per group, mean = 12.3: Crocket and Eisenberg, 1987; see also Chapman and Balcomb, 1998). Small group size was one of the main factors leading to *A. pigra* being assigned species status in the early 1970s (Smith, 1970; Horwich, 1983; Horwich and Johnson, 1986).

The diet of black howlers is described as being as frugivorous as possible and as folivorous as necessary (Silver *et al.*, 1998; Pavelka and Knopff, 2004); however, they are believed to be capable of surviving for long periods on leaves (Silver *et al.*, 2000). Dietary flexibility should make black howlers good candidates for surviving disturbance (Johns and Skopura, 1987). While rare, studies of hurricane-affected invertebrates (Schowalter, 1994; Willig and Camillo, 1991), birds (Askins and Ewert, 1991; Will, 1991), bats (Gannon and Willig, 1994), frogs (Marsh and Pearman, 1997), and primates (Gould *et al.*, 1999; Menon and Poirer, 1996; Ratsimbazafy *et al.*, 2002) suggest that species with more flexible diets will be better able to survive in an environment with limited and/or dramatically altered food production. Black and white ruffed lemurs (*Varecia variegata*) living in a forest damaged by Cyclone Gretelle adjusted to their habitat by eating exotic plant species that were not consumed before the storm (Ratsimbazafy *et al.*, 2002). Lion-tailed macaques (*Macaca fascicularis*) were able to survive in a drought and fire disturbed environment by consuming less of the normally preferred fruit and more of the available insects and leaves (Berenstain, 1986). In addition to threatening individual survival, initially limited production of food in disturbed habitats can prevent females from producing viable offspring because they are unable to obtain adequate nutrients (Gould *et al.*, 1999; Ratsimbazafy *et al.*, 2002).

METHODS

The 52-ha study site is located in southern Belize on the north side of Monkey River, near the coast (16°21′N, 88°29′W). This closely monitored area is part of a larger (approx 100 km^2) forested area along the Monkey River watershed, between the southern highway of Belize and the river mouth. Due to savannah and anthropogenic landscapes to the north and south, and the highway and agricultural development to the west, the monkey population in the watershed forest east of the highway is believed to be discontinuous with monkey populations that may occur further west along the Bladen and Swasey rivers, and into the Maya Mountains (Figure 1). The area receives on average 250 cm of rain annually, which primarily falls from June through December. The most common trees in this seasonally flooded semi-evergreen riparian forest are cohune palms

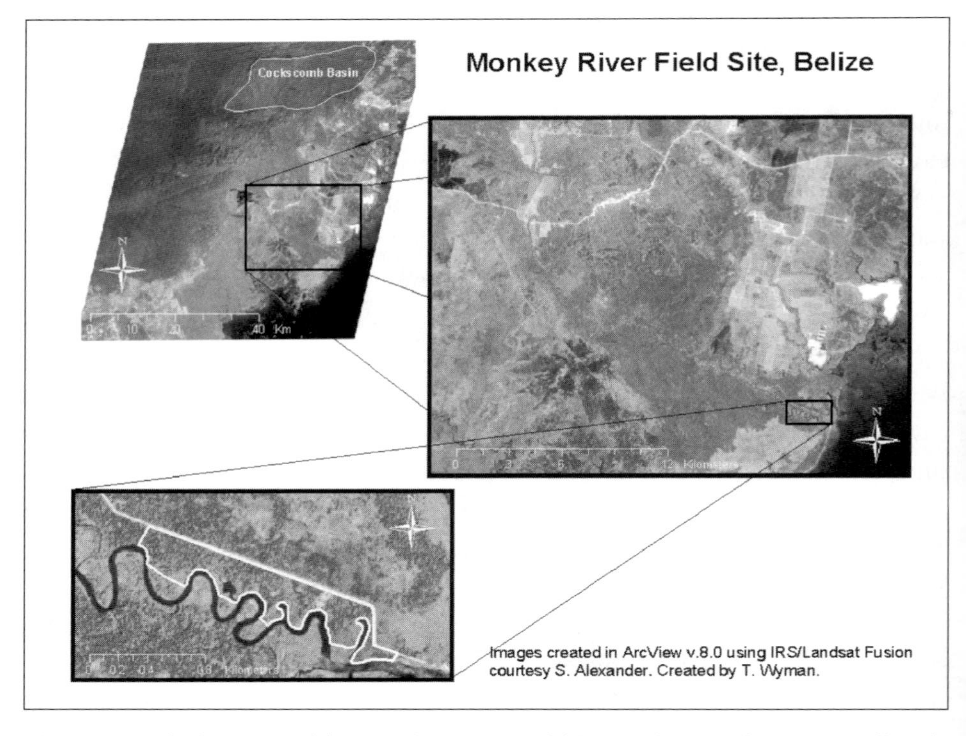

Figure 1. The location of the Monkey River Field Site, Belize. In the most small-scale map, the location of the 52-ha study site where eight groups of black howlers (*A. pigra*) existed prior to Hurricane Iris is indicated by a white line. The body of water in the SE corner is the Caribbean Sea.

(*Attalea* and *Orbigyna*), provision trees (*Pachira*), figs, (*Ficus* spp.), and swamp kaways (*Pterocarpus*) (Pavelka *et al.*, 2003).

Pre-hurricane data on the demography of groups in the 52-ha area were collected between May and August 1999, January and May 2000, January and May 2001, and October 6th and 7th, 2001. On October 6th and 7th we were able to confirm that group compositions were the same as they had been in May. By May 2001, group size, composition, and home range were known for eight groups and detailed behavioral data had been collected on five of them. Groups were recognized by consistent group membership and home range site fidelity.

The hurricane made landfall at 7 pm on October 8, 2001. Since October 16, 2001, the site has been under constant monitoring wherein all monkey groups are generally located every 2–3 days. Thus, it is likely that a change in group composition would have been noted within a day or two of its occurrence. Despite the fact that at least two researchers walk the extensive trail system (18.5 km) almost daily, we have not found a monkey carcass since immediately after the storm. However, with the deadfall and extensive new growth on the forest floor, movement off the trail system was very limited and it is possible that carcasses in the 52-ha area were undiscovered.

RESULTS

General Demographic Changes

Prior to Hurricane Iris, 53 individuals in eight groups lived within the 52-ha study area, and the population density was 102 individuals/km^2. Group size averaged 6.4 individuals, but varied from 2 to 10 individuals (Pavelka, 2003). The initial impact of the hurricane reduced the population by 42%. The surviving animals experienced a period of social disorganization involving transient individuals, high numbers of solitary monkeys, and small fragmentary social groups. Over the course of the first 12 weeks following the hurricane, groups began to reform and we found a decrease in the number of solitary individuals, and an increase in the average group size. By February of 2002, we were able to reestablish the trail system and confirm that 31 monkeys in 5 social groups now inhabited the 52-ha area (Pavelka *et al.*, 2003).

Continuous monitoring of this population for the next 29 months has shown a slow steady decline in total population size, density, number of groups, and

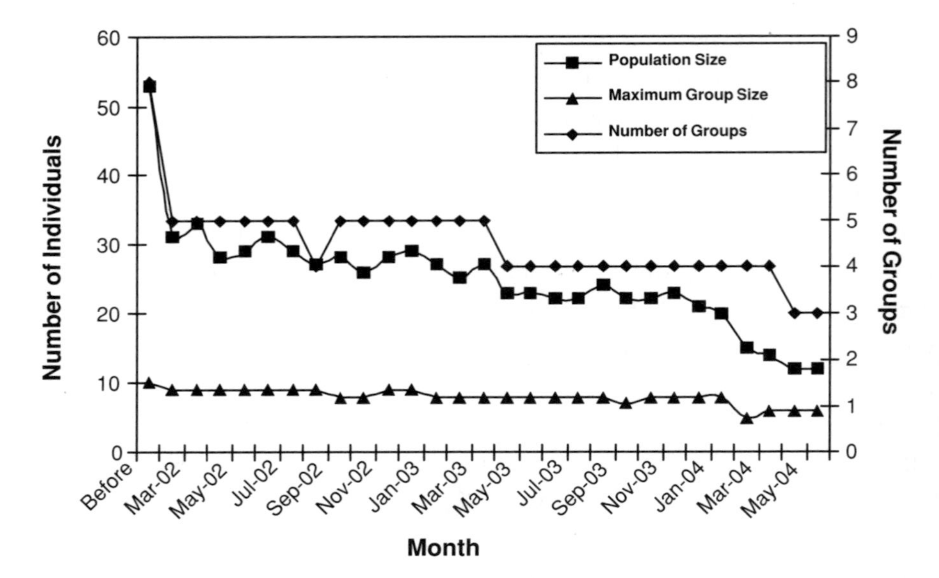

Figure 2. The total population size, number of groups, and maximum group size of the black howler (*A. pigra*) study population prior to Hurricane Iris and for the subsequent 29 months. Group size ($r = -0.657$; $p < 0.001$), number of groups ($r = -0.820$, $p < 0.001$), maximum group size ($r = -0.833$, $p < 0.001$), population size ($r = -0.848$, $p < 0.001$) are related to month and population size and maximum group size are also negatively related ($r = -0.843$, $p < 0.001$, $n = 29$ in all cases).

maximum group size (Figure 2). The population reduction is the result of both smaller and fewer groups. Prior to the hurricane, the maximum group size at Monkey River was 10 animals, with the majority of established groups containing 8 individuals (Pavelka, 2003). Since the storm, maximum group size has steadily declined from nine group members in February 2002 to six by May 2004 (Figure 2). Likewise, the number of groups has also declined from eight before the storm to five shortly after (February 2002) to only three groups as of May 2004. These three groups contain three, two, and six members; thus, the population total within the study area is 12. This constitutes a drop in population density from 102 individuals/km^2 before the storm to 23 individuals/km^2, 29 months after—a loss of 78% of the population in the study area, which is assumed to be reflective of changes in the larger watershed forest area.

All age–sex classes have declined more or less steadily, with the exception of the number of infants which, after an initial decline, has risen (Figure 3). Before the storm, there were slightly more males than females in the population (17 females and 21 males, including two solitary males). The number of males

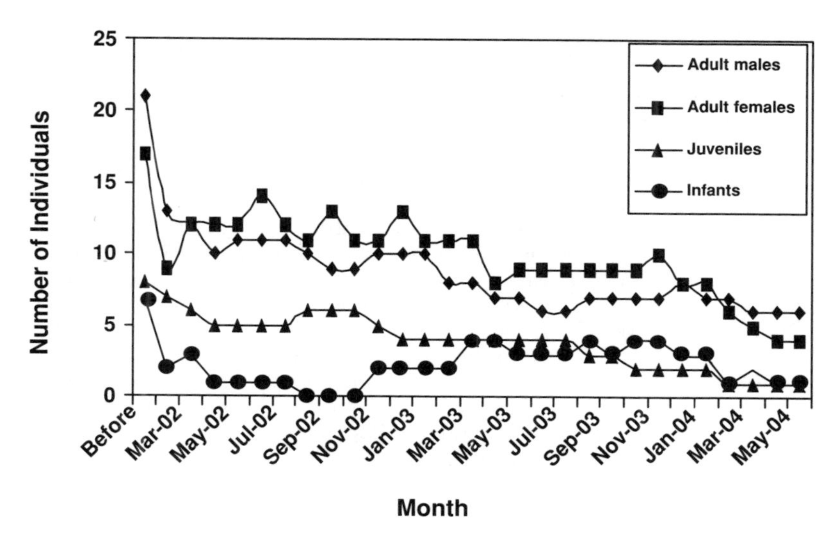

Figure 3. Number of black howlers (*A. pigra*) in each age–sex category for each month following Hurricane Iris.

dropped below that of females after the storm and while both fell steadily, the number of males continued to be lower than the number of females until February 2004 (Figure 3).

Ten infants were born between November 2002 and March 2004 (Table 1). The first infants were born more than a full year after the hurricane. With a gestation length of 6 months (Brockett *et al.,* 2000a), these conceptions must have occurred in June or July 2002, 9 months after the storm. The fact that no infants were born for an entire year after Hurricane Iris suggests that either pregnant females did not carry their pregnancies to term or did not survive the storm.

The temporal distribution of the 10 births suggest 2 birth peaks, in the dry season months of March and April, and in the late wet season months of October and November; however, a larger sample is needed to confirm this suggestion. Six of the 10 infants born were male, 3 were female, and for 1 the sex was unknown, due to its disappearance shortly after birth (Table 1). Survivorship of infants after the storm has been very low (Table 1). As of June 2004, only 1 of the 10 infants was still alive. Four of the 10 infants disappeared with whole or partial groups, and thus may be alive elsewhere; however, we believe this is unlikely, as we have just completed a survey of the larger forested area of the watershed to determine if the data coming from the study area are

Table 1. Infants born into the black howler (*A. pigra*) study site population after Hurricane Iris made landfall in October 2001.

Date of birth	Sex	Fate
November 2002	F	Disappeared with entire N group, April 2003
November 2002	M	Disappeared alone, assumed dead, April 2003; B group
March 2003	?	Disappeared with entire N group, April 2003
March 2003	F	Disappeared with half of Q group, February 2004
April 2004	M	Killed by resident adult male, D group, May 2004
April 2003	F	Disappeared alone, assumed dead, January 2004; A group
August 2003	M	Disappeared after being injured by adult female in A group, November 2003
October 2003	M	Disappeared with half of Q group, February 2004
January 2004	M	Disappeared with mother in April, 2004, after D group breakup in March 2004
March 2004	M	Alive in June 2004

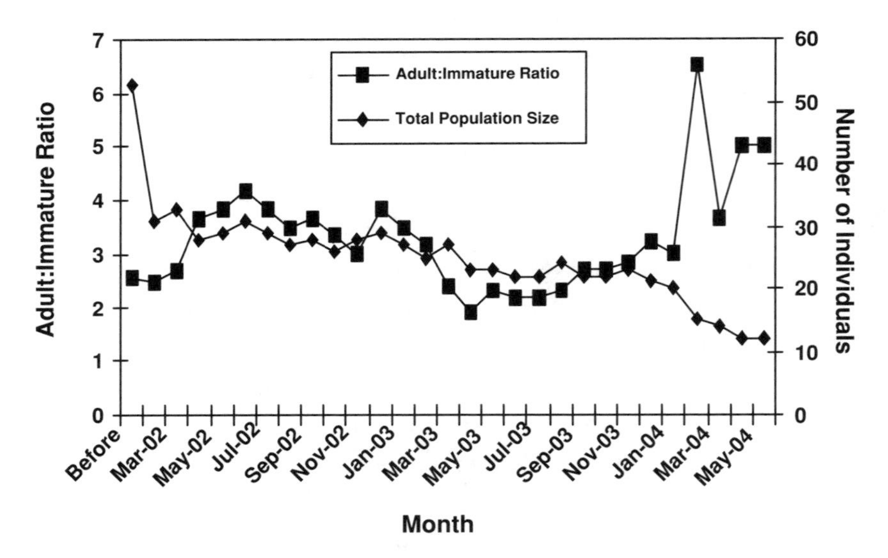

Figure 4. The ratio of adult black howlers (*A. pigra*) to immatures and total population size at Monkey River Field Site following Hurricane Iris.

representative of the larger area. This survey suggests that the low population densities and small social groups (no larger than five or six individuals) found at our study site characterize the whole area, which was equally affected by the hurricane (unpublished data).

Before the hurricane, the adult to immature ratio was 2.5:1 (38 adults and 15 immatures; Figure 4). Over the subsequent 29 months, the number of adults to immatures has varied, climbing to a maximum of 6.5:1 in February 2004, when there were 13 adults and 2 infants, falling back to 5.0:1 by May of 2004 with the loss of 3 of those adults from the study area.

Description of Specific Losses

The immediate cause or sequence of events of the falling population appears to be dispersal events sometimes preceded by intragroup aggression as well as poor infant survival. The population is declining as whole groups or parts of groups suddenly disappear from the study area (Figures 5 and 6). The first major loss to the population after the initial period of decline came with the complete disappearance of N group (six members) in April 2003. The birth of two infants in A and D groups during the same month meant a net loss of four animals, reducing the population from 27 to 23. N group lived at the east end of the study area, closest to the coast, and was one of the groups that appeared to have remained intact following the hurricane. N group inhabited a patch of forest that is fairly discrete, bordered on the east by a road and anthropogenically cleared areas on the coast, the river to the south, a road to

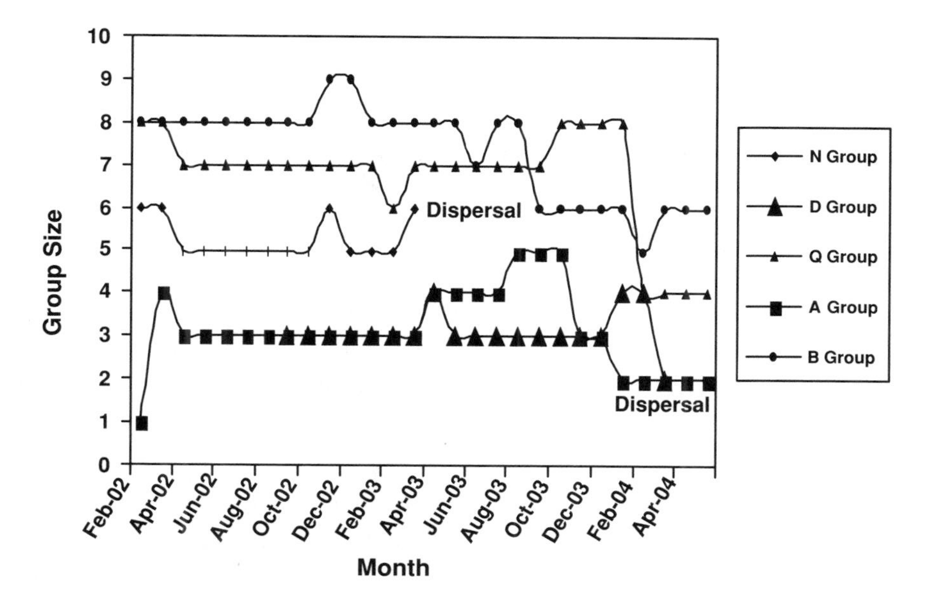

Figure 5. The size of individual groups of black howlers at Monkey River Field Site following Hurricane Iris. Groups N and D dispersed out of the study area and this is indicated by having the size data stop at the time of dispersal.

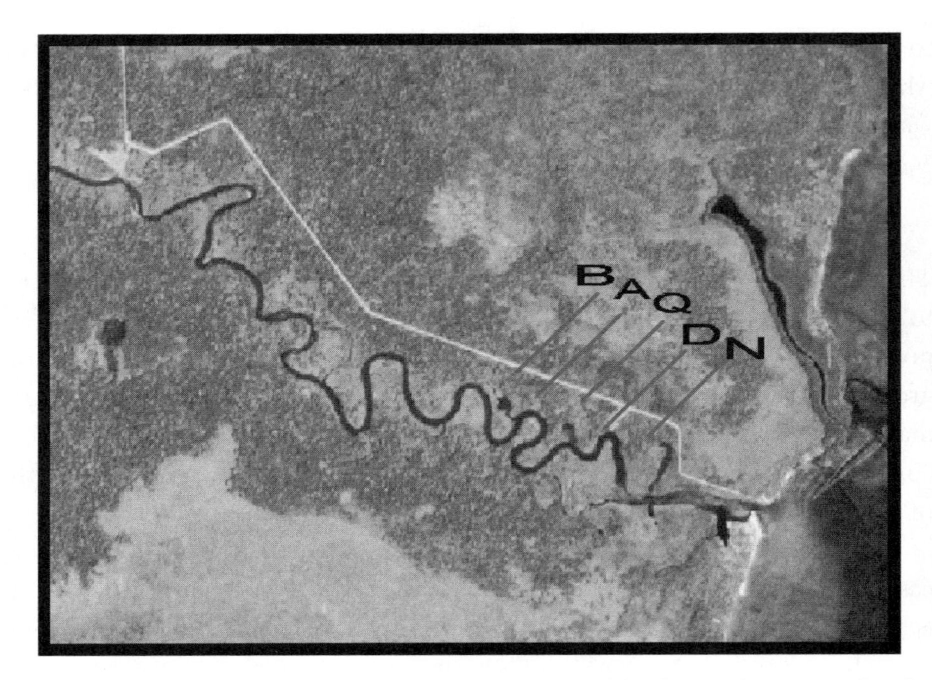

Figure 6. The study site showing the range location of the five study groups after the hurricane. Map prepared by Aaron Osicki.

the north, and the field containing the research camp on the west (Figure 6). Monkeys were infrequently observed crossing the field to the west and the road to the north, but these movements were temporary and rare, and for the most part the group's home range was limited by these boundaries. This patch of forest has had monkeys living in it for as long as local people can recall, and since April 2003 it is vacant.

D group first appeared in the study area in September 2002, almost a year after the hurricane, as an adult male with two adult females (Figure 5). There was an infant born to this group in April 2003; however, it was killed by the resident adult male less than 10 days later (Knopff *et al.*, 2004). In January 2004, they produced another infant; however, in March, 2 months later, the adult male and the adult female without an infant disappeared, leaving the lactating mother and infant on their own. This mother–infant pair stayed in their range for almost a month, and was last seen heading west out of the study area on April 15, 2004. They moved through the ranges of Q, A, and B groups without joining any of these groups.

The adult female and infant remnant of D group did not join the remnant of neighboring Q group, despite the fact that the two groups were familiar with one another, having shared adjacent and sometimes overlapping ranges for 15 months. Q group was one of the largest and most stable groups until late January 2004, when they dropped from eight to four members (Figure 5). One adult female along with a large juvenile female and both infants disappeared, leaving behind two adult males, a subadult male, and one adult female and resulting in an unusual group composition, with an adult male to female sex ratio of 3:1. The actual circumstances of the 50% reduction in Q group are not known, but again in the absence of any evidence of predation or carcasses being found, we assume the adult female, juvenile, and infants left the study area, moving west through the forest along the river. The ultimate fate of these four members of Q group is unknown, however, given that conditions throughout the continuous watershed forest available to them appear to be the same as those in the study area, and that the population throughout the watershed is falling, we suspect that few are surviving. While a merger of the remaining D group adult female and infant pair with the neighboring Q group, three males and one female would have produced a group with a more typical composition, the D group female moved through Q's range without joining them. Perhaps, infanticide risk deterred the female from approaching a group containing three adult males (see Knopff *et al.*, 2004). Also, increasing group size due to merging of small groups is rare in primates and not known to occur in *A. pigra*.

The next group to the west, A group, formed after the hurricane when two adult females and one infant entered the study area and joined a solitary male to form a group of four (Figure 5). The adult females and the infant had been observed on the north side of the road outside of the study area, and the solitary male on the south side of the road within the study area, for several weeks. They formed another small group that showed promise of growing into a larger established group. The infant disappeared within a couple of weeks, and the group remained at three individuals until April of 2003 when an infant was born, then in August a second infant was born. This group of five remained together until November, when one female attacked the other female and infant, injuring the infant, who subsequently disappeared and is presumed to have died. The female continued directing agonistic behavior toward the other female until she left the group and area. Her fate is unknown. The infant of the "aggressive" female disappeared a month later. This group now consists of a male–female pair.

DISCUSSION

The density-dependent factors that are most commonly suggested to influence population size and distribution are predation, food resource availability and quality, disease and parasitism, and social factors (Andrewartha and Birch, 1954; Boutin, 1990; Krebs, 1978; Nicholson, 1933). Although jaguars and tayras are found in the area, we have no evidence to suggest that predators are playing a significant role in determining the density of black howlers at Monkey River. Of course, this absence of evidence is not evidence of the absence of predation. It is possible that damage to the structure of the canopy has interfered with predation avoidance strategies. Like other howlers, *A. pigra* is known to prefer large trees, which are much less common at the site since the storm. Hunting by local farmers, in an effort to protect their livestock from the perceived threat of jaguar predation, does occur, and likely keeps the density low. At this point, it is not possible for us to say what role predation is playing in the continued population decline, only that we have no direct evidence of predation on the monkeys before or after the storm.

The most apparent factor that could explain both the initial drop in population density and the subsequent decline over 29 months is the change in food availability that resulted from hurricane damage. Pavelka and Behie (2005) found that the hurricane caused the mortality of 35% of the major food trees with a circumference of more than 40 cm. If the abundance and quality of available food were lower than needed by the population, one might expect to see low infant survival and an increase in emigration. It is interesting to note that after a hiatus of 1 year, infants continued to be born in this population. Thus, while food stress is undoubtedly part of the explanation for the falling population, it did not translate directly into a failure of females to produce infants (see Glander, discussion of seasonal weight loss and survival/reproduction, this volume). On the other hand, the fact that larger groups at Monkey River did not have longer day ranges or spend less time inactive than smaller groups (after controlling for habitat quality differences; Knopff and Pavelka, submitted) suggests that food resources may not currently be limited for this population. In general, black howler group size may be below the threshold at which limited food would require a behavioral response (Chapman and Pavelka, 2005).

While it is clear that there is less food available, particularly fruits (Pavelka and Behie, 2005), it is also possible that the quality of the available foods have changed. It is well documented that in response to real or simulated herbivory, plants can increase concentrations of secondary compounds (Schultz, 1988) making their leaves less palatable. In contrast, Coley (1983) demonstrated that canopy gaps are typically colonized by climbers and fast-growing pioneer tree species whose leaves generally have more protein, less fiber, and a lower phenolic content than the leaves of persistent canopy tree species. Protein and fiber content of foods are known to be important to leaf-eating monkeys (Milton, 1979, 1998; Chapman and Chapman, 2002; Chapman *et al.*, 2002). Thus, the species of trees that are colonizing and regenerating in the areas opened by Hurricane Iris may be suitable food sources. These two observations present conflicting possibilities for how the quality of the foods available to the black howlers would have changed after the hurricane. We are currently investigating the potential role of food quality and availability by quantifying the secondary compounds in available foods (saponins, alkaloids, and cyanogenic glycosides) and examining the applicability of Milton's protein/fiber model to this black howler population. Milton (1979) proposed that the protein to fiber ratio was a good predictor of leaf choice in mantled howlers (see also McKey, 1978). By measuring overall mature leaf acceptability as the ratio of protein to fiber, several subsequent studies have found positive correlations between colobine biomass and this index of leaf quality at local (Chapman *et al.*, 2002; Ganzhorn, 2002) and regional scales (Oates *et al.*, 1990; Waterman *et al.*, 1988). After determining whether their population was at a level that could be predicted by this model before the hurricane, we can test the prediction that the population should stabilize at a level suggested by the current protein to fiber levels in their foods.

Finding single-factor explanations for complex biological phenomena, such as determinants of black howler abundance, is unlikely. Rather, recent long-term studies have highlighted the importance of multifactoral explanations. For example, based on a 68-month study of howler monkeys (*A. palliata*) and a parasitic botfly (*Alouattamyia baeri*), Milton (1996) concluded that the annual pattern of howler mortality results from a combination of effects including age, physical condition, and larval burden of the parasitized individual, which becomes critical when the population experiences dietary stress. Similarly, Gulland (1992) studied the interactions of Soay sheep (*Ovis aries*) and nematode parasites and demonstrated that at times of population crashes sheep were emaciated, had high nematode burdens, and showed signs of protein-energy malnutrition. In the field, sheep treated with antihelminthics had lower mortality rates, while experimentally infected sheep with high parasite loads, but fed nutritious diets, showed no sign of malnutrition.

It is also likely that the hurricane has altered black howler/parasite relationships in such a way as to negatively impact the howlers. For example, for directly transmitted parasites the reduction in the physical structure of the forest and in the number of food trees may mean that the animals spend more time in one location and thus infection risk increases. Gillespie *et al.* (submitted) demonstrated that selective logging has resulted in higher densities of infective-stage parasites common to red colobus (*Piliocolobus badius*), black-and-white colobus (*Colobus guereza*), and redtail guenons (*Cercopithecus ascanius*). The redtail guenons in logged areas had higher prevalence and richness of gastrointestinal parasites than individuals in unlogged areas.

It is also possible that the initial dietary stress caused by food tree reduction (Pavelka and Behie, 2005) may have adversely affected resistance to parasitic infection by reducing the effectiveness of the immune system (Holmes, 1995; Milton, 1996). This food shortage could have resulted in a higher parasite burden, which in turn could have increased nutritional demands on the howlers and accentuated the effects of food shortages. Thus, nutritional status and parasitism could have had synergistic effects on the host, i.e., the individual effects of each factor would be amplified when co-occurring. The interaction between nutritional stress and parasitism has been examined in a number of laboratories (Crompton *et al.*, 1985; Munger and Karasov, 1989) and in field studies (Gulland, 1992; Murray *et al.*, 1996, 1998; Toque, 1993). These have led to speculation that the interacting effects of food shortage and parasitism may influence vertebrate populations (Holmes, 1995; Keymer and Dobson, 1987). The interactive effects of parasitism and nutritional status have rarely been examined in primates (but see Milton, 1996). Social stress caused by the disruption of the groups' normal composition could also cause stress that could have interacted with both nutritional stress and parasite burden to negatively influence the howler population recovery. Currently we (Alison Behie, Pavelka, and Chapman) are investigating the possible role played by parasites, particularly helminths, by assessing parasite infections through fecal analysis. Since data are not available from before the hurricane, we are comparing the parasite community from Monkey River with those at the Community Baboon Sanctuary, a healthy control population that was not affected by Hurricane Iris.

The hurricane resulted in social disorganization within the area (Pavelka *et al.*, 2003). The possible influence of this social stress on the black howler population dynamics is difficult to assess. However, we have witnessed an adult male killing an infant in his group (Knopff *et al.*, 2004) and a female being so aggressive to another that the recipient left the group. These, along with the dispersal/disappearance of whole and partial groups, suggest that the population is not stable and may be under stress. To evaluate this, we are monitoring stress in general through the quantification of fecal glucocorticoid levels. A considerable body of research on humans and other mammals demonstrates that large and prolonged elevated glucocorticoid levels (cortisol is one type of glucocorticoid) typically reduces survival and reproduction (Bercovitch and Ziegler, 2002; Creel, 2001; Sapolsky, 1986; Wasser *et al.*, 1997). Although data on fitness effects of elevated glucocorticoid levels in the wild are currently limited, the expectation from lab studies is that fitness will decrease as population level stressors become more severe or more prolonged. For specific individuals, we can examine factors coinciding with periods of elevated cortisol levels, be it social stress, food scarcity, or changes in parasite burden.

In conclusion, we have documented that following Hurricane Iris's passing through an area that contained a study population of black howlers, there was a dramatic decline in the population's size and composition. Major disruptions are still occurring some 29 months after the storm and the population is in progressive decline and may be headed for local extinction (see Ford, this volume, for a discussion of population fragmentation and local extinction). We are investigating the possibility that the decline is caused by the reduction of available food trees, and also possible synergistic interactions between this nutritional stress, social disruption, and parasite burden.

SUMMARY

A Central American black howler population in Monkey River, Belize, was monitored from May of 1999 to May of 2001 and was determined to have similar small group size with multi- and single-male groups. Fifty-three monkeys lived in 8 social groups in a 52-ha study area (population density 102 individuals/km^2) that is part of the larger continuous forested area of the Monkey River watershed. On October 8, 2001, the study area was severely damaged by Hurricane Iris, a category four storm that resulted in complete defoliation of the forest along with severe structural damage to those trees not snapped or uprooted. When the area could be accessed again in February 2002, it was determined that the population had dropped by 42%, with 31 monkeys in 5 social groups inhabiting the study area. While initially it was hoped that the population would stabilize at this level, subsequent monitoring through May of 2004 (29 months post hurricane) has revealed a slow but steady decline in the population through the apparent dispersal of whole or parts of social groups, and poor infant survival. We hypothesize that a combination of nutritional and social stress interacting with increased parasite loads (and possibly increased predation) is leading groups and individuals to leave the area, moving west along the river in search of better habitat that is not available. The watershed forest fragment (approximately 100 km^2) was equally damaged by the storm from the southern highway of Belize to the coast, leading us to believe that survival of the animals leaving the study site is unlikely. We are currently investigating phytochemical, hormonal, and parasite contributions to the continued decline of the Monkey River howler monkey population following hurricane Iris.

ACKNOWLEDGMENTS

We would like to thank Tracy Wyman, Aaron Osicki, Laura McKenzie, Andrea Faulkner, Melanie Luinstra, Keriann McGoogan, Travis Steffens, Lisa Corewyn, Kyle Knopff, Aliah Knopff, Shelley Alexander, and the students in the Belize Primatology Field School for help with various aspects of this project. In particular, we wish to acknowledge the ongoing research of Alison Behie, a doctoral student at the University of Calgary, who has collected much of the demographic data described in this paper, as well as all of the fecal samples that will allow us to understand the role of parasite load and stress hormone levels in the population decline following a hurricane. This forms the basis of A. Behie's doctoral research. We are also grateful to Alejandro Estrada, Paul Garber, and Leandra Lueke for organizing the Mesoamerican Primate symposium and the ASP meetings in 2004 and this volume, and Paul Garber, Alejandro Estrada, Tracy Wyman, and an anonymous reviewer for helpful feedback on the manuscript. As always we thank the people of Monkey River, including the Village Council and the Monkey River Tour Guide Association for continued support for this project. Funding was provided by the National Sciences and Engineering Research Council, the National Geographic Society, Conservation

International Margo Marsh Biodiversity Fund, The University of Calgary, and the Calgary Institute for Humanities.

REFERENCES

Andrewartha, H. G. and Birch, L. C. 1954, *The Distribution and Abundance of Animals*. University of Chicago Press, Chicago.

Askins, R. A. and Ewert, D. N. 1991, Impact of Hurricane Hugo of bird populations on St. John, U.S. Virgin Islands. *Biotropica* 23:481–487.

Behie, A. M. and Pavelka, M. S. M. 2005, The short-term effects of a hurricane on the diet and activity of black howlers (*Alouatta pigra*) in Monkey River, Belize. *Folia Primatol.* 835:1–9.

Bercovitch, F. B. and Ziegler, T. E. 2002, Current topics in primate socioendrocrinology. *Annu. Rev. Anthropol.* 31:45–67.

Berenstain, L. 1986, Responses of long-tailed macaques to drought and fire in Eastern Borneo: A preliminary report. *Biotropica* 18(3):257–262.

Boutin, S. 1990, Food supplementation experiments with terrestrial vertebrates: Patterns, problems and the future. *Can. J. Zoolog.* 68:203–220.

Brockett, R. C. and Clark, B. C. 2000, Repatriation of two confiscated black howler monkeys (*Alouatta pigra*) in Belize. *Neotrop. Primates* 8:101–103.

Brockett, R. C., Horwich, R. H., and Jones, C. B. 1999, Disappearance of infants following male takeovers in the Belizean black howler monkey (*Alouatta pigra*). *Neotrop. Primates* 7:86–88.

Brockett, R. C., Horwich, R. H., and Jones, C. B. 2000a, Reproductive seasonality in the Belizean black howler (*Alouatta pigra*). *Neotrop. Primates* 8:136–138.

Brockett, R. C., Horwich, R. H., and Jones, C. B. 2000b, Female dispersal in the Belizean black howling monkey (*Alouatta pigra*). *Neotrop. Primates* 8:32–34.

Chapman, C. A. and Balcomb, S. R. 1998, Population characteristics of howlers: Ecological conditions or group history. *Int. J. Primatol.* 19(3):385–403.

Chapman, C. A. and Chapman, L. J. 2002, Foraging challenges of red colobus monkeys: Influence of nutrients and secondary compounds. *Comp. Biochem. Phys.* 133:861–875.

Chapman, C. A., Chapman, L. J., Bjorndal, K. A., and Onderdonk, D. A. 2002, Application of protein to fiber ratios to predict colobine abundance on different spatial scales. *Int. J. Primatol.* 23:283–310.

Chapman, C. A. and Pavelka, M. S. M. 2005, Group size in folivorous primates: Ecological constraints and the possible influence of social factors. *Primates* 46(1):1–9.

Chapman, C. A. and Peres, C. 2001, Primate conservation in the new millennium: The role of scientists. *Evol. Anthropol.* 10:16–33.

Clark, B. and Brockett, R. C. 1999, Black howler monkey (*Alouatta pigra*) reintroduction program: Population census and habitat assessment. *Neotrop. Primates* 7: 51–53.

Coley, P. D. 1983, Herbivory and defensive characteristics of tree species in a lowland tropical forest. *Ecol. Monogr.* 53:209–233.

Creel, S. 2001, Social dominance and stress hormones. *Trends Ecol. Evol.* 16:491–497.

Crockett, C. M. and Eisenberg, J. F. 1987, Howlers: Variations in group size and demography, in B. B. Smuts, D. L. Cheney, R. M. Seyfarth, R. W. Wrangham, and T. T Struhsaker, eds., *Primate Societies*, The University of Chicago Press, Chicago, pp. 54–68.

Crompton, D. W. T., Arnold, S. E., Walters, D. E., and Whitfield, P. J. 1985, Food intake and body weight changes in mice infected with metacestodes of *Taenia crassiceps*. *Parasitology* 90:449–456.

Estrada, A., Castellanos, L., Garcia, Y., Franco, B., Munoz, D, Ibarra, A., Rivera, A., Fuentes, E., and Jimenez, C. 2002a, Survey of the black howler monkey, *Alouatta pigra*, population at the Mayan site of Palenque, Chiapas, Mexico. *Primates* 43:51–58.

Estrada, A., Llueke, L., Van Belle, S., French, K., Munoz, D., Garcia, Y., Castellanos, L., and Mendoza, A. 2002b, The black howler monkey (*Alouatta pigra*) and spider monkey (*Ateles geoffroyi*) in the Mayan site of Yaxchilan, Chiapas, Mexio: A preliminary survey. *Neotrop. Primates* 10:89–95.

Estrada, A., Mendoza, A., Castellanos, L., Pacheco, R., Van Belle, S., Garcia, Y., and Munoz, D. 2002c, Population of the black howler monkey (*Alouatta pigra*) in a fragmented landscape in Palenque, Chiapas, Mexico. *Am. J. Primatol.* 58:45–55.

Gannon, M. R. and Willig, M. R. 1994, The effects of Hurricane Hugo on bats of the Luquillo Experimental Forest of Puerto Rico. *Biotropica* 23:320–331.

Ganzhorn, J. U. 2002, Distribution of a folivorous lemur in relation to seasonally varying food resources: Integrating quantitative and qualitative aspects of food characteristics. *Oecologia* 131(3):427–435.

Gillespie, T. R., Chapman, C. A., and Greiner, E. C. (submitted). Long-term effects of logging on parasite dynamics in African primate populations. *J. Appl. Ecol.*

Gonzalez-Kirchner, J. P. 1998, Group size and population density of the black howler monkey (*Alouatta pigra*) in Muchukux Forest, Quintana Roo, Mexico. *Folia Primatol.* 69:260–265.

Gould, L., Sussman, R. W., and Sauther, M. L. 1999, Natural disasters and primate populations: The effects of a 2-year drought on a naturally occurring population of ring-tailed lemurs (*Lemur catta*) in Southwestern Madagascar. *Int. J. Primatol.* 20:69–84.

Gulland, F. M. D. 1992, The role of nematode parasites in Soay sheep (*Ovis aries* L.) mortality during a population crash. *Parasitology* 105:493–503.

Holmes, J. C. 1995, Population regulation: A dynamic complex of interactions. *Wildl. Res.* 22:11–20.

Horwich, R. H. 1983, Species status of the black howler monkey, *Alouatta pigra*, of Belize. *Primates* 24:288–289.

Horwich, R. H., Brockett, R. C., James, R. A., and Jones, C. B. 2001a, Population structure and group productivity of the Belizean black howling monkey (*Alouatta pigra*): Implications for female socioecology. *Primates Rep.* 61:47–65.

Horwich, R. H., Brockett, R. C., James, R. A., and Jones, C. B. 2001b, Population growth in the Belizean black howling monkey (*Alouatta pigra*). *Neotrop. Primates* 9:1–7.

Horwich, R. H. and Johnson, E. D. 1986, Geographic distribution of the black howler monkey (*Alouatta pigra*) in Central America. *Primates* 27:53–62.

Johns, A. D. and Skorupa, J. P. 1987, Responses of rain-forest primates to habitat disturbance: A Review. *Int. J. Primatol.* 8(2):157–191.

Keymer, A. E. and Dobson, A. P. 1987, Parasites of small mammals: Helminth population dynamics. *Mammal Rev.* 17:105–116.

Knopff, K. H., Knopff, A. R. A., and Pavelka, M. S. M. 2004. Observed case of infanticide committed by a resident male Central American black howler monkey (*Alouatta pigra*). *Am. J. Primatol.* 24(2):239–244.

Knopff, K. H. and Pavelka, M. S. M. (submitted). Determinants of group size in the Central American black howler monkey (*Alouatta pigra*): Evaluating the importance of ecological constraints.

Krebs, C. J. 1978, A review of the Chitty hypothesis of population regulation. *Can. J. Zool.* 56:2463–2480.

Marsh, D. M. and Pearman, P. B. 1997, Effects of habitat fragmentation of the abundance of 2 species of *Leptodactylid* frogs in an Andean montane forest. *Conserv. Biol.* 11:1323–1328.

McKey, D. B. 1978, Soils, vegetation, and seed-eating by black colobus monkeys, in G. G. Montgomery, ed., *The Ecology of Arboreal Folivores,* Smithsonian Institution Press, Washington, D.C., pp. 423–437.

Menon, S. and Poirer, F. E. 1996. Lion-tailed macaques (*Macaca silenus*) in a disturbed forest fragment: Activity patterns and time budget. *Int. J. Primatol.* 17(6): 969–985

Milton, K. 1979, Factors influencing leaf choice by howler monkeys: A test of some hypotheses of food selection by generalist herbivores. *Am. Nat.* 114:362–378.

Milton, K. 1996, Effects of bot fly (*Alouattamyia baeri*) parasitism on a free-ranging howler (*Alouatta palliata*) population in Panama. *J. Zool. (Lond.)* 239:39–63.

Milton, K. 1998, Physiological ecology of howlers (*Alouatta*): Energetic and digestive considerations and comparison with the Colobinae. *Int. J. Primatol.* 19:513–547.

Munger J. C. and Karasov, W. H. 1989, Sublethal parasites and host energy budgets: Tapeworm infection in white-footed mice. *Ecology* 70:904–921.

Murray, D. L., Keith, L. B., and Cary, J. R. 1996, The efficacy of anthelminitic treatment on the parasite abundance of free-ranging snowshoe hares. *Can. J. Zool.* 74:1604–1611.

Murray, D. L., Keith, L. B., and Cary, J. R. 1998, Do parasitism and nutritional status interact to affect production in snowshoe hares? *Ecology* 79:1209–1222.

National Research Council. 1992, *Conserving Biodiversity: A Research Agenda for Development Agencies.* National Academy Press, Washington, D.C.

Nicholson, A. J. 1933, The balance of animal populations. *J. Anim. Ecol.* 2:132–178.

Oates, J. F., Whitesides, G. H., Davies, A. G., Waterman, P. G., Green, S. M., Dasilva, G. L., and Mole, S. 1990, Determinants of variation in tropical forest primate biomass: New evidence from West Africa. *Ecology* 71:328–343.

Ostro, L. E. T., Silver, S. C., Koontz, F. W., Horwich, R. H., and Brockett, R. 2001, Shifts in social structure of black howler (*Alouatta pigra*) groups associated with natural and experimental variation in population density. *Int. J. Primatol.* 22:733–748.

Ostro, L. E. T., Silver, S. C., Koontz, F. W., and Young, T. P. 2000, Habitat selection by translocated black howler monkeys in Belize. *Anim. Conserv.* 3:175–181.

Ostro, L. E. T., Silver, S., Koontz, F. W., Young, T. P., and Horwich, R. H. 1999, Ranging behavior of translocated and established groups of black howler monkeys *Alouatta pigra* in Belize, Central America. *Biol. Conserv.* 87:181–190.

Pavelka, M. S. M. 2003, Population, group, and range size and structure in black howler monkeys (*A. pigra*) at Monkey River in southern Belize. *Neotrop. Primates* 11(3):187–191.

Pavelka, M. S. M. and Behie, A. M. 2005, The effect of Hurricane Iris on the food supply of Black Howlers (*Alouatta pigra*) in Southern Belize. *Biotropica* 37(1): 102–108.

Pavelka, M. S. M., Brusselers, O. T., Nowak, D., and Behie, A. M. 2003, Population reduction and social disorganization in *Alouatta pigra* following a hurricane. *Int. J. Primatol.* 24:1037–1055.

Pavelka, M. S. M. and Knopff, K. 2004, Diet and activity in *A. pigra* in Southern Belize: Does degree of frugivory influence activity level? *Primates* 45:105–112.

Ratsimbazafy, J. H., Harisoa, V., Ramarosandratana H. V., and Zaonarivelo, R. J. 2002, How do black and white ruffed lemurs still survive in a highly disturbed habitat? *Lemur News* 7:7–10.

Sapolsky, R. M. 1986, Endocrine and behavioral correlates of drought in wild olive baboons (*Papio anubis*). *Am. J. Primatol.* 11:217–227.

Schowalter, T. D. 1994, Invertebrate community structure and herbivory in a tropical rain forest canopy in Puerto Rico following Hurricane Hugo. *Biotropica* 26:312–319.

Schultz, J. C. 1988, Plant responses induced by herbivores. *Trends Ecol. Evol.* 3:45–49.

Silver, S. C. and Marsh, L. K. 2003, Dietary flexibility, behavioral plasticity, and survival in fragments: Lessons from translocated howlers, in L. K. Marsh, ed., *Primates in Fragments: Ecology and Conservation.* Kluwer Academic/Plenum, New York, pp. 251–265.

Silver, S. C., Ostro, L. E. T., Yeager, C. P., and Dierenfeld, E. S. 2000, Phytochemical and mineral components of foods consumed by black howler monkeys (*Alouatta pigra*) at two sites in Belize. *Zoo. Biol.* 19:95–109.

Silver, S. C., Ostro, L. E. T., Yeager, C. P., and Horwich, R. 1998, Feeding ecology of black howler monkeys (*Alouatta pigra*) in northern Belize. *Am. J. Primatol.* 45:163–179.

Smith, J. D. 1970, The systematic status of the black howler monkey, *Alouatta pigra* Lawrence. *J. Mammal.* 51:358–369.

Toque, K. 1993, The relationship between parasite burden and host resources in the desert toad (*Scaphiopus couchii*), under natural conditions. *J. Anim. Ecol.* 62:683–693.

Treves, A., Drescher, A., and Ingrisano, N. 2001, Vigilance and aggregation in black howler monkeys (*Alouatta pigra*). *Behav. Ecol. Sociobiol.* 50:90–95.

Walker, L. R., Lodge, D. J., Walker, N. V. L., and Waide, R. B. 1991, An introduction to hurricanes in the Caribbean. *Biotropica* 23:313–316.

Wasser, S. K., Bevis, K., King, G., and Hanson, E. 1997, Noninvasive physiological measures of disturbance in Northern Spotted Owl. *Conserv. Biol.* 11:1019–1022.

Waterman, P. G., Ross, J. A. M., Bennett, E. L., and Davies, A. G. 1988, A comparison of the floristics and leaf chemistry of the tree flora in two Malaysian rain forest and the influence of leaf chemistry on populations of colobine monkeys in the Old World. *Biol. J. Linn. Soc.* 34:1–32.

Will, T. 1991, Birds of a severely hurricane damaged Atlantic coastal rain forest in Nicaragua. *Biotropica* 23:497–507.

Willig, M. R. and Camillo, G. R. 1991, The effect of Hurricane Hugo on six invertebrate species in the Luquillo Experimental Forest (LEF) of Puerto Rico. *Biotropica* 23:455–461.

CHAPTER SIX

The Effects of Forest Fragment Age, Isolation, Size, Habitat Type, and Water Availability on Monkey Density in a Tropical Dry Forest

Holly Noelle DeGama-Blanchet and Linda Marie Fedigan

INTRODUCTION

The future of many primate species is uncertain. Countries with primate populations are losing 125,140 km^2 of forest annually and this destruction is considered to be a major threat to their survival (Chapman and Peres, 2001; Cowlishaw and Dunbar, 2000). Human disturbance of tropical forest is not

Holly Noelle DeGama-Blanchet • Department of Anthropology, University of Calgary, Calgary, AB, Canada, T2N 1N4. **Linda Marie Fedigan** • Department of Anthropology, University of Calgary, Calgary, Alberta, Canada, T2N 1N4.

New Perspectives in the Study of Mesoamerican Primates: Distribution, Ecology, Behavior, and Conservation, edited by Alejandro Estrada, Paul A. Garber, Mary S. M. Pavelka, and LeAndra Luecke. Springer, New York, 2005.

only leading to habitat destruction but also to massive habitat fragmentation. In many areas of the world, fragmented habitats are becoming the dominant geographic feature (Laurance, 1999). The present study focuses on how the age of forest fragments since disturbance as well as the isolation and size of fragments affect the population density of white-faced capuchins (*Cebus capucinus*), black-handed spider monkeys (*Ateles geoffroyi*), and mantled howlers (*Alouatta palliata*) in tropical dry forest, in Costa Rica. Additionally, the present study examines how the habitat types and dry-season water availability within forest fragments affect the population density of these three primates. Using the line transect method, we censused the three monkey species within the megapark "Area de Conservación Guanacaste" (ACG). We conducted censuses in three parts of ACG where all three primates were present: Sector Santa Rosa (Santa Rosa National Park, SRNP), Sector Cerro el Hacha, and Sector Murciélago. We use these data to draw comparisons with previous primate census studies.

The conservation of primates living in tropical dry forest has received little attention, and there are only a few studies that provide the necessary information to produce conservation management plans for monkey species found in dry forests (Chapman *et al.*, 1989). Such data are needed because tropical dry forest is the most severely threatened of all major tropical habitat types, with less than 0.02% remaining in the world (The Government of Costa Rica, 1998). The near extinction of this forest type and the lack of information on primates living in dry forest fragments indicate that studies on this topic are urgently required. Our study contributes to this body of knowledge.

Background

One major variable we examine is the effect on monkey population densities (hereafter referred to as densities/density) of age since disturbance of the forest fragment. In SRNP, Sorensen and Fedigan (2000) found that capuchins returned to abandoned pastures after 14–25 years, howlers returned after 30–60 years, and spider monkeys returned after 60–80 years. These authors report that densities of all three primates were higher in older patches of forest (Sorensen and Fedigan, 2000). The current study examines whether these findings apply not just to SRNP but to other sectors of ACG, as well.

A second variable we examine is the effect of forest fragment isolation, measured as the distance between forest fragments, on primate density within the fragment. Comparisons are made with the Estrada and Coates-Estrada (1996) study of mantled howlers where the density of primates was influenced by fragment isolation. Estrada and Coates-Estrada found that there were fewer individuals of *A. palliata* in more isolated habitat fragments. We also compare our results to the Onderdonk and Chapman (2000) study where the presence or absence of a primate species was affected by fragment isolation. Onderdonk and Chapman found that fragment isolation affected the presence of Pennant's red colobus monkeys (*Procolobus pennantii*); specifically that these monkeys were more likely to be present in forest fragments that were closer to Kibale National Park, Uganda. We believe that a discussion of the presence/absence of a species is relevant to density values because the absence (or extinction) of a primate population in an area may be the result of small population size, which is expressed as density values (Cowlishaw and Dunbar, 2000).

The third variable we examine is the effect of forest fragment size on primate density. As with isolation, comparisons are made with studies where the density, and presence or absence of primates was found to be related to fragment size. Estrada and Coates-Estrada (1996) found that larger habitat fragments supported larger populations of *A. palliata* in Mexico. Additionally, Kumar *et al.* (1995) determined that the presence of lion-tailed macaques (*Macaca silenus*) and Nilgiri langurs (*Presbytis johni*) in habitat patches was positively related to patch size. Lastly, red colobus monkeys (*Procolobus badius*), crested mangabeys (*Cercocebus galeritus*), and Syke's monkeys (*Cercopithecus mitis*) were usually more likely to be found in larger habitat patches, compared to smaller ones, along the Tana River in Kenya (Cowlishaw and Dunbar, 2000).

The fourth variable we examine is the effect of habitat type on primate density. Prior studies in ACG and SRNP (e.g. Freese, 1976; Fedigan *et al.*, 1998; Sorensen and Fedigan, 2000) have produced conflicting results about what can be considered optimal habitat for mantled howlers, whether it is deciduous or evergreen forest. As for spider monkeys, Freese (1976) found that, in SRNP, they occur in evergreen forests and in small, isolated islands of evergreen trees in semi-deciduous forests. However, Chapman *et al.* (1989) found spider monkeys in all major forest types except for some riverine strips in other sectors of ACG. Several studies of white-faced capuchins in SRNP found that they occur in all forest types (e.g. Freese, 1976; Chapman *et al.*, 1989; Sorensen and Fedigan, 2000). The present study re-examines the issue of which habitat types in ACG contain the highest densities of all three primates.

The fifth and last variable we examine is the effect of dry-season water availability on primate density. White-faced capuchins become central-place foragers near standing sources of water during the dry season in SRNP (Rose and Fedigan, 1995), which may affect their density around such areas. This chapter describes how each of these five variables (forest fragment age, isolation, size, habitat type, and dry-season water availability) affects primate density and compares our results to the two previous primate transect census studies in ACG (Chapman *et al.,* 1989; Sorensen and Fedigan, 2000).

Forest Fragmentation

The vegetation in ACG has been fragmented from the once continuous dry forest into much smaller pieces (Janzen, 1986). The patches of forest under investigation range in size from 0.07 to 95.71 km^2, and we suggest that these patches can be regarded as forest fragments for at least two reasons. First, two other studies have considered patches of forest that are within the size range of those under investigation in the present study, except the largest fragment of 95.71 km^2, as forest fragments (e.g., Goncalves *et al.* (2003) and Ferrari *et al.* (2003)). Additionally, Chiarello (2003) classified habitat patches much larger than 95.71 km^2 as forest fragments. Second, there is often a different microclimate near the periphery of a fragmented habitat, and this "edge effect" may render areas near edges inhospitable, thus leading to an effective reduction in fragment size for some species (Turner, 1996). Therefore, all of the area in the study fragments are not likely to be available to the primates, and the effective size of the study areas may be smaller than reported here.

METHODS

Study Site

In total, ACG is approximately 153,000 ha in size (110,000 ha is terrestrial habitat and 43,000 ha is marine) (Blanco, pers. comm.), and constitutes 2% of Costa Rica's landmass (The Government of Costa Rica, 1998), ranging from sea level to 1916 m (Chapman *et al.*, 1989). Dry successional deciduous forest is the predominant habitat type in the park, varying from 20 to 400 years of regeneration (Janzen, 1986). Regeneration is occurring quickly as vegetation and fauna spread into abandoned pastures and reconstitute themselves (The Government of Costa Rica, 1998).

We collected data in three sectors of ACG, northwestern Costa Rica. The forests in ACG have received nearly four centuries of highly heterogeneous damage due to a number of factors (The Government of Costa Rica, 1998); therefore, each sector has a different history of disturbance. There are little detailed data on the levels of disturbance in each sector, but what is known is reported here. In general, the logging, burning, hunting, ranching, and agriculture that occurred across ACG were not sufficient to completely eliminate any habitat type or any species, except for the green and scarlet macaw (The Government of Costa Rica, 1998).

The first study area is Sector Santa Rosa, or SRNP, which was established in 1971 and is approximately 10,800 ha in size (Fedigan *et al.*, 1985). SRNP is located approximately 35 km northwest of the town of Liberia in Guanacaste province, and is situated between the Pan-American Highway and the Pacific Ocean (Fedigan and Jack, 2001). Map coordinates of the Area Administrative in Sector Santa Rosa are 10 51′ N Lat and 85 37′ W Long (The Government of Costa Rica, 1998). Much of Guanacaste Province was originally covered in tropical dry forest (The Government of Costa Rica, 1998). The landscape is a series of stepped plateaus from the foothills of volcanic mountains down to the Pacific coastal plain (Fedigan *et al.*, 1998) and ranges in elevation from 300 m down to the ocean (Chapman, 1988).

There are two seasons at Santa Rosa, and the vast majority of the rain (approximately 900–2500 mm annually) falls in the wet season, usually between mid-May and mid-December. The non-riparian trees lose their leaves during the dry season, and most waterholes and streams gradually dry up. The strips of riparian and evergreen forests are more likely than other forest types to retain their leaves throughout this time (Fedigan and Jack, 2001).

The second oldest ranch in Costa Rica is located in Sector Santa Rosa (The Government of Costa Rica, 1998). This area was originally covered with dry deciduous forest dominated by clumps of oak forest (*Quercus oleoides*) (Chapman, 1988); but over the past 300 years, half of the upper plateau area was cleared for cattle pasture and planted with the East African jaragua grass (*Hyparrhenia rufa*). The forests in this area were selectively logged primarily for mahogany (*Swietenia macrophylla*) (Fedigan *et al.*, 1998) and fires entered Santa Rosa from runaway grass fires set outside the park by ranchers to keep woody vegetation out of their pastures. Additionally, attempts were made to grow dryland rice and cotton in this area. Specific types of disturbance in the Santa Rosa transects in this study include fire and wind damage to trees, and logging for fencepost material. Since the establishment of Santa Rosa National Park in 1971, this area has been protected from fire, poaching, ranching, farming, and logging (The Government of Costa Rica, 1998).

In Santa Rosa, as in the other sectors of ACG, restoration is occurring rapidly as tree seedlings establish themselves spread onto abandoned fields and pastureland. The establishment of woody vegetation is enhanced by fire prevention, which allows the introduced East African jaragua grass to continue to grow until it chokes itself out, leaving the small tree seedlings (The Government of Costa Rica, 1998). Due to the history of differential disturbance and protection of areas in Santa Rosa, the landscape is a mosaic of regenerating forest. There are fragments of evergreen, riparian, oak, mangrove, and early secondary forest in former pastures (Fedigan and Jack, 2001). No data are available on the proportional representation of each habitat type within this sector.

The second study area is Sector Murciélago, which is 12,200 ha in size (Janzen, 1986), and is located on the western coast bordering the Pacific Ocean. Map coordinates for the guard station in this sector are 1054.074′ N Lat and −85 43.754′ W Long (Medina, pers. comm.). This sector is composed of evergreen forest, semi-deciduous forest, deciduous forest, mangrove forest, and shrub vegetation surrounded by grassland (Medina, pers. comm.). The proportional representation of these forest types is presently unknown. Murciélago was incorporated into ACG in 1980 as an addition to SRNP. The forests in this area were differentially disturbed by centuries of European use (The Government of Costa Rica, 1998), and logging was detected in some of the transects. The forests began to recover in the mid-1980s after the cessation of ranching (The Government of Costa Rica, 1998).

The third study site is Sector Cerro el Hacha, which is 5000 ha in size, and is located on the northern boundary of the park near Nicaragua. Map coordinates for the guard station are 11 01.931′ N Lat and −85 31.691′ W Long (Medina, pers. comm.). This sector is composed of evergreen forest (which has ever-flowing creeks even in the dry season), native grasses, and secondary forest in varying stages of regeneration surrounded by grassland (Janzen, 1986). The proportional representation of these habitat types is unknown. Cerro el Hacha was incorporated into ACG between 1988 and 1989 (Blanco, pers. comm.). Some of the lower slopes in this sector still have virgin forest, while the upper slopes were largely deforested and were covered with native grasses. Some areas were used for growing corn and beans (Janzen, 1986). In this sector, logging and fire damage were detected in the transects. All three sectors can be characterized as a series of forest fragments surrounded by grassland.

Data Collection

We collected data using the line transect method (Brockelman and Ali, 1987). We collected line transect data between January and June 2003 (mostly dry season, and a few weeks of the wet season) on 28 transects (8 in Santa Rosa, 8 in Cerro el Hacha, and 12 in Murciélago). Transects were approximately 600 m in length (mean: 612 ± 56). We walked transects between 18 and 43 times (mean: 31 ± 8) each. Difficulties with transportation and accessing some sites during the rainy season made it impossible to acquire equal sample sizes from all areas.

Line transects were located in a total of six forest fragments for Sectors Murciélago and Santa Rosa (labeled A–F) (Figure 1). Santa Rosa contained Fragments A, B, and E, while Murciélago contained the remaining Fragments

Figure 1. Forest fragments A–F containing line transects in Sectors Murciélago and Santa Rosa, in the megapark Area de Conservación Guanacaste. Fragments A, B, and E are located in Santa Rosa, while Fragments C, D, and F are located in Murciélago.

Table 1. Line transect locations within six forest fragments

	Santa Rosa	Murciélago	Total no. of transects
Fragment A	4 Transects	Not in this sector	4
Fragment B	1 Transect	Not in this sector	1
Fragment C	Not in this sector	6 Transects	6
Fragment D	Not in this sector	1 Transect	1
Fragment E	1 Transect	Not in this sector	1
Fragment F	Not in this sector	1 Transects	1
Not in a fragment	2 Transects	4 Transects	6
Total no. of transects	8	12	20

C, D, and F. Out of the 20 transects in Santa Rosa and Murciélago, four were in Fragment A, one was in Fragment B, six were in Fragment C, one was in Fragment D, one was in Fragment E, one was in Fragment F, and six transects were not located in a fragment (Table 1). We were able to plot the geographical waypoint locations of each transect in Sector Cerro el Hacha (waypoint locations were recorded for each transect in all three sectors), but there were no satellite images available for this sector, so it was not possible to determine the size and isolation of forest fragments in which these transects were found.

We recorded the perpendicular distance, sighting distance, and sighting angle (Brockelman and Ali, 1987) from the geometric center of the *C. capucinus* and *A. palliata* groups (Brockelman and Ali, 1987; Anderson *et al.*, 1979). However, for *A. geoffroyi*, we recorded all measurements for each individual separately (Brockelman and Ali, 1987). We measured distances with a range finder, or we estimated them, when using the rangefinder was not possible because of environmental conditions. We conducted inter-observer reliability testing between the two observers, ensuring 90% accuracy, before data collection began and once or twice each month in the following period.

We recorded the availability of water along a transect during the dry season as a "yes/no" variable. Water was available either from streams or artificial waterholes, such as horse troughs, but not from tree hollows, which evaporate during the dry season. There were nine transects in total that contained water; two transects with water were located in Murciélago, four in Cerro el Hacha, and three in Santa Rosa. Additionally, we hired a local botanist to classify the habitat type of each transect. We grouped habitat types into three categories: Habitat Type 1 consisted of deciduous and/or semi-deciduous forest; Habitat Type 2 was mainly semi-deciduous forest, but at least 100 m of the transect consisted of forest in which up to 50% of the species were evergreen; Habitat Type 3 included transects where there were both semi-deciduous and evergreen forest, or simply evergreen forest. Habitat types were determined based on the majority of species present. For example, if 50% or more of the species present in a transect were evergreen then that forest was classified as evergreen. There were 6 transects classified as Habitat Type 1 (2 in Murciélago, 2 in Cerro el Hacha, 2 in Santa Rosa), 12 classified as Habitat Type 2 (8 in Murciélago, 1 in Cerro el Hacha, 3 in Santa Rosa), and 10 classified as Habitat Type 3 (2 in Murciélago, 5 in Cerro el Hacha, 3 in Santa Rosa). Lastly, the botanist also determined the approximate age since disturbance of each transect based on species composition, canopy height, and history of the area. The forest ages ranged from 14 to 118 years (mean: 33 ± 27).

Forest fragment size and isolation were determined with the Geographical Information System software ArcGIS (v.8.1). Geographical waypoint locations were marked with a Global Positioning System at the beginning and the end of each transect. These points were then plotted onto a satellite image of ACG that had been classified into general categories of forested and non-forested habitat based on 80% canopy closure (Sanchez-Azofeifa and Calvo, 2004). The transects that were classified as being "not in a fragment" were located in areas of vegetation that did not meet this 80% criteria.

We calculated the size of forest fragments surrounding our study transects in Sectors Murciélago and Santa Rosa using a buffer zone of 30 m. Any patch of forest that was located within 30 m of the fragment containing a line transect was included in the size calculation for that study fragment. Forest fragments outside this 30 m buffer zone were used to calculate the isolation distance of each census fragment. The closest distance to the next nearest patch of forest from the edge of any forest fragment within the 30 m buffer zone from a census fragment was considered to be the isolating distance. If there were no forest fragments within 30 m of a census fragment, then the distance from the edge of the census fragment to the next nearest patch of forest was considered to be the isolating distance. A buffer zone of 30 m was chosen because it is a short enough distance that it would not act as a barrier to any of the three species moving terrestrially between forest patches. The size of the census fragments ranged between 0.07 and 95.71 km^2 (mean: 35 ± 44 km^2) (Table 2). Forest fragments in SRNP containing line transects exceeded the boundaries of the park and, therefore, the size for all these fragments combined is larger than the size of the park itself. Additionally, the isolating distances for all census

Table 2. Table showing the size of census fragments (those containing a line transect), the size of forest fragments beyond 30 m from a census fragment (those that were used to determine isolation), and the isolating distances between them

Census fragments in Murciélago and Santa Rosa		Fragments beyond 30 m from a census fragment		
Fragment name	Size (km^2)	Label	Size (km^2)	Distance from census fragment (m)
		1	0.06	40.67
		2	0.12	40.67
A	95.71	3	0.21	40.67
		4	0.39	40.67
		5	0.07	40.67
		6	0.06	40.67
B	47.30	7	0.06	40.67
		8	0.25	40.67
C	19.26	9	3.16	655.84
D	5.28	10	0.06	333.21
E	0.08	11	0.06	626.46
F	0.07	12	0.60	3814.19

fragments ranged from 40.67 to 3814.19 m (mean: 140 ± 239 m) (Table 2). It is important to note that eight fragments were isolated from census fragments (those containing a line transect) by a distance of 40.67 m. This is due to the resolution of the image; distance measurements are only discernable in increments that reflect the size of the pixel, which in this case was 40.67 m.

Density Analysis

We calculated densities as per the National Research Council (1981). Absolute density (individuals/km^2) was calculated as:

$$\text{Density: } \frac{\text{Estimated animal population}}{\text{Census area}} = \frac{\text{Number of animals seen in sample area}}{\text{Sample area}}$$

The sample area was calculated as: Area: $l \times w$; where l is the length of the transect line (multiplied by the number of times the transect was walked) and w is the strip width, or the width on one side of the transect line multiplied by two (since data were taken on both the sides).

We used the maximum distance method (National Research Council, 1981) to designate strip width using sighting distances (the maximum distance of all sightings were used to demarcate the area sampled). We recorded both the perpendicular and sighting distances during data collection but, by comparing density estimates based on both methods in areas of known primate density in SRNP, found that sighting distances yielded more accurate density estimates. We truncated the data by removing 5% of the most distant measurements as the farthest sighting events provide little information about density (Buckland *et al.*, 1996). We excluded solitary capuchins and howlers from analysis to avoid over counting individuals (Fedigan *et al.*, 1998). Few solitaries of these species were seen throughout the course of the study, and, therefore, this was not problematic.

Statistical Analyses

We conducted all statistical analyses using the statistical program SPSS (v.11.5). We performed standard multiple regressions for each species separately to examine the combined influence of forest fragment age since disturbance, isolation, and dry-season water availability on monkey absolute density (individuals/km^2). We eliminated forest fragment size from the multivariate model due to multicollinearity; size was correlated with forest fragment age since disturbance ($r = 0.759$) beyond an acceptable limit (Tabachnick and Fidell, 1996). Therefore, Pearson's product–moment correlations were used to test for the relationship between forest fragment size and the absolute density of monkeys. Lastly, we ran Kruskal–Wallis two-tailed tests to examine the effects of forest fragment habitat type on monkey absolute density. Habitat type was not included in the multiple regression analyses because it is not a continuous or dichotomous independent variable, which is required by the model. Habitat type was grouped into three categories based on the dominant species present, as mentioned above.

RESULTS

Absolute densities for all monkey species for each transect are reported in Table 3. Additionally, transect sample size and forest fragment age, size, isolation, habitat type, and dry-season water availability are reported in this table. These results indicate that the highest densities for capuchins and howlers were in Sector Santa Rosa (capuchins: 34.47 individuals/km^2, howlers: 28.64

Table 3. Primate densities, transect sample size, and forest fragment age, isolation, size, habitat type, and dry-season water availability for all transects

Transect	No. of times walked	Age (years)	Habitat Type	Size (km^2)	Isolation (m)	Water	Capuchin density (per km^2)	Spider density (per km^2)	Spider density (per km^2)
ME1[a]	30	16	2[b]	19.26	655.84	No	0.00	0.00	1.98
ME2	30	23	2	19.26	655.84	No	0.00	0.00	.00
ME3	30	18	2	19.26	655.84	No	0.00	0.00	17.81
ME4	30	16	1[c]	19.26	655.84	No	3.17	0.00	10.42
ME5	30	25	2	19.26	655.84	No	3.47	0.00	1.63
ME6	21	23	3[d]	0.07	3814.19	No	14.43	0.00	5.67
ME7	30	15	2	0.00[e]	0.00[f]	No	0.00	0.00	0.00
ME8	19	18	2	0.00	0.00	No	2.51	0.00	8.12
ME9	30	30	3	19.26	655.84	Yes	9.26	1.68	9.92
ME10	28	14	1	0.00	0.00	No	0.00	0.00	3.92
ME11	32	16	2	5.28	333.21	No	0.00	0.00	4.46
ME12	30	15	2	0.00	0.00	Yes	0.00	0.00	0.00
CH1[g]	30	20	3	Not available	Not available	No	3.05	0.00	2.22
CH2	30	23	3	Not available	Not available	Yes	18.52	5.70	26.50
CH3	30	25	3	Not available	Not available	Yes	11.06	0.00	4.87
CH4	30	16	1	Not available	Not available	No	0.00	0.00	0.00

CH5	19	30	3	Not available	Not available	Yes	31.52	0.00	12.53
CH6	18	18	1	Not available	Not available	No	0.00	0.00	0.00
CH7	22	19	2	Not available	Not available	Yes	0.00	0.00	0.00
CH8	19	20	3	Not available	Not available	No	0.00	0.00	0.00
SR1[h]	43	33	2	0.00	0.00	No	3.75	8.24	0.00
SR2	42	28	2	0.00	0.00	No	8.63	5.75	7.35
SR3	41	48	1	0.08	626.46	Yes	29.51	19.27	1.81
SR4	42	58	2	95.71	40.67	Yes	34.47	0.00	0.00
SR5	40	68	1	95.71	40.67	Yes	4.80	0.00	0.00
SR6	40	103	3	47.30	40.61	No	20.23	0.00	0.00
SR7	40	80	3	95.71	40.67	No	4.81	8.01	13.89
SR8	41	118	3	95.71	40.67	No	19.83	28.64	6.54

[a] Murciélago transects.
[b] Semi-deciduous forest, but at least 100 m of the transect consisted of forest in which up to 50% of the species were evergreen.
[c] Semi-deciduous and/or deciduous forest.
[d] Both semi-deciduous and evergreen forest, or simply evergreen forest.
[e] Zeros indicate that line transects were not located in a fragment, and therefore there were no values for this variable.
[f] Zeros indicate that line transects were not located in a fragment, and therefore there were no values for this variable.
[g] Cerro el Hacha transects.
[h] Santa Rosa transects.

individuals/km^2), whereas the highest densities for spider monkey came from Cerro el Hacha (26.50 individuals/km^2). Throughout ACG, capuchin densities ranged from 0.00 to 34.47 individuals/km^2, howler densities ranged from 0.00 to 28.64 individuals/km^2, and spider monkey densities ranged from 0.00 to 26.50 individuals/km^2. Capuchins were found in all Santa Rosa transects but they were absent in 11 transects in Cerro el Hacha and Murciélago. Howlers were rare throughout ACG as indicated by their absence from 21 transects from all three sectors. Lastly, spider monkeys were absent from 11 transects in all three sectors.

In the multiple regression model, the age of the forest fragment ($T = 3.247$, F (3,16) = 6.838, $p = 0.005$) and dry-season water availability ($T = 3.050$, F (3,16) = 6.838, $p = 0.008$) both made significant contributions to explaining capuchin absolute density, whereas degree of isolation did not. Age made the strongest unique contribution. The model accounts for 56.2% of the variance in capuchin absolute density (Table 4).

Age of the forest fragment ($T = 2.990$, F (3,16) = 3.147, $p = 0.009$) made the only strong, unique, significant contribution to explaining howler monkey absolute density in the multiple regression model. This model explains 37.1% of the variance in this species' density (Table 4). No variable in the model made a strong, unique, significant contribution to explaining spider monkey absolute density. The model accounts for only 3.7% of variance in this dependent variable (Table 4).

We used Pearson's product–moment correlations to test for the relationship between forest fragment size, and the absolute density of monkeys. We found no significant results. There was a positive relationship between size and capuchin absolute density ($r = 0.411$, $df = 18$, $p = 0.071$), but it did not reach statistical significance (Table 4).

We ran Kruskal–Wallis tests to examine the effects of forest fragment habitat type on monkey absolute density. We obtained significant results only for capuchin absolute density ($\chi^2 = 7.274$, $df = 2$, $p = 0.026$). Capuchin density was highest (had the highest rank) in transects where there were both semi-deciduous and evergreen forest, or simply evergreen forest (Habitat Type 3). Although not reaching a level of statistical significance, howler monkey absolute density ($\chi^2 = 1.446$, $df = 2$, $p = 0.485$) and spider monkey absolute density ($\chi^2 = 4.133$, $df = 2$, $p = 0.127$) are all highest in Habitat Type 3, as well (Table 4).

Table 4. Table showing results from statistical tests

Variable	Statistical test	Capuchin density[a]	Howler density[b]	Spider density[c]
Age	Multiple regression			
	P	0.005**[d]	0.009**	0.965
	T	3.247	2.990	0.045
Isolation	*P*	0.112	0.999	0.567
	T	1.683	0.001	0.584
Dry-season water availability	*P*	0.008**	0.981	0.564
	T	3.050	−0.025	0.590
	R^2	0.562	0.371	0.037
	F (3,16)	6.838	3.147	0.208
Size	Pearson's product–moment correlation			
	P	0.071	0.210	0.814
	R	0.411	0.293	0.056
	df	18	18	18
Habitat Type	Kruskal–Wallis			
	K	3	3	3
	P	0.026*[e]	0.485	0.127
	HT 1[f] Rank	12.50	13.67	11.83
	HT 2[g] Rank	11.00	13.33	12.42
	HT 3[h] Rank	19.90	16.40	18.60
	df	2	2	2

[a] $\Upsilon = 0.213 + 3.589\ \text{E} - 03 + 11.290 + -4.383$.
[b] $\Upsilon = 0.147 + 1.952\ \text{E} - 06 + -0.068 + -2.134$.
[c] $\Upsilon = 2.702\text{E} - 03 + 1.140\text{E} - 03 + 1.998 + 3.747$.
[d] Significant at the 0.01 level.
[e] Significant at the 0.05 level.
[f] Semi-deciduous and/or deciduous forest.
[g] Semi-deciduous forest, but at least 100 m of the transect consisted of forest in which up to 50% of the species were evergreen.
[h] Both semi-deciduous and evergreen forest, or simply evergreen forest.

DISCUSSION

Forest Fragment Age

Age of a forest fragment since disturbance is an important variable in explaining capuchin and howler monkey absolute density in ACG. Higher densities of capuchins and howlers were found in transects where there was older forest. These findings are in accordance with a previous study by Sorensen and Fedigan (2000) in SRNP in which they found that capuchin and howler densities were positively related to forest age since disturbance. Sorensen and Fedigan (2000) also found that capuchin food biomass (measured as the combined biomass of fruit and leaves from trees that constituted at least 2% of the species' diet based on published accounts) increased linearly with forest age in SRNP. Therefore, older forests may exhibit higher densities of capuchin monkeys because these areas contain higher food biomass for this species.

Furthermore, forest fragment age is the only significant independent variable in the multiple regression model for explaining howler monkey density. There may be at least two reasons why older forests have higher densities of howlers. First, in SRNP, howlers prefer to forage in larger trees (Larose, 1996), and these trees are found in older areas of forest. Second, food biomass for the howlers (measured as the combined biomass of fruit and leaves from trees that constituted at least 2% of the species' diet based on published accounts) was found to increase linearly with forest age in SRNP, just as it did for the capuchins (Sorensen and Fedigan, 2000). The proportion of evergreen trees in this sector also increased linearly with forest age (Sorensen, 1998). Therefore, more leaf food sources are available in older areas, assuming that evergreen trees produce uniformly palatable leaves that howlers consume (Sorensen and Fedigan, 2000). Then, it is to be expected that howlers would have higher densities in older forests that contain the trees that they prefer as well as high food biomass.

Sorensen and Fedigan (2000) found that in SRNP spider monkeys returned to abandoned pastures after 60–80 years of regeneration and that forest fragment age was significant in explaining their density. The present findings are different from the prior study in that we did not find forest fragment age to affect spider monkey density and we also found spider monkeys in much younger forest than previously reported. We found spider monkeys in Santa Rosa in 28-year-old forest (7.35 individuals/km^2), in Murciélago in 14-year-old forest (3.92 individuals/km^2), and in Cerro el Hacha in 20-year-old forest (2.22 individuals/km^2).

We suggest that since the time of the earlier study, spider monkeys may have simply expanded their ranges to include younger as well as older areas of forest in SRNP. This might explain why our findings are different from those of Sorensen and Fedigan (2000). If spider monkeys have expanded their ranges, then this would indicate that they can forage and travel in much younger areas than previously thought.

In summary, forest fragment age since disturbance was found to make a significant, positive contribution to explaining capuchin and howler monkey density in this study, whereas it made no contribution to explaining the density of spider monkeys. Capuchins and howlers may have been found in older forests in higher densities because these areas contain higher food biomass for these species. Additionally, the large trees howlers prefer are located in older forest, which may also account for their higher densities in such areas. Spider monkey densities were not explained by fragment age, possibly because these primates have simply expanded their ranges to include older and younger areas of forest since earlier studies.

Forest Fragment Isolation

Isolation was not found to make a significant contribution to explaining the density for any of the three primates in this study. It is generally expected that as isolation increases, the probability of colonization, or immigration to a habitat fragment will decrease (Rodriguez-Toledo *et al.*, 2003). This immigration to a fragment could influence population density. We suggest that forest fragment isolation is not an important variable in explaining primate density in our study because of the size of fragments surveyed. Our study fragments ranged in size up to 95.71 km^2. It is possible that immigration occurs between the study fragments and surrounding ones, but that this exchange has little effect on density in most of the fragments surveyed. Most of the study fragments are large and may contain populations sufficient in size that demographic processes rather than rates of immigration regulate them.

Previous studies that have found forest fragment isolation to affect primate density, and presence/absence (e.g., Estrada and Coates-Estrada, 1996; Onderdonk and Chapman, 2000) came from research in forest fragments ranging in size between 0.8 and 1000 ha. Primate populations in such small forest fragments will likely be smaller in size than half the populations and the fragments surveyed in our study (three of the six forest fragments in our study were smaller than 1000 ha, while the remaining three were significantly larger). We suggest that immigration may have a more significant effect on the small populations studied by Estrada and Coates-Estrada (1996) and Onderdonk and Chapman (2000). It has been argued that the addition of individuals through immigration has a more significant effect on smaller populations because they are more vulnerable to extinction (Cowlishaw and Dunbar, 2000). The isolation distances in the two aforementioned studies ranged from 50 to 8000 m, and in our study they ranged from 41 to 3814 m. Since the isolation distances in the present study are within the range of those in the other two studies, this suggests that some other variable is overriding the effects of isolation.

In summary, forest fragment isolation was not found to make a contribution to explaining the density of the three primates in this study. We contend that this is due to the large size of the fragments surveyed. Other demographic processes may be more important than immigration in regulating the study populations.

Forest Fragment Size

Forest fragment size was found to make little contribution to explaining primate density in ACG. Previous primate studies have found higher densities of primates in larger fragments of forest (e.g., Estrada and Coates-Estrada, 1996), or that primates were more likely to occur in larger patches of forest compared to smaller ones (e.g., Kumar *et al.,* 1995; Cowlishaw and Dunbar, 2000). Other studies have not come to this conclusion. For example, Rodriguez-Toledo *et al.* (2003) found the opposite: higher densities of *A. palliata mexicana* were found in smaller forest fragments. Additionally, Kowalewski and Zunino (1999) found that when forest patches in Argentina were reduced in size, a population of *Alouatta caraya* remained the same size. In the present study, no significant relationships were found between primate density and forest fragment size.

Therefore, we conclude that forest fragment size offers limited insight into primate density at our study sites. As with isolation, it may be that the large sizes of most of the forest fragments we surveyed explain why this variable made no explanatory contribution. The large patches that are under analysis here (up to 95.71 km^2) may not constrain primate population sizes the way a 10 ha fragment would. As with isolation, the primate populations living in the large ACG forest fragments may be big enough that fragment size has little or no effect on their population densities.

Gilbert (2003) argues that capuchins and spider monkeys cannot live in fragments under 1 km^2. Additionally, Ferrari *et al.* (2003) found that white-fronted spider monkeys (*Ateles marginatus*) were absent from fragments less than 1 km^2 in size. The results of the present study contradict those of the previous studies; the capuchins and spider monkeys were found in fragments as small as 0.07 km^2. This suggests that, in ACG, a lower size limit is required for capuchins and spider monkeys to exist in forest fragments. Studies of howler monkeys in fragments have found that this species can exist in those as small as 0.01 km^2 [red howlers

(*Alouatta seniculus*) in the Biological Dynamics of Forest Fragments Project in Brazil (Gilbert, 2003)], and 0.70 km^2 [red-handed howlers (*Alouatta belzebul*) along the Santarem–Cuiaba Highway in Brazil (Ferrari *et al.*, 2003)]. In the present study, howlers were found in fragments as small as 0.08 km^2.

In summary, forest fragment size was not important in explaining primate density at our study sights. We attribute this to the large size of the fragments surveyed. The study populations may be big enough that forest fragment size has little influence on their densities. Additionally, capuchins and spider monkeys were found in smaller fragments compared to previous studies, whereas the howlers were found in fragments that were in the size range of those they have been found in previously.

Habitat Type and Dry-Season Water Availability

The presence of evergreen forest is an important explanatory variable for primate species density in ACG. The absolute density of capuchins was significantly higher in transects where there was both semi-deciduous and evergreen forest, or simply evergreen forest (Habitat Type 3) (13.27 individuals/km^2). Additionally, although not reaching the level of statistical significance, howler monkey absolute density and spider monkey absolute density are all highest in Habitat Type 3 (4.40 individuals/km^2 and 8.21 individuals/km^2, respectively), compared to semi-deciduous or deciduous forest (Habitat Type 1) (howlers: 3.21 individuals/km^2, spider monkeys: 2.69 individuals/km^2), or semi-deciduous forest where at least 100 m of the transect consisted of forest in which up to 50% of the species were evergreen (Habitat Type 2) (howlers: 1.17 individuals/km^2, spider monkeys: 3.45 individuals/km^2).

The above finding is in accordance with previous studies from ACG on howler monkeys. Chapman and Balcomb (1998) found that howler monkey densities in ACG were highest in areas of semi-evergreen forest. Freese (1976) found that howler monkeys in ACG were almost completely confined to mature evergreen forest. Additionally, Chapman (1988) found that the core area used by howlers in Sector Santa Rosa (that is the area used in more than 10% of the observations) was wet semi-evergreen forest. Areas outside of this core tended to be dry semi-deciduous forest (Chapman, 1988). A study by Sorensen and Fedigan (2000) contradicts these findings, however. These authors found that, in 1996, howler densities in SRNP were not higher in old evergreen forests than in deciduous ones.

Spider monkeys, like the other two primates in ACG, were found to have higher densities in transects that contained semi-deciduous and evergreen forest, or simply evergreen forest (Habitat Type 3), although this relationship was not significant. However, this species was found in all habitat types we surveyed. These findings are in accordance with previous studies by Freese (1976) and Chapman *et al.* (1989).

Capuchin monkey density was significantly higher in transects where there was semi-deciduous and evergreen forest, or simply evergreen forest (Habitat Type 3). This finding is in accordance with previous studies by Chapman (1988), Chapman *et al.* (1989), and Freese (1976). Freese explained the extensive distribution of capuchins as a result of their diet and locomotor pattern; he argued that capuchins eat primarily fruit and insects, both of which are assumed to be available on a year-round basis in varying abundance in deciduous forests, and insects can also be found year-round in young secondary, deciduous growth. Additionally, these small primates move quadrupedally, which allows for easy movement through weakly structured, dense, short vegetation (Freese, 1976).

Capuchins in ACG may nonetheless reside preferentially in evergreen forests because these areas contain more water resources. During the dry season, capuchins in ACG usually visit areas with standing water daily (Chapman *et al.*, 1989). They try to maintain access to these limited resources (Fedigan and Jack, 2001), presumably because they cannot obtain the water they need from their foods alone (Freese, 1978). The importance of water to this species is evidenced by the fact that dry-season water availability made large contributions to explaining capuchin density in the multiple regression model. During the dry season in ACG, large evergreen trees provide more tree holes with drinking water (Sorensen and Fedigan, 2000), and evergreen vegetation lines the banks of the few springs and ever-flowing watercourses that are left at this time (Janzen, 1986). Thus, we conclude that capuchins can utilize all forest types within ACG, but preferentially inhabit evergreen areas because of water accessibility. This contention is supported by two previous studies. Fedigan *et al.* (1996) argued that, although SRNP capuchins use new secondary forest areas, they are not able to reside in them exclusively because of the difficulty of accessing fruit trees and water resources in these areas. Additionally, Rose and Fedigan (1995) and Chapman (1988), noted that Santa Rosa capuchins become central-place foragers around water resources in the dry season. If water is more abundant in evergreen forests during the dry season, then it is likely that capuchins would restrict much of their activities to these areas.

Dry-season water availability was not found to contribute to explaining howler and spider monkey density. This is consistent with a previous study by Freese (1978). This author found that spider monkeys and howlers in Santa Rosa did not drink from waterholes; he argues that these species probably obtain all the water they need from food. However, a study by Chapman (1988) found that in SRNP spider monkeys and howlers sometimes drank from waterholes. Another study by Gilbert and Stouffer (1989) found that both *A. palliata* and *A. geoffroyi* drank from a standing water source in the dry season in the tropical dry forest of Palo Verde National Park, Costa Rica. Despite these contradictions, it is clear that dry-season water availability is more important to capuchins than to the other two primates in Santa Rosa.

In summary, the presence of evergreen forest is an important explanatory variable for the density of all three primates in this study. Capuchins, howlers, and spider monkeys were all found in higher densities in Habitat Type 3 (semi-deciduous and evergreen forest, or simply evergreen forest), although the relationship was only significant for the capuchins. Capuchins may restrict some of their activities to evergreen forests because these areas contain more dry-season standing water, compared to deciduous forests. The availability of water was not found to contribute to explaining howler and spider monkey density.

SUMMARY

In summary, forest fragment age is an important explanatory variable for capuchin and howler density (higher densities were found in older areas of forest), whereas it makes no contribution to explaining the density of spider monkeys. The presence of evergreen forests in ACG is also important for explaining the absolute density of all three species, as there were higher densities in fragments containing evergreen forest. Transects where water was available in the dry season had higher capuchin densities; water availability appears to be more important for this species than for the spider monkeys and howlers. Forest fragment isolation and size made little contribution to explaining the density of any primate in ACG, probably due to the large size of forest fragments surveyed. Based on these findings, we conclude that older fragments of forest with dry-season standing water, and a substantial amount of evergreen forest should be preferentially protected to enhance the conservation of white-faced capuchins, black-handed spider monkeys, and mantled howlers in Costa Rica.

ACKNOWLEDGMENTS

We are grateful to the National Park Service of Costa Rica for allowing us to work in ACG in 2003, and to the administrators of ACG for their assistance, especially Roger Blanco, the Research Director. We would also like to thank Waldy Medina for supplying maps of ACG, Dr. Arturo Sanchez-Azofeifa for supplying a satellite image of ACG, Tracy Wyman for assistance with GIS analyses, Dr. Tak Fung and Dr. John Addicott for statistical support, Rob O'Malley for assisting with data collection, Daniel Perez for botanical assistance, and Alex Perez for assistance with transportation. Holly DeGama-Blanchet gratefully acknowledges financial support from a University of Calgary Thesis Research Grant, and a University of Calgary Graduate Research Scholarship. Linda Fedigan's research is funded by the Canada Research Chairs Program and by an on-going operating grant from NSERCC (#A7723).

REFERENCES

Anderson, D. R., Laake, J. L., Crain, B. R., and Burnham, K. P. 1979, Guidelines for line transect sampling of biological populations. *J. Wildl. Manage.* 43:70–78.

Brockelman, W. Y. and Ali, R. 1987, Methods of surveying and sampling forest primate populations, in: C. W. Marsh and R. A. Mittermeier, eds., *Primate Conservation in the Tropical Rain Forest.* Alan R. Liss, New York, pp. 23–62.

Buckland, S. T., Anderson, D. R., Burnham, K. P., and Laake, J. L. 1996, *Distance Sampling: Estimating Abundances of Biological Populations.* Chapman and Hall, London.

Chapman, C. A. 1988, Patterns of foraging and range use by three species of neotropical primates. *Primates* 29(2):177–194.

Chapman, C. A. and Balcomb, S. R. 1998, Population characteristics of howlers: Ecological conditions or group history. *Int. J. Primatol.* 19:385–403.

Chapman, C. A., Chapman, L., and Glander, K. E. 1989, Primate populations in northwestern Costa Rica: Potential for recovery. *Primate Conserv.* 10:37–44.

Chapman, C. A. and Peres, C. A. 2001, Primate conservation in the new millennium: The role of scientists. *Evol. Anthropol.* 10(1):16–33.

Chiarello, A. G. 2003, Primates of the Brazilian Atlantic Forest: The influence of forest fragmentation on survival, in: L. K. Marsh, ed., *Primates in Fragments: Ecology and Conservation*, Kluwer Academic/Plenum Publishers, New York, pp. 99–118.

Cowlishaw, G. and Dunbar, R. 2000, *Primate Conservation Biology.* The University of Chicago Press, Chicago.

Estrada, A. and Coates-Estrada, R. 1996, Tropical rain forest fragmentation and wild populations of primates at Los Tuxtlas, Mexico. *Int. J. Primatol.* 17(5):759–783.

Fedigan, L. M., Fedigan, L., and Chapman, C. A. 1985, A census of *Alouatta palliata* and *Cebus capucinus* monkeys in Santa Rosa National Park, Costa Rica. *Brenesia* 23:309–322.

Fedigan, L. M. and Jack, K. 2001, Neotropical primates in a regenerating Costa Rican dry forest: A comparison of howler and capuchin population patterns. *Int. J. Primatol.* 22(5):689–713.

Fedigan, L. M., Rose, L. M., and Morera Avila, R. 1996, Tracking capuchin monkey (*Cebus capucinus*) populations in a regenerating Costa Rican dry forest, in: M. A. Norconk, A. L. Rosenberger, and P. A. Garber, eds., *Adaptive Radiations of Neotropical Primates*, Plenum Press, New York, pp. 289–307.

Fedigan, L. M., Rose, L. M., and Morera Avila, R. 1998, Growth of mantled howler groups in a regenerating Costa Rican dry forest. *Int. J. Primatol.* 19(3):405–432.

Ferrari, S. F., Iwanaga, S., Ravetta, A. L., Freitas, F. C., Sousa, B. A. R., Sousa, L. L., Costa, C. G., and Coutinho, P. E. G. 2003, Dynamics of primate communities along the Santarem–Cuiaba Highway in south-central Brazilian Amazonia, in: L. K. Marsh, ed., *Primates in Fragments: Ecology and Conservation*, Kluwer Academic/Plenum Publishers, New York, pp. 123–142.

Freese, C. 1976, Censusing *Alouatta palliata*, *Ateles geoffroyi*, and *Cebus capucinus* in the Costa Rican dry forest, in: R. W. Thorington and P. G. Heltne, eds., *Neotropical Primates. Field Studies and Conservation*, National Academy of Sciences, Washington, D.C., pp. 4–9.

Freese, C. H. 1978, The behavior of white-faced capuchins (*Cebus capucinus*) at a dry-season waterhole. *Primates* 19(2):275–286.

Gilbert, K. A. 2003, Primates and fragmentation of the Amazon Forest, in: L. K. Marsh, ed., *Primates in Fragments: Ecology and Conservation*, Kluwer Academic/Plenum Publishers, New York, pp. 145–156.

Gilbert, K. A. and Stouffer, P. C. 1989, Use of a ground water source by mantled howler monkeys (*Alouatta palliata*). *Biotropica* 21(4):380.

Goncalves, E. C., Ferrari, S. F., Silva, A., Coutinho, P. E. G., Menezes, E. V., and Schneider, M. P. C. 2003, Effects of habitat fragmentation on the genetic variability of silvery marmosets, *Mico argentatus*, in: L. K. Marsh, ed., *Primates in Fragments: Ecology and Conservation*, Kluwer Academic/Plenum Publishers, New York, pp. 17–27.

Janzen, D. H. 1986, *Guanacaste National Park: Tropical Ecological and Cultural Restoration*. Fundacion de Parques Nacionales, Editorial Universidad Estatal Distancia, San Jose, Costa Rica.

Kowalewski, M. M. and Zunino, G. E. 1999, Impact of deforestation on a population of *Alouatta caraya* in northern Argentina. *Folia Primatol.* 70:163–166.

Kumar, A., Umapathy, G., and Prabhakar, A. 1995, A study on the management and conservation of small mammals in fragmented rain forests in the Western Ghats, South India: A preliminary report. *Primate Conservation* 16:53–58.

Larose, F. 1996, Foraging strategies, group size and food competition in the mantled howling monkey, *Alouatta palliata*. PhD Thesis, University of Alberta, Edmonton, Alberta, Canada.

Laurance, W. F. 1999, Introduction and synthesis. *Biological Conserv.* 91:101–107.

National Research Council 1981, *Techniques for the Study of Primate Population Ecology*. National Academy Press, Washington, D.C.

Onderdonk, D. A. and Chapman, C. A. 2000, Coping with forest fragmentation: The primates of Kibale National Park, Uganda. *Int. J. Primatol.* 21(4):587–611.

Rodriguez-Toledo, E. M., Mandujano, S., and Garcia-Orduna, F. 2003, Relationships between forest fragments and howler monkeys (*Alouatta palliata mexicana*) in southern Veracruz, Mexico, in: L. K. Marsh, ed., *Primates in Fragments: Ecology and Conservation*, Kluwer Academic/Plenum Publishers, New York, pp. 79–97.

Rose, L. M. and Fedigan, L. M. 1995, Vigilance in white-faced capuchins, *Cebus capucinus. Anim. Behav.* 49:63–70.

Sanchez-Azofeifa, G. A. and Calvo, J. 2004, Guanacaste Conservation Area Forest Cover Map 1:250,000. Earth Observation Systems Laboratory, University of Alberta.

Sorensen, T. C. 1998, Tropical Dry Forest Restoration and its Influence on Three Species of Costa Rican Monkeys. MSc Thesis, University of Alberta, Edmonton, AB, Canada.

Sorensen, T. C. and Fedigan, L. M. 2000, Distribution of three monkey species along a gradient of regenerating tropical dry forest. *Biological Conserv.* 92:227–240.

Tabachnick, B. G. and Fidell, L. S. 1996, *Using Multivariate Statistics,* 3rd Edn. HarperCollins, New York.

The Government of Costa Rica, 1998, Area de Conservación Guanacaste. Nomination for inclusion in the world heritage list of natural properties.

Turner, I. M. 1996, Species loss in fragments of tropical rainforest: A review of the evidence. *J. Appl. Ecol.* 33:200–209.

CHAPTER SEVEN

Forest Fragmentation and Its Effects on the Feeding Ecology of Black Howlers (*Alouatta pigra*) from the Calakmul Area in Mexico

Andrómeda Rivera and Sophie Calmé

INTRODUCTION

The endemic Mesoamerican black howler monkey (*Alouatta pigra*) is found in the southern states of Campeche, Quintana Roo, parts of Tabasco and northern Chiapas, in Mexico, northern Guatemala, and Belize (Horwich and Johnson, 1986). Although Mexico harbors about 80% of the geographic distribution of *A. pigra*, it is the least studied of the three primate species that exist in Mexico. Until now, studies on this species in Mexico have consisted primarily of population surveys (Estrada *et al.*, 2002a,b, 2004), and a single study about diet and activity pattern (Barrueta, 2003). In the case of Guatemala,

---

Andrómeda Rivera and Sophie Calmé • El Colegio de la Frontera Sur, Division of Conservation of Biodiversity, AP 424, Chetumal, Quintana Roo, México.

New Perspectives in the Study of Mesoamerican Primates: Distribution, Ecology, Behavior, and Conservation, edited by Alejandro Estrada, Paul A. Garber, Mary S. M. Pavelka, and LeAndra Luecke. Springer, New York, 2005.

population surveys (Coelho *et al.*, 1976 Baumgarten and Hernández, 2002; Estrada *et al.*, 2004), a study on seed dispersal (Ponce-Santizo, 2004) and another on fragmentation (Rosales-Meda, 2003) have been conducted. At present, almost all information on the species comes from Belize, in particular, regarding feeding behavior (Silver *et al.*, 1998, 2000; Pavelka and Knopff, 2004). The restricted distribution of *A. pigra* and the rapid fragmentation and conversion of its natural habitat to pasture lands and agricultural fields place populations of this primate species at risk (Estrada *et al.*, 2004).

This rapid loss of habitat associated with anthropogenic disturbance, such as logging and agriculture, is likely to have a significant impact on howler monkey feeding ecology and patterns of habitat utilization. In particular, the presence of distinct dry and rainy seasons is reported to influence leaf and fruit production by food trees. Typically, fruit production in tropical forests peaks in the late dry season or early rainy season (Janson and Chapman, 1999). This seasonality in food production affects primate behavior, impacting populations most strongly during times of resource limitation (Terborgh, 1986a).

A. pigra has been found to respond to variation in seasonal resource abundance by exploiting leaves from January to March and shifting to mostly fruits from April to July, which corresponds to the late dry season and the beginning of the rainy period (Pavelka and Knopff, 2004). Knowledge of the manner in which different primate species respond to seasonal changes in the availability and distribution of resources is critical for developing conservation and management policies. For example, if important feeding and refuge tree species are left standing in selective logging operations, population declines following logging are likely to be lower and/or the speed of recovery more rapid for those primate species requiring these resources (Chapman *et al.*, 2000).

Several studies have reported that the diet of howlers is comprised mainly of fruits, leaves, and flowers belonging to the Moraceae, Fabaceae, Sapotaceae, and Lauraceae families (Milton, 1980; Gaulin and Gaulin, 1982; Estrada, 1984; Julliot and Sabatier, 1993; Stoner, 1996; Estrada *et al.*, 1999). In fragmented habitats with introduced vegetation, they also consume exotic species like oranges, *Citrus sinensis* (Bicca-Marques and Calegaro-Marques, 1994) and mangos, *Mangifera indica* (Fuentes *et al.*, 2003). However, despite the fact that howlers are reported to consume a wide range of plant species, tradeoffs in the availability, distribution, and nutritional quality (i.e. ratio of protein to fiber and toxicity) of these resources suggest that commonly used resources are not necessarily the highest quality resources. In this regard, the words *use* and *selection* have been often applied interchangeably in the ecological literature (Litvaitis *et al.*, 1996). *Use* only indicates the consumption of food, whereas *selection* implies a choice among alternative foods that are available to the forager. Use is selective if components are exploited disproportional to their availability (Litvaitis *et al.*, 1996). Few primate studies have focused on food selection, i.e. using an index of selectivity (Sourd and Gautier-Hion, 1986; Julliot, 1996; McConkey *et al.*, 2002), and only one of them (Julliot, 1996) concerned a species of the genus *Alouatta*. Each of these studies demonstrated that monkeys are selective in their fruit choice using information on fruit color, fruit or seed size, amount of pulp, and water content in foraging decision.

Most forests in fragmented tropical landscapes offer both a reduced and disturbed space where monkeys are left with few opportunities to choose. If we succeed in understanding how black howler monkeys select the trees on which they feed, it should enable us to assess the quality of a fragmented and disturbed habitat for these monkeys. In this chapter, we compared the feeding ecology of five groups of black howler monkeys (*A. pigra*) existing in habitats that differ in land use patterns. Three troops of *A. pigra* were studied in the protected forest of the Calakmul Biosphere Reserve and two troops living in forest fragments managed as extractive reserves in community-owned land adjacent to the reserve. Specifically, we were interested in examining the effects of fragmentation on howlers' diet and determining the basis for selection of food trees using a selectivity index.

MATERIALS AND METHODS

Study Area

The Calakmul Biosphere Reserve (CBR) is located in the southeastern part of the state of Campeche in the municipality of Calakmul. It is bordered on the east by the state of Quintana Roo and on the south by Guatemala (17°45′–19°15′ N and 89°08′–90°08′ W; Figure 1). The Calakmul Biosphere Reserve protects the largest area of tropical forest in Mexico. It lies within the most important tropical forest region in North America and it forms part of the Mesoamerican Biological Corridor (Galindo-Leal, 1998).

The reserve covers 723,185 ha of largely homogeneous topography ranging in altitude from 260 to 385 m. It has two core areas, one of 147,915 ha in the southern portion where we worked, and another of 100,345 ha in the northern

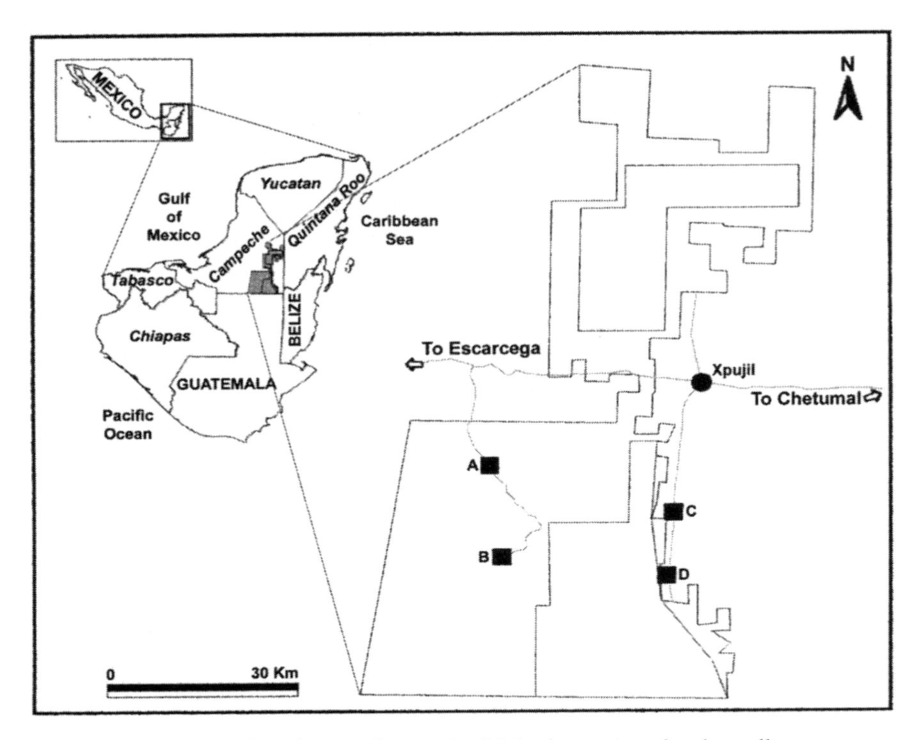

Figure 1. Location of study area (in grey) within the region that broadly encompasses black howler monkey's distribution range. Black squares on the main figure represent the location of the study sites within the study area of Calakmul, and letters refer to each site as illustrated in Figure 3; where A, El Sendero Ecológico; B, Mayan Ruins of Calakmul; C, Cristóbal Colón; and D, Once de Mayo.

portion. The buffer area covers the remaining 474,924 ha. Less than 4% of the buffer area is considered disturbed due to human activities. Surrounding the reserve are communal lands, called *ejidos*. Land use in these farming communities consists mainly of small agricultural plots of mixed crops, such as maize, squash, and beans. Some *ejidos*, especially on the southeastern edge of the CBR, also cultivate jalapeño pepper, and all have some cattle ranching (Klepeis and Roy Chowdhury, 2004). The remaining land cover is similar to CBR, i.e. tropical forest. The climate is warm subtropical, with a mean annual temperature of 22–26°C. Rainfall presents a north–south gradient. Annual precipitation ranges from 1200 to 1500 mm in the central portion of the Reserve, to 1500 to 2000 mm in the southern portion (García-Gil *et al.*, 2002). There are two well-marked seasons: the dry season is from December to May and the rainy

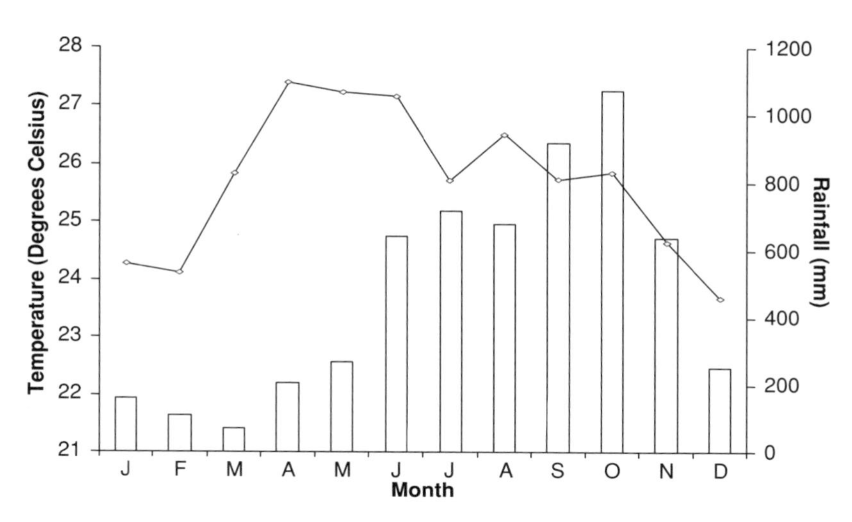

Figure 2. Temperature and rainfall collected by the National Water Commission in four sites at Calakmul for the period 1995–1999. Bars indicate mean rainfall, line indicates mean temperature.

season, which concentrates 81% of total rainfall, occurs from June to November (Figure 2).

Vegetation types in the reserve are: (1) tall semi-deciduous forests which reach heights of over 30 m, and cover a surface area of less than 10,000 ha; (2) medium semi-deciduous forests, which range in height from 15 to 25 m and cover the largest area of the reserve (480,000 ha); and (3) short semi-deciduous forests, with tree heights ranging from 4 to 15 m and covering an area of approximately 85,000 ha (Arriaga *et al.*, 2000).

Study Sites and Focal Troops

Two study sites were located within CBR; that served as controls. The first site is known as El Sendero Ecológico (18°18′58″N, 89°51′23″W), and is situated at 26 km south of the Highway Escárcega-Chetumal (Figures 1 and 3a). The vegetation is medium semi-deciduous forest and tall semi-deciduous forest. When we began the study in February 2003, the troop contained seven individuals (3 females, 1 male, and 3 juveniles). During the second half of the study one infant was born in the troop. The study troop was the only one observed in this site.

The area surrounding the Mayan Ruins of Calakmul (18°06′43″N, 89°48′12″W) is the second study site within CBR (Figures 1 and 3B). It is located 23 km south of the first site, in the center of the southern part of CBR and covers an area of 30 km^2 (Estrada *et al.*, 2004). The vegetation is mainly medium semi-deciduous forest. Estrada *et al.* (2004) report the presence of eight howler monkey troops at this site; we studied two of them in different months, depending on whether we could find the main troop. The main troop had seven individuals at the beginning of the study (2 females, 2 males, 2 juveniles, and 1 infant) and eight individuals by the end (one additional newborn), and the second troop had four individuals (1 female, 1 male, and 2 juveniles).

Two other study sites were located outside CBR. The first is in the farming community *ejido* Cristóbal Colón and is a forest fragment 13.9 ha in size (18°11′25″N, 89°26′12″W; Figures 1 and 3C). The second site is located in the farming community *ejido* Once de Mayo, and consists of a forest fragment 11.6 ha in size (18°07′10″N, 89°27′03″W; Figures 1 and 3D). The distance between these fragments is 8 km. Medium semi-deciduous forest is the predominant vegetation in these sites and both are surrounded by crops (Figure 3). Both forest fragments have trails that people use for timber extraction, which is carried out for domestic purposes such as house construction or maintenance, usually during the dry season when access with a vehicle is possible. In Once de Mayo, the troop had seven individuals at the beginning of the study (2 males, 3 females, and 2 juveniles) and by April one infant was born. In Cristóbal Colón, the troop consisted of four individuals at the beginning of the study (1 male, 2 females, and 1 infant), and one infant was born during the study. Additionally, one adult male immigrated into the troop sometime around June and by December this male was gone. In each of these sites, only one troop was present.

Feeding Behavior

The study was conducted in 2003, from February to May (dry season), and August to November (rainy season). Observations were done over two consecutive days each month, on each study troop, totaling 64 days (3152 records). We observed the monkeys a mean 4.2 ± 0.6 hours per day. To document feeding habits, we used the instantaneous scan-sampling method, recording at 15-minute intervals the activity (feeding, resting, traveling, playing, or vocalizing) displayed by each monkey of the focal troop at the moment they

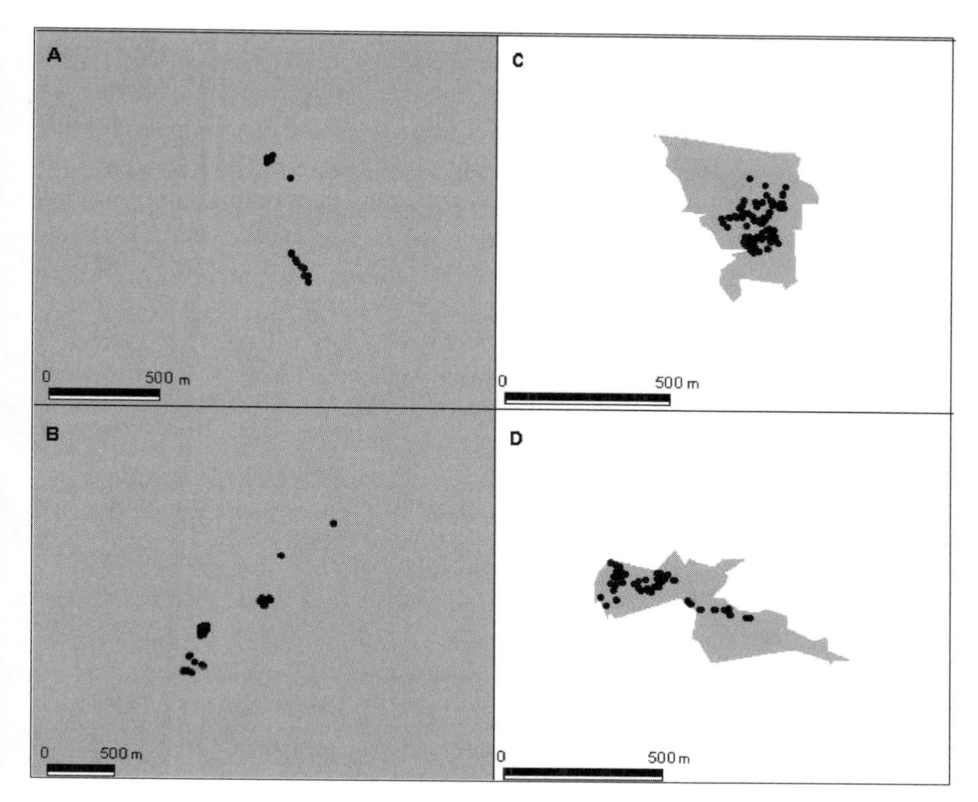

Figure 3. Land cover of the study sites and surroundings (grey represents forest and white represents agriculture). A, El Sendero Ecológico; B, Mayan Ruins of Calakmul; C, Cristóbal Colón; and D, Once de Mayo. Black dots represent locations where howlers were recorded eating and moving from one feeding tree to another.

were observed. If the activity was feeding, we recorded the species and plant part eaten. We chose to use the scan-sampling method because it is well-suited for non-social behavior observations (Altmann, 1974), and it enabled us to collect a large sample of behavioral data. Mitlöhner *et al.* (2001), working with heifers, argued that data collected across 15-min intervals were sufficient to ensure some degree of statistical independence, especially for feeding events, while capturing the whole spectrum of behavior.

Vegetation Sampling

To evaluate how trees used by monkeys for feeding differed from neighboring trees they do not use, we established plots (hereafter named focal plots) of 10 m radius around the focal trees. Plots never overlapped using a 10 m radius, which avoided including the same trees, and thus problems of spatial dependency. We classified trees into three categories: (1) focal trees, where the majority of the monkeys of a given focal troop were feeding; (2) used trees, i.e. trees on which a minority of monkeys of a focal troop were feeding; and (3) non-used trees, i.e. trees not used for feeding. Within each plot, we determined the diameter at breast height (dbh), height, species, and phenology (ripe or unripe fruits, young or mature leaves, leafless, flowers, and buds) of all trees >10 cm dbh. Additionally, we conducted a vegetation census using similar 10-m-radius plots established at random 100–300 m around the areas used by focal troops, in forest stands* where no howler monkey had been observed. We only determined the species of all trees >10 cm dbh, and used this census to assess whether tree species selection is at the stand level by comparing the composition and abundance of tree species between these plots and focal plots.

Statistical Analyses

For the purpose of the present study, we analyzed data corresponding to only adult howlers to avoid any possible age-related bias. To estimate and compare howler monkey's diet composition in each type of forest (continuous, Figures 1A and 1B versus fragments, Figures 1C and 1D), we evaluated the relative percentage of consumption of each species by dividing the number of feeding records of a given species by the total number of feeding records. We did not relate these data to the number of available trees, as we considered that one single tree could provide unlimited resources to a given troop, provided that the tree part they consumed was present (e.g. fruits). For further comparisons between forest fragments and the reserve, we selected only the species that represented >10% of the total number of feeding records, and performed independent likelihood ratio chi-square tests (*G*-tests).

To compare the consumption of the different food items between the reserve and the fragments, we also performed independent *G*-tests. For these analyses,

* Forest stand: A community of trees possessing sufficient uniformity in composition, age, arrangement or condition to be distinguishable from the forest or other growth on adjoining areas, thus forming a temporary silvicultural or management entity. Silvaterm Database, International Union of Forest Research Organizations http://iufro.boku.ac.at/iufro/silvavoc/svdatabase.htm

we selected those species that (1) had various food items consumed, and (2) were used as food sources in both types of forest.

To assess food preference, we compared the proportion of a given tree species in the diet with the proportion of that tree species available in focal plots. We measured diet selection by the howlers using the electivity index (ε) presented by Chesson (1983). The major advantage of this measure of preference is that it is not influenced by food density (i.e. trees in our case), because it is standardized. The Chesson index is based on Manly's alpha selection index (α), which allows to rank plants in order according to frequency in the diet:

$$\alpha_i = \frac{r_i/n_i}{\sum_{j=1}^{m}(r_j/n_j)}$$

where r_i and r_j are the proportions of the tree species i and j, respectively, in the diet; n_i and n_j are the proportions of the tree species i and j, respectively, available in focal plots; and m is the total number of tree species.

Manly's alpha is applicable in situations where the diet plant population can be assumed not to be significantly depleted by feeding activity (Manly, 1974; Chesson, 1983). To obtain results that are comparable between cases in which the number of available tree species varies, we converted Manly's alpha to the selectivity index presented by Chesson (1983):

$$\varepsilon_i = \frac{m\alpha_i - 1}{(m-2)\alpha_i + 1}$$

Chesson's ε potentially ranges between -1 and $+1$. Plant species having negative values are avoided[†] and species with positive values are preferred (Chesson, 1983). If the index value is zero, this represents non-selective feeding on that plant species. We computed the selectivity value for each of the 10 species used for feeding in the CBR and the 16 species used for feeding in the fragments.

To determine if there were structural differences between feeding trees and non-used trees, we compared the dbh and height of the three categories of trees (focal, used, and non-used), using a Tukey–Kramer test for multiple comparisons.

Finally, we compared the frequency distributions of the tree species found in the focal and random plots (where the monkeys did not feed) using the

[†] Avoid: We use this term *sensu* Chesson (1983), who defines that those species less present in the diet than their availability would allow, are avoided.

Kolmogorov–Smirnov two-sample test, after correcting for the unequal number of plots. Then, we selected the most-consumed tree species (>10% of total consumption), and we compared their abundances in focal and random plots using *G*-tests.

RESULTS

Comparison of Diet Composition

In total, 20 tree species were used as sources of food by the howlers, in both the CBR and the forest fragments. Sixteen of these species were consumed in the forest fragments and 10 in CBR (Table 1). Eight out of 10 species consumed in CBR were also a source of food in the fragments.

Table 1. Tree species used for feeding by *Alouatta pigra* at CBR (two sites) and the forest fragments (two sites)

Species	Occurrences (%)			
	Fragments	CBR	*G*	*p*
Aspidosperma megalocarpon (Apocynaceae)	0.8	0.0	1.09	0.295
Brosimum alicastrum (Moraceae)	24.9	31.6	0.80	0.372
Bursera simaruba (Burseraceae)	0.0	1.8	2.52	0.112
Caesalpinia mollis (Leguminosae)	3.1	0.9	1.32	0.250
Celtis trinervia (Ulmaceae)	–*	1.8	–	–
Coccoloba acapulcensis (Polygonaceae)	1.8	1.3	0.08	0.781
Croton arboreus (Euphorbiaceae)	0.3	–	–	–
Eheretia tinifolia (Boraginaceae)	8.7	0.0	12.00	<0.001
Ficus sp (Moraceae)	–	50.5	–	–
Krugiodendron ferreum (Rhamnaceae)	0.0	0.9	1.26	0.262
Lonchocarpus xuul (Leguminosae)	0.5	0.0	0.73	0.392
Manilkara zapota (Sapotaceae)	22.0	8.4	6.31	0.012
Neea choriophylla (Nyctaginaceae)	0.8	0.0	1.09	0.295
Platymiscium yucatanum (Leguminosae)	7.9	2.2	3.30	0.069
Protium copal (Burseraceae)	0.3	0.0	0.42	0.519
Sideroxylon salicifolium (Sapotaceae)	2.4	0.0	3.29	0.070
Tabebuia chrysanta (Bignonaceae)	3.1	–	–	–
Talisia olivaeformis (Sapindaceae)	17.1	0.5	20.10	<0.001
Vitex gaumeri (Verbenaceae)	1.8	–	–	–
Unknown	4.2	0.0	5.85	0.016
Number of consumed species	16	10	1.40	0.237

*A dash in place of a value for occurrence indicated that the species was not present in the focal plots; if the species was not used for feeding but present, the corresponding value was 0. The likelihood ratio chi-square (*G*) is the result of the comparison of the frequency of consumption between fragments and the reserve. The associated probability is noted as *p*.

In CBR, feeding records totaled 440; *Ficus* sp. and *Brosimum alicastrum* contributed approximately 82% of the feeding records. In the fragments, feeding records totaled 381, with *B. alicastrum*, *Manilkara zapota*, and *Talisia olivaeformis* contributing 64% of all feeding records (Table 1).

We found no significant difference in the overall consumption (fruits and leaves) of *B. alicastrum* between fragments and the reserve. This species was the most consumed in fragments and the second most consumed in CBR, contributing to 24.9% and 31.6% of the total consumption, respectively (Table 1). We found significant differences in the consumption of *M. zapota* and *T. olivaeformis* between fragments and CBR. Both the species contributed significantly more to the diet in the forest fragments. Finally, the most important species as source of food (50% of total consumption) in the CBR, *Ficus* sp., was absent in the forest fragments.

Comparison of Tree Parts Consumed

Among the 20 species used for feeding by black howlers, only *B. alicastrum* and *M. zapota* were consumed in both types of forest and for each species a range of different food types were consumed. We found that there were no significant differences in the consumption of young leaves, matures leaves, and fruits of *B. alicastrum* between CBR and the forest fragments (Table 2). However, young leaves represented half the consumption of this species in the fragments, whereas consumption of young leaves (36.7%), mature leaves (28%),

Table 2. Consumption of leaves and fruits of *B. alicastrum* and *M. zapota* in the CBR (two sites) and the forest fragments (two sites). The likelihood ratio chi-square (G) is the result of the comparison of the frequency of consumption between fragments and the reserve. The associated probability is noted as p

	Consumption (%)			
Tree part	Fragments	CBR	G	p
Brosimum alicastrum				
Young leaves	49.47	36.69	1.90	0.170
Mature leaves	29.47	28.06	0.03	0.850
Fruit	21.05	35.25	3.62	0.060
Manilkara zapota				
Young leaves	14.29	40.54	13.10	<0.001
Mature leaves	<0.01	21.62	29.97	<0.001
Fruit	85.71	37.84	19.05	<0.001

and fruits (35.2%) of *B. alicastrum* were almost equally represented in the diet at CBR. For *M. zapota*, howlers ate significantly more young and mature leaves in CBR than in the forest fragments. Conversely, they ate significantly more *Manilkara* fruits in the fragments than in CBR (Table 2).

For both *B. alicastrum* and *M. zapota*, we found that howlers always ate ripe fruits whenever they were available. In the case of *M. zapota*, we registered 14 and 72 feeding records on ripe fruits in the CBR and the forest fragments, respectively; and only in two cases, howlers also ate young leaves. There was no difference in the consumption of unripe fruits of *Manilkara* between CBR and the fragments ($G = 1.03$, $p = 0.3$). However, in both the forest fragments and the CBR, howlers ate significantly more ripe than unripe *Manilkara* fruits (both $G \geq 4.8$, $p \leq 0.02$). We also found that they consumed mature leaves of this species only when fruits were unavailable or immature.

Unlike the case of *M. zapota*, black howlers ate significantly more unripe fruits of *B. alicastrum* in the CBR than in the fragmented sites ($G = 11.3$, $p < 0.001$). However, as for *Manilkara*, howlers ate significantly more ripe than unripe fruits of *B. alicastrum* in both types of forest (both $G \geq 27.7$, $p < 0.001$). In fact, in the forest fragments, when ripe fruits were available, black howlers never ate unripe fruits of *B. alicastrum*. We also found that in fragments they consumed mature and young leaves of this species only when ripe fruits were unavailable. In the CBR, however, black howlers also ate mature or young leaves as well as ripe fruits of *B. alicastrum*.

Dietary Selectivity in Fragments and CBR

We computed the Chesson's electivity index for the 20 tree species that were consumed both in the fragments and the CBR, and only if species were present in the focal plots. We omitted the species not used by monkeys for feeding, as we had no local evidence of their edibility.

Howlers in the forest fragments fed selectively on 12 of the 16 species they consumed, while *T. olivaeformis* and *Caesalpinia mollis* were not used selectively. *Lonchocarpus xuul*, *Protium copal*, *Bursera simaruba*, and *Krugiodendron ferreum* were avoided (Figure 4), although the former two, which represented only 0.8% of the feeding records, were consumed.

Of the 10 tree species used as a source of food in the CBR, 9 species were selected (i.e. had positive electivity indices). Eight species were negatively selected, one of which was consumed however (*T. olivaeformis*; Figure 4).

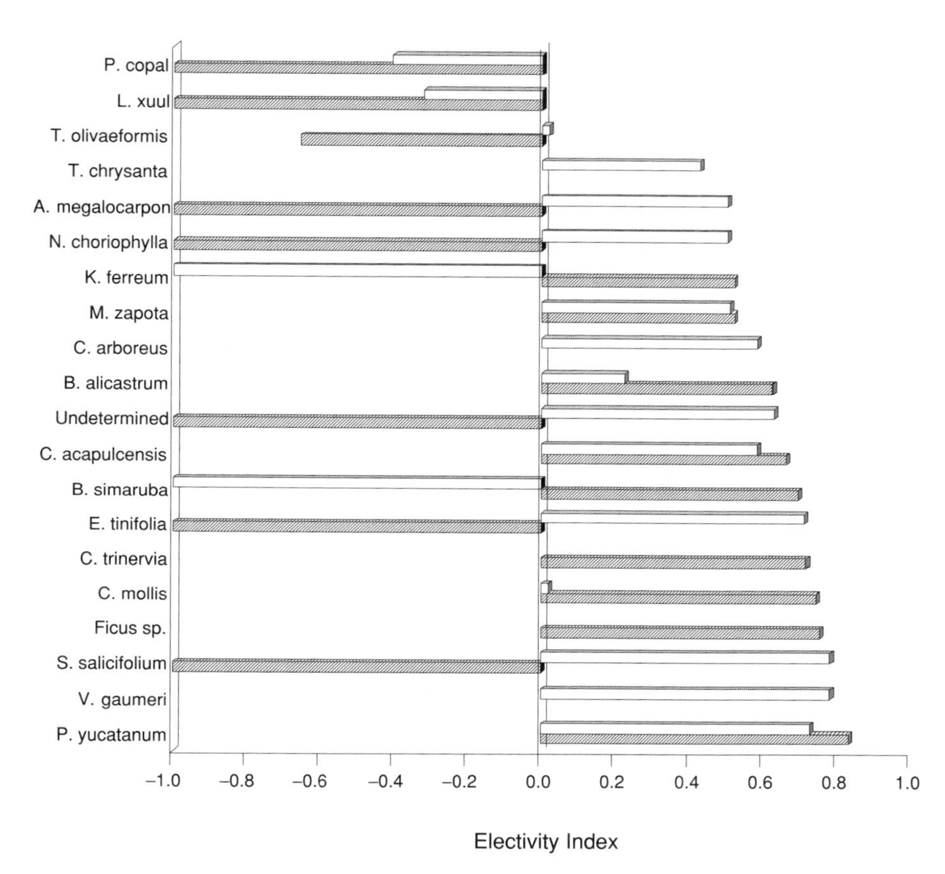

Figure 4. Electivity values (Chesson's ε) for the species used as food sources by black howler monkeys in the Calakmul Biosphere Reserve (hatched bars) and forest fragments (white bars). The more howlers select a species, the higher the value (maximum value is 1). Complete avoidance is denoted by −1, while 0 represents random selection.

Characteristics of Consumed and Non-consumed Trees

In the CBR and the forest fragments, heights of focal and used trees were similar between and among sites. However, in these forests, non-used trees were significantly shorter (by 4–6 m) than both focal and used trees (Tukey–Kramer multiple comparisons, $p < 0.05$ in all the cases; Table 3).

Among the fragments, dbh of focal and used trees also were similar, and greater (by ≥13 cm) than non-used trees (T–K multiple comparison, $p < 0.05$). At CBR, focal trees were greater in diameter than used trees (46 cm), whereas used trees were larger (25 cm) than non-used trees (T–K multiple comparison, $p < 0.05$). Both focal and used trees at CBR were also larger than their

Table 3. Characteristics of used and non-used trees within the Calakmul Biosphere Reserve and in the forest fragments outside the reserve

Tree category	*N*	Height (m)					Diameter at breast height (cm)				
		Mean	SD	T–K	Median	Range	Mean	SD	T–K	Median	Range
Fragments											
Focal tree	21	17.8	3.9	a	18	[14–26.5]	35.2	15.4	a	32	[22.5–65.5]
Used tree	51	16.5	2.9	a	16	[11–27.8]	33.0	14.8	a	31	[12–69]
Non-used tree	488	12.6	3.6	b	12	[6–29]	20.1	9.7	d	17	[10–80.2]
CBR											
Focal tree	24	18.0	4.5	a	18	[10–30]	95.2	91.6	b	54	[17–357]
Used tree	56	16.4	3.5	a	17	[10–25]	48.7	40.1	ac	35	[11–196]
Non-used tree	261	12.0	4.1	b	11	[4–25]	23.3	15.8	ad	19	[10–143]

Table 4. Comparisons of the distribution of frequency of tree species between focal and random plots using the two-sample Kolmogorov–Smirnov test (statistics D). All plots had a 10 m radius

Site	D	P
Fragments		
Once de mayo	0.2449	0.006
Cristóbal Colón	0.1633	0.147
CBR		
Sendero	0.3367	<0.001
Mayan Ruins of Calakmul	0.1429	0.270
All sites	0.2347	0.009

counterparts in the fragments by 60 and 15.7 cm, respectively (significant only for focal trees). On the other hand, non-used trees had similar diameters in the reserve and in the fragments (Table 3).

Tree Species Composition and Abundance in Focal and Random Plots

We counted a mean number of 20.5 trees larger than 10 cm at breast height in the focal plots, and similarly, 18.5 trees in the additional random plots. Considering both focal and random plots, there were a total of 74 tree species in CBR, and 57 in the fragments. Overall, the frequency distribution of tree species in focal and random plots was significantly different (Table 4). However, random plots were different from focal plots only in the ejido Once de Mayo, outside the reserve, and in El Sendero Ecológico, within the reserve.

B. alicastrum was more abundant in the focal plots than in the random plots, but this difference only approached statistical significance ($G = 3.32$, $p = 0.068$). *Ficus* sp. and *M. zapota* were significantly more abundant in the focal plots (both $G > 6.8$, $p \leq 0.009$). For *Ficus*, the analysis was done only for the sites in the reserve, as no *Ficus* tree was detected in the forest fragments. The fourth species most used for feeding in the fragments, *T. olivaeformis*, was equally abundant in focal and random plots in these sites ($G = 0.06$, $p = 0.79$).

DISCUSSION

Feeding Habits in CBR and the Forest Fragments

The diet of black howler monkeys in the protected and extensive forest of CBR and forest fragments outside CBR differed in the number of tree species they used as sources of food, and in dietary composition. For instance, in the extensive forest, 50% of feeding time was devoted to a single species, *Ficus* sp., while the most frequent species in the diet of howlers in the fragments, *B. alicastrum*, represented only one quarter of the total consumption. Thus, howlers were found to have a more narrow-based diet in the reserve, whereas their diet was more broad-based in the fragments. Several factors could account for this, including the fact that *Ficus* sp. was very rare in the fragments (and absent from all plots), and that when available, *Ficus* fruits and leaves are reported to be major food resources for howlers at other sites in Mesoamerica (Silver *et al.*, 1998; Estrada *et al.*, 1999; Serio-Silva *et al.*, 2002; Pavelka and Knopff, 2004).

Our results on diet composition and diversity differ from those of Silver *et al.* (1998) in the Community Baboon Sanctuary (CBS) in Belize. At this site, black howlers are reported to feed on 53 tree species (of a total of 60 tree species identified), with no single species accounting for more than 12.5% of feeding time, although all *Ficus* species together accounted for 31% of feeding time. The top five tree species consumed at CBS contributed 42.8% of feeding time, far from the 80.6% and 96.3% of the top five species consumed in fragments and CBR, respectively, in this study. Black howlers in CBS thus appear to exploit a highly broad-based diet. The habitat at CBS is more disturbed than even the forest fragments in our study, as suggested by the presence of pioneer trees like *Cecropia* spp., and exotics like *C. sinensis.* On the other hand, black howlers of El Tormento, a large managed forest fragment (1400 ha) in Southwestern Campeche, are reported to exploit 19 tree species (Barrueta, 2003), similar to what we report here for fragments.

Moreover, as in CBR and forest fragments in Calakmul, *M. zapota* and *B. alicastrum* also were important species in the diet of howlers at El Tormento. These two species were absent from CBS forest (Silver *et al.*, 1998). Our study suggests that, when possible, howlers may limit their diet to a small number of particular plant species present in sufficient abundance and available over a large time span (and probably highly palatable). In the case of *Ficus*, for example, a single tree may produce a sufficiently large fruit and leaf crop to feed a group for several months. Serio-Silva *et al.* (2002) previously suggested that mantled howlers (*A. palliata*) may concentrate their feeding activities on a single or small number of *Ficus* trees. Such a pattern may reduce time and energy spent traveling, permitting more time and energy for the digestion of a high fiber diet. This is consistent with an energy-minimizing foraging strategy (Milton, 1980).

We hypothesize that in the fragments, howlers compensated for the lack of *Ficus* by eating more *Manilkara* fruits plus "alternative" species, all of them consumed significantly more in the fragments (Table 1). Three of these four species, *Eheretia tinifolia*, *T. olivaeformis*, and the unknown species, were also selected in the fragments, whereas they were strongly avoided in CBR (Figure 4). The four species accounted for 52.1% of howlers feeding time in the fragments. This was not significantly different from the value of 50.5% that *Ficus* represented in the diet of howlers in the reserve ($G = 0.57$, $p = 0.45$), suggesting that howlers might respond to the absence of figs by selecting alternative species and increasing the consumption of other species already present in the diet. The exploitation of *Ficus*, *Brosimum*, and *Manilkara* appear to represent staple resources for many howler populations (Estrada *et al.*, 1999; García del Valle, 2001; Barrueta, 2003). Factors affecting the use of "alternative" species remain unclear and require analyzing the nutritional quality and phytochemical components of these resources.

Howlers ate significantly more ripe fruits than unripe fruits and leaves of *M. zapota* and *B. alicastrum* in both the forest types. Thus, ripe fruits appeared to be a preferred food type when available. In forest fragments, however, leaf consumption of *B. alicastrum* was correlated with a decrease in fruit availability. A similar relationship was found when consuming mature leaves of *M. zapota* in both the fragments and the reserve. Our results are in agreement with that of Silver *et al.* (1998), who mentioned that mature leaves may be secondary or supplemental choices to howlers. These results are consistent with Yeager's (1989) suggestion that increased dietary diversity is associated with food scarcity.

We expected leaf and unripe fruit consumption of *B. alicastrum* and *M. zapota* to be higher in the fragments, because we assumed that trees with ripe fruits should be less numerous in a reduced and disturbed space and both tree species are subject to extraction in the forest fragments. However, we found no differences in the relative consumption of young and mature leaves of *B. alicastrum* between fragments and CBR. In addition, unripe fruits were consumed more in CBR. For *M. zapota*, the relative consumption of young and mature leaves was higher in CBR, and unripe fruit consumption was similar between the two types of forest. In contrast with our results, Fairgrieve and Muhumuza (2003) found that overall consumption of unripe fruits by *Cercopithecus mitis* was higher in a disturbed (logged) forest in Uganda. More detailed analysis of the relationship between habitat disturbance and its effects on forest composition and primate behavior are needed to clearly understand changes in dietary food items.

Food Selection in CBR and the Forest Fragments

If howlers did not behave selectively in their dietary choice, we should expect diet diversity to be broadly similar to that of tree species diversity in the studied sites. Thus, howlers' diet is expected to be more diverse in CBR than in the fragments, as 74 species were identified in the vegetation plots in CBR compared to 57 in the fragments. However, of the 20 species that were used for feeding in both types of forests, 16 species were consumed in the forest fragments and 10 were consumed in the CBR. This is best explained by the predominance of figs (*Ficus* sp.) in howlers' diet in CBR, which was highly selected and represented half of all consumption events. In fact, many fruit-eating primates are reported to preferentially consume figs even when other food is abundant (O'Brien *et al.*, 1998). Figs also have been suggested as keystone resources in tropical forests (Terborgh, 1986b), and in playing a special role in howler conservation (Coates-Estrada and Estrada, 1986; Milton, 1991; Serio-Silva *et al.*, 2002).

In general, most species present in the diet were highly selected. Our data also show that 9 of the 10 species consumed in CBR, and 12 of the 16 species consumed in the fragments were preferentially selected by the howlers. However, it was striking that some species reported to be commonly eaten by howlers at other sites (e.g. Estrada, 1984; Serio-Silva, 1992; Julliot, 1996; Silver *et al.*, 1998; Barrueta, 2003) were not seen being consumed by black howlers during the present study. This was the case of *L. castilloi*, *P. campechiana*, *Spondias mombin*, *Hampea tribolata*, and *Chrysophyllum mexicanum*. Species of the genera *Drypetes*, *Piscidia*, *Pouteria*, *Guettarda*, *Diospyros*, and *Trichilia*, which are known to be consumed by other species of *Alouatta* (Milton, 1980; Julliot, 1996), though present in the fragments and in the reserve also were not consumed. There are several explanations to account for this, first the fact that we did not observe howlers feed on these species does not necessarily mean that they were not consumed. Second, perhaps our observations did not correspond to the period of maximum fruiting of these species (e.g. *P. piscipula* and *Trichilia minutiflora*), and third, in the case of CBR, the extended and asynchronous pattern of fig fruiting and leafing may have enabled the howlers to exploit this species throughout much of the year as it occurred at high enough densities (O'Brien *et al.*, 1998).

Within suitable habitats, the quality of different plants is probably the main factor leading to diet selection (Markkola *et al.*, 2003). The quality of food varies according to energy, protein and water content, soluble carbohydrates, digestibility and toxicity, and several studies on howler diet indicate that selection appears to be based on phytochemical factors rather than the relative availability of potential food (Milton, 1980; Silver *et al.*, 2000). Chemical analyses on the plant species and food items consumed by black howlers in Calakmul would provide greater insight into the criteria used in food selection.

Selection of Feeding Sites at Tree and Stand Levels

The selectivity analysis clearly indicates that howlers are not feeding on tree species based on their availability, but have marked preferences for some species and aversions to others. In general, we found that tree species used by howlers as sources of food were taller and larger than non-used species in both the forest fragments and the reserve. Used trees were usually dominant trees in the canopy, and were either older trees or individuals of species reaching larger dimensions. Non-used trees were usually components of the under-canopy. In fact, all the abundant species clearly avoided by howlers (i.e. abundant in focal plots but not consumed, such as *Drypetes lateriflora*, *Pouteria reticulata*, and *T. minutiflora*) were very small and barely reached the mean values for height and dbh of non-used trees. Other authors who have studied *A. pigra* (Barrueta, 2003) or *A. palliata* (Estrada and Coates-Estrada, 1986; García del Valle, 2001) have found similar mean heights and diameters of the feeding trees in their study sites; however, they did not provide a basis for comparison as they did not measure non-used trees.

In terms of forest management, the characteristics of used trees have clear implications for the conservation of howlers. In effect, selective logging implies the removal of individuals of commercial tree species above a minimum diameter. Most species on which howlers fed are used commercially (e.g. *B. alicastrum*, *C. mollis*, *E. tinifolia*, *M. zapota*, and *P. yucatanum*). Logging decreases the densities of these trees and increases the openness of the canopy from less than 5% to more than 30% in this type of forest (Dickinson, 1998). Thus, in managed forests, howlers have to face both the lack of continuity in the canopy that makes movements more difficult and the loss of many vital trees for feeding. In the forest fragments we studied, the diameter of focal trees was smaller than in the reserve, even after excluding *Ficus* trees from the analyses (mean $dbh_{fragment} = 35.2 \pm 15.4$ cm versus mean $dbh_{reserve} = 45.0 \pm 18.2$ cm). This suggests that the small-scale logging for domestic purposes that local inhabitants practice has an important impact on the trees howlers use for feeding. Chapman *et al.* (2000) reported that a reduction of food availability due to logging leads to increased infant and juvenile mortality in species such as *Macaca sinica* and *Papio cynocephalus*. Therefore, it is necessary to understand how the effects of vegetation changes commonly associated with logging influence primate feeding ecology (Fairgrieve and Muhumuza, 2003); this will help in the implementation of management plans based on conservation of howler food trees.

We also were interested in determining if the forest stands in which howlers were feeding differed from surrounding stands where howlers had not been observed, in order to evaluate the spatial scale at which they select feeding trees. However, focal plots differed from random plots in only two of the four study sites, one within the reserve and the other in a forest fragment, making conclusions unclear. Nevertheless, the four species that accounted for more than 75% of the feeding records overall provide interesting insight. Actively selected species such as *Ficus*, *M. zapota*, and to a lesser extent *B. alicastrum*, were more abundant in the focal plots than in random plots. In contrast, *T. olivaeformis* was equally abundant in focal and random plots, but was not a selected species in fragments and was avoided in the reserve. This suggests that selection at the stand level might be linked to the presence/abundance of the preferred tree species.

In conclusion, black howler monkeys showed strong selection for the consumption of plant food of particular tree species. Forest fragmentation may have served to relax the degree of dietary selectivity in howlers. Howlers depended on a small number of tree species, the lowest ever reported for *A. pigra*, and this was relatively independent of their availability. In fact, the second most abundant tree species in this study, *P. reticulata*, was never used as a source of food by howlers. However, this species is of low stature and small diameter. Black howlers selected the trees on which they fed based at least partially on their stature, dbh, and maturity. Moreover, areas they selected to spend most of their time contained higher abundances of their preferred feeding tree species. As tree species selected by howlers also are commonly exploited, these findings provide a good basis for establishing criteria useful in forest management plans compatible with the conservation of *A. pigra*.

SUMMARY

Information on food selection and feeding habits is critical for species conservation, particularly in the context of forest landscapes heavily transformed by human activities. In this study, we examined the degree to which *A. pigra* feeding habits differed between two sites in the conserved forest of the Calakmul Biosphere Reserve (CBR) and two forest fragments outside the reserve, and how monkeys select the trees on which they feed. Our results suggest that howlers tended to exploit a smaller set of fruit and leaf species in the conserved sites, whereas their diet was more diverse in the fragments. This can be explained probably by the role, at CBR, of *Ficus* sp. in howlers' diet, as it was highly selected and represented half of all feeding events. Chesson's electivity index showed that howlers in fragments selected 12 of the 16 species used for feeding; while in CBR, they selected 9 of the 10 species used for feeding. At both types of forests, feeding trees were taller and were greater in diameter than non-feeding trees. Trees exploited by howlers for feeding have commercial dimensions and most of these species are commercially logged. As a result, in fragments and logged forests, howlers have to face the lack of continuity in the canopy and the loss of many vital trees for feeding. We expect howler monkeys to survive in fragmented sites if tree species important in their diet are conserved.

ACKNOWLEDGMENTS

This study would not have been possible without the help in the field of Oscar Fernando Ramírez, Roberto "Chibebo" Rojo, Johannes Hechenbichler, Rubén Aguilar, Luciano Pérez, Margarito Tuz, and Andrés Sánchez. The identification work of nearly 1000 trees was realized by Demetrio "Coyote" Álvarez. We are indebted to the people of the *ejidos* Once de Mayo and Cristóbal Colón for granting us access to their land. We are also grateful to the Direction of Calakmul Biosphere Reserve for allowing us to study howler monkeys and SEMARNAT for the plant collection license No./SGPA/DGVS/02475. Earlier drafts of this manuscript benefited from helpful comments from Gabriel Ramos Fernández, Jorge León Cortés, Alejandro Estrada, Mary Pavelka, Paul Garber, and an anonymous reviewer. Figures 1 and 3 were produced by Holger Weissenberger. Financial and logistic support for the project was provided by

ECOSUR. AR benefited from a scholarship granted by CONACYT (contract number 169607).

REFERENCES

Altmann, J. 1974, Observational study of behavior: Sampling methods. *Behaviour* 49:227–267.

Arriaga, L., Espinoza, J. M., Aguilar, C., Martínez, E., Gómez, L., and Loa, E. (Coordinators). 2000, Regiones terrestres prioritarias de México. Comisión Nacional para el Conocimiento y Uso de la Biodiversidad, México.

Barrueta, T. 2003, Reconocimiento demográfico y dieta de Alouatta pigra en un fragmento de selva en El Tormento, Campeche, México. MSc Thesis, El Colegio de la Frontera Sur, México, p. 49.

Baumgarten, A. and Hernández, J. F. 2002, Características poblacionales y uso de hábitat del mono aullador negro (*Alouatta pigra*) en la zona de influencia del Parque Nacional Laguna Lachua, Alta Verapaz. Revista Científica del Instituto de Investigaciones Químicas y Biológicas, Universidad de San Carlos de Guatemala 15:1–9.

Bicca-Marques, J. C. and Calegaro-Marques, C. 1994, Exotic plant species can serve as staple food sources for wild howler populations. *Folia Primatol.* 63:209–211.

Chapman, C. A., Balcomb, S. R., Guillespie, T. R., Skorupa, J. P., and Struhsaker, T. T. 2000, Long-term effects of logging on african primate communities: A 28-year comparison from Kibale National Park, Uganda. *Conserv. Biol.* 14:207–217.

Chesson, J. 1983, The estimation and analysis of preference and its relation to foraging models. *Ecology* 64:1297–1304.

Coelho, A. M., Bramblett, C. A., Quick, L. B., and Bramblett S. S. 1976. Resource availability and population density in primates: A socio-bioenergetic analysis of the energy budgets of Guatemalan howler and spider monkeys. *Primates* 17:63–80.

Coates-Estrada, R. and Estrada, A. 1986, Fruiting and frugivores at strangler fig in the tropical rain forest of Los Tuxtlas, Mexico. *J. Trop. Ecol.* 2:349–357.

Dickinson, M. B. 1998, Tree regeneration in natural and logging canopy gaps in a semideciduous forest. PhD Thesis, Florida State University, p. 177.

Estrada, A. 1984, Resource use by howler monkeys (*Alouatta palliata*) in the rain forest of Los Tuxtlas, Veracruz, Mexico. *Int. J. Primatol.* 5:105–131.

Estrada, A. and Coates-Estrada, R. 1986, Use of leaf resources by howling monkeys (*Alouatta palliata*) and leaf-cutting ants (*Atta cephalotes*) in the tropical rain forest of Los Tuxtlas, Mexico. *Am. J. Primatol.* 10:51–66.

Estrada, A., Juan-Solano, S., Ortiz, T., and Coates-Estrada, R. 1999, Feeding and general activity patterns of a howler monkey (*Alouatta palliata*) troop living in a forest fragment at Los Tuxtlas, Mexico. *Am. J. Primatol.* 48:167–183.

Estrada, A., Luecke, L., Van Belle, S., Barrueta, E., and Rosales, M. 2004, Survey of black howler (*Alouatta pigra*) and spider (*Ateles geoffroyi*) monkeys in the Mayan Sites of Calakmul and Yaxchilan, Mexico and Tikal, Guatemala. *Primates* 45:33–39.

Estrada, A., Luecke, L., Van Belle, S., French, K., Muñoz, D., García, Y., Castellanos, L., and Mendoza, A. 2002a, The black howler monkey (*Alouatta pigra*) and spider monkey (*Ateles geoffroyi*) in the Mayan Site of Yaxchilán, Chiapas, Mexico: A preliminary survey. *Neotrop. Primates* 10:89–95.

Estrada, A., Mendoza, A., Castellanos, L., Pacheco, R., Van Belle, S., García, Y., and Muñoz, D. 2002b, Population of the black howler monkey (*Alouatta pigra*) in a fragmented landscape in Palenque, Chiapas, Mexico. *Am. J. Primatol.* 58:45–55.

Fairgrieve, C. and Muhumuza, G. 2003, Feeding ecology and dietary differences between blue monkey (*Cercopithecus mitis stuhlmanni* Matschie) groups in logged and unlogged forest, Budongo Forest Reserve, Uganda. *Afr. J. Ecol.* 41:141–149.

Fuentes, E., Estrada, A., Franco, B., Magaña, M., Decena, Y., Muñoz, D., and García, Y. 2003, Reporte preliminar sobre el uso de recursos alimenticios por una tropa de monos aulladores, *Alouatta palliata*, en El Parque La Venta, Tabasco, México. *Neotrop. Primates* 11:24–29.

Galindo-Leal, C. 1998, El mal congénito de Calakmul. *Biodiversitas* 17:9–15.

García del Valle, Y. 2001, Estudio preliminar de los patrones de alimentación de monos aulladores (*Alouatta palliata*) en semilibertad en el parque Yumka', Tabasco, México. BSc Thesis, Universidad Juárez Autónoma de Tabasco, México, p. 70.

García-Gil, G., Palacio-Prieto, J. L., and Ortíz-Pérez, M. A. 2002, Reconocimiento geomorfológico e hidrográfico de la Reserva de la Biosfera Calakmul, México. Investigaciones Geográficas. Boletín del Instituto de Geografía. Universidad Nacional Autónoma de México 48:7–23.

Gaulin, J. C. and Gaulin, C. K. 1982, Behavioral ecology of *Alouatta seniculus* in cloud forest. *Int. J. Primatol.* 3:1–32.

Horwich, R. H. and Johnson, E. D. 1986, Geographical distribution of the black howler (*Alouatta pigra*) in Central America. *Primates* 27:53–62.

Janson, C. H. and Chapman, C. A. 1999, Resources and primates community structure, in: J. G. Fleagle, C. H. Janson, and K. E. Reed, eds., *Primates Communities*, Cambridge University Press, Cambridge, pp. 237–267.

Julliot, C. 1996, Fruit choice by red howler monkey (*Alouatta seniculus*) in a tropical rain forest. *Am. J. Primatol.* 40:261–282.

Julliot, C. and Sabatier, D. 1993, Diet of the red howler monkey (*Alouatta seniculus*) in French Guiana. *Int. J. Primatol.* 14:527–550.

Klepeis, P. and Roy Chowdhury, R. 2004, Institutions, organizations, and policy affecting land change: Complexity within and beyond the *ejido*, in: B. L. Turner II, J. Geoghegan, and D. R. Foster, eds., *Integrated Land-Change Science and Tropical*

Deforestation in the Southern Yucatán, Oxford University Press, Oxford, pp. 145–169.

Litvaitis, J. A., Titus, K., and Anderson, E. M. 1996, Measuring vertebrate use of terrestrial habitats and food, in: T. A. Bookhout, ed., *Research and Management Techniques for Wildlife and Habitats*, The Wildlife Society, Bethesda, MD, pp. 254–274.

Manly, B. F. J. 1974, A model for certain types of selection experiments. *Biometrics* 30:281–294.

Markkola, J., Niñéemela, M., and Rytkönen, S. 2003, Diet selection of lesser white-fronted geese *Anser erythropus* at a spring staging area. *Ecography* 26:705–714.

McConkey, K. F., Aldy, F., Ario, A., and Chivers, D. J. 2002, Selection of fruit by gibbons (Hylobates muelleri × agilis) in the rain forest of Central Borneo. *Int. J. Primatol.* 23:123–145.

Milton, K. 1980, *The Foraging Strategy of Howler Monkeys: A Study in Primate Economics.* Columbia University Press, New York, p. 165.

Milton, K. 1991, Leaf change and fruit production in six neotropical Moraceae species. *J. Ecol.* 79:1–26.

Mitlöhner, F. M., Morrow-Tesch, J. L, Wilson, S. C., Dailey, J. W. and McGlone, J. J. 2001, Behavioral sampling techniques for feedlot cattle. *J. Anim. Sci.* 79:1189-1193.

O'Brien, T. G., Kinnaird, M. F., Dierenfeld, E. S., Conklin-Brittain, N. L., Wrangham, R. W., and Silver, S. C. 1998, What's so special about figs? *Nature* 392:668.

Pavelka, M. S. M. and Knopff, K. H. 2004, Diet and activity in black howler monkeys (*Alouatta pigra*) in southern Belize: Does degree of frugivory influence activity level? *Primates* 45:105–111.

Ponce-Santizo, G. 2004, Dispersión de semillas por mono araña (*Ateles geoffroyi*), saraguate negro (*Alouatta pigra*) y escarabajos coprófagos en el Parque Nacional Tikal, Guatemala. BSc Thesis, Departamento de Biología, Universidad del Valle de Guatemala, Guatemala, p. 89.

Rosales-Meda, M. M. 2003, Abundancia, distribución y composición de tropas del mono aullador negro (*Alouatta pigra*) en diferentes remanentes de bosque en la ecoregión Lachuá, Guatemala. BSc Thesis, Facultad de Ciencias Químicas y Farmacia, Universidad de San Carlos de Guatemala, Guatemala, p. 105.

Serio-Silva, J. C. 1992, Patrón diario de actividades y hábitos alimenticios de *Alouatta palliata* en semilibertad. BSc Thesis, Facultad de Biología, Universidad Veracruzana, Córdoba, México, p. 66.

Serio-Silva, J. C., Rico-Gray, V., Hernández-Salazar, L. T., and Espinosa-Gómez, R. 2002, The role of *Ficus* (Moraceae) in the diet and nutrition of a troop of Mexican howler monkeys, *Alouatta palliata mexicana*, released on an island in southern Veracruz, Mexico. *J. Trop. Ecol.* 18:913–928.

Silver, S. C., Ostro, L. E. T., Yeager, C. P., and Dierenfeld, E. S. 2000, Phytochemical and mineral components of foods consumed by black howler monkeys (*Alouatta pigra*) at two sites in Belize. *Zoo Biol.* 19:95–109.

Silver S. C., Ostro L. E. T., Yeager C. P., and Horwich, R. 1998, Feeding ecology of the black howler monkey (*Alouatta pigra*) in northern Belize. *Am. J. Primatol.* 45:263–279.

Sourd, C. and Gautier-Hion, A. 1986, Fruit selection by a forest guenon. *J. Anim. Ecol.* 55:235–244.

Stoner, K. E. 1996, Habitat selection and seasonal patterns of activity and foraging of mantled howling monkeys (*Alouatta palliata*) in northeastern Costa Rica. *Int. J. Primatol.* 17:1–30.

Terborgh, J. 1986a, Community aspects of frugivory in tropical forests, in: A. Estrada and T. H. Fleming, eds., *Frugivores and Seed Dispersal*, Dr. W. Junk Publishers, The Hague, The Netherlands, pp. 371–384.

Terborgh, J. 1986b, Keystone plant resources in the tropical rain forest, in: M. Soule, ed., *Conservation Biology*, Sinauer Associates, Sunderland, pp. 330–344.

Yeager, C. P. 1989, Feeding ecology of the proboscis monkey (*Nasalis lavvatus*). *Int. J. Primatol.* 10:497–530.

CHAPTER EIGHT

Intestinal Parasitic Infections in *Alouatta pigra* in Tropical Rainforest in Lacandona, Chiapas, Mexico: Implications for Behavioral Ecology and Conservation

Kathryn E. Stoner and Ana M. González Di Pierro

Kathryn E. Stoner and Ana M. González Di Pierro • Centro de Investigaciones en Ecosistemas, Universidad Nacional Autónoma de México, Apartado Postal 27-3, Morelia, Michoacán, México 58089.

New Perspectives in the Study of Mesoamerican Primates: Distribution, Ecology, Behavior, and Conservation, edited by Alejandro Estrada, Paul A. Garber, Mary S. M. Pavelka, and LeAndra Luecke. Springer, New York, 2005.

INTRODUCTION

Endoparasitic infections are common in nonhuman primates, nevertheless, until recently, the majority of studies have been conducted in captive animals (Kalter, 1989). Primates are particularly vulnerable to parasitic infections because many species live in cohesive social groups characterized by frequent social interactions that facilitate parasite transmission between individuals (Freeland, 1983; Stoner, 1995). Although parasitic infections do not always produce direct pathology affecting host survivorship, in many cases they may increase host susceptibility to predation or decrease the competitive fitness of the individual (for a review, see Scott, 1998). Predation may be greater in individuals infected because they are more obvious to the predator or less likely to escape. Competitive fitness may be reduced in infected individuals because they experience a lower dominance, resulting in the inability to maintain a large territory allowing access to more females.

The study of parasitic infections in nonhuman primates is important for understanding ecological and evolutionary host–parasite relationships and for recommending conservation strategies for endangered species (Stuart and Strier, 1995). Only a few studies document endoparasitic infections in Neotropical primates (Tables 1 and 2). Although intestinal parasitic infections have been

Table 1. Endoparasites identified in Neotropical primates (see Table 2 for *Alouatta* spp.).

Species	Parasite	Reference/Origin
Anthropoidea		Thatcher and Porter (1968)
Callithrichidae	Acanthocephala	
Saguinus oedipus		
Atelidae	Trematoda	Price (1928)
Ateles geoffroyi	*Controchis biliophilus*	
Brachyteles aracnoides	Nematoda	Stuart *et al.* (1993)
	Strongyloides sp.	
	Tripanoxyuris brachytelesi	
Cebidae		
Aotus sp.	Nematoda	Diaz-Ungria (1965)
	Dipetalonema gracile	
	Unidentified Acantocephala	Thatcher and Porter (1968)
Cebus capucinus	Trematoda	Stuart *et al.* (1998)
	Controchis biliophilus	
	Nematoda	
	Strongyloides stercolaris	
Pithecia pithecia	Nematoda	Diaz-Ungria (1965)
	Dipetalonema gracile	
Saimiri sp.	Unidentified Acantocephala	Thatcher and Porter (1968)

Table 2. Endoparsites identified in *Aloutta* spp. For a comprehensive review see Toft (1996) and Stuart *et al.* (1998).

Species	Parasite	Reference/Origin
Anthropoidea	Nematoda	Stiles *et al.* (1929)
Atelidae		
Alouatta belzebul	*Ascaris elongata*	
Alouatta caraya	Nematoda	Santa Cruz *et al.* (2000)
	Oxyuridae	
	Strongyloides sp.	
	Trematoda	
	Bertiella mucronata	
Alouatta fusca	Nematoda	Stiles (1929)
	Tripanoxyuris minutus	Diaz-Ungria (1965)
	Parabronema bonnei	
Alouatta palliata	Acantocephala	Thatcher and Porter (1968)
	Prosthenorchis elegans	
	Cestoda	
	Railleitina sp.	
	Nematoda	
	Parabronema bonnie	
	Nematoda	Stoner (1996)
	Parabronema sp.	
	Unidentified nematode	
	Trematoda	
	Unidentified trematode	
Alouatta seniculus	Acanthocephala	Gilbert (1994b)
	Prosthenorchis sp.	
	Cestoda	
	Railleitina sp.	
	Nematodes	
	Parabronema bonnei	
	Strongyloides sp.	
	Tripanoxyuris sp.	

documented in *Alouatta* in wild populations, only five of eight species have been studied so far (Table 2) and no studies have documented intestinal parasites in the regionally endemic black howler, *Alouatta pigra*.

Interspecific, intraspecific, and inter-individual variations in parasitic infections in nonhuman primates may be related to many factors including demographic factors (i.e., density), environmental factors (i.e., seasonality or humidity), social interactions, age–sex class, reproductive condition, habitat fragmentation, and diet. Below we summarize some of the evidence for each of these factors.

Intraspecific variation in intestinal parasitic infection in wild populations is positively related to population density because it increases the chance of infection (Alexander, 1974; Gilbert and Dodds, 1987; Scott, 1988; Nunn *et al.*,

2003; but see Chapman and Gillespie, in press). For example, Stuart *et al.* (1990) found a higher incidence of infection of *A. palliata* in the tropical dry forest of La Pacifica, Costa Rica (12–37% depending on the parasite species) compared to the tropical dry forest site of Santa Rosa, Costa Rica (0–11% depending on the parasite species) and suggested that this was due to the higher population density of howlers at La Pacifica (73.3 individuals per km^2) compared to Santa Rosa (4.7 individuals per km^2). Similarly, Gilbert (1994a) found in tropical rainforest in Manaus, Brazil, a higher incidence of infection with increased population density for *A. seniculus.* Nevertheless, the study of Stoner (1995) in the tropical rainforest of northwestern, Costa Rica, concluded that density was not necessarily the most important factor in determining parasitic infections in *A. palliata*. She argued that many other factors including ranging patterns and humidity influence the incidence of infection. Several other studies have shown that environmental factors such as humidity influence the incidence of parasitic infections in primates since parasite eggs and larva survive longer in warm humid environments contributing to a greater chance of infection (Stuart *et al.*, 1990, 1993; Jones, 1994; Stoner, 1996).

Behavior and social interactions, and age–sex class of the host may affect both the period the host is exposed to possible infection and their immune response once they are infected by a parasite (Campillo *et al.*, 1999). Social primates that live in large groups are generally more susceptible to parasitic infections because they have contact with more individuals than primates that live in smaller groups (Alexander, 1974; Gilbert, 1994b). The amount of social behavior depends on the species, age, sex, and social rank of the individual (Hausfater and Watson, 1976). For most vertebrate species, males have a higher incidence and intensity of infection than females for many parasite species largely due to the greater immunological capacity of females (for a review, see Klein, 2004). In particular, in females, innate responses, antibody-mediated responses, and cellular responses are higher than in males (Zuk and McKean, 1996; Schuurs and Verheul, 1990). Finally, age-class appears to affect parasitic infections since several studies with vertebrate animals have shown that younger individuals are more often parasitized than older ones (Ramírez-Barrios *et al.*, 2004).

Forest fragmentation and disturbed habitats facilitate contact between humans and wild primates that may result in increased opportunities of infection between them (Stuart and Strier, 1995; Gillespie, 2004), especially if the populations are concentrated in small fragments (Stoner, 1995). For this reason, parasitic infections appear to be an important factor to consider when designing corridors or when implementing translocation programs for wild primates (McGrew *et al.*, 1989; Stoner, 1995).

Finally, the type of food ingested by animals also may affect their susceptibility to some parasites. In particular, if the host animal consumes plants that contain secondary compounds, this may have a negative effect on the establishment and survival of some intestinal parasites such as nematodes (Freeland, 1983; Huffman and Wrangham, 1994; Krief, 2003). Selective foraging strategies by some primates include the ingestion of plant species that are known to have medicinal value (Phillips-Conroy, 1986; Huffman and Seifu, 1989). For example, Phillips-Conroy (1986) suggested that baboons found in areas with high levels of *Schistosoma* sp. contamination consumed leaves of *Balanites aegyptica*, a bush known to be toxic to this parasite.

It is well documented that howler monkeys, in general, selectively forage to maximize nutrients and minimize the quantity of secondary compounds ingested (Glander, 1978; Milton, 1980; Serio-Silva *et al.*, 2002). Several studies have shown that howlers consume a large amount of *Ficus* spp. leaves and fruit (Stoner, 1996; Estrada *et al.*, 1999; Serio-Silva *et al.*, 2002). This Pantropical genus contains as many as 2000 species and is characterized by latex that contains secondary compounds including alkaloids, steroids, flavonoids, and terpenoides. Furthermore, this genus has been used for a variety of pharmaceutical applications, including the use as a medication to eliminate intestinal parasites (Schultes and Raffauf, 1990). In our study area, the Reserva de la Biosfera Montes Azules, Chiapas, the latex in the fruits, leaves, and bark of *Ficus* is used by the local population to make a tea as an anti-parasite medicine (per. obs.). No studies have evaluated the possible relationship between *Ficus* ingestion and intensity of intestinal parasitic infections in howler monkeys.

The objectives of the present study include: (1) Identify the species of endoparasites in *A. pigra*; (2) determine if infection prevalence and intensity are affected by group size, age–sex class, or season; and (3) determine if the intensity of infection is negatively correlated to the time dedicated to consuming *Ficus*.

METHODS

Study Species and Site

The black howler monkey (*A. pigra*) is one of the eight species in the genus *Alouatta* and is found in Mexico (east of Tabasco, south to Chiapas, and in the Yucatán Penisula), Guatemala, and Belize (Nowak, 2000). This species is classified as in danger of extinction and has been listed by CITES on Appendix I since 1975 (Inskipp and Wells, 1979). Howler monkeys are classified as behavioral folivores, but depending upon the season and site, they also may consume many fruits and flowers (Milton, 1980; Stoner, 1996; Estrada *et al.*, 1999). Troop size varies considerably depending on the species, with *A. pigra* having some of the smallest troops (4–6 individuals; Crocket and Einsberg, 1987). As for most other Mesoamerican primates, habitat destruction is the main threat to the survival of this species; however, several other natural factors contribute to their mortality in the wild including predation, intraspecific aggression, disease, and parasites (Stoner, 1995; Milton, 1996; Scott, 1998). Parasites and disease are likely to be among the main forces responsible for howler monkey mortality in the wild since both predation (Sherman, 1991) and intraspecific aggression are rare (Milton, 1980; Neville *et al.,* 1988; Crockett, 1998).

The study was conducted in a tropical rainforest in Chiapas, Mexico, in the southern part of the Reserva de la Biósfera Montes Azules in the Lacandona forest. This large reserve encompasses an area of approximately 3000 km^2 (16°07′ 58″N, 90°56′36″W; Medellín, 1994). Average annual temperature is 22°C with the coldest month greater than 18°C (Herrera-MacBryde and Medellín, 1997). Average annual rainfall is 3000 mm with approximately 88% of this falling in the rainy season between June and November (Medellín and Equihua, 1998).

Data Collection

Foraging data and fecal samples were collected from three groups of *A. pigra* from January through October 2003 (except for April). The groups were selected to represent three different sizes and to be separated from each other by at least 1.5 km. We also attempted to select groups that contained adults and juveniles whenever possible. The largest group was the Station group consisting of two adult males, three adult females, two juvenile females, and one infant. The intermediate size group was the River group consisting of one adult male, two adult females, one juvenile female, one juvenile male, and one infant. The smallest group was the Tablero group consisting of one adult male and two adult females. Once a month, data were collected for 3 days from each group from 7:00 h to 18:00 h. Individuals were identified by recognizing unique marks by sex–age class. Foraging data were collected on adults and juveniles using 5-min focal animal observations (Altmann, 1974); focal animals were randomly changed after each 5-min observation. Plant species was identified and young leaves, mature leaves, petiole, immature fruit, mature fruit, flower, or bark was consumed. When the species was unknown, we collected a sample for later identification using the herbarium at the Instituto de Biologia, Universidad Nacional Autònoma de Mèxico.

One to five fecal samples were collected from each individual each month. Samples were collected immediately after defecation from identified individuals. Samples were placed in 30 ml vials with 10 ml of 10% formalin; the same quantity of feces was placed in each vial to reach 15 ml (Brooke and Goldman, 1949).

Fecal Analysis

Since it often takes several samples to isolate parasitic infections, fecal sampling was repeated for each individual to assure that parasitic infections were identified. Qualitative analysis using direct smears was used to determine the presence of ova and larvae of different parasite species. Quantitative analysis using a sedimentation technique with ethyl acetate/formalin centrifugation (Long *et al.*, 1985) was used to estimate intensity of infection. After centrifugation, two drops of concentrate and one drop of iodine were placed on a microscope slide and all ova and larvae under a 22 mm × 30 mm glass coverslip were counted. Three slides were systematically scanned for each fecal sample and the total number of ova and larvae for each parasite species was summed to represent the intensity of infection for each sample. This estimate is taken to represent a minimum intensity of infection since parasites that produce few eggs will be underrepresented. We assume here, as has been done with other primate field studies (Stoner, 1996; Knezevich, 1998), that the number of parasites shed was a general reflection of the severity of parasitic infection. Actual adult parasite numbers cannot be estimated without necropsy. Parasite eggs and larvae were identified based on size and morphology (Markell and Voge, 1984; Ash and Orihel, 1997). Because not all the parasite life stages were collected, identification of parasites based on ova and larvae is tentative.

Data Analysis

To determine if the frequency of individuals infected varied between groups, sexes (adults only), or age-classes (adult versus juvenile), we used generalized linear models applying the GENMOD procedure (SAS, 2000) for repeated measurements where individuals were the repeated factor in the model. The categorical independent variable was group, sex, or age-class, for each of the three analyses, respectively, and the dependent variable was the number of samples that contained each parasite. This analysis uses a binomial distribution and a logit link function. The same types of analyses were used to compare intensity of infection between groups, sexes, and age-class, respectively, but a Poisson distribution was employed. Similarly, to determine if seasonality affects parasitic infections, generalized linear models were used to compare the frequency of individuals affected during the wet (June–November) and dry season (December–May) and to compare the intensity of infection during the wet and dry season. Finally, a linear regression using the procedure PROCREG (SAS, 2000) was used to determine if time devoted to feeding on *Ficus* affects intensity of infection of individuals. It should be noted that sample sizes were relatively small for all statistical comparisons (3 groups, 4 adult males versus 7 adult females, and 11 adults versus 4 juveniles). Because of the small sample sizes, we adopt a p-value <0.1 as significant (Effective Clinical Practice, 2001).

RESULTS

Species Identified in *Alouatta pigra*

One hundred and fifty-one fecal samples were collected: 69 from the Station group, 30 from the Tablero group, and 52 from the River group. Eight species of parasites were identified based on morphology and size (Figure 1): (1) Protozoan: *Blastocystis* sp. (8.0 ± 0.6 μm $\times$ 10.0 ± 0.6 μm, $n = 14$), *Entamoeba* sp. (19.0 ± 0.7 μm $\times$ 18.0 ± 0.7 μm, $n = 11$), *E. coli* (25.0 ± 0.8 μm $\times$ 19 ± 0.5 μm, $n = 7$), and *Isospora* sp. (20.0 ± 0.6 μm $\times$ 11.0 ± 0.6 μm, $n = 3$; (2) Nematodes: *Enterobius* sp. (49.0 ± 0.7 μm $\times$ 20.0 ± 0.01 μm, $n = 2$), *Strongyloides* sp. (49.0 ± 0.7 μm $\times$ 36.0 ± 0.7 μm, $n = 11$), and *Trichostrongyloides* sp. (75.0 ± 0.6 μm $\times$ 0 ± 0.6 μm, $n = 3$); and (3) Unidentified Trematode: a dark staining digenean egg (Platyhelminthe: Digenea) measuring 35.0 ± 1.0 μm $\times$ 20.0 ± 0.05 μm, $n = 4$. Flukes are particularly difficult to identify based on eggs since many species are quite similar (Ash and Orihel, 1997). The incidence of infection of the observed parasites in the samples was: *Blastocystis* sp. 65%, *Strongyloides* sp. 45%, *Entamoeba* sp. 42%, *E. coli* 38%, *Isospora* sp. 10%, digenean 6%, and *Trichostrongyloides* sp. and *Enterobius* sp. 2%. No statistical analyses were preformed for *Isospora*

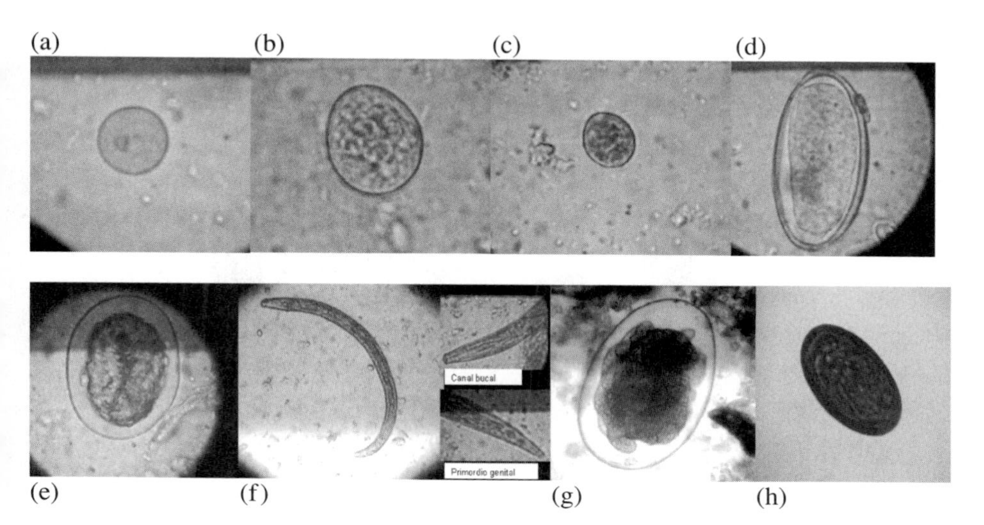

Figure 1. Parasites found in *Alouatta pigra*. (a) Egg of the protozoa *Entamoeba* sp. (Lens power 10×). Identifying characteristics include the two nuclei present and size (20 μm × 18 μm). (b) Egg of the protozoa *Entamoeba coli* (Lens power 40×). Identifying characteristics include the presence of granular chromatin and size (25 μm × 19 μm). (c) Egg of the protozoa *Isospora* sp. (Lens power 10×). Identifying characteristics include the thin cell wall, granular sporoblaysts, and size (20 μm × 10 μm). (d) Egg of the nematode *Enterobius* sp. (Lens power 40×). Identifying characteristics include the long shape, thick cell wall, partially embryonated state, and size (50 μm × 20 μm). (e) Egg of the nematode *Strongyloides* sp. (Lens power 40×). Identifying characteristics include the thin cell wall and size (49 μm × 36 μm). (f) Rhabditoide larva from *Strongyloides* sp. (Lens power 40×). Identifying characteristics include the short buccal canal, prominent primordial genital, and size (180 μm × 14 μm). (g) Egg from the nematode *Trichostrongyloides* sp. (Lens power 40×). Identifying characteristics include the thin clear color cell wall, wrinkled internal membrane, and size (75 μm × 40 μm). (h) Unidentified fluke egg (Trematode: Platyhelminthe: Digenea) (Lens power 40×). Identifying characteristics include the dark color, cell wall of medium thickness, and size (35 μm × 19 μm).

sp., digenean, *Trichostrongyloides* sp., or *Enterobius* sp. due to their low prevalence.

Effect of Group and Sex on the Incidence and Intensity of Infection

The Station group had a significantly higher incidence of infection of *Blastocystis* sp. than the other two groups ($X^2 = 3.7$, df $= 2$, $p = 0.10$; Figure 2a). There

(a)

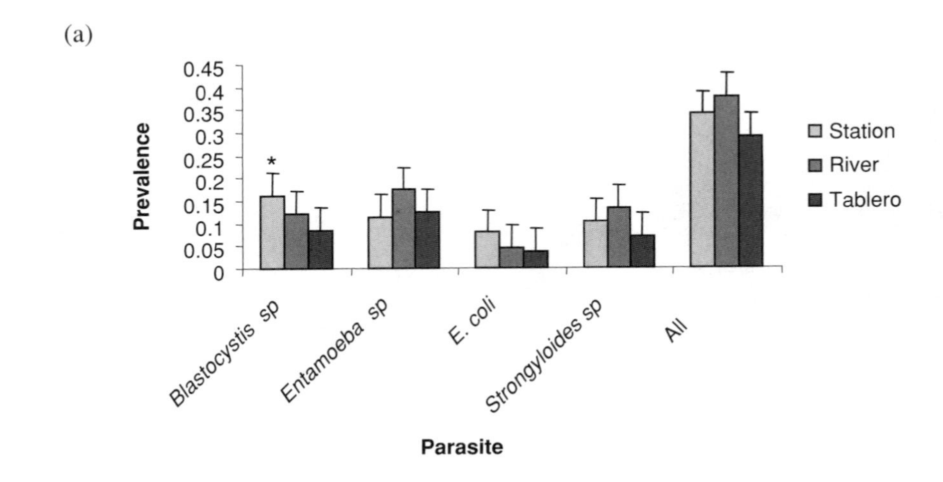

(b)

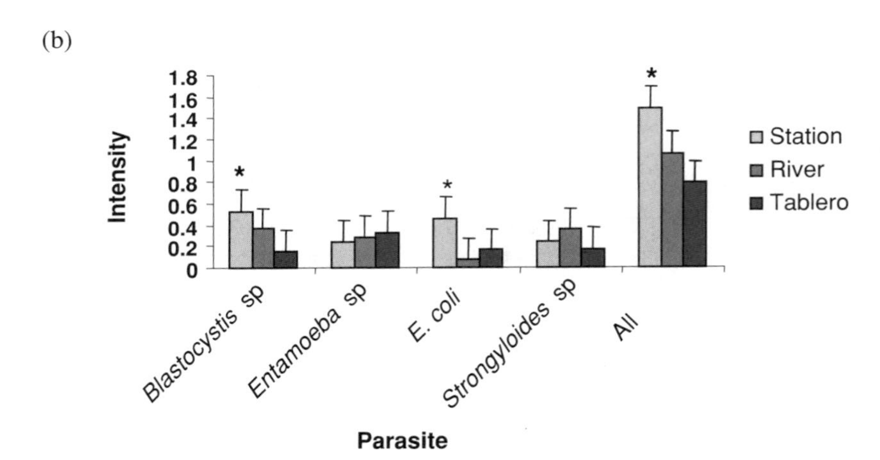

Figure 2. Prevalence (a) and intensity (b) of infection of endoparasites in the three groups of *Alouatta pigra*. Bars indicate standard error ($^{*}p < 0.1$, $^{**}p < 0.05$).

was no significant difference in the frequency of individuals infected with the other parasites between the three groups: *Entamoeba* sp. ($X^2 = 2.05$, df $= 2$, $p = 0.35$), *E. coli* ($X^2 = 1.53$, df $= 2$, $p = 0.46$), and *Strongyloides* sp. ($X^2 = 1.11$, df $= 2$, $p = 0.57$). Similarly, no differences were found between groups in the frequency of individuals infected when analyzing parasites as a whole ($X^2 = 0.73$, df $= 2$, $p = 0.69$). The intensity of infection of intestinal parasites was significantly different between groups with the Station group displaying a greater intensity of infection of *Blastocystis* sp. ($X^2 = 4.72$, df $= 2$, $p = 0.09$; Figure 2b), *E. coli* ($X^2 = 4.13$, df $= 2$, $p = 0.10$), and when all parasites were

(a)

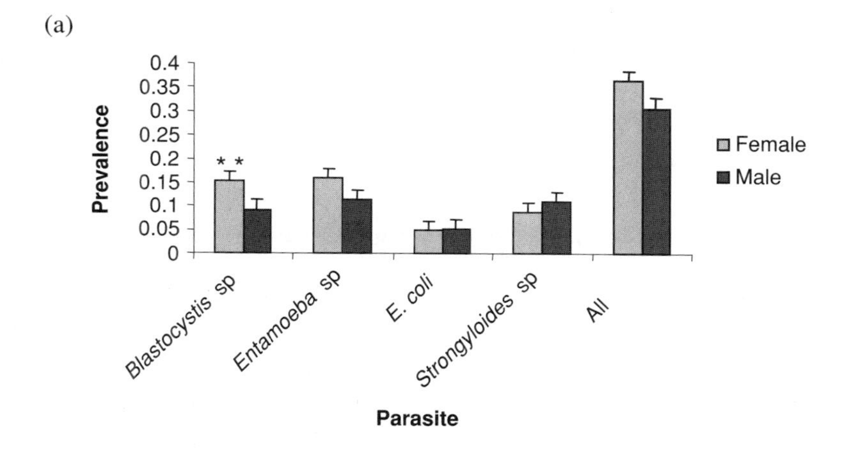

(b)

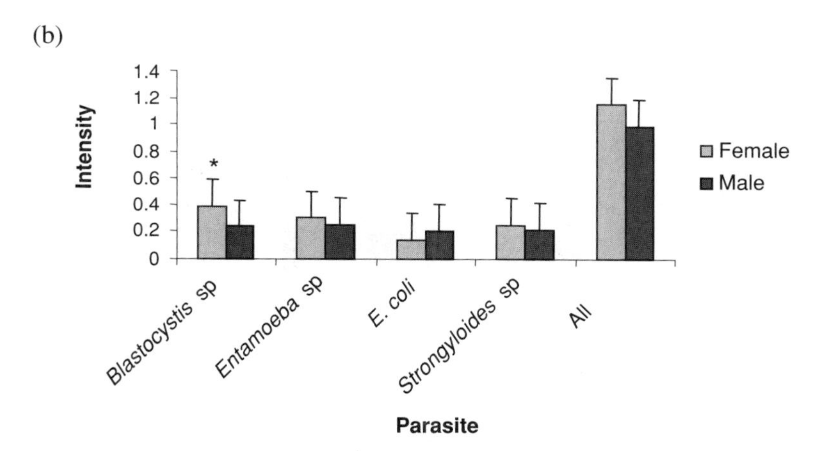

Figure 3. Prevalence (a) and intensity (b) of infection of females and males. Bars indicate standard error (* $p < 0.1$, ** $p < 0.05$).

analyzed as a whole ($X^2 = 4.95$, df $= 2$, $p = 0.08$). No differences were found between groups for the other parasites: *Entamoeba* sp. ($X^2 = 0.35$, df $= 2$, $p = 0.83$) and *Strongyloides* sp. ($X^2 = 0.8$, df $= 2$, $p = 0.66$).

Females had a greater frequency of infection than males for *Blastocystis* sp. ($X^2 = 3.53$, df $= 1$, $p = 0.05$; Figure 3a). No differences were observed between sexes for the other parasites: *Entamoeba* sp. ($X^2 = 0.74$, df $= 1$, $p = 0.39$), *E. coli* ($X^2 = 0.03$, df $= 1$, $p = 0.86$), *Strongyloides* sp. ($X^2 = 0.42$, df $= 1$, $p = 0.51$), or when analyzing all parasites as a whole ($X^2 = 0.79$, df $=$ 1, $p = 0.37$). Similarly, females had a significantly higher intensity of infection than males for *Blastocystis* sp. ($X^2 = 2.21$, df $= 1$, $p = 0.10$). No differences were found between females and males for the other parasites: *Entamoeba* sp. ($X^2 = 0.18$, df $= 1$, $p = 0.67$), *E. coli* ($X^2 = 0.51$, gl $= 1$, $p = 0.57$), and *Strongyloides* sp. ($X^2 = 0.07$, gl $= 1$, $p = 0.79$), or when analyzing all parasites as a whole ($X^2 = 1.00$, df $= 1$, $p = 0.31$).

Effect of Age on the Incidence and Intensity of Infection

Juveniles had a significantly greater frequency of infection than adults for *Blastocystis* sp. ($X^2 = 4.99$, df $= 1$, $p = 0.02$) and *Strongyloides* sp. ($X^2 = 3.49$, df $= 1$, $p = 0.06$), and for all parasites as a whole ($X^2 = 5.71$, df $= 1$, $p = 0.01$; Figure 4a). No differences were observed between adults and juveniles for *Entamoeba* sp. ($X^2 = 0.12$, df $= 1$, $p = 0.9$) or *E. coli* ($X^2 = 0.24$, df $= 1$,

(a)

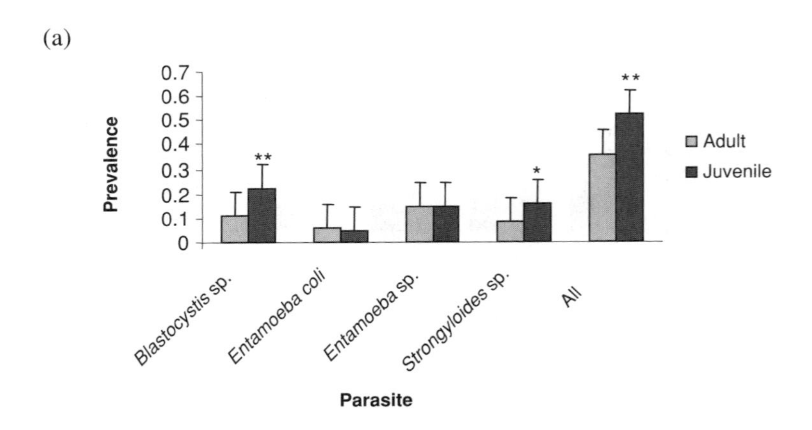

(b)

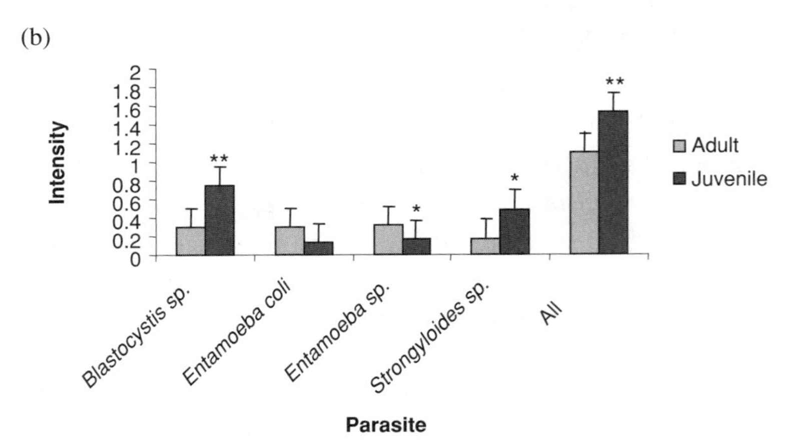

Figure 4. Prevalence (a) and intensity (b) of infection of adults and juveniles. Bars indicate standard error ($^{*}\,p < 0.1$, $^{**}\,p < 0.05$).

$p = 0.6$). Similarly, juveniles showed a greater intensity of infection than adults for *Blastocystis* sp. ($X^2 = 5.2$; df = 1; $p = 0.02$), *Strongyloides* sp. ($X^2 = 2.7$, df = 1, $p = 0.09$), *Entamoeba* sp. ($X^2 = 2.3$, df = 1, $p = 0.10$), and when analyzing all parasites as a whole ($X^2 = 2.63$, df = 1, $p = 0.10$; Figure 4b). No significant difference was observed for *E. coli*. ($X^2 = 1.2$, df = 1, $p = 0.2$).

Effect of Season on the Incidence and Intensity of Infection

A significantly higher incidence of infection was observed during the dry season for *Blastocystis* sp. ($X^2 = 34$, df = 1, $p = 0.0001$), *E. coli* ($X^2 = 4.43$, df = 1, $p = 0.03$), *Entamoeba* sp. ($X^2 = 2.46$; gl = 1; $p = 0.10$), and when analyzing all parasites as a whole ($X^2 = 5.21$, df = 1, $p = 0.022$; Figure 5a). In

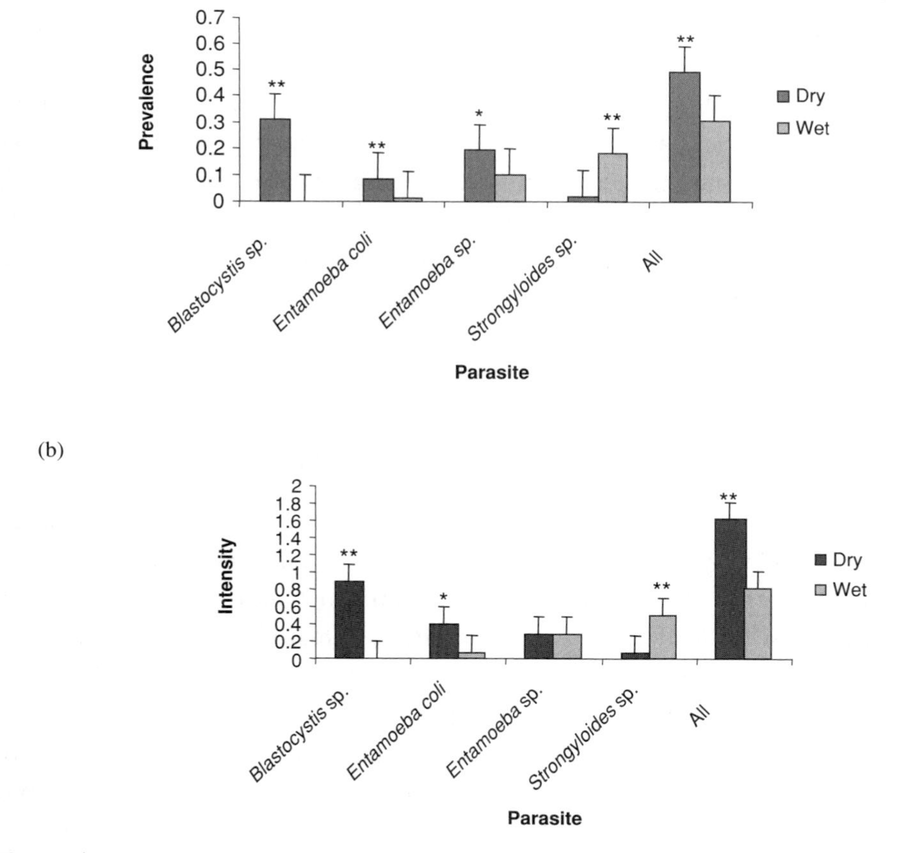

Figure 5. Prevalence (a) and intensity (b) of infection during the dry and rainy season. Bars indicate standard error (* $p < 0.1$, ** $p < 0.5$).

contrast, a significantly higher incidence of infection was observed during the wet season for *Strongyloides* sp. ($X^2 = 14.65$, df $= 1$, $p = 0.0001$). Similarly, intensity of infection was significantly greater in the dry season for *Blastocystis* sp. ($X^2 = 65.41$, df $= 1$, $p = 0.0001$), *E. coli* ($X^2 = 3.32$, df $= 1$, $p = 0.06$), and when analyzing all parasites as a whole ($X^2 = 9.14$, df $= 1$, $p = 0.002$), while intensity of infection was greater in the rainy season for *Strongyloides* sp. ($X^2 = 4.09$, df $= 1$, $p = 0.04$; Figure 5b). No difference was observed between seasons for *Entamoeba* sp. ($X^2 = 0$, df $= 1$, $p = 0.9$).

Relationship Between Intensity of Infection and Feeding on *Ficus*

During 172 h of observation (River group 58 h, Tablero group 52 h, Station group 62 h), we collected 48.6 h of foraging data (Table 3). We recorded feeding from 24 plant species, with the River group consuming 5 species (3 families), the Tablero group consuming 6 species (5 families), and the Station group 19 species (12 families). Both the River group and the Tablero group spent a considerable amount of their time consuming *Ficus tecolutensis* fruits and leaves, 42% and 45%, respectively, while the Station group only spent 3% of their time eating this species, in spite of the fact that it was observed in their foraging area (pers. obs.). We found a significant negative relationship between the intensity of infection of parasites and time spent feeding on *Ficus tecolutensis* fruits and leaves ($R^2 = 0.4$, $p = 0.02$; Figure 6).

DISCUSSION

Alouatta pigra Compared to Other Species

One of the eight species of intestinal parasites found in our study population (the nematode *Trichostrongyloides* sp.) has not been reported previously for the genus *Alouatta*, or for other Neotropical primate species (Tables 1 and 2). Trichostrongyles are a diverse group of bursate nematodes mainly found in the gastrointestinal tract. They have been found in all vertebrates except fish and are especially common in ruminants such as cattle and sheep (Noble *et al.*, 1989). The life cycle is similar for most members of this group. Eggs are passed out in feces where the rhabditiform larvae feed on fecal microflora and develop into the infective third stage larvae that attach to plants and are ultimately eaten by herbivores or omnivores.

Table 3. Foraging time dedicated to different species and plant parts consumed by each troop.

Species	River Troop Total 995 min (%) Part consumed %	Tablero Troop Total 824 min (%) Part consumed %	Station Troop Total 1100 min (%) Part consumed %
Acanthaceae			
Bravaisia sp.	0	0	69 (6) 5YL, 1P
Arrabideae	0	0	14 (1) 1YL
Bignoneaceae	0	109 (13) 13YL	14 (1) 1ML
Caesalpinioideae			
Dialium guianense	0	0	18 (2) 2YL
Schizolobium sp.	0	0	1 (0.1) 0.1P
Cecropiaceae			
Cecropia obtusifolia	0	82 (10) 10YL	59 (5) 3YL, 2MF
Chrysobalanaceae			
Licania platypus	0	0	176 (16) 11YL, 5B
Faboideae			
Lonchocarpus sp.	0	0	48 (4) 4ML
Macaerium sp.	0	0	78 (7) 6YL 1FL
Platymiscium sp.	0	0	338 (31) 31YL
Magnoliaceae			
Talauma sp.	0	129 (16) 16P	0
Malpigiaceae	0	0	6 (1) 0.1FL
Mimosoideae			
Acacia usumacintensis	0	1 (0.1) 0.1YL	40 (4) 4YL
Albizia Leucocalyx	155 (16) 5YL, 9ML, 1MF	0	0
Cojoba arborea	0	0	10 (1) 1YL
Inga sp.	0	0	15 (1) 1YL
Moraceae			
Brosimum alicastrum	275 (28) 24YL, 4IF	0	48 (4) 3YL, 1MF
Castilla elastica	0	0	20 (2) 2MF
Fícus yoponensis	0	133 (16) 16YL	0
Ficus tecolutensis	413 (42) 21YL, 21MF	370 (45) 6YL, 39MF	33 (3) 3MF
Maclura tinctoria	42 (4) 4YL	0	0
Rutaceae			
Zanthoxylum riedelianum	109 (11) 11YL	0	0
Sapindaceae			
Paulinia fibrígera	0	0	73 (7) 7YL
Vitaceae			
Cissus microcarpa	0	0	40 (4) 1YL, 3P
Total	**100**	**100**	**100**

YL = Young leaves, ML = Mature leaves, IF = Immature fruit, MF = Mature fruit, P = petiole, B= Bark and FL = flower.

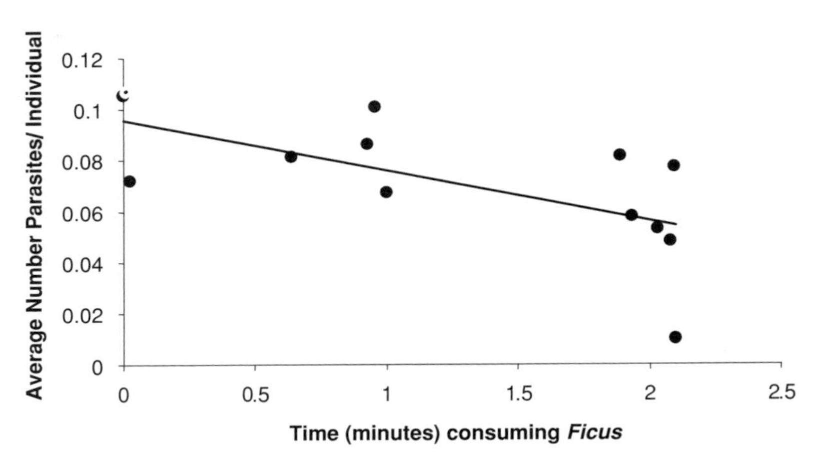

Figure 6. Relationship between individual intensity of infection and feeding time devoted to *Ficus tecolutensis.*

Seven of the parasites we found have been previously reported for *Alouatta. Entamoeba* spp. has been described for *A. caraya* and *A. palliata* (Stiles *et al.*, 1929; Hegner, 1935). *Isospora* sp. has been reported for *A. palliata* in northwestern Costa Rica (Stuart *et al.*, 1990) and in Panama (Hendricks, 1977). *Enterobius* sp. was found in *A. fusca* in Brazil (Kopper *et al.*, 2000). A nematode from the genus *Strongyloides* was reported for *A. palliata* in Costa Rica (Stuart *et al.*, 1990) and for *A. caraya* in Argentina (Santa Cruz *et al.*, 2000). An unidentified digenean was identified in *A. palliata* in Costa Rica (Stoner, 1996).

In spite of the fact that we found a relatively high number of parasites in our population compared to other studies on *Alouatta* (Table 2), the intensity of infection was rather low compared to other studies that provide this information for wild primates (see references from Tables 1 and 2). This may not be surprising since protozoans and many nematodes infect hosts through accidental ingestion of contaminated water or food (Ash and Orihel, 1997), and our study site included a large well-preserved area with little opportunity for contamination. The low intensity of infection of trematodes is likely a result of the low percentage of invertebrates that howlers include in their diet, since many trematodes require invertebrates as intermediate hosts before arriving at their definitive vertebrate host; the first host for several trematode species are ants or snails (Ash and Orihel, 1997). Since these items are not typical of the diet of howlers, it is likely that they may only become infected when inadvertently ingesting these items while foraging on fruits, leaves, or drinking contaminated water from tree holes (Gilbert, 1994b). Another possible explanation for the low intensity of infection observed compared to other studies of *A. palliata* is that *A. pigra* is routinely found in smaller groups than *A. palliata* (Crockett and Eisenberg, 1987).

Some behaviors of howlers may partially explain low intensity of parasitic infections observed. Since they are largely arboreal, they minimize their risk of coming into contact with infectious agents on the ground (Gilbert, 1994b). In addition, howlers display few social behaviors such as grooming (Crockett and Eisenberg, 1987), which also reduce the opportunity of passing parasitic infections between individuals because individuals do not come in contact as often with possibly contaminated hair or skin. Finally, it has been suggested that howlers' behavior of specific defecation sites that occur in areas free of underlying vegetation may help them avoid parasitic infections (Gilbert, 1997).

Effect of Group

Significant differences in intensity of infection were found between groups for two of the parasites infecting *A. pigra* at our study site and for overall intensity. In all cases, the largest group, the Station group, had the highest intensity of infection. These results suggest that, in spite of relatively little variation in the size of the three troops studied (i.e., 3, 5, and 7), group size may play a significant role in intensity of infection. The only other sympatric primate at our study site, *Ateles geoffroyi*, was frequently observed in the area of the Station troop, while they were never seen in the area of the other two troops. This may be another factor affecting the differences in parasites between the troops. The two parasites that showed greater intensity of infection in the Station troop, *Blastocystis* sp. and *Entamoeba coli*, are acquired through direct ingestion of contaminated food or water and can be transmitted between individuals by contact with fecal material (Noble *et al.*, 1989; Ash and Orihel, 1997). On several occasions, we observed individuals scratching themselves in the perineal area and rubbing their behind on branches; however, we have no quantitative data on the frequency of these behaviors. Another factor that may have contributed to the higher intensity of infection observed in the Station troop is their close proximity to the Chajul Biological Station. Some humans camp in the area, resulting in fecal contamination. Although howlers do not go down to the ground often, they have been observed to do this on occasion (Gilbert and Stouffer, 1989), and this might provide a new source of parasite contamination.

Differences Between Sexes

No differences were found between males and females in the incidence and intensity of infection for most of the parasites analyzed; nevertheless, in contrast to our prediction that males would be more infected than females, females had a significantly higher incidence and intensity of infection for the most common parasite observed, *Blastocystis* sp. This may be a result of the hormonal changes that occur in females during pregnancy and lactation. Progesterone, which plays a critical role in maintaining pregnancies, is typically regarded as an immunosuppressive that may ultimately allow parasites to infect and proliferate in the host (Klein, 2004). Once female adult howler monkeys reach maturity, they spend most of their adult life lactating or pregnant (Crockett and Eisenberg, 1987).

Differences Between Adults and Juveniles

Our study shows that juveniles were significantly more affected by intestinal parasites than adults, both in incidence and intensity of infection. In general, this may result because the age of an individual influences the capacity of parasites to invade and survive in the host (Schalk and Forbes, 1997). In many cases, juveniles do not yet have specific immunity to parasites, which makes them more susceptible to infections (Stuart *et al.*, 1998).

In addition, the behavior of juveniles may influence their greater incidence and intensity of intestinal parasites acquired through direct contamination or contact with each other. Although howlers in general have been described as not spending much time on social behaviors such as grooming (Crockett and Eisenberg, 1987; Wang and Milton, 2003), juveniles represent the age group that most often make contact with each other while resting and also while playing (Gilbert, 1994b).

Seasonal Differences

We found a higher incidence and intensity of infection in the dry season for parasites that are acquired by direct ingestion of contaminated food or water. We suggest that this may have occurred because protozoans such as *Blastocystis* sp. and *Entamoeba* spp. have a durable membrane that allows them to remain in the cyst stage in dry environments; the cyst is still infective if ingested (Noble *et al.*, 1989). Another important factor that may have influenced the higher incidence and intensity of infection by protozoans during the dry season is that howlers were observed routinely drinking water from tree holes, especially during the dry season. Since water in tree holes is not often washed out by rains in the dry season, it stagnates and may allow for the accumulation of feces from howlers and other mammals and birds. Our findings of significantly higher incidence and intensity of infection of the nematode *Strongyloides* sp. in the wet season agrees with previous studies (Stuart *et al.*, 1990, 1993; Stoner, 1996). This may occur because *Strongyloides* sp. do not have a cyst stage and humid environments allow the larvae to live longer allowing contamination (Stuart *et al.*, 1990, 1993; Jones, 1994; Stuart and Strier, 1995; Stoner, 1996).

Ficus Consumption and Parasitic Infections

We found a significant negative relationship ($R^2 = 0.49$, $p = 0.01$) between the amount of time an individual devoted to consuming *Ficus tecolutensis* and the intensity of infection. In the Amazon Basin in Brazil, some indigenous populations use the genus *Ficus* as an intestinal parasite medicine. Pharmacological studies with the genus *Ficus* in this area have shown that, indeed, this plant genera works in eliminating some intestinal parasites from the host (Schultes and Raffauf, 1990). We suggest that the consumption of *Ficus* by howlers may work as parasite medicine and contribute to the low incidence and intensity of parasites observed. Nevertheless, this is speculative at this point, since it is impossible to separate the importance of group size from the importance of consuming *Ficus*. The Station group was the largest, but it was also the group that spent the least amount of time consuming *Ficus*. Future bromatological studies need to be conducted on several of the plants that howlers consume to isolate and recognize the compounds responsible for allowing *Ficus* to work in eliminating intestinal parasites. More detailed studies on howler foraging and intensity of parasitic infections should be conducted to further evaluate this possible relationship.

Implications for Conservation

Our study has important implications for the conservation of howler monkeys and possibly other endangered primate species. The ever-increasing rate of deforestation in tropical regions (Myers, 1989) forces primates into smaller areas and translocation programs are becoming more common (Estrada and

Coates-Estrada, 1988; Horwich *et al.*, 1993; Rodríguez-Luna and Cortés-Ortiz, 1994; Rodríguez-Luna *et al.*, 1993, Vié and Richard-Hansen, 1997). Because our results show that many factors affect the incidence and intensity of intestinal parasitic infections of howler monkeys in the wild, we feel that it is important to consider all of these factors when planning primate translocation programs. In particular, it is important to consider the primate density that will result after translocation, the age–sex class structure of the group to be translocated, the season of translocation, and the species available for food.

One of the novel results of our work is the suggestion that individual diet may partially determine the intensity of parasitic infection. We suggest that future studies should concentrate on detailed foraging analysis combined with bromatological analysis of species consumed to identify plants that may function as anti-parasitic agents. In particular, *Ficus* should be further studied to determine the potential medicinal properties of this genus as a natural anti-parasite medicine in primates. When planning management programs in fragmented landscapes or translocation programs, not only the variety of plant species should be considered, but also the specific species that may have medicinal value to the primates.

SUMMARY

Only a few studies have documented intestinal parasites in populations of wild Neotropical primates, and many of these have focused on captive animals. Intestinal parasites have been documented in five of the eight species of howler monkeys (*Alouatta*). The black howler monkey (*A. pigra*) has not yet been studied. We documented the prevalence of parasitic infections in three troops of *A. pigra* in the tropical rainforest in Chiapas, Mexico. We evaluated the importance of demographic factors (i.e., group size), age–sex class, environmental factors (i.e., seasonality or humidity), and diet in determining parasitic infections.

Eight species of parasites were identified in *A. pigra*: (1) Protozoan: *Blastocystis* sp., *Entamoeba* sp., *E. coli*, and *Isospora* sp.; (2) Nematodes: *Enterobius* sp., *Strongyloides* sp., and *Trichostrongyloides* sp.; and (3) Trematode: Unidentified dark staining digenean egg (Platyhelminthe: Digenea). Similar to several other studies with wild animals, we found a higher incidence and intensity of infection in the largest troop. We also found a higher incidence and intensity of infection in females and juveniles compared to males and adults, respectively. We found that parasitic infections depended upon the season. Protozoans that form protective hard cysts and can live a relatively long time were more common in the dry season. The nematode *Strongyloides* sp. was more common during the rainy season, probably due to the fact that more humid conditions allowed infective larvae to live longer. Finally, we found a significant negative relationship between intensity of infection of an individual and time spent foraging on *Ficus tecolutensis.* We suggest that the consumption of this species may function as an anti-parasite medicine. Based on our results, we recommend that primate management and translocation programs consider the age–sex class structure of the groups and the species available for food. In addition, translocation programs should also consider the season and humidity conditions when undertaking translocation. Taking these factors into account when designing management programs in fragmented landscapes and making decisions about translocating primate groups will help deter the negative effects of intestinal parasitic infections and contribute to the successful conservation of primates.

ACKNOWLEDGMENTS

We thank A. Estrada for inviting us to contribute to this book. We are grateful to Rafael Lombera Estrada for his valuable assistance in the field, R. Medellín for allowing us to use his boat for transportation during fieldwork, and D. Silva Aguilar for laboratory assistance. We thank H. Ferreria and G. Sanchez Montoya for technical assistance and three anonymous reviewers for helpful comments on a previous draft of this manuscript. We thank the Centro de Investigaciones en Ecosistemas, UNAM, for providing financial assistance to conduct this research and we thank the Consejo International de Ciencia y Tecnologia, Mexico, for providing a graduate student scholarship to A. M. Gonzalez Di Pierro.

REFERENCES

Alexander, R. D. 1974, The evolution of social behavior. *Annu. Rev. Ecol. Syst.* 5:325–383.

Altmann, J. 1974, Observational study of behavior: Sampling methods. *Behavior* 49:227–267.

Ash, L. R. and Orihel, T. C. 1997, *Atlas of Human Parasitology,* 4th Edn. American Society of Clinical Pathologists, Chicago University Press, Chicago.

Brooke, M. M. and Goldman, M. 1949, Polyvinyl alcohol fixative as a preservative and adhesive for protozoa in dysenteric stools, and other liquid materials. *J. Lab. Clin. Med.* 34:1554–1560.

Campillo, M. C., Vázquez, F. A. R., Fernández, A. R. M., Acedo, M. C. S., Rodríguez, S. H., López-Cozar, I. N., Baños, P. D., Romero, H. Q., and Varela, M. C. 1999, *Parasitología Veterinaria.* Editorial McGraw-Hill-Interamericana, Madrid, Spain.

Chapman, C. A. and Gillespie, T. R. Gastrointestinal parasite prevalence and richness in sympatric colobine populations in relation to a dramatic increase in host density: Contrasting density-dependent effects. *Am. J. Primatol.*, in press.

Crockett, C. M. 1998, Conservation biology of the genus *Alouatta. Int. J. Primatol.* 3:549–578.

Crockett, C. M. and Eisenberg, J. G. 1987, Howlers: Variations in group size and demography, in: B. B. Smuts, D. L. Cheny, R. M. Seyfarth, R. W. Wrangham, and T. T. Struhsaker, eds., *Primate Societies*, University of Chicago Press, Chicago, p. 5468.

Diaz-Ungria, C. 1965, Nematodos de primates Venezolanos. *Boletín de la Soc. Venez. de Cienc. Naturales* 25:393–398.

Effective Clinical Practice 2001, Primer on Statistical Significance and *P* Values. July/August 2001. http://www.acponline.org/journals/ecp/julaug01/primer.htm

Estrada, A. and Coates-Estrada, R. 1988, Tropical rainforest conversion and perspectives in the conservation of wild primates (*Alouatta* and *Ateles*) in Mexico. *Am. J. Primatol.* 14:315–327.

Estrada, A., Juan Solano, S., Ortíz Martínez, T., and Coates-Estrada, R. 1999, Feeding and general activity patterns of a howler monkey (*Alouatta palliata*) troop living in a forest fragment at Los Tuxtlas, Mexico. *Am. J. Primatol.* 48:167–183.

Freeland, W. J. 1983, Parasites and the coexistence of animal host species. *Am. Nat.* 2:223–236.

Gilbert, F. F. and Dodds, D. G. 1987, *The Philosophy and Practice of Wildlife Management.* Robert E. Krieger Publishing, Malabar, Florida.

Gilbert, K. A. 1994a, Parasitic infection in red howling monkeys in forest fragments. *Neotrop. Primates* 2:10–12.

Gilbert, K. A. 1994b, Endoparasitic infection in red howling monkeys *Alouatta seniculus* in the Central Amazonian basin. A cost of sociality? PhD Thesis, Rutgers University, New Brunswuick, New Jersey.

Gilbert, K. A. 1997, Red howling monkey use of specific defecation sites as a parasite avoidance strategy. *Anim. Behav.* 54:451–455.

Gilbert, K. A. and Stouffer, P. C. 1989, Use of a ground water source by mantled howler mlkneys (*Alouatta palliata*). *Biotropica* 21:380.

Gillespie, T. R. 2004, Effects of human disturbance on primate–parasite dynamics. PhD Dissertation, University of Florida, Gainesville, Florida.

Glander, K. E. 1978, Howling monkey feeding behavior and plant secondary compounds: A study of strategies, in: G. G. Montgomery, eds., *Ecology of Arboreal Folivores*, Smithsonian Institution Press, Washington, D.C., pp. 561–574.

Hausfater, G. and Watson, D. F. 1976, Social and reproductive correlates of parasite ova emissions by baboons. *Nature* 262:668–669.

Herrera-MacBryde, O. and Medellín, R. A. 1997, Lacandona rain forest region, México, in: S. D. Davis, V. H. Heywood, O. Herrera-Macbryde, J. Villa-Lobos, and A. C. Hamilton, eds., *Centers of Plant Diversity, A Guide and Strategy for their Conservation*, IUCN Publications Unit, Cambridge, U.K., pp. 125–129.

Horwich, R. H., Koontz, F., Saqui, H., and Glander, K. 1993, A reintroduction program for the conservation of the black howler monkey in Belize. *Endang. Species Updat.* 10:1–6.

Huffman, M. A. and Seifu, M. 1989, Observations on the illness and consumption of a possibly medicinal plant *Vernonia amygdalina* (Del.), by a wild chimpanzee in the Mahale Mountains National Park, Tanzania. *Primates* 30:51–63.

Huffman, M. A. and Wrangham, R. W. 1994, Diversity of medicinal plants use by chimpanzees in the wild, in: R. W. Wrangham, W. C. McGrew, F. B. de Wall, and P. G. Heltne, eds., *Chimpanzee Cultures*, Harvard University Press, MA, pp. 129–148.

Inskipp, T. and Wells, S. 1979, *International Trade in Wildlife*. Earthscan Publication, London.

Jones, C. 1994, Injury and disease of the mantled howler monkey in fragmented habitats. *Neotrop. Primates* 4:4–5.

Kalter, S. 1989, Infectious diseases of nonhuman primates in a zoo setting. *Zoo Biol.* 1:61–76.

Knezevich, M. 1998, Geophagy as a therapeutic mediator of endoparasitismin a free-ranging group of rhesus macaques (*Macaca mulatta*). *Am. J. Primatol.* 44:71–82.

Klein, S. L. 2004, Hormonal and immunological mechanisms mediating sex differences in parasite infection. *Parasite Immunol.* 26:247–264.

Kopper, G., Krambeck, A., Braga, Z., and Da Silva, H. 2000, Levantamento preliminar de endoparasitas do tubo digestivo de bigios *Alouatta guarida clamitans. Neotrop. Primates* 8:107–108.

Krief, S. 2003, Métabolites secondaires des plantes et comportement animal: Surveillance sanitaire et observations de l'alimentation des chimpanzés (*Pan troglodytes schweinfurthii*) en Ouganda. Activités biologiques et étude chimique des plantes consommées. Thèse. Muséum National d'Histoire Naturelle, Paris.

Long, E. G., Tsin, A. T., and Robinson, B. A. 1985, Comparison of the FeKal CON-Trate system with the formalin-ethyl acetate technique for detection of intestinal parasites. *J. Clin. Microbiol.* 22:210–211.

Markell, E. K. and Voge, M. 1984, *Parasitología: Diagnóstico, Prevención y Tratamiento*, 5thEdn. Editorial El Manual Moderno, México, D.F.

McGrew, W. C., Tutin, C. E. G., Collins, D. A., and File, S. K. 1989, Intestinal parasites of sympatric *Pan troglodytes* and *Papio* spp. at two sites: Gombe, Tanzania and Mt. Assirik, Senegal. *Am. J. Primatol.* 2:147–155.

Medellín, R. A. 1994, Mammal diversity and conservation in the Selva Lacandona, Chiapas, México. *Conserv. Biol.* 3:780–799.

Medellín, R. A. and Equihua, M. 1998, Mammal species richness and habitat use in rainforest and abandoned agricultural fields in Chiapas, México. *J. Appl. Ecol.* 35:13–23.

Myers, N. 1989, *Deforestation Rates in Tropical Forest and their Climatic Implications.* Friends of the World, London.

Milton, K. 1980, *The Foraging Strategy of Howler Monkeys: A Study in Primate Economics.* Columbia University Press, New York.

Milton, K. 1996, Effects of bot fly (*Alouattamyia baeri*) parasitism on free ranging howler monkey (*Alouatta palliata*) populations in Panama. *J. Zool. Lond.* 239:39–63.

Neville, M. K., Glander, K. E., Braza, F., and Rylands, A. B. 1988, The howling monkeys, genus *Alouatta*, in: A. Mittermeier, A. B. Rylands, A. Coimbra-Filho, and G. A. B. Fonseca, eds., *Ecology and Behaviour of Neotropical Primates*, World Wildlife Fund, Washington, D.C.

Noble, E. R., Noble G. A., Schad, G. A., and MacInnes, A. J. 1989, *Parasitology: The Biology of Animal Parasites.* Lea and Febiger, Philadelphia, Pennsylvania.

Nowak, R. M. 2000, *Walker's Primates of the World.* Johns Hopkins University Press, Baltimore, Maryland.

Nunn, C. L., Altizer, S., Jones, K. E., and Sechrest, W. 2003, Comparative tests of parasite species richness in primates. *Am. Nat.* 162:597–614.

Phillips-Conroy, J. E. and Knopf, P. M. 1986, The effects of ingesting plant hormones on schistosomiasis in mice: An experimental study. *Biochemical Systematics and Ecology* 6:637–645.

Price, E. 1928, Preliminary note on the identification of gastrointestinal helminth parasites of a wild troop of Howler Monkeys *Alouatta palliata* in southern Mexico, in: Pastor-Nieto, ed., *New Helminth Parasites from Central American Mammals. Proceeding US National Museum* 73:1–7.

Rodríguez-Luna, E. and Cortés-Ortiz, L. 1994, Translocacion y seguimiento de un grupo de monos *Alouatta palliata* librado en una isla (1988–1994). *Neotrop. Primates* 2:1–5.

Rodríguez-Luna, E., García-Orduña, F., Canales-Espinoza, D., and Serio-Silva, J. C. 1993, Introduction of howler monkey: A four year record. *Am. Assoc. Zool. Parks Aquar. Region Conference Proceedings*, 767–774.

Ramírez-Barrios, R. A., Barboza-Mena, G., Muñoza, J., Angulo-Cubillán, F.,

Hernández, E., González, F., and Escalona, F. 2004, Prevalence of intestinal parasites in dogs under veterinary care in Maracaibo, Venezuela. *Vet. Parasitol.* 121:11–20.

Santa Cruz, A., Borda, J., Patiño, E., Gómez, L., and Zunino, G. 2000, Habitat fragments and parasitism in howler monkeys (*Alouatta caraya*). *Neotrop. Primates* 8: 146–148.

SAS Institute, Inc. 2000, *SAS User's Guide: Statistic*, vol. 1, Version 6, 4th Edn. SAS Institute, Inc., Cary, North Carolina.

Schalk, G. and Forbes, M. R. 1997, Male biases in parasitism of mammals: Effects of study type, host age and parasite taxon. *Oikos* 78:67–74.

Schultes, R. E and Raffauf, R. F. 1990, *The Healing Forest. Medicinal and Toxic Plants of the Northwest Amazonia*. Dioscorides press, Portland, Oregon.

Schuurs, A. H. and Verheul, H. A. 1990, Effects of gender and sex steroids on the immune response. *J. Steroid. Biochem.* 35:157–172.

Scott, M. E. 1998, The impact of infection and disease on animal populations: Implications for conservation biology. *Conser. Biol.* 2:40–56.

Serio-Silva, J. C., Rico-Gray, V., Hernández-Salazar, M. T., and Espinosa-Gómez, R. 2002, The Role of *Ficus* (Moraceae) in the diet and nutrition of a troop of mexican howler monkeys *Alouatta palliata mexicana*, released on an island in southern Veracruz, Mexico. *J. Trop. Ecol.* 18:913–928.

Sherman, P. T. 1991, Harpy eagle predation on a red howler monkey. *Folia Primatol.* 56:53–56.

Stiles, C. W., Hassall, A., and Nolan, O. 1929, Key-catalogue of parasites reported for primates (monkeys and lemurs) with their possible public health importance and key-catalogue of primates for which parasites are reported. U. S. Treasury Department, Public Health Service, *Hygienic Lab. Bull.* 152:406–601.

Stoner, K. 1995, Prevalence and intensity of intestinal parasites in mantled howling monkeys (*Alouatta palliata*) in Northeastern Costa Rica: Implications for conservation biology. *Conser. Biol.* 2:539–546.

Stoner, K. E. 1996, Habitat preferences and seasonal patterns of activity and foraging in two troops of mantled howling monkeys (*Alouatta palliata*) in a rainforest in northeastern Costa Rica. *Int. J. Primatol.* 17:1–30.

Stuart, M. D., Greenspan, L. L., Glander, K. E., and Clarke, M. 1990, A coprological survey of parasites of wild mantled howling monkeys, *Alouatta palliata palliata. J. Wildl. Dis.* 26:547–549.

Stuart, M. D., Strier, K. B., and Pierberg, S. M. 1993, A coprological survey of parasites of wild muriquis, *Brachyteles arachnoides*, and brown howling monkeys, *Alouatta fusca. J. Helminthol. Soc. Wash.* 60:111–115.

Stuart, M. D. and Strier, K. B. 1995, Primates and parasites: A case for a multidisciplinary approach. *Int. J. Primatol.* 4:577–593.

Stuart, M., Pendergast, V., Rumfelt, S., Pierberg, S., Greenspan, L., Glander, D., and Clarke, M. 1998, Parasites of wild howlers (*Alouatta* spp.). *Int. J. Primatol.* 3:493–512.

Thatcher, V. E. and Porter, J. A. Jr. 1968, Some helminth parasites of Panamian primates. *Trans. Am. Microsc. Soc.* 87:186–196.

Vié, J. -C. and Richard-Hansen, C. 1997, Primate translocation in French Guiana—A preliminary report. *Neotrop. Primates* 5:1–3.

Wang, E.and Milton, K. 2003, Intragroup social relationships of male *Alouatta palliata* on Barro Colorado Island, Republic of Panama. *Int. J. Primatol.* 24:1227–1243.

Zuk, M. and McKean, K. A. 1996. Sex differences in parasite infections: Patterns and processes. *Int. J. Parasitol.* 26:1009–1023.

PART THREE

Behavior and Ecology

Introduction: Behavior and Ecology

M. S. M. Pavelka, A. Estrada, and P. A. Garber

Long-term studies of primates in Mesoamerica have contributed significantly to our understanding of the behavior, ecology, social organization, and life history strategies of neotropical monkeys. This section of the book highlights recent advances in this area. Long-term projects have focused almost exclusively on white-faced capuchin monkeys and black and mantled howlers. Not surprisingly, the chapters in this section deal with *Cebus* and *Alouatta*—Mesoamerican taxa that, while exhibiting very different adaptive patterns, represent the most geographically widespread and most well studied of New World primates.

The first chapter in this section highlights an important and often overlooked problem in the behavior and ecology literature—that is, the use of average body weight for a species as a predictive tool. These predictions, however, are only as reliable as the body weight data used, and average weights vary widely from

M. S. M. Pavelka • Department of Anthropology, University of Calgary, 2500 University Drive NW, Calgary, Alberta, Canada T2N 1N4. **A. Estrada** • Estación Biológica Los Tuxtlas, Instituto de Biología, Universidad Nacional Autónoma de México, Apartado Postal 176, San Andrés Tuxtla, Veracruz, México. **P. A. Garber** • Department of Anthropology, University of Illinois at Urbana-Champaign, 109 Davenport Hall, Urbana, IL 61801, USA.

New Perspectives in the Study of Mesoamerican Primates: Distribution, Ecology, Behavior, and Conservation, edited by Alejandro Estrada, Paul A. Garber, Mary S. M. Pavelka, and LeAndra Luecke. Springer, New York, 2005.

one sample time to another for the same individual, and from site to site. Ken Glander demonstrates this variability by examining average body mass for mantled howling monkeys (*Alouatta palliata*) from two sites in Costa Rica and one in Panama, including 34 years of body mass fluctuations in one population of mantled howlers. This examination reveals significant variability that is typically hidden in "average" body mass values. In her chapter on developmental plasticity in Costa Rican mantled howlers, Clara Jones further explores the variation expressed within individuals over their lifetimes. Building on the numerous studies that show behavioral plasticity in howlers, particularly with respect to diet, timing of reproduction, bisexual dispersal patterns, and number of males in a social group, Jones describes the results of an exploratory study of female chest circumference variation in response to habitat differences. This functional ecology approach is relatively new in primatology, and Jones argues that it will have important implications for primate and other mammalian development, energetics, life history, evolution, and conservation because it involves an understanding of growth, survival, and reproduction relative to environmental regimes. Paul Garber takes a cognitive ecology approach, using feeding behavior to assess cognitive abilities in mantled howlers (with P. Jelinek) and in white-faced cebus monkeys (with E. Brown). In mantled howlers, Garber and Jelinek examine foraging strategies and travel patterns in a group of Nicaraguan mantled howler monkeys, documenting howler path sequences and use of consecutive feeding and resting sites in order to address questions about the degree to which mantled howlers represent spatial information as a route-based or as a metric-based cognitive representation. The degree to which howlers use topological (route-based) spatial representations or geometric (coordinate-based) spatial representations to locate and revisit these feeding and resting sites has been unclear, but in this study howlers were found to take direct routes to feeding and resting sites and reuse the same route segments on several occasions. On the basis of data on canopy visibility, tree distribution, and the distance between sequential feeding sites, the authors conclude that howlers rely on particular landmark cues to orient and reorient their direction of travel, maintaining information about the locations of numerous intersecting routes of travel and landmarks within their home range. Reuse of these route segments is consistent with a typological or route-based spatial representation.

An experimental field study of wild capuchin monkeys explored the extent to which these larger brained omnivores, reported to rely on complex spatial information to locate distant feeding sites, actually use landmark cues to locate feeding sites. In this study, Garber and Brown use field experimental design to assess the ability of *Cebus capucinus* to use landmarks singly and relationally to compute the location of baited feeding sites in small-scale space. Their results indicate that capuchins quickly learn to attend to the spatial positions of landmark arrays and use this information to compute the location of reward platforms. Further, given that foragers arrived at the feeding station from different directions and encountered alternative views of the landmarks, wild capuchins may be able to mentally rotate the configuration of the landmark array to efficiently solve this foraging problem.

Two chapters in this section explore ontogeny in *Cebus* and *Alouatta*. First, Michelle Bezanson examines evidence of age-related or ontogenetic patterns of locomotion in white-faced capuchins and mantled howling monkeys, two species that share a pattern of growth in which limb proportions and body mass increase with age reducing postnatal hand and foot dominance, yet which show important differences in the timing of life-history events and in patterns of growth. *Cebus* appear to show adult-like locomotor behavioral patterns at an earlier age than do the howlers. Bezanson concludes that ontogenetic differences in linear growth, body mass, and life history timing did not predictably influence locomotion in either species. Next, Katherine MacKinnon expands our understanding of juvenile foraging behavior by examining how the diets of smaller and larger juvenile capuchin monkeys (among the most altricial at birth of all neotropical monkeys) vary with seasonal changes in a tropical dry forest, as compared with adults. She argues that ontogenetic factors such as physiological constraints, the effects of experience, and differing nutritional requirements play an important role in food choice and food acquisition in young capuchins. The data also suggest that the dietary profiles of smaller juveniles, larger juveniles, and adults are quite similar.

In two chapters from the long-term project at Santa Rosa National Park in Costa Rica, researchers employ new methods of field genetics and field endocrinology to significantly advance our understanding of mating strategies of male and female capuchin monkeys. First, Katharine Jack and Linda Fedigan explore relationships between dominance and reproductive success, and reveal much about the advantages of being an alpha male. Using recent advances in paternity testing in field situations (using DNA obtained from hair and feces amplified using PCR - Polymerase Chain Reaction), these authors were able to perform paternity exclusions on 15 of 19 infants genotyped. The analysis showed that alpha males in two study groups sired significantly more offspring (a minimum of 13 and maximum of 17) than did subordinates (who sired between two and six of the infants born). Despite an egalitarian mating system without overt mate guarding by dominant males, these data indicate a strong reproductive advantage for alpha males in this species.

How are alpha males obtaining these reproductive advantages without overt male-male competition for access to mates? Carnegie, Fedigan, and Ziegler provide an answer to this question in their chapter on postconceptive mating in the same population of capuchin monkeys. They examined sexual and affiliative behaviors of females who were determined, by hormonal assays, to be cycling, noncycling, or pregnant, in an effort to determine if there are behavioral indicators of reproductive condition in wild female capuchins. They discovered that cycling and pregnant females copulated and received more courtship displays from adult males than did noncycling females, and that pregnant females mated primarily with subordinate males whereas cycling (fertile) females mated with alpha males. Pregnant females appeared to actively court subordinate males—possibly an infanticide avoidance strategy—and cycling females accepted the increased advances they received from alpha males. This combination of genetic, hormonal, and behavioral information highlights the value of integrating new methodologies in piecing together the puzzle of primate reproductive strategies.

CHAPTER NINE

Average Body Weight for Mantled Howling Monkeys (*Alouatta palliata*): An Assessment of Average Values and Variability

Kenneth E. Glander

INTRODUCTION

Body weight is a universally accepted morphological descriptor for most organisms and is frequently given as a group or population average. This average is then used as a predictive tool in many fields such as ecology (Kendeigh *et al.*, 1977); behavior (Clutton-Brock and Harvey, 1977a,b); physiology (Pedley, 1977; Schmidt-Nielsen, 1979); and paleontology (Gould, 1966; Rensch, 1960).

Kenneth E. Glander • Biological Anthropology and Anatomy, Duke University, Durham, NC 27708.

New Perspectives in the Study of Mesoamerican Primates: Distribution, Ecology, Behavior, and Conservation, edited by Alejandro Estrada, Paul A. Garber, Mary S. M. Pavelka, and LeAndra Luecke. Springer, New York, 2005.

These predictions are only as reliable as the average body weight chosen because often there are several available. Frequently, this is not apparent to the reader of a published paper as no mention is made of the fact that there was more than one "average" available. Sometimes an average of the means from different sites is given without the original weights, standard deviations, or sample size specified (Froelich and Thorington, 1982). Further, weakening the validity of a cited average body weight is that the original source was an unpublished source.

In this paper, I present average body mass for mantled howling monkeys (*Alouatta palliata*) from two sites in Costa Rica and one in Panama, examine what a given "average" body mass number hides, and offer detailed analyses of 34 years of body mass fluctuations in one population of mantled howlers. The geographic relationship of the three study sites is given in Figure 1.

METHODS

Sites

Hacienda La Pacifica, Costa Rica

Hacienda La Pacifica (LP) (10°28′N and 85°07′W) is a 1990-ha ranch containing a mosaic of farmland, dry tropical forest, evergreen forest, and reclaimed pasture in varying degrees of secondary succession (Glander and Nisbett, 1996). The ranch experiences distinct wet (May to November) and dry (December to April) seasons with 99% of the average rainfall of 1553 mm (records for 81 years) occurring during the wet season of mid-May to mid-December.

Santa Rosa National Park, Costa Rica

Santa Rosa (SR) (10° 45′ to 11° 00′ N and 85° 30′ to 85° 45′ W) is a 10,700-ha National Park created in 1971 (Chapman *et al.*, 1995). The park experiences two distinct seasons with 98% of the 1527-mm yearly average rainfall occurring from late May to mid-December (Chapman *et al.*, 1995).

Barro Colorado Island, Panama

Barro Colorado Island (BCI) (9°09′N, 79°51′W) is a 1500-ha nature reserve created in 1914 when Lake Gatun was produced during the building of the Panama Canal (Leigh, 1982). The average rainfall for the island is 2600 mm

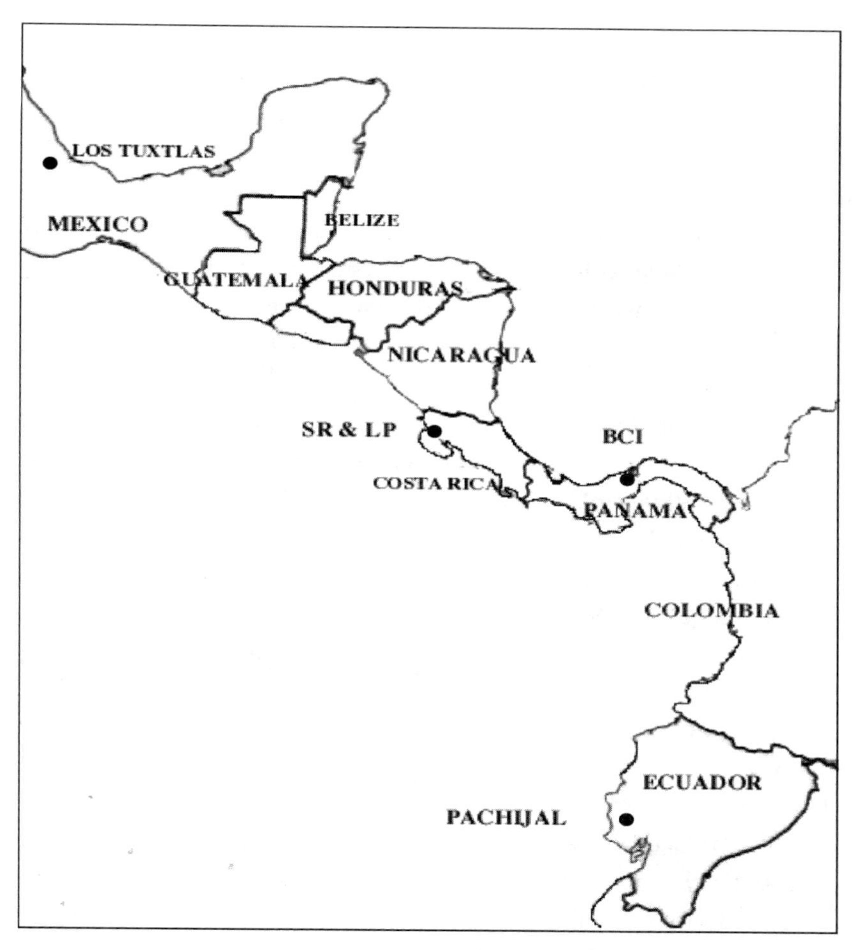

Figure 1. Location of the Costa Rican and Panamanian study sites plus sites from the northern and southern extremes of mantled howler geographic distribution. Map is modified version of the one supplied by courtesy of the General Libraries, The University of Texas at Austin. http://www.lib.utexas.edu/maps/americas/middleamerica.jpg, http://www.lib.utexas.edu/usage_statement.html

(records for 50 years) with 90% of that falling during the wet season from May through December (Rand and Rand, 1982).

Procedure

Individuals were captured using the Pneu-Dart^{TM} system (Pneu-Dart, Inc., HC 31, Williamsport, PA 17701,USA) as part of long-term studies at each of the study sites (for a complete description of the capture methodology, see

Glander *et al.*, 1991). Once captured, the monkeys were weighed, measured, and marked for positive identification (unique collars plus tattoos as well as Avid chips (AVID, Norco, CA) for the LP individuals). Body weight in kg (to the nearest 100 g) was obtained using a 20 kg Pesola® scale.

All body weights are for fully adult individuals. Females were judged to be adult at 48 months and males at 60 months (based on known ages of 50 individuals of each sex). Weights for pregnant females were excluded after 3 months of pregnancy. The fetus weighs less than 100 g until after the fourth month. Gestation for mantled howlers is 6 months (Glander, 1980). All captured females were palpated and fetuses 3–4 weeks old can be detected.

The body weights for LP have been collected as part of a continuing long-term study begun in 1970 (Clarke, 1990; Clarke *et al.*, 2002; Clarke and Zucker, 1994; Glander, 1980, 1992; Glander and Nisbett, 1996; Glander *et al.*, 1991; Stuart *et al.*, 1998; Teaford and Glander, 1997; Ungar *et al.*, 1995; and references in these papers). The weights for SR were obtained in 1985, 1986, and 1992 (Chapman *et al.*, 1995; Glander *et al.*, 1991). The weights for BCI were collected in 1986 and 2000 (Glander and Milton, in prep.).

RESULTS

The body mass of howling monkeys from three locations (two in Costa Rica and one in Panama) was significantly different (Table 1). All three populations are sexually dimorphic with the males being significantly heavier than the females (BCI: $F = 65.71$, $p < 0.00001$); SR: $F = 65.60$, $p < 0.00001$; LP: $F = 621.72$, $p < 0.00001$). Males from BCI were significantly heavier than males from both SR ($F = 23.33$, $p < 0.00001$) and LP ($F = 295.50$,

Table 1. Average body weight for mantled howlers at three sites. BCI = Barro Colorado Island, Panama; SR = Santa Rosa, Costa Rica; LP = La Pacifica, Costa Rica. See Figure 1 for locations

	Males	Females
BCI	7562 g (±731) $N = 38$	6445 g (±553) $N = 49$
SR	6573 g (±483) $N = 15$	5161 g (±537) $N = 21$
LP	5790 g (±578) $N = 288$	4726 g (±616) $N = 663$

The values in the parentheses are SD.

$p < 0.00001$). The SR males were significantly heavier than the LP males ($F = 26.54$, $p < 0.0001$). BCI females were significantly heavier than females from both SR ($F = 80.48$, $p < 0.00001$) and LP ($F = 360.50$, $p < 0.00001$). The SR females were significantly heavier than the LP females ($F = 10.25$, $p < 0.001$).

Females were significantly heavier during the wet season while males did not demonstrate a seasonal fluctuation in body weight (Table 2). Both females and males from Riparian (River) forests were significantly heavier than females and males living in Upland (Dry) forests (Table 3).

When season and habitat were considered together by sex, there were no seasonal differences in body weight either for females (Table 4) or for males (Table 5), but there was a habitat effect. Females living in Riparian Forests were significantly heavier in both dry and wet seasons than those living in Upland Forests (Table 4) just as the males living in Riparian Forests also were significantly heavier in both seasons than those in the Upland Forests (Table 5).

For individuals, the amount and percentage of change in body weight were much greater than the average. Females fluctuated 616 g on average or 13% of their body weight (Table 1), but it was as much as 1506 g for a riparian female (Purple) or 23% of her body weight (Figure 2) and 1500 g for an Upland female (Trinka) or 30% (Figure 2). Male body weights fluctuated less, both on average

Table 2. Seasonal differences in weights for La Pacifica howlers

	Dry season	Wet season	*F*	*p*-Value
Females	4598 g (±658) *N* = 106	4750 g (±607) *N* = 539	5.18	0.02
Males	5691 g (±630) *N* = 48	5807 g (±571) *N* = 233	1.58	NS

The values in the parentheses are SD.

Table 3. Body weight comparisons for La Pacifica howlers by habitat

	Upland Forest	Riparian Forest	*F*	*p*-Value
Females	4478 g (±521) *N* = 262	5006 g (±630) *N* = 243	105.91	<0.00001
Males	5695 g (±472) *N* = 134	6163 g (±598) *N* = 63	35.24	<0.00001

The values in the parentheses are SD.

Table 4. Female body weight by season and habitat at La Pacifica. Habitats within seasons affect body weights (columns) while season within the habitat does not (rows)

	Dry season	Wet season	*F*	*p*-Value
Upland Forest	4387 g (±543) $N = 66$	4509 g (±510) $N = 196$	2.70	NS
Riparian Forest	4994 g (±683) $N = 36$	5008 g (±622) $N = 207$	0.01	NS
F	24.18	77.07		
p-Value	<0.00001	<0.00001		

The values in the parentheses are SD.

Table 5. Male body weight by season and habitat at La Pacifica. Habitats within seasons affect body weights (columns) while season within habitats does not (rows)

	Dry season	Wet season	*F*	*p*-Value
Upland Forest	5606g (±570) $N = 34$	5725g (±434) $N = 100$	1.61	NS
Riparian Forest	6086g (±662) $N = 7$	6172g (±595) $N = 56$	.13	NS
F	3.92	28.98		
p value	0.05	<0.0001		

The values in the parentheses are SD.

and for specific individuals. On average, all males fluctuated 578 g or 10% of their body weight (Table 1) while a Riparian male (Bandit) oscillated 1300 g or 18% of his body weight and an Upland male (Chief) varied 1000 g or 16% (Figure 3).

DISCUSSION

It is clear from these results that there is no "average" body weight for mantled howlers. Populations of mantled howlers from three different locations in Central America differed significantly in body mass even though the two Costa Rica populations (LP and SR) are only 70 km apart. The SR males are 14% larger than the LP males while the SR females are 9% larger in average body weight (Table 1). The difference in average body weight is much greater between Panama and Costa Rican populations with BCI males being 15% larger than SR males and 31% heavier than LP males while BCI females are 25% larger than SR females and 36% heavier than LP females (Table 1).

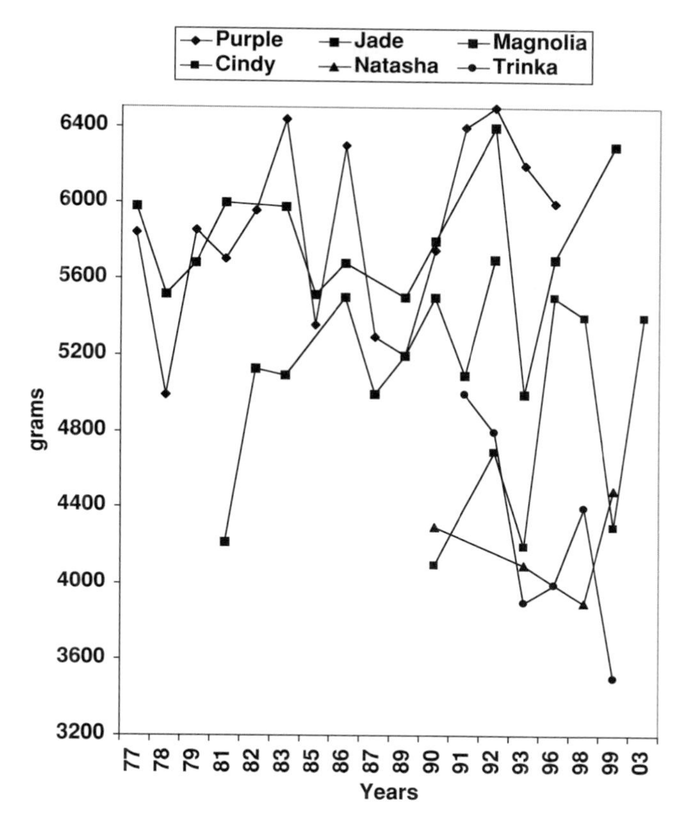

Figure 2. Weight fluctuations for females at La Pacifica. Purple, Jade, and Magnolia are from Riparian Forest while the other three are from Upland Forest.

Variation is even greater when the actual range for each population is considered rather than the range between "average" body mass. The body weight range of the LP population was 3320 g for adult females and 3000 g for adult males (Table 6). The smallest range was found in the SR population, but it was still 2000 g for females and 1500 g for males. Estrada (1982) reported similar ranges in both sexes for a Mexican population of mantled howlers (Table 6). In all documented populations, there is overlap between the sexes in body weight, that is, there were some females as large or larger than some males (Table 6).

Discussing "average" weights for various populations or even the actual ranges in body weights of a population does not value what fluctuations in body weight signify for individuals. Weight change presumes biological consequences, especially when a female loses 23–30% of her body mass (Purple and Trinka, Figure 2). Going from 5 kg to 3.5 kg (Trinka) may have a major impact

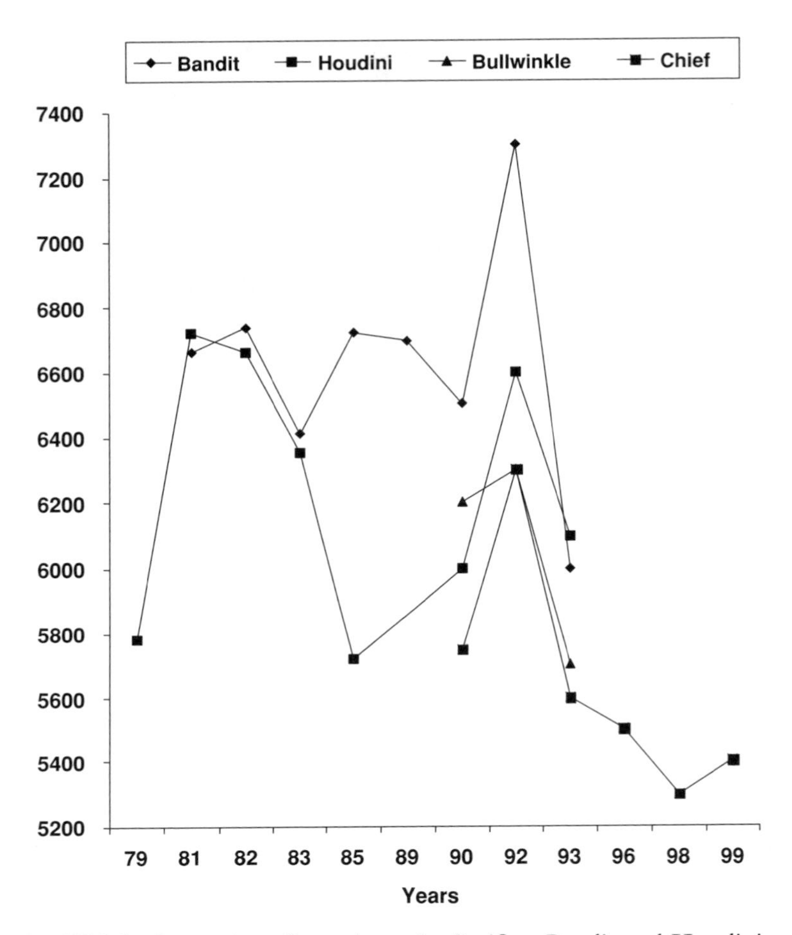

Figure 3. Weight fluctuations for males at La Pacifica. Bandit and Houdini are from Riparian Forest while the other two are from Upland Forest.

Table 6. Range of body weight from North to South in the geographic range of mantled howlers. See Figure 1 for site locations

Site	Males (g)	Females (g)	Source
Los Tuxtlas, Mexico	6430–9007 $N = 5$	5000–8000 $N = 7$	Estrada (1982)
Santa Rosa, Costa Rica	5750–7250 $N = 15$	4000–6000 $N = 21$	Glander *et al.* (1991)
La Pacifica, Costa Rica	4300–7300 $N = 288$	3178–6500 $N = 663$	This paper
BCI, Panama	6000–8750 $N = 38$	5300–7900 $N = 49$	Glander and Milton (in prep.)
Pachijal, Ecuador	7000–7150 $N = 2$	na	Glander (unpublished)

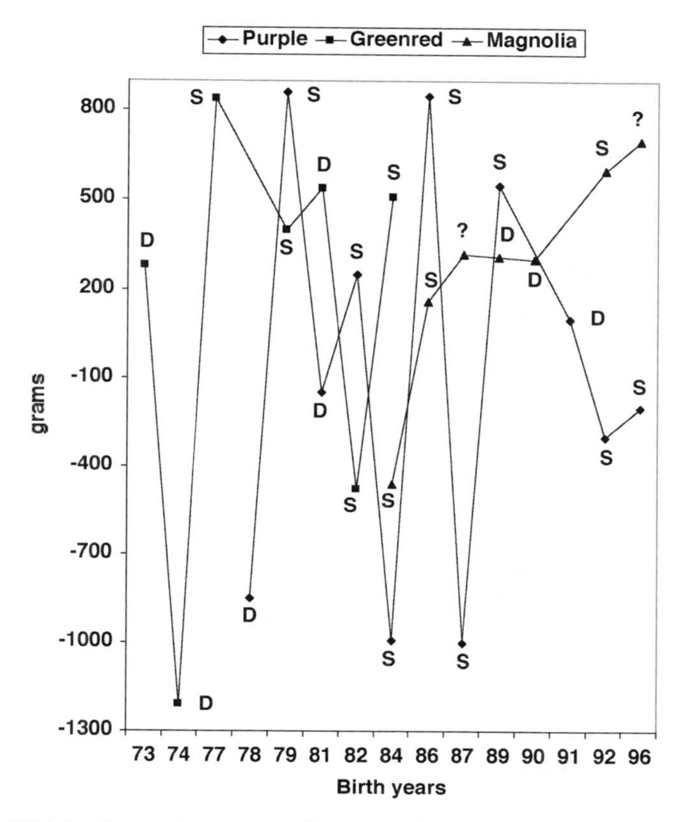

Figure 4. Weight fluctuations and infant mortality for three riparian females at La Pacifica. The first data point is the difference (plus or minus) from the previous year. The gain or loss occurred within ±6 months of the birth, thus impacts both gestation and lactation. D = died; S = survived; ? = unknown fate (D, S, and ? pertain to the infant's fate).

on life expectancy and reproduction (but see next paragraph). The males in this population experienced less actual body weight loss (900–992 g) and percentage loss (11%) (Figure 4). One reason for this may be that males are dominant and are not limited in their access to high quality food (Glander, 1992) as are the lower ranked females (Purple and Trinka were both low ranked in their groups). Also, males do not bear the burden of pregnancy or lactation that may contribute to a female's weight change.

Many models of within-group feeding competition assume that even slight differences in feeding efficiency or access to resources have a major impact on individual survivorship and fitness. Yet my data show that individuals can lose

10–30% of their body mass within a given year and still survive for many years (Figure 2). For example, Purple lost 1000 g or 15% of her body mass between July 1986 and August 1987, but lived for another 11 years. Trinka lost 900 g or 19% of her body weight between July 1992 and February 1993, but lived another 9 years. Both of these females had other dramatic weight fluctuations, as did all other females at LP without affecting their life span.

The same models suggest that a reduction in body mass should result in a reduction in fertility or infant survivorship. Again, my data contradict this assumption (Figure 4). Purple suffered weight losses of 990 and 1000 g but her infants born during and/or after these changes survived. Greenred experienced a weight loss of 480 g, but her infant also survived. These are just a few examples of many similar occurrences in the LP population. Figure 4 clearly shows that weight gain or loss is not correlated with infant mortality when viewed over the long-term.

These extreme variations in body mass of individuals or populations did not occur between members of different species but in individuals of the same species, i.e., *A. palliata*, or more commonly known as mantled howling monkeys (also referred to as howler monkeys, but I prefer the grammatically correct phrase "howling monkey" or "howlers" used by C. R. Carpenter in 1934). Mantled howlers have one of the largest geographic distributions, from southern Mexico to Ecuador (Rowe, 1996; Wolfheim, 1983). Within this geographic range, the "average" male howler weighs 4300–9007 g and the "average" female weighs 3178–8000 g (Table 6). The largest of each sex is more than double the smallest, yet all of these individuals are *A. palliata* (Ellsworth, 2000).

This tremendous range in body mass should raise a red flag for taxonomists, paleontologists, ecologists, and behaviorists who rely on body size to distinguish species, allocate species designation to fossils, study niche separation, and even predict behavior. Body size is one of the factors used to differentiate species (Groves, 2001) and differences in size are used by paleontologists to assign fossils to different species (Ciochon *et al.*, 2001). In ecological theory, body size is believed to be critical in foraging (Temerin *et al.*, 1984) as well as in community structure (Schoener, 1984; Terborgh, 1983). Vervaecke *et al.* (1999) used body size to estimate relative fighting abilities in *Pan paniscus* and predict dominance hierarchies.

The astonishing variation for each sex described here obviously calls into question the use of "average" body weights in any of these theories and models. The generally held theory is that reduced competition should select for larger individuals (Lomolino, 2004). If this were the case, then LP howlers should be heavier than either the SR or BCI howlers since there are no other primates occupying the LP habitat while *Cebus capucinus* and *Ateles geoffroyi* share both the SR and BCI habitats with howlers. A similar line of reasoning suggests that smaller members of a feeding guild are usually heavier where competitors are missing (Grant, 1965; Schoener, 1970; McNab, 1971). The reality is the reverse of the expected.

Using habitat variations to explain these divergences in body mass for the mantled howlers is tempting since mantled howler geographic distribution is extensive with habitats including deciduous, riparian, evergreen, and montane forests (Crockett and Eisenberg 1987). One of the primary sources of habitat variation is food availability. And, established theory holds that reduced food availability should select for smaller body size (Hessee *et al.*, 1951; Lawlor, 1982). The BCI habitat is certainly different from the LP and SR habitats in rainfall and tree species composition and the BCI howlers are significantly heavier than either of the Costa Rican populations. However, the LP and SR habitats are similar in terms of tree species and rainfall (Chapman *et al.*, 1995; Glander and Nisbett, 1996) and the howlers at SR are also significantly heavier than those at LP despite similar habitats and similar food availability (Chapman *et al.*, 1995; Glander and Nisbett, 1996).

BCI does have more tree species and many more *Ficus* tree species and individuals (Milton, 1980) than either Costa Rican site (Chapman *et al.*, 1995; Glander and Nisbett, 1996). Certainly, the abundance of figs could help in explaining the larger body size of the BCI howlers despite the competition from two other primate species (Milton, 1982). But, there is no similar difference in food availability between the LP and the SR to explain the variation in body weights between these two Costa Rican populations.

It is also tempting to use population density as an explanation for body size variation, but the howler density of 72.7/km^2 for LP, Costa Rica (Clarke *et al.*, 2002), is less than the 91.7/km^2 on BCI, Panama (Milton, 1982). Further diluting this argument are the population densities of 4.9/km^2 at Santa Rosa National Park, Costa Rica (Fedigan *et al.*, 1985) and 23/km^2 at Los Tuxtlas in Mexico (Estrada, 1982). The largest "average" body size for both sexes is found in the habitat with the highest density of howlers (BCI) and this does not consider the added presence of two other primate species on BCI. Santa Rosa with the lowest population density has howlers that are significantly heavier than LP, but significantly lighter than BCI with the highest population density.

Based on these numbers, there is no relationship between howler population density and body size. The population density of the other primate species at some of these sites is not relevant because the competition explanation is not applicable as discussed earlier.

An argument could be made that the weight differences are due to microhabitat differences, but these differences (if they exist) have not been discovered after long-term studies at both sites (21 years at SR and 34 years at LP).

These 34 years of research at LP and 1564 individual captures of uniquely marked animals provided an opportunity to examine body weight changes in the LP howlers over their life span (30 years), seasonally, and by habitat. Only the females showed a seasonal effect by weighing more in the wet season (Table 2). Both males and females living in the riparian forests were significantly heavier than those in the Upland forests (Table 3). It is not clear why only the females demonstrated a seasonal effect because fruit is primarily a wet season occurrence (Glander and Nisbett, 1996) and the males are dominant to all females (Glander, 1980). The habitat variation effect may be due to the availability of fruit since there are more fruiting species in the riparian habitats (Glander and Nisbett, 1996).

Despite the fact that females were significantly heavier in the wet season while the males showed no seasonal difference (Table 2), there were season and habitat effects by sex. There were no seasonal effects for females and males in either habitat. Tables 4 and 5 clearly show that habitat has a greater impact on body size than season. This points out the importance of looking at more than just overall seasonal and habitat differences. In fact, the overall seasonal impact disappears when broken down by season within habitats. The weight of females and males living in both types of habitat does not change seasonally, but both sexes living in the Riparian habitats are heavier than those living in the Upland forests.

CONCLUSION

Mantled howling monkeys (*A. palliata*) from two Costa Rican populations (SR and LP) and the island population of BCI were significantly different in body size with the "average" body weight of females ranging from 6445 g for BCI to 4726 g for LP. The "average" male body weight ranged from 7562 g for BCI to 5790 g for LP. The BCI females and males were significantly heavier than either of the Costa Rican populations while both SR sexes were significantly heavier than their LP counterparts.

As expected, the sexes were significantly different in body size; however, there was much greater variation in body weight between the populations than is usual when considering the same species. Especially, when body weights within populations exceeded that between populations. There also were seasonal and habitat variations in the weights of LP individuals and extreme individual fluctuations that did not correspond to reproductive condition nor to infant mortality.

These extreme weight differences are not explained by competition or by food availability for the Costa Rican populations. The type and variety of food in BCI may be a factor in the case of BCI howlers of both sexes being significantly heavier than the Costa Rican populations, even though the BCI population has the highest density and greater potential competition from other primate species. The heaviest individual howler of each sex is found at Los Tuxtlas, Mexico; a very different habitat from that found in BCI in terms of rainfall and tree species composition.

An "average" body mass is a myth for mantled howling monkeys. Anyone using mantled howlers as models must carefully consider the season, forest type, and population rather than simply using one average body size to represent all *A. palliata*. Unfortunately, detailed analyses of season, forest type, and populations are seldom available, but if they are, the "average" weight used for predictive purposes should be matched to the finest grained conditions available.

SUMMARY

A comparison of mantled howling monkey (*A. palliata*) body weights from the two Costa Rican populations at SR and LP plus the island population of BCI yielded average body weights of 6445 g for BCI females ($N = 49$), 5161 g for SR females ($N = 21$), and 4726 g for LP females ($N = 663$). Average male body weight for these same three populations was 7562 g for BCI ($N = 38$), 6573 g for SR ($N = 15$), and 5790 g for LP ($N = 288$). All three populations are sexually dimorphic with the males being significantly heavier than the females (BCI: $F = 65.71$, $p < 0.00001$); SR: $F = 65.60$, $p < 0.00001$; LP: $F = 621.72$, $p < 0.00001$).

The BCI females are significantly heavier than the SR females ($F = 80.48$, $p < 0.00001$) and LP females ($F = 360.50$, $p < 0.00001$). The BCI males are significantly heavier than the SR males ($F = 23.23$, $p < 0.00001$) and the

LP males ($F = 295.50$, $p < 0.00001$). The SR females are significantly heavier than the LP females ($F = 10.25$, $p < 0.001$) and the SR males are significantly heavier than the LP males ($F = 26.54$, $p < 0.0001$).

Female weight at LP showed a seasonal difference while season had no effect on male weight. There were habitat effects on both female and male weights. Individual body weights at LP oscillated from 10% to 30% within and between years. These dramatic changes in body mass did not reduce life span nor affect infant survivorship.

ACKNOWLEDGMENTS

Thanks to Stephan Schmidheiny and the Board of Directors of La Pacifica for their permission to work on La Pacifica and for their continued support and help. Special thanks goes to Lily and Werner Hagnauer and Verena and Tony Leigh. The success of the La Pacifica Howler Research Project is due to the contributions of many researchers and students. There are too many to mention individually but my gratitude goes to each one with notable thanks to the many Earthwatch volunteers who have supported this project. This research has been funded in part by NSF grants GS-31733 and BNS 8819733 plus an REU Supplement, Duke University Research Council grants, Duke University Biomedical Research Support grants, and COSHEN-Pew grants.

REFERENCES

Carpenter, C. R. 1934, A field study of the behavior and social relations of howling monkeys (*Alouatta palliata*). *Comp. Psychol. Monogr.* 10:1–168.

Chapman, C. A., Wrangham, R. W., and Chapman, L. J. 1995, Ecological constraints on group size: An analysis of spider monkey and chimpanzee subgroups. *Behav. Ecol. Soc.* 36:59–70.

Ciochon, R. L., Gingerich, P. D., Gunnell, G. F., and Simons, E. L. 2001, Primate postcrania from the late middle Eocene of Myanmar. *Proc. Natl. Acad. Sci.* 98:7672–7677.

Clarke, M. R. 1990, Behavioral development and socialization of infants in a free-ranging group of howling monkeys (*Alouatta palliata*). *Folia primatol.* 54:1–15.

Clarke, M. R., Crockett, C. M., Zucker, E. L., and Zaldivar, M. 2002, Mantled howler population of Hacienda La Pacifica, Costa Rica, between 1991 and 1998: Effects of deforestation. *Am. J. Primatol.* 56:155–163.

Clarke, M. R. and Zucker, E. L. 1994, Group takeover by a natal male howling monkey (*Alouatta palliata*) and associated disappearances and injures of immatures. *Primates* 35:435–442.

Clutton-Brock, T. H. and Harvey, P. H. 1977a, Species differences in feeding and ranging behavior in primates, in: T. H. Clutton-Brock, ed., *Primate Ecology*, Academic Press, New York, pp. 557–584.

Clutton-Brock, T. H. and Harvey, P. H. 1977b, Primate ecology and social organization. *J. Zool.* 183:1–39.

Crockett, C. M. and Eisenberg, J. F. 1987, Howlers: Variation in group size and demography, in: B. B. Smuts, D. L. Cheney, R. M. Seyfarth, R. W. Wrangham, and T. T. Struhsaker, eds., *Primate Societies*, University of Chicago Press, Chicago, pp. 54–68.

Ellsworth, J. A. 2000, Molecular evolution, social structure, and phylogeography of the mantled howler monkey (*Alouatta palliata*). Ph.D. Dissertation, University of Nevada, Reno.

Estrada, A. 1982, Survey and census of howler monkeys (*Alouatta palliata*) in the rain forest of "Los Tuxtlas", Veracruz, Mexico. *Am. J. Primatol.* 2:363–372.

Fedigan, L. M., Fedigan, L., and Chapman, C. 1985, A census of *Alouatta palliata* and *Cebus capucinus* monkeys in Santa Rosa National Park, Costa Rica. *Brenesia* 23:309–322.

Froelich, J. W. and Thorington, R. W. Jr. 1982, The genetic structure and socioecology of howler monkeys (*Alouatta palliata*) on Barro Colorado Island, in: E. G. Leigh, Jr., A. S. Rand, and D. M. Windsor, eds., *The Ecology of a Tropical Forest*, Smithsonian Institution Press, Washington, D.C., pp. 291–305.

Glander, K. E. 1980, Reproduction and population growth in free-ranging mantled howling monkeys. *Am. J. Phys. Anthropol.* 53:25–36.

Glander, K. E. 1992, Dispersal patterns in Costa Rican mantled howling monkeys. *Int. J. Primatol.* 13:415–436.

Glander, K. E., Fedigan, L. M., Fedigan, L., and Chapman, C. 1991, Capture techniques and measurements of three monkey species in Costa Rica. *Folia Primatol.* 57:70–82.

Glander, K. E. and Nisbett, R. A. 1996, Community structure and species density in tropical forest associations in Guanacaste Province, Costa Rica. *Brenesia* 45–46:113–142.

Gould, S. J. 1966, Allometry and size in ontogeny and phylogeny. *Biol. Rev. Camb. Philos. Soc.* 41:587–640.

Grant, P. R. 1965, The adaptive significance of some size trends in island birds. *Evolution* 19:353–367.

Groves, C. 2001, Why taxonomic stability is a bad idea, or why are there so few species of Primates (or are there?). *Evol. Anthropol.* 10:192–198.

Hessee, R., Allee, W. C., and Schmidt, K. P. 1951, *Ecological Animal Geography*. Wiley, New York.

Kendeigh, S. C., Dol'nik, V. R., and Govrilov, V. M. 1977, Avian energetics, in: J. Pinowski and S. C. Kendeigh, eds., *Granivorous Birds in Ecosystems.* Cambridge University Press, Cambridge, pp. 127–204.

Lawlor, T. E. 1982, The evolution of body size in mammals: Evidence from insular populations in Mexico. *Am. Nat.* 119:54–72.

Leigh, E. G., Jr. 1982, Introduction, in: E. G. Leigh, Jr., A. S. Rand, and D. M. Windsor, eds., *The Ecology of a Tropical Forest*, Smithsonian Institution Press, Washington, D.C., pp. 11–17.

Lomolino, M. V. 2004, Body size of mammals on Islands: The island rule reexamined. *Am. Nat.* 125:310–316.

McNab, B. K. 1971, On the ecological significance of Bergmann's rule. *Ecology* 52:845–854.

Milton, K. 1980, *The Foraging Strategies of Howler Monkeys.* Columbia University Press, New York.

Milton, K. A. 1982, Dietary quality and demographic regulation in a howler monkey population, in: E. G. Leigh, Jr., A. S. Rand, and D. M. Windsor, eds., *The Ecology of a Tropical Forest*, Smithsonian Institution Press, Washington, D.C., pp. 273–289.

Pedley, T. J. 1977, *Scale Effects in Animal Locomotion.* Academic Press, New York.

Rand, A. S. and Rand, W. M. 1982, Variation in rainfall on Barro Colorado Island, in: E. G. Leigh, Jr., A. S. Rand, and D. M. Windsor, eds., *The Ecology of a Tropical Forest*, Smithsonian Institution Press, Washington, D.C., pp. 47–59.

Rensch, B. 1960, *Evolution Above the Species Level.* Columbia University Press, New York.

Rowe, N. 1996, *The Pictorial Guide to the Living Primates.* Pogonian Press, New York.

Schmidt-Nielsen, K. 1979, *Animal Physiology: Adaptations and Environment*, 2nd Edn. Cambridge University Press, Cambridge.

Schoener, T. 1970, Size patterns in West Indian *Anolis* lizards. II. Correlation with the sizes of particular sympatric species-displacement and convergences. *Am. Nat.* 104:155–174.

Schoener, T. W. 1984, Size differences among sympatric bird-eating hawks, in: D. R. Strong, Jr., D. Simberloff, L. G. Abele, and A. B. Thistle, eds., *Ecological Communities: Conceptual Issues and the Evidence*, Princeton University Press, Princeton, pp. 254–281.

Stuart, M. V., Pendergast, S., Rumfelt, S., Pierberg, L., Greenspan, L., Glander, K., and Clarke, M. 1998, Parasites of wild howlers (*Alouatta* spp.) *Int. J. Primatol.* 19:493–512.

Teaford, M. R. and Glander, K. E. 1997, Dental microwear and diet in a wild population of mantled howlers (*Alouatta palliata*), in: M. A. Norconk, A. L. Rosenberger, and P. A. Garber, eds., *Adaptive Radiations of Neotropical Primates*, Plenum Press, New York, pp. 433–449.

Temerin, L. A., Wheatley, B. P., and Rodman, P. S. 1984, Body size and foraging in primates, in: P. S. Rodman and J. G. H. Cant, eds., *Adaptations for Foraging in Nonhuman Primates*, Columbia University Press, New York, pp. 215–248.

Terborgh, J. 1983, *Five New World Primates: A Study in Comparative Ecology*. Princeton University Press, Princeton.

Ungar, P. S., Teaford, M. F., Glander K. E., and Pastor, R. F. 1995, Dust accumulation in the canopy: A potential cause of dental microwear in primates. *Am. J. Phys. Anthropol.* 97:93–99.

Vervaecke, H., de Vries, H., and van Elsacker, L. 1999, An experimental evaluation of the consistency of competitive ability and agonistic dominance in different social contexts in captive bonobos. *Behaviour* 136:423–442

Wolfheim, J. H. 1983, *Primates of the World: Distribution, Abundance, and Conservation*. University of Washington Press, Seattle.

CHAPTER TEN

An Exploratory Analysis of Developmental Plasticity in Costa Rican Mantled Howler Monkeys (*Alouatta palliata palliata*)

Clara B. Jones

INTRODUCTION

The genus *Alouatta* (howler monkeys) is the most widely distributed platyrrhine genus, occupying a broad range of biogeographic regimes (Groves, 2001; Crockett, 1998; Curdts, 1993). It has been postulated by several authors that the ecological success of howlers is in part a function of phenotypic plasticity (phenotypic variation expressed by reproductive individuals throughout their lifetimes: Crockett and Eisenberg, 1987; Jones, 1995a,b,c, 1997a, 2002,

Clara B. Jones • Department of Psychology, Fayetteville State University, Fayetteville, NC 28301-4298, USA.

New Perspectives in the Study of Mesoamerican Primates: Distribution, Ecology, Behavior, and Conservation, edited by Alejandro Estrada, Paul A. Garber, Mary S. M. Pavelka, and LeAndra Luecke. Springer, New York, 2005.

2003a, 2005; Crockett, 1998; Horwich *et al.*, 2000; Clarke *et al.*, 2002a; Silver and Marsh, 2003; Pavelka *et al.*, 2003; Kowalewski and Zunino, 2004; also see Jones, 1978, 1981; Kinzey and Cunningham, 1994; Strier, 1992, 1996; Brockmann, 2001; Jones and Agoramoorthy, 2003; Reader and Laland, 2003). Phenotypic plasticity is thought to be favored in response to environmental heterogeneity (changes in abiotic or biotic events over time and space), optimizing genotypic and phenotypic success in conditions of uncertainty and/or risk (Meyers and Bull, 2002; Lewontin, 1957; also see West-Eberhard, 1979, 2003).

Numerous studies document plasticity in the feeding responses exhibited by howlers. Glander (1975), for example, showed within- and between-season differences in plant selectivity by Costa Rican mantled howlers (*A. palliata palliata*). Studying *A. seniculus* and *A. pigra*, de Thoisy and Richard-Hansen (1997) and Ostro *et al.* (2000), respectively, reported changes in food and site selectivity before and after translocation. These studies support Crockett's (1998: 549) suggestion that the success of howlers is facilitated by "their ability to exploit folivorous diets" and a broad range of habitat types.

Research documenting seasonal peaks in births for some howler species in some habitats also provides evidence for phenotypic plasticity in these monkeys. Jones (1980a,b) showed that mantled howler groups in deciduous habitat of Costa Rican tropical dry forest environment (Frankie *et al.*, 1974) exhibited birth seasonality, but that birth seasonality was not evident for groups occupying riparian habitat. Similarly, Fedigan *et al.* (1998) documented birth seasonality in mantled howlers occupying deciduous habitat of tropical dry forest in Guanacaste, Costa Rica. Studying *A. pigra* (the black howling monkey), Brockett *et al.* (2000) provided evidence for birth seasonality at one semideciduous forest site in Belize where a significant proportion of females appeared to adjust the timing of gestation with peaks in preferred food. In their report on *A. caraya* (the black and gold howler) in Argentina, Kowalewski and Zunino (2004) documented birth seasonality in riparian forests of Argentina and an absence of birth seasonality on a nearby island; and Crockett and Rudran (1987) showed a peak in births for red howlers (*A. seniculus*) in the more heterogeneous of two Venezuelan habitats. Strier *et al.* (2001) did not detect birth peaks for brown howler monkeys (*A. fusca clamitans*) in the Atlantic Forest of Brazil. Brockett *et al.* (2000) concluded that the reported patterns of howler gestation and birth were likely to be a function of differential patterns of rainfall, possibly as condition-dependent and/or facultative responses to the availability of limiting resources, as suggested by Kowalewski and Zunino (2004).

Crockett (1998) stressed the importance of bisexual dispersal as a plastic response to local conditions for howlers. Mesoamerican *A. palliata*, for example, exhibit variations in dispersal patterns as a function of dominance rank (Jones, 1980a) and habitat perturbation (Clarke *et al.*, 2002a; see Jones, 1999, 2004; Estrada *et al.*, 2002). Additional features of howler species reflecting phenotypic plasticity are demonstrated by results showing effects of group and/or population size (density) on relative reproductive success (the mean number of immatures:females per female group size (*A. palliata*: Jones, 1996a; *A. pigra*: Horwich *et al.*, 2001a)) and variations in the number of males in a group (Horwich *et al.*, 2001b).

Recently, within-species variation in howler behavior and social organization was highlighted by Wang and Milton's (2003) work showing that characteristics of the dominance hierarchy of mantled howlers (*A. palliata aequatorialis*) in Panamanian semideciduous lowland tropical forest may differ from the same features in howlers inhabiting tropical dry forests of Costa Rica (Jones, 1978, 1980a,b; Glander, 1980). Wang and Milton's (2003) study documents a relaxed dominance hierarchy in their subjects on Barro Colorado Island, contrasting with the linear hierarchies documented for the mantled howlers in Costa Rican tropical dry forest (Jones, 1978, 1980a; Glander, 1980). These results strengthen the interpretation that differences in habitat (e.g. dispersion and/or quality of limiting food resources; rainfall) may explain observed differences in morphology, behavior, and sociosexual organization within howler species. This chapter describes the results of an exploratory study of habitat differences in chest circumference for female mantled howler monkeys (*A. palliata palliata*) and proposes that the findings are a result of developmental plasticity, a component of phenotypic plasticity whereby between-individual variation(s) in fixed traits result(s) from differences in environments encountered during development.

METHODS

The concepts employed in this chapter have recently been reviewed by Piersma and Drent (2003), West-Eberhard (2003), and Meyers and Bull (2002) (also see Sultan and Spencer, 2002; Kingsolver *et al.*, 2002). As pointed out by Piersma and Drent (2003), definitions for terms and concepts related to phenotypic plasticity are not standardized in the literature, and different fields may utilize different meanings for the same words or phrases. One factor retarding standardization between the social and biological sciences is that definitions in the latter disciplines are generally derived from population genetics, a field that few social scientists have studied, and a field promoting analysis at the population rather than the individual level. Concepts related to quantitative genetics, however, are common to both behavioral genetics (a field studied by many social scientists, especially psychologists) and population genetics. Thus, potential exists for a common vocabulary in this domain of investigation.

Study Site: Hacienda La Pacífica

The study site was Hacienda La Pacífica, Cañas, Guanacaste, Costa Rica, a lowland cattle ranch comprising approximately 13.3 km^2 of pastureland, agricultural fields, and forest fragments at the time of the surveys reported in the present chapter (early- to mid-1970s; see Malmgren, 1979). Details of the study site can be found elsewhere (e.g. Clarke *et al.*, 2002b; Clarke and Zucker, 1994; Clarke *et al.*, 1986; Malmgren, 1979; Glander, 1975). Hacienda La Pacífica is situated within tropical dry forest environment whose natural components include riparian and deciduous forest habitats (Frankie *et al.*, 1974; see Jones, 1996b). Riparian and deciduous habitats are seasonal with flower and fruit activity occurring primarily during the dry season, November through April (Frankie *et al.*, 1974). In the deciduous forest, leaf fall is synchronized for most trees during early to mid dry season. Most trees in the riparian forest retain their leaves throughout the year, displaying a phenological pattern similar to wet forest sites in Costa Rica (Frankie *et al.*, 1974). Riparian habitat, with higher humidity and greater proportion of evergreen vegetation, is most likely characterized by a higher level of primary productivity compared to deciduous habitat (G. W. Frankie, pers. comm., 2004), although quantitative data are lacking.

A third habitat, irrigation, is discussed in this chapter. Irrigation habitat is a degraded secondary deciduous habitat surrounding irrigation ditches at the ranch. Irrigation ditches were constructed consequent to anthropogenic perturbation for the purposes of farming and cattle ranching (see Clarke *et al.*, 1986). To my knowledge, irrigation habitat has not previously been discriminated in other reports based upon research at Hacienda La Pacífica. In this report, irrigation habitat is presumed to be more stressful than riparian or deciduous habitats for mantled howlers based upon the lower proportion of leaf cover and presumed desiccating effects. These assumptions, although untested, are consistent with assumptions made by other primatologists reporting from the field (e.g. Ravosa *et al.*, 1993; Hunt and McGrew, 2002).

The climatological features (e.g. patterns of temporal and spatial autocorrelations of rainfall: see Jones, 1997b) throughout Central America are very similar (Rand and Rand, 1982). These characteristics are a component of (abiotic) environmental heterogeneity which is thought to be a major force in the selection of phenotypic plasticity (see, for example, Sultan and Spencer, 2002). It is argued in this chapter, however, that local rather than global features of the environment are most likely to influence developmentally plastic features of the phenotype, a position consistent with recent discussions (see Kingsolver *et al.*, 2002; Piersma and Drent, 2003; West *et al.*, 2002).

Mantled howlers have been systematically studied at Hacienda La Pacífica since early 1970s, most notably by faculty and students of the Organization for Tropical Studies (OTS). When the present data were collected, approximately 16 howler groups occupied the ranch on variably sized forest fragments (Malmgren, 1979), and no other non-human primate species inhabited the ranch with the exception of the occasional *Cebus* vagrant. The organismic data on which this paper are based are extracted from the censuses conducted in the early- to mid-1970s by Dr. Norman J. Scott, Jr. (US Fish and Wildlife Service, Retired) and his assistants, including the present author.

Animals: *Alouatta palliata*

Mantled howlers, with a maximum body weight of approximately 7 kg (Wolfheim, 1983), are distributed throughout the forests of Middle America and the Pacific coast of northern South America (Groves, 2001). Populations are generally structured into highly communal, polygynandrous (multimale–multifemale) groups, though social organization may include polygynous and "age-graded" varieties of sociosexual architecture (Crockett and Eisenberg, 1987). Howlers are classified as diurnal, arboreal folivores (primary consumers), and are herbivorous primates, preferring new leaves, fruit, and flowers (Glander, 1975; Milton, 1980; Crockett and Eisenberg, 1987; Jones, 1996b). This chapter emphasizes data for adult female mantled howling monkeys.

Field Procedures

Morphometric data (weight (g), length of body (mm), length of tail (mm), pubis width (mm), length of arm (mm), and chest circumference (mm)) were collected from marked and aged (see Scott *et al.*, 1976; Malmgren, 1979; also see Glander *et al.*, 1991; Glander, 1993; Jones, 1980a) animals (127 adult females and 36 adult males). Age was determined by tooth wear (Scott *et al.*, 1976), whereby age class 1 was estimated to be 5–7 years old; age class 2, 7–10 years old; age class 3, 10–15 years old; and age class 4, >15 years old. Subjects were censused and measured (Malmgren, 1979; Scott *et al.*, 1976) in three discriminable habitats on the ranch: riparian (canopy cover estimated at 65–100%), deciduous (canopy cover 40–75%), and irrigation (canopy cover 10–45%). Some animals were followed by radio-tracking (AVM Instrument Company, 810 Dennison Drive, Champaign, IL 61820, USA), necessitated by the extrusion of lime stone aggregated upon a rough landscape, features of the deciduous habitat interfering with location of study groups. Some sample sizes differ in the present report ($N = 127$) and that of Jones (2003b; $N = 120$), because the number of valid cases (cases without missing data) was not the same for all the analyses. Data were analyzed with EcStatic software (Chalmer, 1990), and all tests are two-tailed with α set at 0.05.

PREVIOUS RESULTS FOR THE RELATIONSHIP BETWEEN CHEST CIRCUMFERENCE AND HABITAT IN MANTLED HOWLERS AT HACIENDA LA PACIFICA

A previous report (Jones, 2003b) on the present sample of adult male and female mantled howlers showed that there was no significant difference between habitats in the proportion of each of four age classes represented in the sample. For males, there was a significant negative correlation coefficient between *habitat* and *weight* ($r = -0.4224$, $p = 0.004$, $N = 35$) and a significant negative correlation between *habitat* and *chest circumference* ($r = -0.3273$, $p = 0.024$, $N = 35$). For both the comparisons, *weight* and *chest circumference* were smallest in the irrigation habitat. For females, a significant correlation coefficient was found between *habitat* and *chest circumference* ($r = -0.1851$, $p = 0.021$, $n = 119$). An analysis of variance (ANOVA) comparing male weight with habitat yielded a nearly significant finding (riparian > deciduous > irrigation), but female weight did not differ with habitat (Table 1).

Table 2 displays the means and standard deviations of chest circumference (CC) in all three habitats for both sexes. An ANOVA for the data in Table 2 showed a significant between-habitat difference for females (Table 3) but not for males, and a Newman–Keuls post-test demonstrated that CC was significantly smaller for females in irrigation habitat relative to CC for females in riparian or deciduous habitats (irrigation < riparian, deciduous;

Table 1. Weights (g) of adult male and female mantled howler monkeys in three habitats in the present survey. Null hypothesis was riparian = deciduous = irrigation

Sex	Riparian	Deciduous	Irrigation
Males[a]	5912.00 ± 594.53, $n = 10$	5755.45 ± 586.33, $n = 11$	5333.13 ± 621.17, $n = 15$
Females[b]	4530.91 ± 419.45, $n = 44$	4554.57 ± 407.18, $n = 37$	4439.44 ± 396.05, $n = 39$

See Jones (2003b) and text for further discussion.
[a] $F_{2,23} = 3.1413$, $p = 0.056$.
[b] $p > 0.05$.

Table 2. Means and standard deviations of chest circumference (CC: mm) for adult males and females in the present survey. The null hypothesis was riparian = deciduous = irrigation

Sex	Riparian	Deciduous	Irrigation
Males[a]	328.40 ± 12.14, $n = 10$	328.64 ± 15.67, $n = 11$	316.64 ± 13.45, $n = 14$
Females[b]	289.59 ± 13.94, $n = 44$	291.03 ± 13.13, $n = 37$	283.28 ± 13.45, $n = 39$

See Jones (2003b) and text for further discussion.
[a] $p > 0.05$.
[b] See Table 3.

Table 3. A source table (ANOVA) of adult female chest circumference (CC: dependent variable) × habitat (independent variable)

Source	SS	df	MS	F	p
Habitat[a]	1318.46	2	659.23	3.5986	0.0304
Residual	21,433.5068	117	183.1921		
Total	22,751.9667	119	191.1930		

See text and Jones (2003b) for further information.
[a] Irrigation × riparian (♀♀): $p < 0.05$ (Newman–Keuls test (Chalmer, 1990)). Irrigation × deciduous (♀♀): $p < 0.05$ (Newman–Keuls test (Chalmer, 1990)).

riparian = deciduous; Figure 1). This finding, the only significant comparison yielded by all morphometric analyses for females, may be indicative of differential (energy) investment to cardiovascular function(s) as has been reported for Indian children (Sundaram *et al.*, 1995). Comparable analyses for males yielded no significant results.

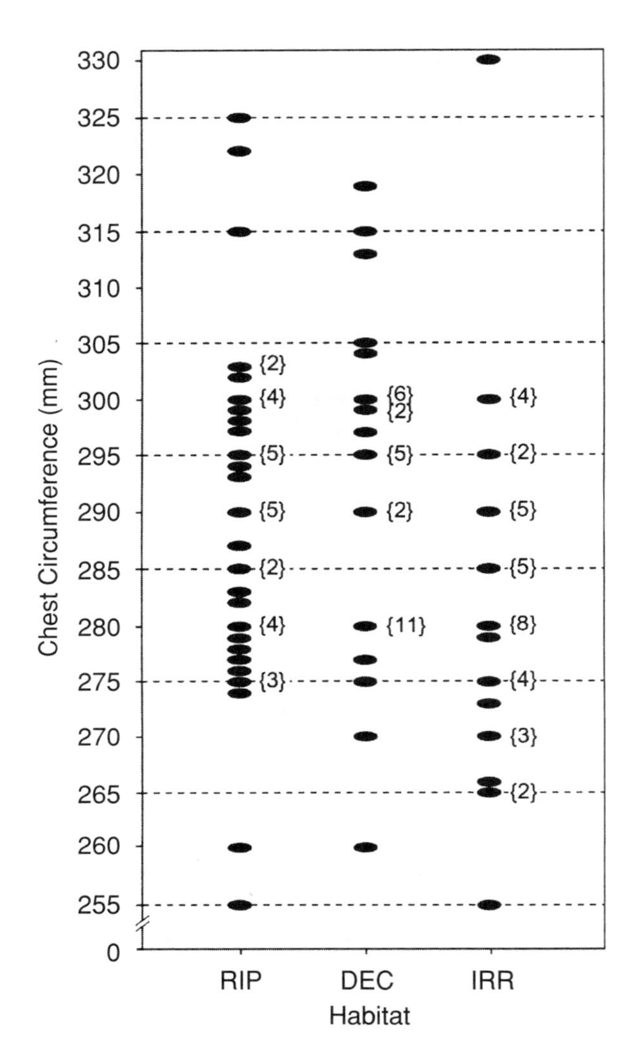

Figure 1. Distribution of chest circumference (CC: mm) × habitat (riparian = RIP; deciduous = DEC; irrigation = IRR) for adult female subjects in the present study ($N = 127$). Numbers in parenthesis = number of females in the sample with the specified chest circumference. See text for further explanation.

CORRELATIONS IN MORPHOMETRIC CHARACTERS AND BETWEEN-HABITAT DIFFERENCES FOR ADULT FEMALE CHEST CIRCUMFERENCE: PUBIS WIDTH RATIO

Table 4 displays correlation coefficients for CC relative to four other morphometric characters for adult females. The strongest correlation is a negative non-significant one shown for CC and pubis width (P). A further test evaluated the

Table 4. Correlation coefficients (r), sample sizes (N), and significance level (p) comparing female chest circumference (mm) with four other morphometric characters for females, pubis width (mm), arm length (mm), total body length (mm), and tail length (mm). Note that total sample size is divided between three habitats (riparian, deciduous, and irrigation)

	Pubis	Arm	Body	Tail
Chest	$r = -0.1225$, $N = 120$, $p = 0.091$	$r = -0.0829$, $N = 120$, $p = 0.184$	$r = 0.0340$, $N = 120$, $p = 0.356$	$r = -0.0376$, $N = 120$, $p = 0.342$

See Jones (2003b) and text for further discussion.

possibility that a tradeoff exists between CC and P in primate females inhabiting irrigation habitat. A test of this possibility showed no significant correlation coefficients between CC and P in riparian ($r = 0.0758$, $p = 0.313$, $n = 44$) and deciduous ($r = -0.0063$, $p = 0.485$, $n = 37$) habitats. In irrigation habitat, however, the correlation coefficient between CC and P was highly significant ($r = -0.3895$, $p = 0.005$, $n = 39$), indicative of a tradeoff.

The ratio between chest circumference and pubis width (CC:P; Figure 2) was calculated for each female subject. The resulting ANOVA comparing CC:P by habitat showed no significant relationships. Based on the results displayed in Tables 2 and 3, it is expected that, in future studies with larger sample sizes and correction for sources of error (discussed below), CC:P in irrigation habitat

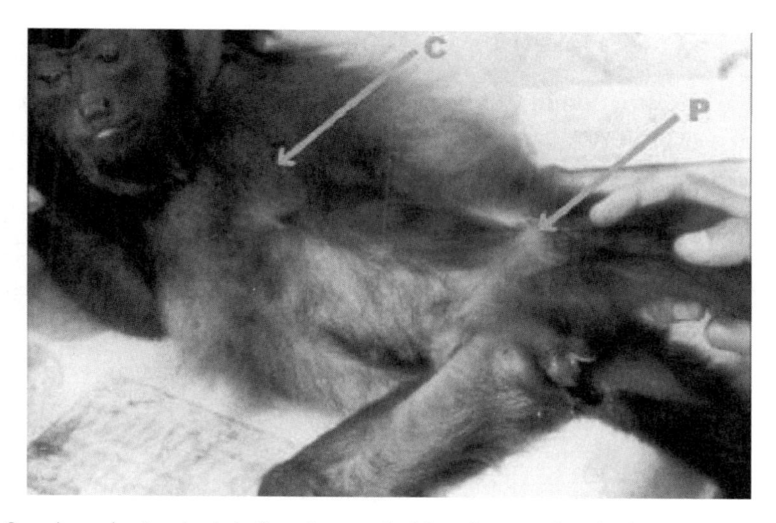

Figure 2. Anesthetized adult female mantled howler monkey (*Alouatta palliata mexicana*) showing approximate location of chest (C) and pubis (P). Juan Carlos Serio Silva©.

(Mean = 0.6962, SD = 0.06, $n = 39$) will be shown to be significantly smaller than the same ratio in the remaining habitats, riparian (Mean = 0.7045, SD = 0.05) and deciduous (Mean = 0.7193, SD = 0.06).

The above expectations rest upon two assumptions. The first assumption is that energy is limiting for an individual so that an increase in energy investment to one structure or function implies a decrease in energy investment to one or more alternative structures or functions. The second assumption is that significant differences exit between the three habitats discriminated in this study, possibly differences in primary productivity. A tentative test of this idea using Malmgren's (1979) estimates of adult density (adults/km^2) for 10 groups showed that mean adult density in riparian habitat was 312.5 ($n = 2$), for deciduous habitat, 159.4 ($n = 5$), and for irrigation habitat, 211.33 ($n = 3$) ($F_{2,7} = 5.3641$, $p < 0.0387$). A Newman–Keuls post-test (Chalmer, 1990) showed that adult density in both deciduous and irrigation habitats was significantly smaller than adult density in riparian habitat ($p < 0.05$) but did not differ from each other ($p > 0.05$). These limited findings indicate that the riparian habitat supports a higher density of adults and may be more productive. Additional research is required to determine the phytogeochemical differences among these three habitats and to test their proposed ontogenetic consequences for mantled howler females.

WITHIN-HABITAT VARIATION IN CC:P RATIO FOR ADULT FEMALE MANTLED HOWLERS

Within-habitat variation in CC:P ratio was assessed for adult females. For these treatments, an ANOVA compared CC:P for each group in each habitat. Statistical analysis showed within-habitat variation to be highly significant for each habitat analyzed separately (riparian: $F_{5,120} = 7.67$, $p = 0.00001$, 6 groups; deciduous: $F_{5,120} = 4.67$, $p = 0.0006$, 6 groups; irrigation: $F_{3,122} = 4.80$, $p = 0.0034$, 4 groups). Interestingly, then, within-habitat variation in CC:P ratio was greater than between-habitat variation, suggesting that, for adult females, positively assortative habitat selection (i.e. a female moving from one riparian habitat to another) is not occurring or is not marked at La Pacífica.

DISCUSSION

The present findings suggest that allocation of energy to CC is more plastic than allocation of energy to P, presumably because functions associated with

P (locomotion, birth) contribute more than pulmonary function, on average, to survival and/or reproduction (see Lloyd, 1987). The negative trend found in the present results cannot be explained by variations in P as a function of habitat since Jones (2003b) found no statistically significant habitat × character comparisons for any morphometric measurement except CC. Adult females in irrigation habitat, then, had a smaller, though not statistically significant, chest circumference relative to pubis than the same ratio for adult females in the other habitats, suggesting a developmental tradeoff between these two body parts and indicative of the relative importance (conservation) of pubis size for females.

Are CC and CC:P Endogenously or Exogenously Induced?

Environmental heterogeneity will prevent individuals from responding *optimally* to any set of conditions since heterogeneity will decrease the accuracy of responses, on average (Meyers and Bull, 2002; Piersma and Drent, 2003). Several authors (e.g. West-Eberhard, 2003; Sih, 2004) have pointed out that developmental plasticity, including tradeoffs in the relationships between growth and development of body parts, may represent facultative adjustment to local conditions (e.g. diet: see Emlen, 1997). Since howlers demonstrate a significant degree of plasticity in their feeding tactics and strategies (e.g. Silver and Marsh, 2003; Fuentes *et al.*, 2003; Zunino *et al.*, 2001; Milton, 1980; Glander, 1975; also see Kowalewski and Zunino, 2004), it will be important to investigate in future the extent to which variations in body mass and sizes of body parts reflect autonomous (endogenous) factors (e.g. geneotype, physiology) and/or exogenous ones (e.g. climate, competition for limiting food of varying dispersion, and/or quality).

Both endogenous (e.g. somatic or physiological perturbations) and/or exogenous (e.g. abiotic or social perturbations) induction of plastic responses may lead to differential allocation of an organism's resources (energy). A tradeoff between chest circumference and pubis width assessed in the present chapter may represent such a case. Although the present report advances the interpretation that CC and CC:P vary as a function of developmental plasticity, future studies will need to identify the functional relationship, if any, between CC and P (F. Nijhout, pers. comm., 2004). For example, if CC and P vary as a function of developmental plasticity, then we expect that changes in endogenous and/or exogenous factors, possibly food dispersion, quality, and/or nutritional status, will be causally related to variations in these morphometric characters. However, if CC and P vary as a function of some third variable, then any association between them is expected to be purely correlational, not causal. The "original Darwinian dilemma" (F. Nijhout, pers. comm., 2004) in interpreting structures that correlate negatively requires resolution for the present observations.

As a partial test of alternative explanations for the pattern of results presented here, it will be necessary to survey and measure the population at discrete intervals over time (C. P. Groves, pers. comm., 2004). Other important *caveats* to the methods and interpretations of the present results entail possible sampling error introduced by inter-investigator error (i.e. measurements were recorded by more than one researcher: Scott *et al.*, 1976), error introduced by surveys taken over a several year period (Scott *et al.*, 1976), and error resulting from the observation that most home ranges of these monkeys overlap more than one habitat type. Future studies need to eliminate, to control, or to correct for these potentially confounding effects and to confidently evaluate hypothesized differences in primary productivity between riparian, deciduous, and irrigation habitats. Furthermore, in order to employ statistical regression on morphometric factors as a function of habitat, it will be instructive in future to measure habitat (food dispersion and quality) with a continuously distributed variable, such as amount of forest cover and resource productivity *for each group*. Such an assessment seems particularly important since a high level of within-habitat variability was shown. Finally, the potential for habitat selection by female mantled howlers is in need of empirical support by radio-tracking dispersing females in order to determine rates of dispersal for irrigation habitat relative to these rates for riparian and deciduous habitats and the fates of dispersing females. Alternatively, within-individual (e.g. variations in genotype) and/or other within-habitat effects (e.g. local competition for limiting food resources) may explain the tendency for females with relatively smaller chest circumference to be clustered in irrigation habitat.

Competition among Body Parts in the Development and Evolution of Mantled Howler Monkeys

Studies with invertebrates (Stern and Emlen, 1999; Nijhout and Emlen, 1998; Emlen and Nijhout, 2001; also see Plaistow *et al.*, 2004) have shown that changes in the relative growth of bodily structures represent competition among body parts for energy and that these responses are apparently under hormonal control. Stern and Emlen (1999) pointed out that there are parallels between insects and vertebrates in the proposed mechanisms for control of body part growth. These authors also pointed out that, similar to insects, some vertebrate body parts grow relative to overall body size. For female mantled howler monkeys in the present sample, Jones (2003b) found no significant relationships between weight, on average, and habitat (see Glander, this volume, for a discussion of individual variation in body mass). Thus, overall body size alone (weight) did not account for the results described in this paper, supporting arguments that assessment of mortality in addition to assessment of body size is required for valid estimation of life-history features (see Stearns, 1984, 1992; Jones, 1998).

Stearns (1984: 694) suggested that life-history evolution is a function of "extrinsic age- and size-specific shifts in mortality rates that interact with . . . the intrinsic constraints and potentials of organisms." Pubis width, then, may be conserved because of costs to reproduction or survival occasioned by a smaller (threshold) pubis size (see Ridley, 1995). Research on a number of vertebrate species has demonstrated ontogenetic changes in body parts in association with the utilization of new habitats (see Shubin and Dahn, 2004). Future studies of female mantled howlers should measure differential reproductive success of individuals of different size, body proportions, genotype, and behavior within and between habitats. These considerations lead to the hypothesis that pubis size for female mantled howlers is a conservative character, presumably due to the constraints of birth and/or locomotion (see Fleagle, 1999: 34) and that smaller chest circumference of females in the irrigation habitat represents a plastic response to local conditions. The effects of chest circumference on reproductive output or mortality in riparian and deciduous habitat and of decreased chest circumference on these life-history parameters in irrigation habitat remain unclear.

Thresholds of Patch Quality as Generators of Dispersal and Phenotypic Plasticity

Results of the work presented in the present report and in Jones (2003b) demonstrate that metapopulation effects (i.e. genetic and/or phenotypic effects of habitat fragmentation) between the variably sized forest fragments at Hacienda La Pacífica have not prevented females residing on patches of irrigation habitat from exhibiting, on average, CC or CC:P smaller than females residing in either riparian or deciduous habitats. Extending a previous theoretical treatment by Moran (1992), Sultan and Spencer (2002; see Kingsolver *et al.*,

2002) have formulated an elegant model showing that phenotypic plasticity in morphological and other traits may be favored where dispersal occurs at sites differing in the relative frequencies of "environments" (e.g. relative frequencies of riparian, deciduous, and irrigation habitats).

Interestingly, Sultan and Spencer (2002: 279–280) show that, in the presence of dispersal, fixation of the plastic genotype may occur even when its fitness is lower than that of other genotypes as long as its costs are small and responsive to local conditions (e.g. competition for food or other limiting resources). As pointed out above, it will be important for primatologists to assess the costs as well as the benefits of phenotypic plasticity (see, for example, Sih, 2004). If a tradeoff is occurring between CC and P for adult female mantled howlers, one would expect that cardiopulmonary function is being compromised for these individuals in the most stressful habitat (i.e. smaller CC in irrigation habitat relative to adult females in riparian or deciduous habitats, Figure 1). This putative cost to females may represent a cost for a female residing in irrigation habitat. If females are "energy maximizers" (Schoener, 1971), then female mantled howlers in irrigation habitat at Hacienda La Pacífica may be exhibiting plasticity in resource (energy) allocation at a measurable cost to survival or reproduction. As Kingsolver *et al.* (2002) conclude, research is needed to verify the predictions of Sultan and Spencer's (2002) model. These authors' conclusions suggest that it will be important to obtain estimates of dispersal rates for primates (see Pope, 1992; Hanski, 2001), to compare and contrast populations varying in levels of plasticity, to evaluate the extent to which thresholds of plastic responses are sensitive to local compared to global conditions, and to conduct simulations and field experiments to manipulate the sizes and connectivity of habitat patches.

SUMMARY

The topic of developmental plasticity is fundamentally related to life-history evolution (West-Eberhard, 2003), in particular, patterns of survival and reproduction. Jones (1997b) employed matrix analysis (see Alberts and Altmann, 2003) of Scott's census data with age structure for mantled howlers at Hacienda La Pacífica to estimate life-history parameters including survivorship, fecundity, and mortality. The suite of life-history traits described by this author (e.g. low survivorship in more than one age class, iteroparity, relatively small reproductive effort) is consistent with the view that mantled howlers, and possibly other members of the genus, express tactics and strategies minimizing costs to fecundity. Since changes in CC and/or CC:P are irreversible morphological changes, it is proposed that female mantled howlers are capable of responding to local conditions with mechanisms of developmental plasticity, a within-individual strategy compatible with the life-history strategy of mantled howlers (Meyers and Bull, 2002; Table 1; see Ravosa *et al.*, 1993). Further research is required to test alternate hypotheses for the present results (e.g. natural selection (C. P. Groves, pers. comm.; F. Nihout, pers. comm.)) and to examine the possibility that there is a threshold of response to locally stressful conditions in irrigation habitat exhibited by female howlers and manifested as developmental plasticity in CC and CC:P.

The present report is consistent with the program of Stearns *et al.* (2003: 311) expressed in the following statement: "Alternative explanations for characteristic male and female growth schedules, and the consequences of the patterns seen in each species . . . all call for investigation across the spectrum of primate social systems." The study of the functional ecology, including physiological ecology and developmental plasticity, of primates is in its early stages (Milton, 1998; also see Strier, 1992; Ravosa *et al.*, 1993; Crockett, 1998; Reader and Laland, 2003: 20–21; Jones, 2005), investigations which are likely to occupy laboratory and field investigators for many years. This body of research will have important implications on primate and other mammalian development, energetics, life history, evolution, and conservation, as it involves an understanding of growth, survival, and reproduction relative to environmental regimes.

ACKNOWLEDGMENTS

I am very grateful to Dr. Norman J. Scott, Jr. for sharing with me the morphometric data upon which the present analyses are based. I thank the Werner Hagnauer family for allowing me to conduct studies at Hacienda La Pacífica intermittently from 1973 to 1980. I appreciate Dr. Alejandro Estrada for inviting me to contribute to this volume and all of the book's editors, especially Dr. Paul A. Garber and Dr. Alejandro Estrada, for providing constructive criticisms on prior drafts of my chapter. Dr. Fred Nijhout and, especially Dr. Colin P. Groves, made very helpful comments on the first version of this chapter which significantly improved the manuscript. Dr. Walter L. Ellis and Dr. David Wallace provided me with expert discussion of my statistical analyses, and Dr. Juan Carlos Serio Silva generously gave permission to use his photograph of an anesthetized adult female mantled howler (*A. p. mexicana*; Figure 2).

REFERENCES

Alberts, S. C. and Altmann, J. 2003, Matrix models for primate life history analysis, in: P. M. Kappeler and M. E. Pereira, eds., *Primate Life Histories and Socioecology*, The University of Chicago Press, Chicago, pp. 66–102.

Brockett, R. C., Horwich, R. H., and Jones, C. B. 2000, Reproductive seasonality in the Belizean black howling monkey (*Alouatta pigra*). *Neotrop. Primates* 8:136–138.

Brockmann, H. J. 2001, The evolution of alternative strategies and tactics. *Adv. Stud. Behav.* 30:1–51.

Chalmer, B. 1990, *EcStatic: A Statistical Package*. SomeWare in Vermont, Inc., Montpelier, VT.

Clarke, M. R., Collins, D. A., and Zucker, E. L. 2002a, Responses to deforestation in a group of mantled howlers (*Alouatta palliata*) in Costa Rica. *Int. J. Primatol.* 23:365–381.

Clarke, M. R., Crockett, C. M., Zucker, E. L., and Zaldivar, M. 2002b, Mantled howler population of Hacienda La Pacífica, Costa Rica, between 1991 and 1998: Effects of deforestation. *Am. J. Primatol.* 56:155–163.

Clarke, M. R. and Zucker, E. L. 1994, Survey of the howling monkey population at La Pacífica: A seven-year follow up. *Int. J. Primatol.* 15:61–73.

Clarke, M. R., Zucker, E. L., Scott, N. J. Jr. 1986, Population trends of the mantled howler groups of La Pacífica, Guanacaste, Costa Rica. *Am. J. Primatol.* 11:79–88.

Crockett, C. M. 1998, Conservation biology of the genus *Alouatta*. *Int. J. Primatol.* 19:549–578.

Crockett, C. M. and Eisenberg, J. F. 1987, Howlers: Variations in group size and demography, in: B. B. Smuts, D. L. Cheney, R. M. Seyfarth, R. W. Wrangham, and T. T. Struhsaker, eds., *Primate Societies*, The University of Chicago Press, Chicago, pp. 54–68.

Crockett, C. M. and Rudran, R. 1987, Red howler monkey birth data I: Seasonal variation. *Am. J. Primatol.* 13:347–368.

Curdts, T. 1993, Distribución geografica de las dos especies de mono zaraguate que habitan en Guatemala; *Alouatta palliata* y *Alouatta pigra*, in: A. Estrada, E. Rodriguez-Luna, R. Lopez-Wilchis, and R. Coates-Estrada, eds., *Estudios Primatológicos en México*, Vol. I, Biblioteca Universidad Veracruzana, Veracrizana, pp. 317–329.

de Thoisy, B. and Richard-Hansen, C. 1997, Diet and social behaviour changes in a red howler monkey (*Alouatta seniculus*) troop in a highly degraded rain forest. *Folia Primatol.* 68:357–361.

Emlen, D. J. 1997, Diet alters male horn allometry in the beetle *Onthophagus acuminatus* (Coleoptera: Scarabaeidae). *Behav. Ecol. Sociobiol.* 41:335–341.

Emlen, D. J. and Nijhout, H. F. 2001, Hormonal control of male horn length dimorphism in *Onthophagus taurus* (Coleoptera: Scarabaeidae): A second critical period of sensitivity to juvenile hormone. *J. Insect Physiol.* 47:1045–1054.

Estrada, A., Mendoza, A., Castellanos, L., Pacheco, R., Van Belle, S., Garcia, Y., and Munoz, D. 2002, Population of the black howler monkey (*Alouatta pigra*) in a fragmented landscape in Palenque, Chiapas, Mexico. *Am. J. Primatol.* 58:45–55.

Fedigan, L. M., Rose, L. M., and Avila, R. M. 1998, Growth of mantled howler groups in a regenerating Costa Rican dry forest. *Int. J. Primatol.* 19:405–432.

Fleagle, J. G. 1999, *Primate Adaptation and Evolution*, 2nd Edn. Academic, San Diego.

Frankie, G. W., Baker, H. G., and Opler, P. A. 1974, Comparative phenological studies of trees in tropical wet and dry forests in the lowlands of Costa Rica. *J. Ecol.* 62:881–919.

Fuentes, E., Franco, B., Magaña-Alejandro, M., Decena, Y., Muñoz, D., García, Y., and Estrada, A. 2003, Plantas utilizadas como alimento por una tropa de monos aulladores *Alouatta palliata*, en al Parque La Venta, Tabasco, Mexico. *Universidad y Ciencia (MX)* 37:43–52.

Glander, K. E. 1975, Habitat and resource utilization: An ecological view of social organization in mantled howling monkeys. Doctoral Dissertation, University of Chicago, Illinois.

Glander, K. E. 1980, Reproduction and population growth in free-ranging mantled howling monkeys. *Am. J. Phys. Anthropol.* 53:25–36.

Glander, K. E. 1993, Capture and marking techniques for arboreal primates, in: A. Estrada, E. Rodriguez-Luna, R. Lopez-Wilchis, and R. Coates-Estrada, eds., *Estudios Primatológicos in México*, Vol. I. Biblioteca Universidad Veracruzana, Veracruzana, pp. 299–304.

Glander, K. E., Fedigan, L. M., Fedigan, L., and Chapman, C. 1991, Field methods for capture and measurement of three monkey species in Costa Rica. *Folia Primatol.* 57:70–82.

Groves, C. P. 2001, *Primate Taxonomy*. Smithsonian Institution Press, Washington, D.C.

Hanski, I. 2001, Population dynamic consequences of dispersal in local populations and in metapopulations, in: J. Clobert, E. Danchin, A. A. Dhondt, and J. D. Nichols, eds., *Dispersal*, Oxford University Press, Oxford, pp. 283–298.

Horwich, R. H., Brockett, R. C., James, R. A., and Jones, C. B. 2001a, Population structure and group productivity of the Belizean black howling monkey (*Alouatta pigra*): Implications for female socioecology. *Primate Rep.* 61:47–65.

Horwich, R. H., Brockett, R. C., James, R. A., and Jones, C. B. 2001b, Population growth in the Belizean black howling monkey (*Alouatta pigra*). *Neotrop. Primates* 9:1–7.

Horwich, R. H., Brockett, R. C., and Jones, C. B. 2000, Alternative male reproductive behaviors in the Belizean black howler monkey (*Alouatta pigra*). *Neotrop. Primates* 8:95–98.

Hunt, K. D. and McGrew, W. C. 2002, Chimpanzees in the dry habitats of Assirik, Senegal and Semliki Wildlife Reserve, Uganda, in: C. Boesch, G. Hohmann, and L. F. Marchant, eds., *Behavioural Diversity in Chimpanzees and Bonobos*, Cambridge University Press, Cambridge, pp. 35–51.

Jones, C. B. 1978, Aspects of reproduction in the mantled howler monkey (*Alouatta palliata* Gray). PhD Dissertation, Cornell University, Ithaca, NY.

Jones, C. B. 1980a, The functions of status in the mantled howler monkey, *Alouatta palliata* Gray: Intraspecific competition for group membership in a folivorous Neotropical primate. *Primates* 21:389–405.

Jones, C. B. 1980b, Seasonal parturition, mortality, and dispersal in the mantled howler monkey, *Alouatta palliata* Gray. *Brenesia* 1:1–10.

Jones, C. B. 1981, The evolution and socioecology of dominance in primate groups: A theoretical formulation, classification, and assessment. *Primates* 22:70–83.

Jones, C. B. 1995a, Howler monkeys appear to be preadapted to cope with habitat fragmentation. *Endang. Species Updat.* 12:9–10.

Jones, C. B. 1995b, Howler subgroups as homeostatic mechanisms in disturbed habitats. *Neotrop. Primates* 3:7–9.

Jones, C. B. 1995c, Alternative reproductive behaviors in the mantled howler monkey (*Alouatta palliata* Gray): Testing Carpenter's hypothesis. *Bol. Primatol. Lat.* 5:1–5.

Jones, C. B. 1996a, Relative reproductive success in the mantled howler monkey: Implications for conservation. *Neotrop. Primates* 4:21–23.

Jones, C. B. 1996b, Predictability of plant food resources for mantled howler monkeys at Hacienda La Pacífica, Costa Rica: Glander's dissertation revisited. *Neotrop. Primates* 4:147–149.

Jones, C. B. 1997a, Subspecific differences in vulva size between *Alouatta palliata palliata* and *A. p. mexicana*: Implications for assessment of female receptivity. *Neotrop. Primates* 5:46–48.

Jones, C. B. 1997b, Life history patterns of howler monkeys in a time-varying environment. *Bol. Primatol. Lat.* 6:1–8.

Jones, C. B. 1998, Book review of the evolving female: A life-history perspective. *Polit. Life Sci.* 17:97–99.

Jones, C. B. 1999, Why both sexes leave: Effects of habitat fragmentation on dispersal behavior. *Endang. Species Updat.* 16:70–73.

Jones, C. B. 2002, Genital displays by adult male and female mantled howling monkeys (*Alouatta palliata*): Evidence for condition-dependent compound displays. *Neotrop. Primates* 10:144–147.

Jones, C. B. 2003a, Urine-washing behaviors as condition-dependent signals of quality by adult mantled howler monkeys (*Alouatta palliata*). *Lab. Prim. News* 42:12–14.

Jones, C. B. 2003b, Chest circumference differs by habitat in Costa Rican mantled howler monkeys: Implications for resource allocation and conservation. *Neotrop. Primates* 11:22–24.

Jones, C. B. 2004, The number of adult females in groups of polygynous howling monkeys (*Alouatta* spp.): Theoretical inferences. *Primate Rep.* 68:7–25.

Jones, C. B. 2005, *Behavioral Flexibility in Primates: Causes and Consequences.* Springer, New York.

Jones, C. B. and Agoramoorthy, G. 2003, Alternative reproductive behaviors in primates: Towards general principles, in: C. B. Jones, ed., *Sexual Selection and Reproductive Competition in Primates: New Perspectives and Directions*, American Society of Primatologists, Norman, OK, pp. 103–139.

Kingsolver, J. G., Pfennig, D. W., and Servedio, M. R. 2002, Migration, local adaptation and the evolution of plasticity. *Trends Ecol. Evol.* 17:540–541.

Kinzey, W. G. and Cunningham, E. P. 1994, Variability in platyrrhine social organization. *Am. J. Primatol.* 34:185–198.

Kowalewski, M. and Zunino, G. E. 2004, Birth seasonality in *Alouatta caraya* in Northern Argentina. *Int. J. Primatol.* 25:383–400.

Lewontin, R. C. 1957, The adaptations of populations to varying environments. *Cold Spring Harb. Symp. Quant. Biol.* 22:395–408.

Lloyd, D. G. 1987, Parellels between sexual strategies and other allocation strategies, in: S. C. Stearns, ed., *The Evolution of Sex and Its Consequences*, Birkhauser Verlag, Basel, pp. 263–281.

Malmgren, L. A. 1979, Empirical population genetics of golden mantled howling monkeys (*Alouatta palliata*) in relation to population structure, social dynamics, and evolution. PhD Dissertation, The University of Connecticut, Storrs.

Meyers, L. A. and Bull, J. J. 2002, Fighting change with change: Adaptive variation in an uncertain world. *Trends Ecol. Evol.* 17:551–557.

Milton, K. 1980, *The Foraging Strategy of Howler Monkeys: A Study in Primate Economics.* Columbia University Press, New York.

Milton, K. 1998, Physiological ecology of howlers (*Alouatta*): Energetic and digestive considerations and comparison with the Colobinae. *Int. J. Primatol.* 19:513–548.

Moran, N. A. 1992, The evolutionary maintenance of alternative phenotypes. *Am. Nat.* 139:971–989.

Nijhout, H. F. and Emlen, D. J. 1998, Competition among body parts in the development and evolution of insect morphology. *Proc. Natl. Acad. Sci. USA* 95:3685–3689.

Ostro, L. E. T., Silver, S. C., Koontz, F. W., and Koontz, T. P. 2000, Habitat selection by translocated black howler monkeys in Belize. *Anim. Conserv.* 3:175–181.

Pavelka, M. S. M., Brusselers, O. T., Nowak, D., and Behie, A. M. 2003, Population reduction and social disorganization in *Alouatta pigra* following a hurricane. *Int. J. Primatol.* 24:1037–1055.

Piersma, T. and Drent, J. 2003, Phenotypic flexibility and the evolution of organismal design. *Trends Ecol. Evol.* 18:228–233.

Plaistow, S. J., Lapsley, C. T., Beckerman, A. P., and Benton, T. G. 2004, Age and size at maturity: Sex, environmental variability and developmental thresholds. *Proc. R. Soc. Lond. B* 271:919–924.

Pope, T. R. 1992, The influence of dispersal patterns and mating system on genetic differentiation within and between populations of the red howler monkey (*Alouatta seniculus*). *Evolution* 46:1112–1128.

Rand, A. S. and Rand, W. M. 1982, Variations in rainfall on Barro Colorado Island, in: E. G. Leigh, Jr., A. S. Rand, and D. M. Windsor, eds., *The Ecology of a Tropical Forest*, Smithsonian Institution Press, Washington, D.C., pp. 19–60.

Ravosa, M. J., Meyers, D. M., and Glander, K. E. 1993, Relative growth of the limbs and trunk in sifakas: Heterochronic, ecological, and functional considerations. *Am. J. Phys. Anthropol.* 92:499–520.

Reader, S. M. and Laland, K. N. 2003, Animal innovation: An introduction, in: S. A. Reader and K. N. Laland, eds., *Animal Innovation*, Oxford University Press, Oxford, pp. 3–35.

Ridley, M. 1995, Brief communication: Pelvic sexual dimorphism and relative neonatal brain size really are related. *Am. J. Phys. Anthropol.* 97:197–200.

Schoener, T. W. 1971, Theory of feeding strategies. *Ann. Rev. Ecol. Syst.* 2:369–404.

Scott, N. J. Jr., Scott, A. F., and Malmgren, L. A. 1976, Capturing and marking howler monkeys for field behavioral studies. *Primates* 17:527–534.

Shubin, N. H. and Dahn, R. D. 2004, Lost and found. *Nature* 428:703–704.

Sih, A. 2004, A behavioral ecological view of phenotypic plasticity, in: T. J. DeWitt and S. M. Scheiner, eds., *Phenotypic Plasticity: Functional and Conceptual Approaches*, Oxford University Press, Oxford, pp. 112–125.

Silver, S. C. and Marsh, L. K. 2003, Dietary flexibility, behavioral plasticity, and survival in fragments: Lessons from translocated howlers, in: L. K. Marsh, ed., *Primates in Fragments: Ecology and Conservation*, Kluwer, New York, pp. 251–266.

Stearns, S. C. 1984, Adaptations of colonizers. *Science* 223:694–695.

Stearns, S. C. 1992, *The Evolution of Life Histories.* Oxford University Press, Oxford.

Stearns, S. C., Pereira, M. E., and Kappeler, P. M. 2003, Primate life histories and future research, in: P. M. Kappeler and M. E. Pereira, eds., *Primate Life Histories and Socioecology*, The University of Chicago Press, Chicago, pp. 301–312.

Stern, D. L. and Emlen, D. J. 1999, The developmental basis for allometry in insects. *Development* 126:1091–1101.

Strier, K. B. 1992, Ateline adaptations: Behavioral strategies and ecological constraints. *Am. J. Phys. Anthropol.* 88:515–524.

Strier, K. B. 1996, Male reproductive strategies in New World primates. *Human Nat.* 7:105–123.

Strier, K. B., Mendes, S. L., and Santos, R. R. 2001, Timing of births in sympatric brown howler monkeys (*Alouatta fusca clamitans*) and Northern muriquis (*Brachyteles arachnoides hypoxanthus*). *Am. J. Primatol.* 55:87–100.

Sultan, S. E. and Spencer, H. G. 2002, Metapopulation structure favors plasticity over local adaptation. *Am. Nat.* 160:271–283.

Sundaram, K. K., Seth, V., Jena, T. K., and Shukla, D. K. 1995, Age at which chest circumference overtakes head circumference in children. *Indian J. Pediatr.* 62:89–94.

Wang, E. and Milton, K. 2003, Intragroup social relationships of male *Alouatta palliata* on Barro Colorado Island, Republic of Panama. *Int. J. Primatol.* 24:1227–1243.

West, S. A., Pen, I., and Griffin, A. S. 2002, Cooperation and competition between relatives. *Science* 296:72–75.

West-Eberhard, M. J. 1979, Sexual selection, social competition, and evolution. *Proc. Am. Philos. Soc.* 123:222–134.

West-Eberhard, M. J. 2003, *Developmental Plasticity and Evolution*. Oxford University Press, Oxford.

Wolfheim, J. H. 1983, *Primates of the World: Distribution, Abundance, and Conservation*. University of Washington Press, Seattle.

Zunino, G. E., González, V., Kowalewski, M. M., and Bravo, S. P. 2001, *Alouatta caraya*: Relations among habitat, density and social organization. *Primate Rep.* 61:37–46.

CHAPTER ELEVEN

Travel Patterns and Spatial Mapping in Nicaraguan Mantled Howler Monkeys (*Alouatta palliata*)

Paul A. Garber and Petra E. Jelinek

INTRODUCTION

Studies of foraging and ranging behavior in several primate species indicate evidence of goal-directed travel and relatively straight-line movement between distant or out-of-view feeding sites (Carpenter, 1934; Menzel, 1973; Altmann, 1974; Milton, 1980, 2000; Sigg and Stolba, 1981; Chapman *et al.*, 1989; Garber, 1989; Gallistel and Cramer, 1996; Menzel, 1996; Janson and Di Bitetti, 1997; Janson, 1998; Menzel *et al.*, 2002). Although in some instances group members may deviate from straight-line travel in order to monitor the productivity and phenological state of potential feeding sites or detour to avoid

Paul A. Garber • Department of Anthropology, University of Illinois, Urbana, Illinois 61801.
Petra E. Jelinek • University of Illinois, Urbana, Illinois 61801.

New Perspectives in the Study of Mesoamerican Primates: Distribution, Ecology, Behavior, and Conservation, edited by Alejandro Estrada, Paul A. Garber, Mary S. M. Pavelka, and LeAndra Luecke. Springer, New York, 2005.

areas of high predation risk, in general many researchers have concluded that prosimians, monkeys, and apes maintain detailed internal representations or "mental maps" of the spatial locations and distributions of major feeding and resting trees within their home range. What is less clear, however, is what types of information are contained in these maps, how detailed these representations are, whether landmarks or other physical features of the environment are encoded singly or as an array, and whether spatial information is represented in a route-based system or a coordinate-based system (Poucet, 1993; Tomasello and Call, 1997; Shettleworth, 1998; Byrne, 2000; Garber, 2000; Bicca-Marques and Garber, 2004). Foragers using a route-based representation are expected to encode spatial information as a learned network of fixed points (nodes), travel routes, landmarks, and other salient features of the environment (Poucet, 1993). In contrast, a forager using a metric-based representation is expected to encode spatial information in a coordinate framework in which true distances and directions between multiple landmarks and goals are used to compute efficient routes of travel (Dyer, 1991; Byrne, 2000; Gibson and Kamil, 2001).

Many species of primates spend the majority of their daily activity budget searching for and consuming plant resources such as fruits, flowers, leaves, and seeds (Garber, 1987). Compared to mobile prey such as flying insects and vertebrates, these food patches represent fixed points on the landscape. Although the specific locations and distributions of food-bearing trees change seasonally, foragers may encode and associate the locations of important feeding sites with particular topographical features of the environment or other landmark cues to aid in foraging success (Biegler and Morris, 1996; Kamil and Cheng, 2001; Roberts, 2001). Landmarks are reference points and may be represented as distance, direction, and/or distance and direction vectors to a target or goal (Cheng and Spetch, 1998; Roberts, 2001). A single landmark may be used as a beacon to navigate to a target resting or feeding tree, or serve as a point of orientation to redirect travel (i.e., switchpoint or node) (Dyer, 1991; Poucet, 1993; Garber, 2000). In other cases, spatial representations may include a configuration of several landmarks encoded independently or together and take the form of regularly used routes of travel (Gibson and Kamil, 2001; Garber and Brown, 2005).

Given species differences in dietary patterns, size and cohesion of foraging groups, daily path length, speed of travel, and the number of feeding and resting sites visited per day, cognitive strategies used by primate foragers to effectively encounter and relocate feeding sites are likely to vary considerably (Janson and Di Bitetti, 1997; Milton, 2000).

For example, spider monkeys (*Ateles* spp.) forage in small fission–fusion parties, exploit home ranges of many hundreds of hectares, and travel up to 3–5 km in a single day (Milton, 1981, 2000; van Roosmalen, 1985). It has been suggested that spider monkeys maintain a mental representation of a large number of feeding and resting trees in their home range, and select resting sites based on their proximity to major feeding trees (Milton, 1981; Chapman *et al.*, 1989; van Roosmalen, 1985). This ranging and grouping pattern enables spider monkeys to efficiently exploit a set of widely scattered feeding sites (over 20 individual feeding trees per day, van Roosmalen, 1985). van Roosmalen (1985: 152) points out, however, that whereas some individual spider monkeys (i.e., leading females) appear to possess complex spatial knowledge "and used the shortest possible connections between consecutive food sources", other individuals (i.e., solitary or nonleading females) are "incapable of planning an economic route". In species exhibiting a fission–fusion social system, individual community members have access to very different types of information and foraging experiences and therefore may rely more on self-generated or ego-based information than group-based information.

It has been suggested that the selection pressures associated with encoding self-generated information has played a major role in the evolution of enhanced cognitive capabilities in some primate lineages (Milton, 1981, 2000). In comparison, mantled howler monkeys (*Alouatta palliata*) are reported to travel in single line progression as a cohesive social unit, have small day ranges (100–450 m; Carpenter, 1934; Milton, 1980), small home ranges (2–40 ha; Milton, 1980; Crockett, 1998), and exploit an average of 1.5 primary feeding trees per day (i.e., trees that account for ≥20% of feeding time; Milton, 2000). Over a period of 1–2 weeks, these howlers concentrate their feeding, resting, and ranging activities on a small set of pivotal trees that are visited and revisited several times daily (Milton, 1980). Given mantled howler feeding and ranging patterns, most, if not all, group members have access to the same ecological information (i.e., public information, see Valone, 1989). Carpenter (1934: 36) notes that howlers in his study group traveled "roughly over the same route, which is the shortest distance between the two places". The degree to which mantled howler monkeys at other sites also exhibit a route-based pattern of travel remains unclear.

In this paper, we examine foraging strategies and travel patterns in a group of Nicaraguan mantled howler monkeys (*Alouatta palliata*). Specifically, we document howler path sequences and use of consecutive feeding and resting sites in order to address a series of questions concerning the degree to which these primates represent spatial information as a route-based or as a metric-based cognitive representation.

METHODS

Data on mantled howler foraging and ranging patterns were collected at Estación Biológica de Ometepe, a privately owned research facility dedicated to scientific investigation, teaching, and conservation located on Isla de Ometepe, Nicaragua (11°40′N and 85°50′W; Figure 1). Isla de Ometepe is a large volcanic island (256 km^2) situated in the southeastern edge of Lake Nicaragua and is characterized by disturbed secondary forest, agricultural fields, and undisturbed cloud forest (Garber *et al.*, 1999). In addition to mantled howling monkeys (*Alouatta palliata*), white-faced capuchins (*Cebus capucinus*) also naturally inhabit the island.

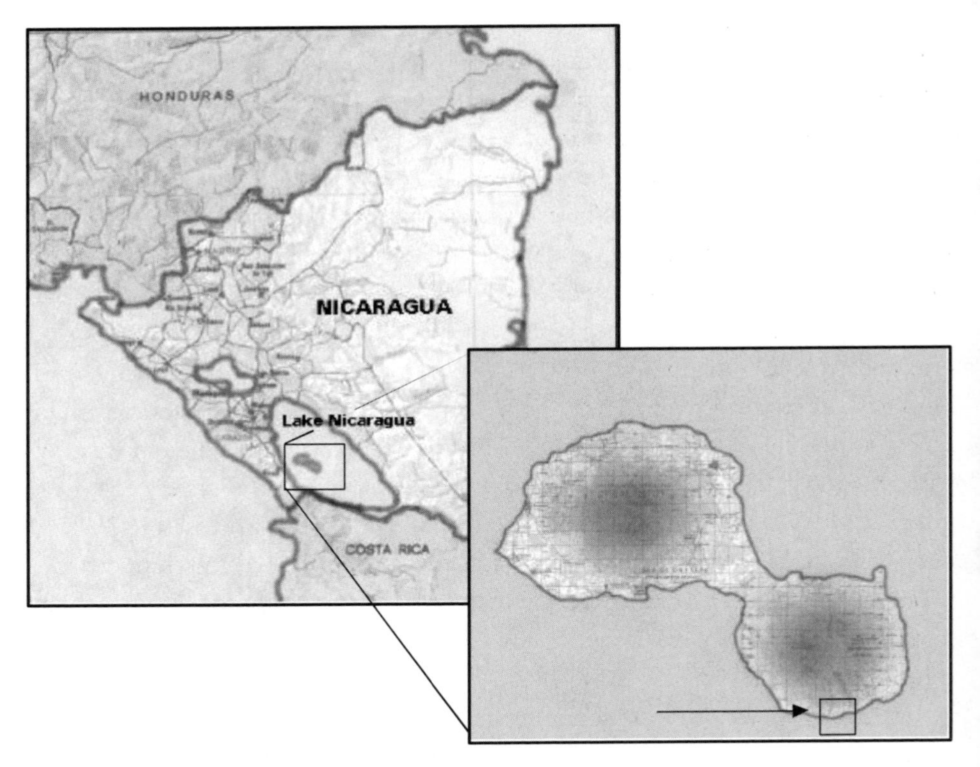

Figure 1. Map of Nicaragua showing Isla de Ometepe and the location of the field site (Estación Biológica de Ometepe).

The home range of our study troop was located in a zone of dry tropical forest situated near the small town of San Ramon at the foothills of Volcan Maderas (Figure 1). Rainfall in this region of Nicaragua is estimated between 1200 and 1800 mm per year (Salas Estrada, 1993). The howler study troop was composed of 26–29 individuals including 10 adult females and 12 adult males. All group members were fully habituated to the presence of the observers. Between July and August 2002 (wet season), a total of 102.7 h of data were collected on the behavior, ecology, and ranging patterns of mantled howler monkeys. This troop has been the focus of previous field studies (Garber *et al.*, 1999; Bezanson *et al.*, 2001).

Behavioral Data

Behavioral data on howler activity budget, social interactions, and ranging behavior were collected on focal animals using a 2-min instantaneous time sampling technique (Altmann, 1974). Feeding was defined as handling and ingesting plant material. Foraging was defined as localized movement in the crown of a feeding tree that was for the specific goal of encountering a food resource. Traveling involved movement between the crowns of trees or within the crown of a tree that was not directly food related. Resting was defined as a period of inactivity. Social interactions included vocalizations directed at other group members, physical contact, displacement, threats, huddling, and grooming.

During all data collecting periods, a team of two researchers were in the forest. One researcher was responsible for collecting behavioral data and the other researcher was responsible for marking every tree visited by the focal animal. These trees were scored as travel trees, feeding trees, and/or resting trees. When traveling between feeding and resting sites, mantled howlers tend to move in a single line progression (Carpenter, 1934; Milton, 2000), and therefore, we feel confident that the path taken by the focal animal was similar to the path taken by other group members.

Several of the howlers in our study group were marked with identification collars or anklets facilitating recognition in the field. The first animal sighted on a given morning was selected as the focal animal and followed continuously. On those occasions when the focal animal was lost from view for a period of $\geq$ 4 min, the next animal sighted became the focal animal. On average, the howlers were followed continuously for 6.5 h per day for 15 consecutive days. We use this 15-day period to illustrate and explain short-term travel patterns, ranging, and patch choice in our mantled howler study group.

Tree Use and Distribution

A trail system and map of the study group's home range was constructed using a Brunton Transit mounted on a tripod. The trail system consisted of 154 marked and mapped points. In addition, each tree the focal animal was observed to travel, feed, and rest in ($N = 250$) was marked with surveyors flagging, assigned an identification number, and distances and angles between trees were measured. The locations and sequential use of trees were plotted daily on the field map. Thus, we were able to record the spatial position of individual howlers in relation to 404 mapped points within their small home range. Daily path lengths were calculated by measuring the straight-line distance between sequential feeding, resting, foraging, and traveling trees directly off the field map and summed. Trees were considered major feeding and resting sites if they accounted for at least 1% of total activity time.

Distribution of Feeding and Resting Sites and Ranging Patterns

In order to determine the spatial distribution of major feeding and resting sites, a grid divided into 25 m × 25 m squares ($N = 63$) was overlaid on the field map. We then tallied the number of major feeding and resting trees located in each of the 63 quadrats. In addition, we tallied the number of times the focal animal entered each quadrat in order to determine patterns of habitat utilization, ranging, and the degree to which individual path segments between major feeding and resting sites were reused during the same day, reused on different days, or whether howlers revisited major feeding sites using a variety of different travel routes. A route segment was defined as a travel path in which the same three or more contiguous quadrats were visited and revisited in sequence (i.e., A–B–C = C–B–A). We also identified the order in which major feeding and resting trees were visited over the course of each day, and compared this with the most efficient tree sequences (in terms of minimizing travel distance) to visit these sites. An index of circuitry (actual distance/most efficient route distance) was calculated to determine the degree to which howler travel patterns were consistent with taking the most efficient route to visit all trees on a daily itinerary.

Habitat Visibility: Field-of-View

Given that howlers have small day ranges, use a core area extensively over a limited period of time, and tend to visit nearby trees, we collected ecological information on tree density and stature as an estimate of habitat visibility or the ability of a forager to sight directly from one major feeding/resting tree to another. Data on tree density were compiled in January 2004 by tallying the number of trees located in a hemisphere with radius of 20 m (area of 628 m^2) around 25 of the 27 major feeding and resting trees (two trees had been damaged or could not be relocated from the previous field season). These sampled plots represent approximately 40% of the core area used by the howlers during our study. For each tree located in the plot with a height of greater than 3 m and a diameter at breast height (DBH) of greater than 15 cm (>45 cm circumference at breast height), we estimated tree height and crown height and measured DBH.

For each of the 25 major feeding and resting trees, we measured crown diameter in two cardinal directions, DBH, as well as estimated tree height and crown height. We also scored the shape of the tree crown (i.e. circular, ellipsoid, rectangle, square, cone) and used the appropriate volumetric formula to determine crown volume from crown height and diameter

In order to obtain an estimate of a howler's "field-of-view", we used the ecological information obtained in each plot to reconstruct visibility at the top, middle, or bottom of the crown for each major feeding or resting tree. The field-of-view was based on the tree's height and crown dimensions, as well as the number, height, and crown dimensions of other trees in each plot that obstruct the foragers' view at a particular location in the canopy. For example, if the crowns of 6 of 12 of the surrounding trees overlapped with the height at the top of the crown of the major tree, a 50% field-of-view score was assigned for that height. If in that same tree, 10 of 12 of the surrounding trees obstructed the view when a howler was located in the middle of the crown, then a 16% field-of-view score was assigned for that height. Although our field-of-view index is at best only a crude and relative measure of the degree to which howlers could sight directly to a subsequent feeding/resting site, we use these data to determine whether howler traveling paths were different when traveling between less visible and more visible targets or goals.

RESULTS

Over the course of 15 consecutive observation days, members of our howler study group exploited a home range of 3.9 ha, and fed, rested, and traveled in 250 trees. The mean number of trees visited per day was 40.6 (±15.7) and mean

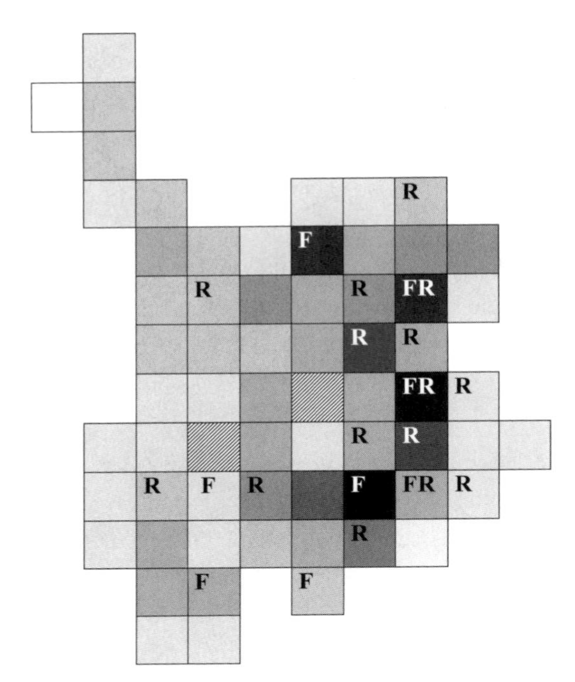

Figure 2. Schematic representation of the study troop's range divided into 63 (25 m × 25 m) quadrats. Lightest gray indicates areas of infrequent use (one time). Quadrats darken with progressive use. The most frequently used quadrat (black) was visited on 10 occasions. Diagonally hatched squares indicate zero-use quadrats. F indicates the location of quadrats containing a major feeding tree(s), R indicates the location of quadrats containing a major resting tree(s), and FR indicates quadrats containing both major feeding and major resting trees.

daily path length was 381 m (±151). As in other field studies (e.g., Milton, 1980), resting (73%), feeding (16.9%), and traveling (9.1%) dominated the howler activity budget.

As indicated in Figure 2 and Table 1, the howlers centered their activities around a small set of important or major feeding and resting trees. We defined a major tree as one in which the howlers spent at least 1% of their total activity time (range 1–9.6%). Twenty-seven individual trees accounted for 69.5% of the howler activity budget. Eight of these trees were used principally for feeding, and the remaining 19 trees were used principally as resting trees. On average, 8.8 (±2.8, range 3–14, 3.3 feeding trees and 5.6 resting trees) of these major trees were visited each day, with 2 of these trees revisited twice on the same day. One feeding tree and one resting tree were each visited 11 times during

Table 1. Daily pattern and frequency of visits to major feeding and resting trees (15-day study period)

Tree ID	% Total frequency of use	Focal activity	Day 1	Day 2	Day 3	Day 4	Day 5	Day 6	Day 7	Day 8	Day 9	Day 10	Day 11	Day 12	Day 13	Day 14	Day 15	Total visits
2.4	1.36	Resting	2		1		1			1		1						6
2.5	3.25	Resting	1	1	2		1	1		1		1	2	1				11
2.7	6.36	Feeding	1	1	1	1	3	1				1	1	1				11
2.9	1.66	Resting	1		2								1					4
2.16	3.60	Resting	1				1											2
2.17	2.24	Feeding	2	1	1		1	1				1	1	1				9
2.20	1.82	Feeding	1		2			1		1	1	1		1		1		9
2.30	1.27	Feeding	1							1	1							3
3.16	1.46	Resting		2				1	1				1					5
3.17	2.79	Resting		2					1									3
4.3	6.39	Resting			1		2					1	1	1		1		7
5.5	2.21	Resting				1			1			1		1				4
5.11	1.40	Resting				1						1		3	2			7
5.12	1.10	Feeding				1						1	2		1			5
5.13	1.49	Resting				1								2	1			4
5.26	9.67	Resting				2			2						2			6
5.29	1.88	Feeding				1			3						4			8
6.3	1.01	Resting					1						1					2

(Continued)

Table 1. Daily pattern and frequency of visits to major feeding and resting trees (15-day study period) (*Continued*)

Tree ID	% Total frequency of use	Focal activity	Day 1	Day 2	Day 3	Day 4	Day 5	Day 6	Day 7	Day 8	Day 9	Day 10	Day 11	Day 12	Day 13	Day 14	Day 15	Total visits	
6.4	2.21	Resting					1				1		1	1	1	1		6	
7.10	2.63	Resting						1								1		2	
7.3	2.04	Feeding						1		1								2	
8.9	1.27	Resting							1						2			3	
10.16	2.50	Feeding								1							1	2	
10.24	1.88	Resting								1	1							2	
11.1	1.40	Resting									4						1	5	
15.15	2.30	Resting												1				1	
16.18	2.37	Resting													1	2	1	4	
		F tree visits	5	2	4	3	5	4	3	4	2	3	4	3	5	1	1	3.267	Mean
		R tree visits	5	5	6	5	6	3	6	3	6	6	7	10	9	5	2	5.6	Mean
		Individual tree visits	8	5	7	7	8	7	6	7	5	9	9	10	8	5	3	6.93	Mean
		Total visits/ day	10	7	10	8	11	7	9	7	8	9	11	13	14	6	3	8.867	Mean
		Revisits per day	2	2	3	1	3	0	3	0	3	0	2	3	6	1	1	2	Mean

The %total frequency of use from column 2 does not equal 100% (see text).

the 15 consecutive day study period. Trees revisited on the same day were used primarily as resting sites. We direct the remainder of our analysis to examine resource use, distribution, and ranging patterns associated with these 27 trees.

Forest Profile and Tree Density

Based on measurements taken in our 25 sample plots, the density of trees greater than 3 m in height and 15 cm DBH in the study troops home range was 148.7 trees/ha. Major feeding and resting trees were characterized by a mean height of 25.8 m (±5.2), a mean DBH of 48.5 cm (±17.1), and a mean crown diameter of 13.4 m (±4.9). Table 2 provides data on the crown height,

Table 2. Characteristics of major feeding and resting trees used by the mantled howler study troop

Tree #	DBH	Crown diameter	Tree height (m)	Crown height (m)	Number of trees in quadrat	Crown volume
2.5	89.1	30.7	35	27	24	6193.2
2.4	33.1	11.7	22	11	14	268
2.7	55.7	16.5	24	14	22	2854.5
2.9	90.7	21.6	31	20	12	1465.5
2.2	51.2	19.1	31	17	16	1403.7
2.17	37.2	15.8	25	14	14	822.4
2.16	44.5	17.2	25	16	22	1098
3.17	50.3	23.9	32	22	12	3479
3.16	22.6	11.0	19	10	24	463.2
4.3	64.0	22.7	31	22	12	735.4
5.12	36.3	15.9	18	15	18	449.6
5.13	53.1	20.0	32	19	14	802.2
5.11	40.1	9.4	18	8	24	217.9
5.26	38.5	13.7	22	13	12	407.7
5.29	60.5	11.5	28	7	22	3208
5.5	45.2	10.0	25	7	14	374.7
2.3	54.4	9.8	22	8	26	281.8
7.1	55.7	17.0	29	15	14	1043.3
6.4	36.3	19.0	23	18	10	724.5
5.3	41.4	11.6	22	10	12	370.7
15.15	51.5	12.6	25	12	12	211.2
16.18	55.7	14.8	31	13	28	1563
10.24	30.8	11.7	18	11	16	184.3
10.16	47.4	23.1	33	22	22	1210.9
7.3	26.1	12.2	21	12	26	69.4
Mean	48.4	16.1	25.6	14.5	17.6	1196.0
SD	16.3	5.3	5.2	5.3	5.6	1399.1

Table 3. Number of trees and vertical height profile of trees in sample vegetation plots

Plot number	Number of trees	% 0–15 m	% 16–20 m	% 21–25 m	% >25 m
1	24	16.6	33.3	33.3	16.6
2	14	42.8	28.5	28.5	0
3	22	18.1	0	54.5	27.2
4	12	0	50.0	33.3	16.6
5	16	14.2	37.5	62.5	0
6	14	0	0	28.5	57.1
7	22	0	36.3	45.4	18.1
8	12	0	83.3	16.6	0
9	24	24.9	33.3	41.6	0
10	12	0	66.6	0	33.3
11	18	33.3	44.4	11.1	11.1
12	14	14.2	28.5	28.5	28.5
13	24	16.6	41.6	33.3	8.3
14	12	16.6	66.6	0	16.6
15	22	27.2	27.2	27.2	18.1
16	14	14.2	28.5	57.1	0
17	26	7.6	15.3	30.7	0
18	14	14.2	28.5	57.1	0
19	10	0	0	80.0	20.0
20	12	33.3	33.3	0	33.3
21	12	0	0	66.6	33.3
22	28	7.1	64.2	21.4	7.1
23	16	0	75.0	12.5	0
24	22	9	45.4	36.3	9
25	26	0	53.8	30.7	15.3
Mean	17.6	12.3	36.8	33.4	14.7
SD	5.6	12.4	23.3	21.1	14.6

crown diameter, and crown volume of major feeding and resting trees. There were no statistically significant differences in any of these features between major feeding trees and major resting trees. A comparison (Table 3) with other trees measured in these quadrats indicates that major feeding and resting trees were of significantly greater stature ($t = 1.65$, df $= 244$, $p < 0.0002$), DBH ($t = 1.96$, df $= 244$, $p < 0.0008$), and crown diameter ($t = 1.97$, df $= 244$, $p < 0.0003$). In 64% of sample plots, a major tree was the tallest or second tallest tree in the area.

Based on our 25 sample vegetation plots, there was evidence that the size and number of trees were not uniformly distributed throughout the groups' home range. The number of trees per plot ranged from 10 to 28 (mean $= 17.6 \pm 5.6$). The vertical height profile of trees in these plots also varied, with eight plots having no trees over 25 m and four plots with at least one-third of all trees greater than 25 m. Given these differences in tree density and tree stature, it is likely that certain parts of the forest offer howlers greater opportunities for continuous and relatively straight-line travel routes than others.

Visibility in the Canopy

In addressing questions of spatial memory, it is important to obtain a measure of the likelihood that a forager can sight directly from one goal to the next (see "Methods"). A howler with a direct field-of-view is not required to maintain a complex spatial representation of the landscape in order to take least distance routes when traveling between sequential feeding and resting sites. Assuming that major trees were the targets or goals, we determined the degree to which a forager had an unobstructed view over a minimum distance of 20 m if positioned in the lower, middle, or upper zones of the tree crown.

Our results indicate that when located in the lowest third of the crown of a major feeding or resting tree, on average 54.7% of the surrounding tree crowns obstructed the howlers' field-of-view (range: 0–100%). When positioned in the middle third of the crown, 68.4% (range: 12–100%) of the view was obstructed, and when positioned in the top of the crown of a major tree 22.9% (range 0–100%) of the view was obstructed. Values for field-of-view in the lowest levels of the tree crown are likely to represent minimum values given that we did not include in our ecological sample trees less than 15 cm DBH and 3 m in height. Such trees are common in the howlers' home range and many have height profiles that extend into and filter visual information in the lowest levels of the major tree crowns.

Given that howlers are reported to spend the majority of their time in the upper and middle levels of the tree crown (Carpenter, 1934), we compared the number of days howlers visited major resting and feeding trees with the field-of-view at the top and middle zones of the crown for that tree. We found that trees with high field-of-view scores (70–100% visibility) in the upper crown were visited on average 4.9 ± 2.8 times over the course of 15 consecutive days, whereas trees with lower visibility (37–50% visibility) were visited only 2.8 ± 1.1 times ($t = 2.5$, $df = 25$, $p = .009$). A similar situation occurred using field-of-view scores at the mid-level of the tree crown. Trees with low visibility (12–50%) were visited less frequently (3.5 times per tree) than trees with higher visibility scores (57–100%, visited 4.75 times per tree). This latter comparison, however, did not reach statistical significance ($t = 1.14$, $df = 23$, $p = 0.13$).

Distribution of Feeding and Resting Sites

In order to determine the distribution of howler resting and feeding sites, we divided the home range into 63 (25 m × 25 m) quadrats. Thirteen of these (20.6%) contained one major tree, four (6.3%) contained two major trees, and two (3.1%) each contained three major trees. In these latter two quadrats, one contained three resting trees, while the other included two resting and one major feeding tree. Seventy-one percent of quadrats did not contain a major feeding or resting tree. Each of the eight major feeding trees was distributed in a different quadrat and spatially restricted to an area of 1.1 ha within the 3.9 ha core area used during this study (see Figure 2). However, given the small size of this core area, and the fact that mean distances between nearest neighbor feeding and resting trees was only 24.7 m (±17.6), these trees were found to best approximate a uniform distribution ($R = 1.23$, $Z = 2.38$, $p = .0084$).

Ranging Pattern and Use of Major Feeding and Resting Sites

To examine howler foraging strategies and travel patterns, we counted the number of times the howlers visited each of the major feeding trees, and how frequently visits to major feeding trees were from different previous trees or directions (Table 4). Of the 31 visits to major feeding trees, 38.7% ($n = 12$) were from the same previous tree, but 61.3% ($n = 19$) were from different previous trees and directions. The mean distance traveled when selecting new tree sequences and directions was 44.7 m (±29.3), which was significantly greater (two-tailed t-test, $t = 2.05$, df $= 25$, $p < .02$) than the distance traveled when reusing the same tree sequences (22.5 ± 10.8 m).

Table 4. Data on the frequency with which howlers revisited individual feeding trees and the frequency those visits were associated with different travel routes

Feeding tree ID	Number of total visits to feeding tree	Number of times visited from different previous tree
2.7	6	4
2.17	5	3
2.20	7	5
2.30	3	3
5.12	4	3
5.29	3	1
7.3	2	2
10.16	1	1
Total	31	22

In order to examine the degree to which howlers reused route segments to reach quadrats with major feeding and resting sites, we examined the number of days in which particular route segments (defined as travel in three or more contiguous quadrats) were used on more than 1 day. Given that any single quadrat was bounded by eight adjacent quadrats, howlers could reach the same target or goal using a range of different quadrat sequences. Our findings indicate that on 11 of 15 days the howlers reused particular route segments of 75–100 m (3–4 quadrats) to reach major feeding and resting sites. For example, on Day 11, our focal howler used a travel path that included seven quadrats. Four of these quadrats (comprising a route segment) were also visited on Day 4 and on Day 12. Similarly, four contiguous quadrats visited by the howlers on Day 8 also were visited in sequence on Day 14.

The Traveling Salesperson Problem

To examine the degree to which howler daily ranging patterns were consistent with a distance-limiting travel itinerary, we examined the sequence of visits to major feeding and resting trees (average 8.8) over the course of each day. Our results (see Table 5) indicate that on 12 of 15 days, the howlers visited these trees in a highly efficient sequence, traveling 0–4% more than the most efficient

Table 5. Comparison of the actual distance traveled by howlers compared to the least distance routes between feeding and resting sites (index of circuity is defined in text)

Day	Number of quadrats	Distance traveled (m)	Most efficient route (m)	Index of circuitry
1	8	193.5	193.5	1
2	3	41.5	41.5	1
3	6	150	150	1
4	8	151	151	1
5	9	138.5	138.5	1
6	8	224.5	166	1.352
7	6	125	125	1
8	7	200.5	192.5	1.041
9	5	131	131	1
10	6	152.5	124	1.225
11	9	271	231	1.173
12	11	230	230	1
13	9	213	213	1
14	4	144.5	142.5	1.014
15	3	173.5	173.5	1
	Sum	2540	2403	1.05
	Mean	169.3	160.2	1.09

route. Over the entire 15-day sample period, the circuity index, or the actual distance traveled by the howlers compared to the most efficient daily route, was 1.05–1.09. Thus, the howlers visited trees in a sequence that resulted in an increase in travel distance of only 5–9% more than that required by selecting the most efficient tree sequence. These data indicate that at our study site, howlers have access to and consistently use and reuse travel paths that offer them direct or relatively straight-line routes to important feeding and resting sites.

DISCUSSION

Recent studies of primate cognitive ecology have examined the kinds of information primates use in decision-making and the spatial strategies primates employ in traveling between distant feeding sites or what has been referred to as large-scale space (Menzel, 1973; Garber, 1989, 2000; Poucet, 1993; Menzel, 1996; Dyer, 1998; Janson, 1998; Menzel *et al.*, 2002). In general, foragers are expected to track their movements in space using a number of different cognitive mechanisms (Garber and Hannon, 1993; Dyer, 2000; Garber, 2000; Roberts, 2001). These may include path integration (an egocentric frame of reference in which an individual tracks its movements from a starting point to a goal and then returns by computing a direct or short distance route), use of systematic foraging rules to reduce opportunities for backtracking or recrossing paths (for example, travel in a straight-line for some distance and then turn right, if food is not found, continue in same direction and then turn left), sighting directly to a feeding site or a landmark in spatial association to a feeding site, navigating using olfactory or acoustic information, or piloting (encoding the spatial positions of multiple landmark cues to find a target). These mechanisms may be hierarchical or act in concert, depending on the cognitive abilities of the forager and the particular spatial problem encountered (Roberts, 2001).

Although it is possible that individuals in some nonhuman primate species maintain an accurate spatial representation of salient features of the environment as a set of geometric coordinates in large-scale space (real distance and direction vectors, i.e., vector map defined by Byrne (2000) or mental map as defined by Dyer (2000)), at present there is little empirical evidence to support this (Byrne, 2000; Garber, 2000). In this regard, Poucet (1993) has proposed a model of spatial cognition in which landmarks and travel routes in large-scale space are encoded as topological or route-based representations (i.e., network map, Byrne, 2000). In this model, place is encoded as a learned network of travel routes, fixed points (nodes), landmarks, and topological features of the environment. Within such a network, there may be multiple routes of travel between sequential feeding sites, but foragers are constrained to "known" routes and path sequences (Poucet, 1993; Garber, 2000). This description of a topological-based spatial representation is consistent with patterns of travel and foraging in mixed species troops of wild moustached (*Saguinus mystax*) and saddleback tamarins (*Saguinus fuscicollis*) in Peru. Garber (2000) reports that within the 40 ha home range of the study troop there existed a set of frequently used turning areas or switch points (18 of 144–50 m × 50 m quadrats) that appeared to function in reorienting travel. The tamarins used these quadrats to reach parts of their home range that contained productive feeding sites. Travel patterns consistent with network or route-based "maps" have been reported in other primates including hamadryas baboons (Sigg and Stolba, 1981), proboscis monkeys (Boonratana, 2000), orangutans (MacKinnon, 1974), spider monkeys (van Roosmalen, 1985; Milton, 1988), and Panamanian mantled howler monkeys (Milton, 1988).

In the present study, we found that Nicaraguan mantled howler monkeys use efficient routes to visit and revisit major feeding and resting sites. During the course of our study we identified 27 individual trees that served as major feeding and resting sites. These trees were the focus of howler ranging patterns and activity budget. Nicaraguan howlers commonly visited these trees using a distance-limiting travel itinerary. Given that many of these trees were distributed within a circumscribed area and were among the tallest trees in the groups' home range, in some cases the howlers had access to visual cues that enabled them to sight directly from their present location to a potential feeding site. We found, for example, that feeding and resting trees that were characterized by high field-of-view indices were visited more frequently than trees characterized by lower visibility. In some cases, howlers as well as other primates may prefer to feed, rest, and/or travel in the upper parts of the canopy because this provides them with a less obstructed field-of-view in locating resources, spatial landmarks, and possibly predators.

We also present evidence that howlers commonly visited the same target feeding and resting sites from several different previous directions and distances. In these cases, howler movement patterns provide information concerning how spatial information is likely represented in a mental map. Our results indicate that travel was principally route-based, with certain path segments reused several times and across different days during our study. This is consistent with data collected by Milton (1980) in a long-term study of ranging patterns of mantled howler monkeys on Barro Colorado Island, Panama. As in our study, Milton (1980: 105) identified particular arboreal pathways that spanned "a distance of approximately 100 m or more, connected important pivotal trees . . . ", and were reused by howlers on numerous occasions.

Using a limited number of pathways to reach a goal offers an advantage because single landmarks provide their most precise spatial information when viewed from the same perspective or orientation (Byrne, 2000). When approached from a different perspective, a forager may fail to recognize the correct distance, angle, or view between the landmark and the goal. This may explain, in part, why many primate species are reported to reuse paths and path segments during travel. However, in both our study and Milton's (1980) study mantled howlers also were found to take direct but alternate routes to reach the same target feeding or resting site. This suggests that mantled howlers (a) individually encode many different views of a single landmark to relocate each important feeding and resting sites, (b) encode views from several different individual landmarks to relocate a given goal, or (c) views using two or more landmarks are encoded together to relocate the goal (i.e., piloting) (Cheng and Spetch, 1998). Although the use of multiple landmarks to encode spatial information may require an increased capacity to store and integrate a relatively large number of points in the environment, Kamil and Cheng (2001: 107) argue that the use of "multiple landmarks functions to increase dramatically the precision of searching in the face of errors in the estimation of distance or direction". Although we lack sufficient information to identify salient features of the landscape that may be used by howlers as landmarks to orient and reorient the direction of travel, it is likely that members of our howler study troop maintained information of the locations of numerous intersecting routes of travel and landmarks within their home range. Reuse of these route segments is consistent with a typological spatial representation (i.e., network map).

Finally, in the current study we monitored howler ranging activities, foraging behavior, and use of feeding and resting trees during a period of only 15 consecutive days. It is likely that howlers exhibit a wider range of spatial strategies than we observed; however, the patterns documented in Nicaraguan mantled howlers were extremely similar to patterns documented in more long-term studies of mantled howlers on Barro Colorado Island, Panama (Carpenter, 1934; Milton, 1980, 1988, 2000). Isla de Ometepe (dry tropical forest) and Barro

Colorado Island (tropical rainforest) represent forests of very different stature, tree species composition, and level of human disturbance. Despite the fact that mantled howlers on Ometepe occupied small home ranges of 3–4 ha and the mantled howlers on BCI exploited home ranges of 30–40 ha, in each habitat individuals appeared to rely on a complex set of cognitive mechanisms and spatial memory to efficiently travel between major resting and feeding sites. In this regard, studies of cognition in other New World primates offer important comparative models for examining whether species that exploit larger home ranges, larger day ranges, and/or more highly dispersed or ephemeral resources (i.e., other *Atelines*, *Cebus*, *Saimiri*, *Saguinus*, *Callimico*, *Leontopithecus*, and *Cacajao*) or species that exploit small ranges and focus their daily activities around a limited set of productive feeding sites (i.e., some gum-feeding marmosets such as *Callithrix jacchus* and *Cebuella pygmaea* and other species of *Alouatta*) rely on different types of social, spatial, and ecological information and mental maps in locating resources.

SUMMARY

In this study, we examined questions of spatial memory and the travel routes taken by Nicaraguan mantled howler monkeys (*Alouatta palliata*) when moving between major feeding and resting sites. Studies of primate foraging and ranging behavior indicate evidence of goal-directed travel and relatively straight-line movement between sequential feeding sites. In the case of mantled howlers on BCI, Panama, Milton (1980) has argued that over the course of several weeks, group members center their feeding, resting, and ranging activities on a small set of pivotal trees that are visited several times daily. The degree to which howlers use topological (route-based) spatial representations or geometric (coordinate-based) spatial representations to locate and revisit these feeding and resting sites remains unclear.

In order to address questions concerning goal-directed travel and spatial memory, we mapped the travel routes taken by a troop of 26–29 mantled howlers inhabiting Estación Biológica de Ometepe located on Isla de Ometepe, Nicaragua. Behavioral data were collected during July and August 2002. Over the course of 15 days and 103 h of observation, all trees the howlers were observed to travel, feed, and rest in were marked ($N = 250$), and distances and angles between trees were measured and mapped. Travel routes were identified by following a focal individual for 6–8 h per day.

The mean number of trees visited per day was 40.6 and mean daily path length was 381 m. Overall, howlers were found to take direct routes to feeding and resting sites, and reuse the same route segments on several occasions. Based on data on canopy visibility, tree distribution, and the distance between sequential feeding sites, howlers appear to rely on particular landmark cues to orient and reorient their direction of travel. These results are consistent with the hypothesis that members of our howler study troop maintained information of the locations of numerous intersecting routes of travel and landmarks within their home range. Reuse of these route segments is consistent with a typological or route-based spatial representation.

ACKNOWLEDGMENTS

The authors would like to acknowledge Michelle Bezanson, Anneke DeLuycker, Teague O'Mara, Bernardo Urbani, and Melissa Raguet-Schofield for critical assistance in collecting behavioral and ecological data in the field. PEJ thanks Rudolfo and his two sons for their invaluable assistance in tracking the howlers. We thank Don Rene Molina, Doña Lilian Molina, and Alvaro Molina for their generosity, support, and efforts in maintaining Estación Biológica de Ometepe. Sara Garber provided assistance in the field and calculated crown volumes of major feeding and resting trees. As always PAG thanks Sara and Jenni for always being Sara and Jenni.

REFERENCES

Altmann, J. 1974, Observational study of behaviour: Sampling methods. *Behaviour* 49:227–265.

Altmann, S. A. 1974, Baboons: Space, time, and energy. *Am. Zool.* 14:221–248.

Bezanson, M. F., Garber, P. A., and DeLuycker, A. M. 2001, Patterns of subgrouping and spatial affiliation in a community of mantled howling monkeys (*Alouatta palliata*). *Am. J. Phys. Anthropol.* 32(Suppl): 39–40.

Bicca-Marques, J. C. and Garber, P. A. 2004, The use of visual, olfactory, and spatial information during foraging in wild nocturnal and diurnal anthropoids: A comparison among *Aotus*, *Callicebus*, and *Saguinus*. *Am. J. Primatol.* 62:171–187.

Biegler, R. and Morris, R. G. M. 1996, Landmark stability: Further studies pointing to a role in spatial learning. *Q. J. Exp. Psychol.* 49:307–345.

Boonratana, R. 2000, Ranging behavior of proboscis monkey (*Nasalis larvatus*) in the lower Kinabatanjan, Northern Borneo. *Int. J. Primatol.* 21:497–518.

Byrne, R. W. 2000, How monkeys find their way: Leadership, coordination, and cognitive maps of African baboons, in: S. Boinski and P. A. Garber, eds., *On the Move: How and Why Animals Travel in Groups*, University of Chicago Press, Chicago, pp. 491–518.

Carpenter, C. R. 1934, A field study of the behavior and social relations of howling monkeys. *Comp. Psychol. Monogr.* 10:1–168.

Chapman, C. A., Chapman, L. J, and McLaughlin, R. L. 1989, Multiple central place foraging by spider monkeys: Travel consequences of using many sleeping sites. *Oecologia* 79:506–511.

Cheng, K. and Spetch, M. L. 1998, Mechanisms of landmark use in mammals and birds, in: S. D. Healy, ed., *Spatial Representation in Animals*, Oxford University Press, Oxford, pp. 1–17.

Crockett, C. M. 1998, Conservation biology of the genus *Alouatta. Int. J. Primatol.* 19:549–578.

Dyer, F. C. 1991, Bees acquire route-based memories but not cognitive maps in a familiar landscape. *Anim. Behav.* 41:239–246.

Dyer, F. C. 1998, Cognitive ecology of navigation, in: R. Dukas, ed., *Cognitive Ecology*, University of Chicago Press, Chicago, pp. 201–260.

Dyer, F. C. 2000, Group movement and individual cognition: Lessons from social insects, in: S. Boinski and P. A. Garber, eds., *On the Move: How and Why Animals Travel in Groups*, University of Chicago Press, Chicago, pp. 127–164.

Gallistel, C. R. and Cramer, A. E. 1996, Computations on metric maps in mammals: Getting oriented and choosing a multi-destinational route. *J. Exp. Biol.* 199:211–217.

Garber, P. A. 1987, Foraging strategies among living primates. *Ann. Rev. Anthropol.* 16:339–364.

Garber, P. A. 1989, Role of spatial memory in primate foraging patterns: *Saguinus mystax* and *Saguinus fuscicollis. Am. J. Primatol.* 19:203–216.

Garber, P. A. 2000, Evidence for the use of spatial, temporal, and social information by primate foragers, in: S. Boinski and P. A. Garber, eds., *On the Move: How and Why Animals Travel in Groups*, University of Chicago Press, Chicago, pp. 261–298.

Garber, P. A. and Brown, E. 2005, Use of landmark cues to locate feeding sites in wild capuchin monkeys (*Cebus capucinus*): an experimental field study. In: New Perspectives in the Study of Mesoamerican Primates: Distribution, Ecology, Behavior, and Conservation, edited by Alejandro Estrada, Paul A. Garber, Mary S. M. Pavelka, and LeAndra Luecke. Springer, New York, pp. 311–322.

Garber, P. A. and Hannon, B. 1993, Modeling monkeys: A comparison of computer generated and naturally occurring foraging patterns in 2 species of neotropical primates. *Int. J. Primatol.* 14:827–852.

Garber, P. A., Pruetz, J. D., Lavallee, A. C., and Lavallee, S. G. 1999, A preliminary study of mantled howling monkey (*Alouatta palliata*) ecology and conservation on Isla de Ometepe, Nicaragua. *Neotrop. Primates* 7:113–117.

Gibson, B. M. and Kamil, A. C. 2001, Test for cognitive mapping in Clark's nutcrackers (*Nucifraga columbiana*). *J. Comp. Pyschol.* 115:403–407.

Janson, C. 1998, Experimental evidence for spatial memory in wild capuchin monkeys, *Cebus apella*. *Anim. Behav.* 55:1229–1243.

Janson, C. H. and Di Bitetti, M. S. 1997, Experimental analysis of food detection in capuchin monkeys: Effects of distance, travel speed, and resource size. *Behav. Ecol. Sociobiol.* 41:17–24.

Kamil A. C. and Cheng, K. 2001, Way-finding and landmarks: The multiple-bearings hypothesis. *J. Exp. Biol.* 204:103–113.

MacKinnon, J. 1974, The behaviour and ecology of wild orang-utans (*Pongo pygmaeus*). *Anim. Behav.* 22:3–74.

Menzel, C. 1996, Structure-guided foraging in long-tailed macaques. *Am. J. Primatol.* 38:117–132.

Menzel, C. R., Savage-Rumbaugh, E. S., and Menzel, E. W. Jr. 2002, Bonobo (*Pan pansicus*) spatial memory and communication in a 20-hectare forest. *Int. J. Primatol.* 23:601–620.

Menzel, E. W. 1973, Chimpanzee spatial memory organization. *Science* 182:943–945.

Milton, K. 1980, *The Foraging Strategy of Howler Monkeys: A Study in Primate Economics*. Columbia University Press, New York.

Milton, K. 1981, Diversity of plant foods in tropical forests as a stimulus to mental development in primates. *Am. Anthropol.* 83:534–548.

Milton, K. 1988, Foraging behavior and the evolution of primate cognition, in: R. W. Byrne and A. Whiten, eds., *Machiavellian Intelligence: Social Expertise and the Evolution of Intellect in Monkeys, Apes, and Humans*, Clarendon Press, Oxford, pp. 285–305.

Milton, K. 2000, Quo vadis? Tactics of food search and group movements in primates and other animas, in: S. Boinski and P. A. Garber, eds., *On the Move: How and Why Animals Travel in Groups*, University of Chicago Press, Chicago, pp. 375–417.

Poucet, B. 1993, Spatial cognitive maps in animals: New hypotheses on their structure and neural mechanisms, *Psychol. Rev.* 100:163–182.

Roberts, W. A. 2001, Spatial representation and the use of spatial codes in animals, in: M. Gattis, ed., *Spatial Schemas and Abstract Thought*, MIT Press, Cambridge, MA, pp. 15–44.

Salas Estrada, J. B. 1993, *Arboles de Nicaragua*. Instituto Nicaraguense de Recursos Naturales Y del Ambiente, Managua, Nicaragua.

Shettleworth, S. J. 1998, *Cognition, Evolution, and Behavior.* Oxford University Press, Oxford.

Sigg, J. and Stolba, J. 1981, Home range and daily march in a hamadryas baboon troop. *Folia Primatol.* 36:40–75.

Tomasello, M. and Call, J. 1997, *Primate Cognition.* Oxford University Press, Oxford.

Valone, T. J. 1989, Group foraging: Pubic information, and patch estimation. *Oikos* 56:357–363.

van Roosmalen, M. G. M. 1985, Habitat preferences, diet, feeding strategy and social organization of the black spider monkey (*Ateles paniscus paniscus* Linnaeus 1758) in Surinam. *Acta Amazonia* 15:7–238.

CHAPTER TWELVE

Use of Landmark Cues to Locate Feeding Sites in Wild Capuchin Monkeys (*Cebus capucinus*): An Experimental Field Study

Paul A. Garber and Ellen Brown

INTRODUCTION

Recent studies of navigation in insects, birds, and mammals indicate that foragers use a range of cognitive mechanisms to locate and travel between distant feeding sites (Biegler and Morris, 1996; Dyer, 1991, 2000; Gallistel, 1990; Garber, 2000; Janson and Di Bitetti, 1997; Kamil and Cheng, 2001; Shettleworth, 1998; Tomasello and Call, 1997). These mechanisms include solar navigation, magnetic compasses, olfactory cues, visual information, and various forms of spatial memory. In particular, researchers have focused on

Paul A. Garber • Department of Anthropology, University of Illinois, Urbana, IL 61801. **Ellen Brown** • Yale University School of Forestry and Environmental Studies, M.E.M, New Haven, CT 06511.

New Perspectives in the Study of Mesoamerican Primates: Distribution, Ecology, Behavior, and Conservation, edited by Alejandro Estrada, Paul A. Garber, Mary S. M. Pavelka, and LeAndra Luecke. Springer, New York, 2005.

how individuals internally represent spatial information and integrate and recall predictable features in the environment as landmarks (Biegler and Morris, 1996; Kamil and Cheng, 2001; Roberts and Pearce, 1998). Landmarks are reference points and may be represented as distance, direction, or distance and direction vectors from a target or goal (Cheng and Spetch, 1998). A single landmark may be used as a beacon and associated with the location of a den, nest, or feeding tree (Dyer, 1991). In other cases, spatial representations may include a configuration of several landmarks encoded individually or together (Gibson and Kamil, 2001; Kamil, *et al.*, 1999; Vander Wall, 1990). The use of multiple landmarks can significantly increase search accuracy by allowing individuals to orient to or travel to an area between a set of features (i.e. piloting) and thus avoid errors in estimating the precise distance or direction to the goal (Kamil and Cheng, 2001).

Based on extensive natural field observations and limited experimental field research, there is evidence that primate foragers often take direct routes to out-of-sight food patches, select nearest-neighbor trees of target species, orient to features of the landscape in order to locate resources, form subgroups under conditions of low food availability, incorporate quantity information and temporal information in regulating return times to renewable feeding sites, and use conspecifics as guides to encounter food rewards (Garber, 1988, 1989, 2000; Garber and Dolins, 1996; Janson, 1998; Janson and Di Bitetti, 1997; Menzel, 1991, 1996; Menzel *et al.*, 2002). This implies an ability to integrate a complex set of social and ecological information during foraging (Bicca-Marques and Garber, 2005). Several questions remain unanswered, however, including how detailed prosimian, monkey, and ape spatial representations are, and how primates internally represent landmarks and other cues in order to relocate feeding sites.

Poucet (1993) proposed a model of spatial cognition in which landmarks and travel routes are encoded as metric representations in small-scale space and as topological or route-based representations in large-scale space. He argues that the recurrent use of specific core areas within a home range, or what can be termed "small-scale space" (nest sites, aggregates of feeding trees, water holes) provides a forager with the opportunity to encounter and integrate different "views" of the same landmarks and features of the environment relative to a set of goals. These "views" might provide the forager with sufficient information to build a mental representation that includes accurate directions and distances between multiple landmarks and goals, and to use this metric information to compute efficient routes of travel (Gibson and Kamil, 2001). Overtime, a forager may maintain detailed metric representations of several such core areas.

In large-scale space or when moving between core areas, however, the forager's ability to construct a metric map is limited by factors of distance and visibility which make it difficult to obtain direct "views" of the location of the goal and single or multiple landmarks near-to-the-goal. Poucet (1993) argues therefore, that in large-scale space, place is encoded as a learned network of fixed points (nodes), travel routes, landmarks, and salient topological features of the environment rather than as a true metric representation. In this regard, compass bearings or directions between landmarks may be more salient or represented more accurately than distances between landmarks (Kamil and Cheng, 2001). Within such a network, there may be multiple routes of travel between sequential feeding sites. An animal possessing computational skills may select the shorter of two or more fixed routes to reach the target, but nonetheless is restricted to reusing particular path segments and to orienting to particular landmarks as steps in the process of locating the goal (Garber, 2000; Poucet, 1993).

In this study, we examine the ability of wild white-faced capuchin monkeys (*Cebus capucinus*) to use landmarks singly and relationally to compute the location of baited feeding sites in small-scale space. Compared to other New World primates, capuchins are characterized by large-brain size, enhanced manual dexterity, frequent object use (i.e. break open hard fruits by striking them against tree trunks and branches; Boinski *et al.*, 2001; Panger, 1998), and are reported to rely on complex spatial information to locate distant feeding sites (Janson, 1996, 1998). The degree to which capuchins use landmarks to locate feeding sites remains unclear.

METHODS

From September 2000 to November 2000, we conducted an experimental field study of spatial cognition and foraging strategies in a group of 15 habituated wild white-faced capuchins (*C. capucinus*) inhabiting La Suerte Biological Research Station in northwestern Costa Rica (10°26′N, 83°47′W). This area receives approximately 4000 mm of rainfall per year. The Research Station is a 700 ha area of tall secondary rainforest and regenerating pasture. In addition to white-faced capuchins, mantled howler monkeys (*Alouatta palliata*) and black-handed spider monkeys (*Ateles geoffroyi*) inhabit the site.

Research Design

The field experiment involved the construction of a feeding station comprising eight visually identical wooden platforms located in the home range of the study group. The platforms were arranged in a circle with a diameter of 8 m (Figure 1). Each platform was positioned at an angle of 45° and a distance of 3.1 m from neighboring platforms. Platforms resembled square tables, and were approximately 1.5 m in height and measured 62 cm × 62 cm (length and width).

The problem presented to the capuchins tested their ability to use a series of experimental landmark cues (identical yellow and pink colored poles measuring 2 m in height) to predict which of the eight feeding platforms contained a hidden food reward (bananas). The spatial configuration of poles was systematically manipulated across experiments (see below). In all the experiments, six platforms each contained two plastic (sham) bananas and two platforms each contained two real bananas (however, neighboring platforms did not each contain a food reward). Real and sham bananas were covered with leaves to eliminate visual cues, and banana skins were placed with plastic bananas to equalize odor cues. The identity of the two reward platforms during each test

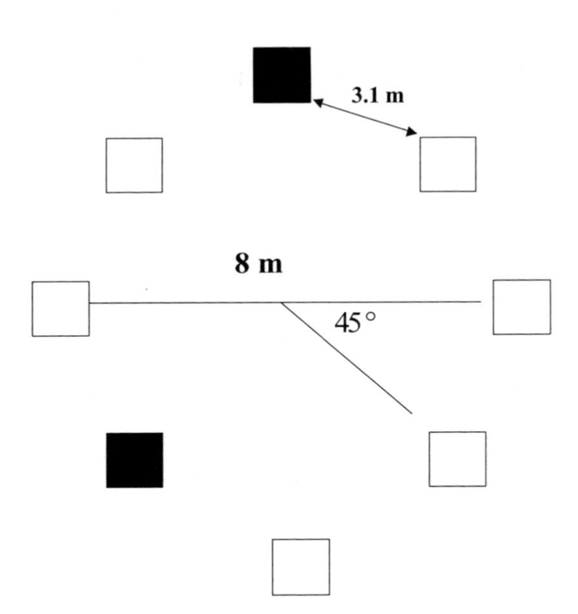

Figure 1. A schematic of the spatial configuration of the eight visually identical feeding platforms that formed the experimental Feeding Station. Dark squares represent baited (reward) feeding sites. White squares represent sham (non-reward) feeding sites.

session was random (place not predictable). Previous studies of these capuchins (Garber, 2000; Garber and Brown, 2004; Garber and Paciulli, 1997) indicated that individual group members used both natural landmarks (features of their habitat) and experimental spatial/associative cues (yellow blocks placed on a platform when it contained a food reward) to locate baited feeding sites.

We conducted a series of six experiments in which the relative spatial positions of artificial landmark cues (identical yellow and pink colored poles) were the only information available to foragers to efficiently locate reward feeding sites. Data were collected between the hours of 5:15 am and 4:00 pm on each of 60 consecutive days (5-day habituation period and 55 days of field experiments). The number of test sessions or trials per experiment was dependent on the number of times the capuchins visited the feeding station per day, as well as the capuchins' performance in locating reward platforms. A trial began when a capuchin was observed within 10 m of the experimental feeding station. The trial concluded when the last group member left the vicinity of the feeding station. Individual group members approached the feeding platforms from a variety of different cardinal directions and descended onto the platforms using a number of different arboreal pathways.

A visit was scored when a capuchin arrived on a feeding platform. Once on a platform, the forager quickly manipulated or picked up the leaf covering, exposing the real or plastic bananas. If the capuchin was successful in obtaining a food reward, it would grab one or both the bananas and rapidly climb on a nearby tree to feed. If the forager was unsuccessful, it would typically climb or jump from the platform to a nearby tree and then travel to another platform. Less commonly, an unsuccessful capuchin forager would run quadrupedally on the ground to a nearby platform, climb on the wooden structure, and search the platform for food. Platforms were rebaited after all capuchins had left the area, and a new trial began when the capuchins arrived in the trees within 10 m of the feeding station. All observations were made with the observer concealed in a blind located approximately 5 m from the nearest platform. The use of a blind insured that the presence of the observer had no effect on the platform choice or behavior of the capuchins.

Data were collected on (a) the time the first capuchin arrived at the feeding station, (b) latency from arrival at the feeding station to the time the first platform was visited, (c) whenever possible the identity of the capuchin arriving at a platform, (d) whether the individual was successful in obtaining a food reward, (e) the total number of platforms visited, and (f) the sequence of platform visits. Although repeated visits to a platform were recorded, only data on the first visit by a capuchin to a reward or sham platform were used in the analyses. This is due to the fact that once a platform was visited, subsequent foragers could use ecological information (sight of real banana or plastic banana, or leaf covering removed and no banana present) or social information (observe the behavior of a group member feeding at a given platform) in selecting or avoiding a feeding platform. On many occasions, several capuchins visited the platforms in the span of a few seconds, and therefore it was not possible to accurately identify each individual forager. In addition, there were occasions when a member of the study group arrived at the feeding station alone and visited feeding platforms in a predictable sequence (i.e. P1–P2–P3–P4). These cases ($N = 23$) were omitted from the analyses because it was apparent that either because of rank or experience at the platforms (consistently arriving after all food had been consumed) that some individual group members did not learn to attend to the landmark cues (see below).

The first 5 days of the study were used to habituate the capuchins to the feeding platforms. During days 1–3, all the platforms were baited with two real bananas. The capuchins visited the feeding station several times per day and fed on all experimental platforms. On the last 2 days of the habituation period, two plastic bananas and two real bananas were placed on all the feeding platforms. The capuchins continued to visit and feed on all the platforms. Over the next 55 days, we then conducted a series of six experiments. Experiment 1 tested the ability of capuchins to associate the spatial position of either one or two landmarks with the spatial position of the reward platforms (search platform nearest a pole or search between the cues). Experiment 2 tested the ability of capuchins to use the spatial relationship between two landmark cues and the platform to locate the food reward (search between the cues only). Experiment 3 tested the ability of the capuchins to compute the spatial relationships between three landmarks to locate one (the reward platform most distant to the landmark cues) of the two reward platforms. Experiment 4 was used as a control for Experiment 3 to determine whether foragers were using the spatial position of the third pole in relation to the spatial position of the other two landmarks to locate the reward platform most distant to the landmark cues. Experiment 5 was a repeat of Condition 3 in order to make certain that the results from that experiment were reliable. Experiment 6 was the control condition in which all the landmark cues were removed and therefore the monkeys' performance should not exceed chance levels (2/8 or 25%). A complete description of each experimental condition is presented in Figure 2.

EXPERIMENT	CONDITION	TASK
1	Place random, absence of visual and olfactory cues. Two 2-meter poles placed on each side of a reward platform	Associate the spatial position of 2 poles with the spatial position of the reward platforms (**search platform nearest a single pole or search between the cues**)
2	Place random, absence of visual and olfactory cues. Two 2-meter poles each placed at a distance of 2 meters behind and at an angle equidistant from its two neighboring platforms. Each single pole has identical spatial relationship to reward and nonreward platforms.	Use the spatial position of array of two distant landmarks to compute the location of a food reward (**search between poles – requires use of two cues**)
3	Place random, absence of visual and olfactory cues. Two 2-meter poles placed at a distance of 2 meters behind and at an angle equidistant from its two neighboring platforms. An imaginary circle with a radius of 1 meter was constructed at the center of the sphere of platforms. Each point on this circle was 3 meters distant from the nearest platform. The location of the 3rd landmark cue was placed in a line between the 2 initial poles and the opposite platform. The distance from the 3rd landmark cue to the reward platform was 5 meters.	Use the spatial position of an array of two landmarks and an array of three landmarks to compute the location of food rewards 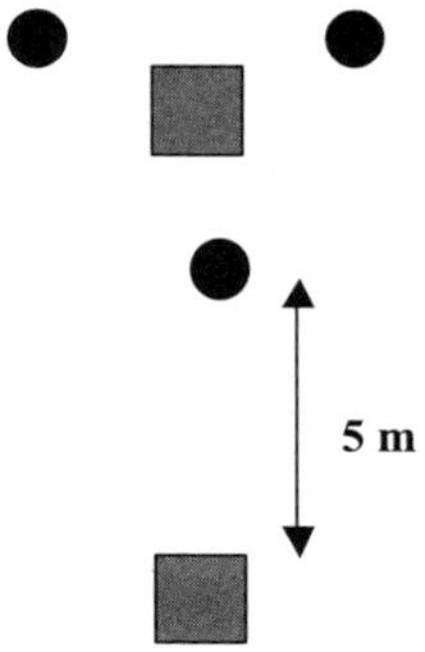 5 m

Figure 2. A description of each of the six Experimental conditions (see text). Only reward platforms are shown (dark squares). Yellow and pink poles that serve as landmark cues are presented as black circles. Note that the spatial locations of the landmark cues changed randomly each time the capuchins arrived at the Feeding Station. However, the spatial relationships between the landmarks and the reward platforms were consistent during a given experiment.

EXPERIMENT	CONDITION	TASK
4	Place random, absence of visual and olfactory cues. Same as Condition 3 except the 3rd cue (single pole) was removed.	Compare capuchins' performance with Condition 3 to determine whether foragers were using the spatial position of the 3rd pole in relation to the spatial positions of the other 2 landmark cues to triangulate and locate baited feeding site. 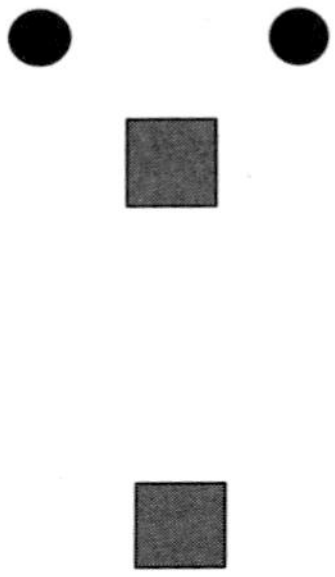
5	Same as Experiment 3	Associate the spatial positions of 3 landmark cues relationally to locate a food reward. (**compute spatial relationships of 3 points in space-repeat Condition 3**) 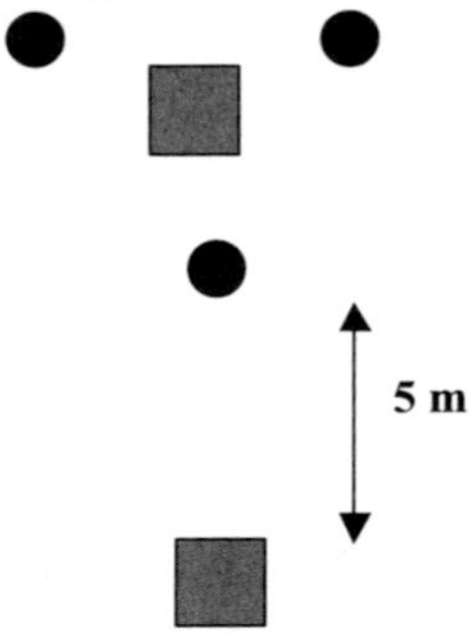

5 m

Figure 2. *(Continued)*

Data on the ability of the capuchins to use landmark information were analyzed based on the probability of selecting two reward platforms during their first four platform selections. This was used as a threshold of efficiency because (a) by chance capuchins were expected to locate one reward platform in the first four platform choices (25%) and (b) individuals who arrived at the Feeding Station after one or two foragers had already begun to explore the platforms, often searched platforms, positioned most distant to the current foragers. This

EXPERIMENT	CONDITION	TASK
6	Place random, absence of visual and olfactory cues. Absence of landmarks. This is the control condition.	In the absence of any spatial cues to predict the location of baited feeding sites, the monkeys' performance should not exceed chance (2/8 = 25% correct, **Control Condition**)

Figure 2. *(Continued)*

was an artifact of dominance, but could result in a capuchin locating a reward platform at a rate greater than expected by chance due to the fact that adjacent platforms were never both reward platforms.

Given that the order in which the two reward platforms were selected was irrelevant to the foraging problem, the data were evaluated as a Bernoulli problem with the selection of each platform as an independent trial according to the formula

$$b(x, n, p) = (n!/(n-x)\, x!)\, p^x q^{n-x}$$

where

b = probability of capuchin performance during a given experiment;

p = probability that two reward platforms are selected by chance during first four platform visits ($p = 0.214$);

q = probably that less than two reward platforms are selected by chance during the first four platform visits ($q = 0.786$);

x = number of trials in which two reward platforms were selected during first four platform visits;

n = number of trials in which two reward platforms were not selected during first four platform visits.

A Student's *t*-test (2-tail probability) was used to examine differences in capuchin behavior (latency to first platform visit, time spent at feeding platforms, and time spent at the feeding station) over the course of the field experiments. For all the analyses, probability was set at 0.05.

RESULTS

General Behavior

Over the course of 55 consecutive days and 227 test trials (six experimental conditions), members of our capuchin study group visited the feeding platforms a total of 3262 times. Experiments lasted from 6 to 15 days and the number of trials per experiment ranged from 20 to 67 (Table 1). Given that only two of eight platforms contained a food reward and the number of capuchin foragers totaled 15, accuracy in locating reward platforms was a major factor in individual feeding success.

The number of experimental trials per day was determined by the frequency with which capuchins returned to the feeding station. Over the course of the study, there was a steady increase in the number of daily visits capuchins' made to the feeding station. This ranged from an average of 1.5 times per day in Experiment 1 to an average of 7.4 times per day in Experiment 6 (Table 1), and appeared to reflect a process of both increased habituation of particular individuals to the test setting, and changes in strategies used by subordinate group members to obtain access to reward platforms.

Based on our observations of capuchin behavior at the Feeding Station, there is evidence that social dominance played an important role in the identity of the first individual to visit a platform and in latency in time of arrival at the Feeding Station to time of first platform visit. For example, in the first two experimental conditions, the highest-ranking adult male, Mr. Cool, was the first capuchin to explore feeding platforms during 75% of group visits (18/24 identified individuals). On average, he explored a platform within 2–3 min of arriving at the Feeding Station. In the last experimental condition, Mr. Cool was the first individual to arrive at a feeding platform during only 11% of group visits (7/62 identified individuals). This resulted from the fact that other group members, in particular, adult females with infants, adult females without infants, and juveniles adopted a behavioral tactic of arriving at the feeding station 5–10 min in advance to the rest of the group. Although these individuals were characterized by increased latency in the time interval between arriving at the Feeding Station

Table 1. Number of experimental trials and summary of time spent by capuchins at the feeding platforms

Experiment (No. of trials)	Duration of experiment (days)[a]	Trials per day	Latency to first platform visit (min)	Total time spent at feeding platforms (min)	Total time spent at feeding station (min)
1 (20)	13	1.5 ± 0.77	2.2 ± 1.6	19.6 ± 28.6	28.5 ± 30.0
2 (20)	9	2.2 ± 1.0	2.4 ± 2.7	13.3 ± 8.4	25.1 ± 14.4
3 (45)	15	3.0 ± 1.6	2.8 ± 2.4	15.4 ± 12.5	27.9 ± 18.7
4 (30)	6	5.0 ± 1.5	3.7 ± 3.2	14.7 ± 9.4	24.9 ± 9.3
5 (45)	7	6.4 ± 1.3	5.3 ± 6.6	15.0 ± 8.3	26.7 ± 12.9
6 (67)	9	7.4 ± 2.0	5.3 ± 6.5	11.5 ± 8.4	27.5 ± 10.4

[a] Capuchins were exposed to the test conditions over 55 consecutive days. However, a given experiment would end at noon on 1 day and the next experiment would begin on the same day. Therefore, the total in column 2 (duration of experiment) exceeds 55 days.

and visiting a feeding platform (5–6 min; Table 1; comparing latency in time to exploring first feeding platform in Experiments 1, 2, and 3 with Experiments 4, 5, and 6, $t = 3.77$, $df = 208$, $p = 0.0002$), animals visiting feeding platforms first had greater opportunities to learn to use experimental landmark cues to predict the location of reward platforms than did later arriving animals. This was the case because although later arriving animals had access to the same landmark array as did earlier arriving animals, the positive reinforcement of locating a food reward was eliminated (reward had already been removed by an earlier forager). A comparison of time group members spent at the feeding platforms (time from first visit to last visit; $t = 1.5$, $df = 223$, $p = 0.13$) and time group members spent in the immediate vicinity of the Feeding Station (time of arrival to time of departure; $t = 0.38$, $df = 169$, $p = 0.069$) during Experimental trials 1–3 versus Experimental trials 4–6 indicates no significant differences.

Field Experiments

In Experiment 1, a 2 m long yellow and pink striped pole was placed directly on each side of a reward platform (a total of four associative cues, two per reward platform). These associative cues provided the only information available to the capuchins to predict the location of feeding sites. The results (Table 2) indicate that the capuchins quickly learned to associate the presence of one or both landmarks with the location of a food reward ($p < 0.01$). In 9 of these first 20 experimental trials, the capuchins located both reward platforms in four or fewer platform choices.

In the second experiment, landmarks were relocated at a distance of 2 m behind a reward platform and at an angle mid-way between neighboring platforms. In this case, the proximity of a single pole was not sufficient to solve the foraging problem because each pole was positioned at the same distance and angle to the reward platform as it was to the unbaited platforms on either side of the reward platform. An understanding of the spatial positions of the two landmarks and the reward platform was required to solve this foraging problem. In this experiment, neighboring platforms were never reward platforms. This was done to eliminate the possibility that all the four landmark cues were adjacent to each other. The capuchins attended to this new cue arrangement rapidly and successfully used the presence of both the landmarks to predict the location of reward platforms. In 13 of 17 trials, individuals selected both the reward platforms in four or fewer platform choices (Table 2, $p < 0.0001$).

In Experiment 3, we tested the ability of capuchins to use an arrangement of three identical landmarks to locate reward platforms. As in the previous experiment, two landmarks were located 2 m behind one reward platform. An imaginary circle with a radius of 1 m was constructed at the center of the sphere of platforms. Each point on this circle was 3 m distant from the nearest platform. The location of the third landmark was placed in a line between the two initial poles and the opposite platform. The distance from the third landmark cue to the reward platform opposite the poles was 5 m. This arrangement is illustrated in Figure 2. The task presented to the capuchins therefore involved the use of spatial relationships of three identical landmarks to compute the trajectory (direction) to the reward platform. Although the specific locations of the three landmarks changed randomly from trial to trial, the spatial relationship between the three landmarks and the reward platforms remained constant throughout this experiment.

The results indicate that in 15 of 25 trials, the capuchins located both reward platforms in four or fewer platform choices (Table 2, $p = 0.0001$). In another nine cases, only a single forager arrived at the feeding station and this individual located one reward platform in two or fewer platform choices. Initially, the capuchins selected the reward platform bounded by the two landmark cues (13/17 trials). In the final 17 trials, however, the capuchins were as likely to select first the reward platform requiring the use of three landmarks (9/17 = 52.9%) as the platform bounded by two landmarks (8/17 = 47.0%) as their first reward platform choice.

In the fourth experiment, we removed the third landmark to determine the degree to which the monkeys were using this cue in conjunction with the other two landmarks to locate baited feeding sites. The results indicate that over the course of 18 test trials, the capuchins located two reward platforms in four or fewer platform choices, significantly greater than expected by chance (7/18, Table 2, $p = 0.046$). In three additional test trials, a single capuchin arrived at the feeding station and this individual located one reward platform in three or fewer platform choices. However, in 76.1% of cases, the capuchins visited the reward platform associated with the two landmark cues first (16/21). This pattern was identical to that observed during the initial phase of Experiment 3, but differed from that found in the latter half of that experiment in which the reward platforms requiring the use of two landmarks and three landmarks were visited first equally. This is consistent with the hypothesis that the presence of the third landmark cue was attended to by the capuchins in Experiment 3 to locate the more distant reward platform.

Table 2. Pattern of reward and sham platform selections by white-faced capuchins under different experimental conditions

	Capuchin performance									
Experiment	0/4	1/4	2/4	1/3	2/3	1/2	2/2	1/1	Total[a]	*p*-Value
1	1	10	4	0	3	0	2	0	20	0.01
2	1	3	2	1	8	0	3	1	19	0.0001
3	1	9	10	1	3	7	2	2	35	0.0001
4	1	10	5	1	1	1	1	1	21	0.046
5	0	12	14	3	3	4	3	3	42	0.0001
6	4	46	6	2	5	2	2	0	67	0.122

[a] Twenty three cases were omitted from the analyses. In each of these cases, the capuchin forager arrived at the feeding station alone and visited the platforms in a predictable sequence (i.e. P1–P2–P3–P4). See "Methods" section.

In Experiment 5, we repeated the conditions of the third experiment to verify the accuracy of our results. We found that in 20 of 32 trials, the capuchins located both reward platforms in four or fewer platform choices (Table 2, $p = 0.0001$) and that in seven additional trials, foragers arriving alone located one reward platform in two or fewer platform choices. In three cases, one reward platform was selected in three platform choices. In the first 22 trials, the capuchins first selected the reward platform that was bounded by two landmark cues 68.1% of the time (15/22). In the final 20 trials, the capuchins were equally likely to select first the reward platform requiring the use of three landmarks (9/20 = 45%) as the platform bounded by two landmarks (11/20 = 55%).

Experiment 6 represented the control condition. All experimental landmarks were removed. Two of eight platforms had real bananas, banana skins were placed with plastic bananas to minimize olfactory information, leaves covered the platforms to eliminate visual cues, and the place was random. In this experiment, the capuchins were successful in locating both the reward platforms in four or fewer platform choices in only 13 of 63 trials (Table 2; $p = 0.122$; in four additional trials one reward platform was selected in two or three platform choices). In 50 of the trials, the capuchins located either 0 ($n = 4$) or 1 ($n = 46$) of the banana rewards in their first four platform choices. The results of this experiment indicate that in the absence of reliable landmark cues, the capuchins were unable to accurately predict the location of reward platforms.

DISCUSSION

A major challenge that animals face in exploiting their environment is the ability to relocate ephemeral, widely scattered, and productive feeding sites. Resources in many tropical forests are patchily distributed in time and space (Johnson *et al.*, 2002), and therefore the search effort required to locate feeding sites can account for a significant proportion of an individuals' daily activity budget. In the case of white-faced capuchins (*C. capucinus*), for example, adult males and females devote between 50% and 65% of their day foraging and traveling (Fedigan, 1993; Rose, 1998) and have a day range of approximately 3000 m (Rose, 1998). Similarly, Robinson (1986: 26) reports that wedge-capped capuchins (*C. olivaceus*) "spend almost 70% of their active hours . . . searching for and processing food." Wedge-capped capuchins travel on average 2141 m per day (Robinson, 1986). Given this commitment of time and energy, it is likely that when exploiting resources that are temporarily renewing and fixed or predictable in space (i.e. individual trees bearing fruits, flowers, young leaves, and colonial insect nests), capuchins and other primates are likely to rely on spatial search strategies that enhance their ability to revisit previously rewarded locations.

In the present study, we examined the ability of wild white-faced capuchins to use the spatial relationship of an array of experimental landmarks to predict the location of baited feeding sites in small-scale space. Previous studies on this capuchin group indicate that individuals use associative spatial cues (yellow block on platform when it contained bananas), temporal information, and quantity information in foraging decisions (Garber, 2000; Garber and Paciulli, 1997). In our current field experiments, the spatial locations of landmarks moved randomly from trial to trial; however, the spatial relationships between landmarks and goals were constant throughout a given experiment. Thus, an understanding of the relative spatial relationships of landmarks to each other was required by the forager to predict the location of hidden food rewards. In addition, given that landmarks presented to the capuchins were visually identical (except in Experiment 1 in which although the landmarks were identical, the foraging problem could be solved by using a single landmark, i.e. search platform nearest to a landmark), individuals could not solve the foraging problem by encoding the spatial relationship between a single landmark and the reward. Rather, in order to find the goal, the forager had to associate the spatial relationship between two or more landmarks in order to efficiently locate reward platforms. As suggested by Spetch *et al.* (1996: 67), when using three identical landmarks, "the spatial relationship between any individual landmark and the goal is not defined except with reference to its position within an array. Therefore, the configuration of the array must be used in spatial localization."

Our results indicate that individual capuchins quickly and flexibly integrated different sets of spatial information under different experimental conditions. For example, after initial exposure to the test conditions in the first experiment, the capuchins rapidly learned to associate the spatial position of local or near-to-site landmarks to successfully locate reward platforms. However, it must be pointed out that not all group members had equal access to the feeding platforms or learned to associate the spatial position of landmarks with a food reward. It appears that dominance and social foraging strategies played a critical role in the types of information individual capuchins used in deciding which platforms to search.

In Experiment 2, capuchins were found to use the relative spatial positions of two landmarks, each displaced 2 m apart from the goal to solve the foraging problem. In Experiment 3, the research design was changed by introducing a third landmark located 5 m from one of the reward platforms. In order to locate both the reward platforms in this experiment, the capuchins were required to encode the relative positions of three landmarks as an array and use this information to compute the direction of the goal. Given that the precise location of the reward platforms varied randomly between trials, the capuchins are likely to have formed a mental representation of the geometric relationships of landmarks to the reward platforms in small-scale space and used this representation to predict the location of hidden food rewards. However, based on conditions of the experiment, it remains possible that the capuchins located the platform distant to the third landmark cue by encoding the spatial relationship between the nearer reward platform, the third pole, and distant reward platform (but see below).

During the initial set of trials in Experiments 3 and 5, the first reward platform identified by the capuchins tended to be the platform bounded by the two nearer-to-site landmarks. This is consistent with data on spatial search strategies in a range of animal species indicating that landmarks nearer to a goal generally are more salient or overshadow landmarks more distant to the goal (Cheng, 1989; Kamil and Cheng, 2001; Spetch *et al.*, 1996). Over the course of each of these experiments, however, there was evidence that the capuchins altered their search strategy and were equally likely to first select the reward platform distant to the third pole as they were to select the reward platform that required using only two landmarks. In addition, given that foragers arrived at the feeding station from different directions and encountered alternative views of the landmarks as their relative spatial locations was changed randomly, it is possible that the capuchins were able to mentally rotate the configuration of an array of three landmark cues to efficiently solve the foraging problem. The mental rotation of spatial information is defined as an ability to match "a previously presented visual sample stimulus and the display of the same stimulus depicted in different orientations" (Vauclair *et al.*, 1993: 99), and has been reported in tamarins, baboons, and humans (Tomasello and Call, 1997; Vauclair *et al.*, 1993).

In Experiment 4, the absence of the third landmark cue resulted in the capuchins principally orienting to the platform with the two nearby landmark cues, and having less success in locating the second reward platform. This supports the contention that in Experiments 3 and 5, the capuchins simultaneously used spatial information from three landmarks to increase their probability of predicting the location of a hidden food reward. In a recent paper, Kamil and Cheng (2001) outlined a series of hypotheses concerning multiple landmark use by foragers. Given evidence that miscalculations in estimating distance result in a greater error in locating a target than a miscalculation in estimating direction, these authors argue that the use of multiple landmarks significantly reduces the search space within which a target can be found. Based on a series of computer simulations using a 1% and a 2% random search error, they found that compared to a search strategy using two landmarks, a three landmark search strategy reduced the error in localizing the goal by 25% and using four landmarks reduced the error by 38%. These authors suggest that if a forger has precise spatial knowledge, it can rely on a single landmark to locate its goal accurately. However, in the absence of a "highly accurate biological compass," the use of three or more landmarks can significantly improve search efficiency (Kamil and Cheng, 2001). Based on the results of our field experiments, there is evidence that capuchins were more successful in locating both reward platforms when using the spatial relationships of three landmarks to compute the position of a goal than when using only two landmarks.

In conclusion, non-human primates navigate across forested landscapes that vary in size from less than 1 ha (e.g. pygmy marmoset) to several square kilometers (baboons and great apes). In order to relocate important feeding and resting sites within their home range, foragers are likely to encode the spatial relationships of salient features of the environment as landmarks, and mentally represent this information in some form of spatial map. A major question addressed in this chapter is whether monkeys use a single landmark as a reference point or whether the relative spatial positions of several landmarks are used to relocate a goal. Based on the results of this experimental field study, there is evidence that wild white-faced capuchins attended to an array of two and three landmarks to predict the location of baited feeding sites. Given that the precise spatial positions of the landmarks changed after each trial, at least some capuchins attended to the spatial configuration or geometry of the landmarks to solve this foraging problem. The design of the experiment also insured that the capuchins obtained alternative "views" of the landmark array. In this context, the spatial skills displayed by the capuchins were consistent with the ability to mentally rotate landmarks and possibly to form a geometric map in a small-scale space. Building on the work of Janson (1996, 1998) and Janson and Di Bitetti

(1997), we intend to develop field experiments to study spatial cognition and use of landmark cues by capuchins in large-scale space.

SUMMARY

Studies on spatial cognition in primates and other animals indicate that landmarks serve as reference points, and may be encoded as distance and/or direction vectors in navigating to concealed or out-of-sight goals (dens, nests, feeding sites). In the present research, we conducted an experimental field study of spatial cognition and foraging strategies in a group of 15 wild white-faced capuchins (*C. capucinus*) in northeastern Costa Rica (10°26′N, 83°47′W). Specifically, we examined the ability of wild capuchins to use the geometric relationships of an array of two and three landmark cues to predict the location of baited feeding sites. The research design involved the construction of eight visually identical feeding platforms arranged in a circle with a diameter of 8 m. We then conducted a series of six experiments in which the relative spatial positions of experimentally manipulated landmarks (yellow- and pink-colored poles measuring 2 m in height) were the only information available to the forager to efficiently distinguish the location of reward and sham feeding sites. Our results indicate that over the course of 55 consecutive days and 227 experimental trials, the capuchins visited the feeding platforms 3262 times. Group members quickly learned to attend to the spatial positions of landmark arrays and use this information to compute the location of reward platforms. In addition, given that foragers arrived at the feeding station from different directions and encountered alternative views of the landmarks, it is possible that the capuchins were able to mentally rotate the configuration of the landmark array to efficiently solve this foraging problem.

ACKNOWLEDGMENTS

We wish to thank Federico Molina and Alvaro Molina for providing logistical support and permission to conduct this research at Estacion Biologica La Suerte. Israel Mesen Rubi provided assistance in the field and in making certain that we had an endless supply of bananas for the capuchins. Dr. Leslea Hlusko provided suggestions for data analysis. As always, PAG thanks Sara and Jenni for their love, and for allowing me to spend time with the animals and plants that inhabit the rainforests of Central America. Funds to conduct this research were provided by the University of Illinois, Urbana-Champaign.

REFERENCES

Bicca-Marques, J. C. and Garber, P. A. 2005, The use of social and ecological information in primate foraging decisions. *Int. J. Primatol.* 26:

Biegler, R. and Morris, R. G. M. 1996, Landmark stability: Further studies pointing to a role in spatial learning. *Q. J. Exp. Psychol.* 49:307–345.

Boinski, S., Quatrone, R. P., and Swarts, H. 2001, Substrate and tool use by brown capuchins in Suriname: Ecological contexts and cognitive bases. *Am. Anthropol.* 102:741–761.

Cheng, K. 1989, The vector sum model of pigeon landmark use. *J. Exp. Psychol. Anim. Behav. Process.* 15:366–375.

Cheng, K. and Spetch, M. L. 1998, Mechanisms of landmark use in mammals and birds, in: S. D. Healy, ed., *Spatial Representation in Animals*, Oxford University Press, Oxford, pp. 1–17.

Dyer, F. C. 1991, Bees acquire route-based memories but not cognitive maps in a familiar landscape. *Anim. Behav.* 41:239–246.

Dyer, F. C. 2000, Group movement and individual cognition: Lessons from social insects, in: S. Boinski and P. A. Garber, eds., *On the Move: How and Why Animals Travel in Groups*, University of Chicago Press, Chicago, pp. 127–164.

Fedigan, L. M. 1993, Sex differences and intersexual relations in adult white-faced capuchins, *Cebus capucinus. Int. J. Primatol.* 14:853–877.

Gallistel, C. R. 1990, *The Organization of Learning*. MIT Press, Cambridge, MA.

Garber, P. A. 1988, Foraging decisions during nectar feeding by tamarin monkeys (*Saguinus mystax* and *Saguinus fuscicollis*, Callitrichidae, Primates) in Amazonian Peru. *Biotropica* 20:100–106.

Garber, P. A. 1989, Role of spatial memory in primate foraging patterns: *Saguinus mystax* and *Saguinus fuscicollis. Am. J. Primatol.* 19:203–216.

Garber, P. A. 2000, Evidence for the use of spatial, temporal, and social information by primate foragers, in: S. Boinski and P. A. Garber, eds., *On the Move: How and Why Animals Travel in Groups*, University of Chicago Press, Chicago, pp. 261–298.

Garber, P. A. and Brown, E. 2004, Wild capuchins (*Cebus capucinus*) fail to use tools in experimental field study. *Am. J. Primatol.* 62:165–170.

Garber, P. A. and Dolins, F. L. 1996, Testing learning paradigms in the field: Evidence for use of spatial and peceptual information and rule-based foraging in wild moustached tamarins, in: M. Norconk, A. L. Rosenberger, and P. A. Garber, eds., *Adaptive Radiations in Neotropical Primates*, Plenum Press, New York, pp. 201–216.

Garber, P. A. and Paciulli, L. M. 1997, Experimental field study of spatial memory and learning in wild capuchin monkeys (*Cebus capucinus*). *Folia Primatol.* 68:236–254.

Gibson, B. M. and Kamil, A. C. 2001, Test for cognitive mapping in Clark's nutcracerks (*Nucifraga columbiana*). *J. Comp. Pyschol.* 115:403–407.

Janson, C. H. 1996, Towards an experimental socioecology of primates: Examples for Argentine brown capuchin monkeys (*Cebus apella nigritus*), in: M. Norconk, A. L. Rosenberger, and P. A. Garber, eds., *Adaptive Radiations in Neotropical Primates*, Plenum Press, New York, pp. 309–325.

Janson, C. H. 1998, Experimental evidence for spatial memory in foraging wild capuchin monkeys, *Cebus apella. Anim. Behav.* 55:1129–1143.

Janson, C. H. and Di Bitetti, M. S. 1997, Experimental analysis of food detection in capuchin monkeys: Effects of distance, travel speed, and resource size. *Behav. Ecol. Sociobiol.* 41:17–24.

Johnson, D. P., Kays, R., Blackwell, P. G., and Macdonald, D. W. 2002, Does the resource dispersion hypothesis explain group living? *Trends Ecol. Evol.* 17:563–570.

Kamil, A. C., Balda, R. P., and Good, S. 1999, Patterns of movement and orientation during caching and recovery by Clark's nutcrackers (*Nucifraga columbiana*). *J. Exp. Psychol. Anim. Behav. Process.* 16:162–168.

Kamil A. C. and Cheng, K. 2001, Way-finding and landmarks: The multiple-bearings hypothesis. *J. Exp. Biol.* 204:103–113.

Menzel, C. R. 1991, Cognitive aspects of foraging in Japanese monkeys. *Anim. Behav.* 41:397–402.

Menzel, C. R. 1996, Structure-guided foraging in lion-tailed macaques. *Am. J. Primatol.* 38:117–132.

Menzel, C. R., Savage-Rumbaugh, E. S., and Menzel, E. W. Jr. 2002, Bonobo (*Pan pansicus*) spatial memory and communication in a 20-hectare forest. *Int. J. Primatol.* 23:601–620.

Panger M. A. 1998, Object-use in free-ranging white-faced capuchins (*Cebus capucinus*) in Costa Rica. *Am. J. Phys. Anthropol.* 106:311–321.

Poucet, B. 1993, Spatial cognitive maps in animals: Hew hypotheses on their structure and neural mechanism. *Psychol. Rev.* 100:163–182.

Roberts, A. D. L. and Pearce, J. M. 1998, Control of spatial behavior by an unstable landmark. *J. Exp. Psychol. Anim. Behav. Process.* 24:172–184.

Robinson, J. G. 1986, Seasonal variations in use of time and space by wedge-capped capuchin monkeys, *Cebus olivaceus*: Implications for foraging theory. *Smithsonian Control. Zool.* 431:1–60.

Rose, L. M. 1998, Behavioral ecology of white-faced capuchins (*Cebus capucinus*) in Costa Rica. PhD Thesis, Washington University, St. Louis.

Shettleworth, S. J. 1998, *Cognition, Evolution, and Behavior*. Oxford University Press, New York.

Spetch, M. L., Cheng, K., and MacDonald, S. E. 1996, Learning the configuration f a landmark array: I. Touch-screen studies with pigeons and humans. *J. Comp. Psychol.* 110:55–58.

Tomasello, M. and Call, J. 1997, *Primate Cognition*. Oxford University Press, New York.

Vander Wall, S. B. 1990, *Food Hoarding in Animals*. University of Chicago Press, Chicago.

Vauclair, J., Fagot, J., and Hopkins, W. D. 1993, Rotation of mental images in baboons when the visual input is directed to the left cerebral hemisphere. *Psychol. Sci.* 4:99–103.

CHAPTER THIRTEEN

Leap, Bridge, or Ride? Ontogenetic Influences on Positional Behavior in *Cebus* and *Alouatta*

Michelle F. Bezanson

INTRODUCTION

During growth and development, ontogenetic changes in body mass, limb proportions, and motor skills are likely to influence locomotion through the arboreal canopy. The arboreal canopy is a discontinuous substrate characterized by branches of varying size, shape, and mass-bearing capacity (Cant, 1992; Ripley, 1967). Gaps in the forest canopy present a set of problems that arboreal primates must solve to forage and travel efficiently both between and within tree crowns. These include absence of continuous routes of travel, difficulties in bridging or leaping across gaps in the tree crown, and problems of maintaining balance and weight support on terminal branches (Ripley, 1967; Cant and Temerin, 1984; Grand, 1984; Cant, 1992). In this regard, factors such as

Michelle F. Bezanson • Department of Anthropology, Northern Arizona University, Campus Box 15200, Flagstaff, AZ 86011-5200 and Department of Anthropology, University of Arizona, Tucson, AZ 85721.

New Perspectives in the Study of Mesoamerican Primates: Distribution, Ecology, Behavior, and Conservation, edited by Alejandro Estrada, Paul A. Garber, Mary S. M. Pavelka, and LeAndra Luecke. Springer, New York, 2005.

body mass and limb length are likely to influence patterns of substrate use and positional behavior (posture and locomotion, Prost, 1965). Moreover, given limitations in strength, coordination, and motor control during development, young primates may be forced to alter travel routes and gap crossing patterns.

In this study, I examine evidence of age-related or ontogenetic patterns of locomotion in two species of prehensile-tailed primates, *Cebus capucinus* (white-faced capuchin) and *Alouatta palliata* (mantled howling monkey). Like other primates, these taxa share a pattern of growth in which limb proportions and body mass increase with age reducing postnatal hand and foot dominance (Lumer and Schultz, 1947; Stahl *et al.*, 1968; Jungers and Fleagle, 1980; Hurov, 1991). *Cebus* and *Alouatta* also show important differences in the timing of life-history events and in patterns of growth. For example, in the genus *Cebus*, limb segments (including the tail) are shorter at birth and grow at a slower rate relative to the trunk when compared to *Alouatta* (Lumer and Schultz, 1947; Stahl *et al.*, 1968). Additionally, recent research on *C. capucinus* in Santa Rosa National Park in Costa Rica suggests that adult males do not reach full body mass until approximately 10 years of age (Jack and Fedigan, 2004), whereas in the larger-bodied *Alouatta*, males attain full adult body mass at approximately 4 years of age (Glander, 1980). Therefore, given species-specific variation in rates of growth, one might expect *Cebus* and *Alouatta* to exhibit differences in the patterns of positional behavior and substrate use during development, whereby juvenile *Alouatta* should resemble the adult pattern relatively earlier than juvenile *Cebus*.

Positional behavior in the genus *Alouatta* has been well studied and is described as cautious, slow arboreal quadrupedal locomotion dominated by above-branch travel and below-branch feeding postures (Richard, 1970; Mendel, 1976; Fleagle and Mittermeier, 1980; Schön Ybarra, 1984; Cant, 1986; Schön Ybarra and Schön, 1987; Gebo, 1992; Bicca-Marques and Calegaro-Marques, 1995; Youlatos, 1993, 1998; Bergeson, 1996, 1998; Bezanson, 1999). In contrast, leaping and running locomotor behaviors are frequent in capuchins and serve to distinguish adult *Cebus* from adult *Alouatta* (Gebo, 1992; Bergeson, 1996; Garber and Rehg, 1999). For example, when crossing gaps in the canopy, adult *C. capucinus* engage in leaping (47.3%) more frequently than *A. palliata* (6.7%), while *Alouatta* adults have been observed to bridge more frequently (31.5%) than *Cebus* adults (11.3%) (Bergeson, 1996). However, it remains unclear when adult-like locomotor patterns first emerge in these platyrrhines and whether *Cebus* and *Alouatta* exhibit similar ontogenetic locomotor trajectories (i.e. develop adult-like patterns during a comparable stage of development). In addition, given their smaller body size, infants and juveniles experience larger and more frequent gaps in the forest canopy than do adults. Leaping, therefore is expected to occur more commonly in younger individuals unless limitations in motor skills cause juveniles to travel in areas of the canopy characterized by a greater degree of branch overlap.

This chapter examines ontogenetic patterns of positional behavior in sympatric populations of wild *A. palliata* and *C. capucinus* in Costa Rica. Specifically, I address the following questions:

(1) What are the ontogenetic patterns of locomotor behavior in *A. palliata* and *C. capucinus*?
(2) Given their smaller body mass and shorter limb lengths, do juvenile *Alouatta* and *Cebus* use different locomotor patterns than adults to cross gaps in the canopy?
(3) Given their more rapid pattern of growth, does *Alouatta* attain an adult-like pattern of positional behavior at an earlier age than *Cebus*?

METHODS

Research was conducted at La Suerte Biological Research Station in northeastern Costa Rica (10°26′N, 83°47′W). This site is a wet tropical lowland rainforest that includes primary forest, advanced secondary forest, and areas that have been selectively logged in the past. Rainfall in this region averages 3962 mm per year (Sanford *et al.*, 1994). Field observations took place from March 2002 to November 2003 and May 2003 to August 2003.

Systematic, quantitative behavioral data were collected utilizing 1-min instantaneous focal animal sampling on habituated groups of howlers and capuchins (Altmann, 1974; Martin and Bateson, 1993). This method allows an observer to analyze the percentage of time devoted to particular positional behaviors while recording the ecological and social context of these behaviors (Dagosto and Gebo, 1998). An instantaneous focal animal sampling method commonly is used to collect quantitative data on primate positional behavior (Garber, 1984; Doran, 1992; Bergeson, 1996; Walker, 1996; McGraw, 1996; Dagosto and Gebo, 1998).

One-minute instantaneous sampling allows a researcher to collect 60 individual activity records (IARs) or point samples of behavior per hour. In the present study, each instantaneous record contained information on the identity of the focal animal, activity pattern (feed, forage, rest, and travel), positional behavior,

Table 1. Positional definitions. Definitions are based on Hunt *et al.* (1996)

Bimanual pull-up	"A typically horizontal support is grasped by both hands and the body is lifted by retracting the humerus and flexing the elbow; the spine may be flexed to aid bringing the hindlimb on top of the support." (379)
Bridge	Gaps are closed by maintaining contact with the hindlimbs and tail of the starting point (gap) and reaching out with the forelimbs across the gap and pulling the body across the gap.
Drop	"...falling after releasing a support." (381)
Vertical climb	Includes flexed-elbow vertical climb, ladder climb, vertical scramble, pulse climb. Ascent on descent on vertical supports.
Quadrupedal run	"Fast locomotion using asymmetrical or irregular gaits and with a period of free flight." (377)
Quadrupedal walk	Walking on four limbs involving a diagonal sequence and a symmetrical gait. The torso is parallel to the substrate.
Leap	Movements across open gaps that require hindlimb propulsion, free flight, and quadrupedal, suspensory, hindlimb, or forelimb landing.
Ventral/side ride	Infant rides by grasping onto the front or side of the torso of the mother.
Dorsal ride	Infant rides by grasping on the back or neck of the mother or in rare cases, other individuals (observed in *Cebus*).

For details on limb placement, flexion, and extension and mass bearing see Hunt *et al.* (1996).

branch size, branch angle, the placement of the limbs, crown location, and diet. During samples in which the animal was leaping or moving from one support to another, the size and orientation of the take-off and landing support were recorded along with the size of the gap traversed. Positional categories used in this analysis are based on definitions proposed by Hunt *et al.* (1996) (Table 1).

Data on the ontogeny of positional behavior were obtained by examining two capuchin groups, each with 11–16 individuals. Three *Alouatta* groups were observed ranging from 9 to 14 individuals per group. Group sizes in each species changed during the two field seasons as several infants were born and individuals left the group, joined the group, or disappeared due to unknown reasons. Five developmental age categories are compared. These include two infant stages, two juvenile stages, and the adult stage (Table 2). Position in the canopy was described in relationship to the location of the focal animal within a particular part of the tree crown (peripheral versus core and lower, middle, and upper portions).

Descriptive statistics were used to examine proportions of observations of positional behavior among the five age classes of *Cebus* and *Alouatta* (Table 1). In addition, statistical analyses were performed using two-sample randomization

Table 2. Age categories are based on information from Oppenheimer (1968), Freese and Oppenheimer (1981) and Glander (1980)

	Age–*Cebus*	Age–*Alouatta*	Behavior
Infant 1	Birth to 2 months	Birth to 3 months	Carried dorsally or ventrally, always on mother. All observed sustenance is achieved through nursing.
Infant 2	2–5 months	3–6 months	Carried dorsally with some independent movement, always associated with mother.
Juvenile 1	5–12 months	6–12 months	Most locomotion and foraging are independent (occasional help crossing gaps) usually associated with mother during travel.
Juvenile 2	12–36 months	12–24 months	All locomotion and foraging is independent while maintaining some proximity with the mother during resting and social behaviors such as grooming.

Infant 1, Infant 2, and Juvenile 1 individuals were known and identified from birth. Juvenile 2 categories were identified based on size of the individual, forehead coloration patterns (*Cebus*), and pelage (*Alouatta*). In this analysis, the youngest age classes are compared to adults to avoid confusion between older juveniles/preadults and small adults. Therefore, this table does not include all of the age categories from birth to adult that were observed at La Suerte.

procedures in Manly's RT version 2.1 statistical software package (Manly, 1997). The use of traditional statistical analyses can be problematic in behavioral studies as many behavioral datasets violate assumptions of random sampling and sample independence (Dagosto, 1994; Dagosto and Gebo, 1998). In addition, many primate studies involve observations of unmarked animals that make the unambiguous identification of individuals difficult and comparison between individuals of different age or sex class difficult. Randomization tests (including Monte Carlo and bootstrap techniques) represent a robust analytical tool used to compare patterns of behavior derived from non-random and non-independent datasets that contain hundreds or thousands of related behavioral data points. Randomization techniques have become increasingly common in behavioral studies (Dagosto, 1994; Adams and Anthony, 1996; Dagosto and Gebo, 1998; Bergeson, 1996; Bejder *et al.*, 1998). Randomization tests involve creating a test statistic or the mean differences in the comparisons for each randomization loop through multiple randomizations of the data (in this case 5000). In all statistical analyses, significance was determined if $p \leq 0.05$.

RESULTS

Ontogenetic Patterns of Locomotion in Howlers and Capuchins

A total of 380 h (22,800 IAR or point samples) of data were collected on *C. capucinus* and 452 h (27,120 point samples) of data were collected on *A. palliata*. Quadrupedal walking was the most common locomotor mode in juvenile and adult *C. capucinus* and in juvenile and adult *A. palliata* (Tables 3 and 4).

Howler Infant 1 and Infant 2 ride ventrally, dorsally, or on the side of their mothers during group travel and foraging. Infant 1 were observed off of their mothers during only 5.3% of total observations. Infant 2 were observed off their mothers during 24.6% of total observations. During play and exploration, quadrupedal walk, leap, and climb were the most frequent locomotor modes in Infant 2. Climbing modes (ladder climb, pulse climb, vertical scramble, and flexed elbow vertical climb) were observed most frequently in Juvenile 1 (10.8%) and Juvenile 2 (9.2%). This behavior accounted for only 5.9% of the adult observations. As in the case of climbing, leaping was observed more frequently in *Alouatta* juveniles (Infant 2: 3.4%, Juvenile 1: 6.1%, Juvenile 2: 7%) than in adults (2.4% of the total observations).

In *Cebus*, climbing was observed most frequently in Juvenile 1 (9.2%) while Juvenile 2 and adults were observed to climb at equal frequency (6.4% and

Table 3. Frequencies of locomotor behavior in *Alouatta*

Positional mode	Infant 1	Infant 2	Juvenile 1	Juvenile 2	Adult
Bimanual pull-up	–	–	2.8	2.1	1.1
Bridge	–	1.3	3.1	12.2	10.2
Drop	–	–	<1	1.4	1.3
Vertical climb	–	1.5	10.8	9.2	5.9
Quadrupedal run	–	–	7.9	5.7	1.8
Quadrupedal walk	5.3	18.4	64.7	62.4	77.3
Leap	–	3.4	6.1	7	2.4
Ventral/side ride	86.3	1.2	–	–	–
Dorsal ride	8.4	74.2	3.6	<1	–

Table 4. Frequencies of locomotor behavior in *Cebus*

Positional mode	Infant 1	Infant 2	Juvenile 1	Juvenile 2	Adult
Bimanual pull-up	–	1.8	1.1	1.1	<1
Bridge	–	<1	<1	<1	<1
Drop	–	<1	<1	<1	<1
Vertical climb[a]	4.5	6.2	9.2	6.4	6.8
Quadrupedal run	–	5.4	17.2	11.7	4.3
Quadrupedal walk	3.4	27.3	55.9	66.7	76.6
Leap	–	3.1	15.5	10.8	10.3
Ventral/side ride	2.8	<1	–	–	–
Dorsal ride	89.3	55.6	<1	–	–

6.8%, respectively). In *Cebus*, leaping modes accounted for 15.5% of Juvenile 1 positional modes but decreased in Juvenile 2 (10.8%) and adults (10.3%). *Cebus* Infant 1 were rarely observed leaping (3.1%) during exploration and play.

Ontogenetic Patterns of Gap Crossing in Howlers and Capuchins

The most common method for gap crossing in all age groups of *Cebus* was leaping (Infant 2: 8.1%, when locomoting independently; Juvenile 1: 86.9%; Juvenile 2: 81.1%; Adult: 86.2%) (Table 5). Bridging gaps increased slightly through the life stages (Infant 2: <1%, when locomoting independently; Juvenile 1: 3.8%; Juvenile 2: 4.7%; Adult: 6.1%). However, overall no significant differences in gap crossing behaviors were found among the Juvenile 1, Juvenile 2, and adult categories ($p > 0.5$ in all comparisons). Howlers were more variable in their methods of gap crossing (Table 6). Both Juvenile 1 and Juvenile 2 were observed to leap significantly more often than adults (Juvenile 1: 43.8%;

Table 5. Frequencies of locomotor modes during gap crossing in *Cebus*

Positional mode	Infant 1	Infant 2	Juvenile 1	Juvenile 2	Adult
Bimanual pull-up	–	4.6	5.8	8.4	2.7
Bridge	–	<1	3.8	4.7	6.1
Drop	–	1.6	2.3	5.8	5.0
Leap	–	8.1	86.9	81.1	86.2
Ventral/side ride	–	<1	–	–	–
Dorsal ride	100	85.2	<1	–	–

No significance in all comparisons between juveniles and adults. 5000 permutations.

Table 6. Frequencies of locomotor modes during gap crossing in *Alouatta*

Positional mode	Infant 1	Infant 2	Juvenile 1	Juvenile 2	Adult
Bimanual pull up	–	–	14.3	11.0	5.3
Bridge	–	–	33.3	51.8	67.1
Drop	–	–	8.6	7.3	8.5
Leap	–	–	43.8	29.9	19.9
Ventral/side ride	94.8	<1	–	–	–
Dorsal ride	5.2	99.8	3.6	<1	–

Statistical significance in comparisons among Juvenile 1, Juvenile 2, and adults—Bimanual pull-up: Juvenile 1 versus Juvenile 2 ($p = 0.006$), Juvenile 1 versus Adult ($p < 0.001$), Juvenile 2 versus Adult ($p < 0.001$); Bridge: Juvenile 1 versus Juvenile 2 ($p = 0.003$), Juvenile 1 versus Adult ($p < 0.001$), Juvenile 2 versus Adult (ns); Drop: ns—all comparisons; Leap: Juvenile 1 versus Juvenile 2 ($p < 0.001$), Juvenile 1 versus Adult ($p < 0.001$), Juvenile 2 versus Adult ($p = 0.002$). 5000 permutations.

Juvenile 2: 29.9%; Adult: 19.9%) ($p < 0.01$ in all comparisons). Bridging behaviors significantly increased from the Juvenile 1 (33.3%) to Juvenile 2 (51.8%) categories ($p < 0.01$). Adults were observed to bridge in 67.1% of all gaps crossed, which was not significantly greater than Juvenile 2.

DISCUSSION

Body Size and Locomotion in White-Faced Capuchins and Mantled Howlers

This research examines the ontogeny of positional behavior and arboreal locomotion in sympatric populations of *C. capucinus* and *A. palliata* inhabiting a tropical rainforest in Costa Rica. These species represent good models for understanding how growth and development influence positional behavior and habitat utilization for several reasons. First, *A. palliata* and *C. capucinus* are found to coexist across a range of habitats and geographical areas, and an understanding of ontogenetic patterns of positional behavior in sympatric species offers a framework for examining the manner in which species differences in growth and development influence locomotion, manipulative abilities, and foraging patterns in the same environment. Second, because *C. capucinus* and *A. palliata* differ in body mass (adult body mass of *C. capucinus* is 3.8 kg in males and 2.6 kg in females and adult body mass of *A. palliata* is 7.1 kg in males and 5.4 kg in females) and growth patterns, they provide additional information on how changes in mass and limb proportions may influence positional behavior (Fleagle and Mittermeier, 1980; Cant, 1992; Ford, 1994; Gebo and Chapman, 1995; McGraw, 1996). Finally, there exist several studies of positional behavior in adult white-faced capuchins and mantled howlers inhabiting a range of forest types (Gebo, 1992; Bergeson, 1996, 1998; Garber and Rehg, 1999). These studies provide a strong comparative framework for examining questions of intraspecific variation in positional behavior and patterns of habitat utilization in these species.

Results obtained in this study indicate that when compared to mantled howlers, juvenile white-faced capuchins exhibited adult-like patterns of locomotion early in development. For example, juveniles that have reached 5 months of age were found to engage in similar proportions of leaping, bridging, and climbing behaviors as adults. *Cebus capucinus* are the most dimorphic capuchin species with males approximately 30% heavier than females (Ford, 1994). A study by Gebo (1992) found no evidence of sex-based differences in positional behavior in a population of *C. capucinus* in a dry forest in Costa Rica. Similarly, I found no differences in positional behavior and substrate utilization among adult male capuchins, female capuchins, and females with infants inhabiting this Costa Rican rain forest (Bezanson, unpublished data.). Several studies on positional behavior in New World monkeys, Old World monkeys, and apes report limited evidence of sex-based differences in positional behavior (Gebo, 1992; Gebo and Chapman, 1995; Garber, in press). These data suggest that changes in limb and trunk proportions that accompany the growing and developing juvenile have only a minor influence on positional behavior in *Cebus* (Lumer and Schultz, 1947; Jungers and Fleagle, 1980) despite the fact that Juvenile 1 and Juvenile 2 weigh approximately 25–50% of adult body mass.

In the case of howlers, there was greater evidence of age-based patterns of positional behavior. Young howlers were characterized by significant increases in leaping behavior relative to adults, and adult howlers were characterized by significant increases in bridging relative to younger juveniles.

A pattern of ontogenetic change in locomotion has been reported in several primate taxa. For example, Doran (1997) reports that mountain gorillas were characterized by developmental shifts from infancy to adulthood that differed in the frequency of quadrupedalism, climbing, and suspensory behaviors. Younger individuals (age 0–23 months) engaged in locomotor modes such as climbing and suspension more frequently than adults. Ontogenetic effects on locomotor behavior also have been described in *Macaca mulatta* (Wells and Turnquist, 2001). Juvenile macaques engaged in more arboreal positional modes in an arboreal setting (adults: 35% and juveniles: 49.5%) and in less quadrupedal walking (adults: 70.85% and juveniles: 31.54%) than adults. However, these patterns were argued to reflect risk-sensitive behavior and predator avoidance more than mechanical problems associated with locomotion.

Locomotor Patterns and Fine Manipulative Capabilities

Young white-faced capuchins resembled adults in their locomotor repertoire earlier in development than young mantled howlers. A similar pattern has been observed in squirrel monkeys. Research on *Saimiri oerstedii* (Boinski and Fragaszy, 1989) indicates that Costa Rican squirrel monkeys locomote independently by 4 weeks of age, began foraging on their own by 7 weeks, were weaned by 16 weeks of age, and "by 4–5 months of age, most of the foraging patterns were indistinguishable from those of older individuals" (Boinski and Fragaszy, 1989: 423). In addition, recent research on *Saimiri sciureus* indicates that infants were as successful in capturing insects as their adult counterparts by 6 months of age (Stone, 2004).

Wild *Cebus olivaceus* show a similar pattern of adult-like manipulative efficiency early during the juvenile period (Fragaszy and Boinski, 1995). Once past infancy, wedge-capped capuchin juveniles were as efficient as adults in the amount of fruit and insects ingested and in the way they foraged for insects and fruits (manipulative activities). Additionally, research on captive groups of *Cebus apella* suggests that captive capuchins exhibited adult-like dextrous manipulative abilities long before the weaning process is complete (Fragaszy and Adams-Curtis, 1997). *Cebus capucinus* inhabiting a dry forest showed a similar pattern (MacKinnon, 1995). Young juveniles acquired an adult-like foraging repertoire (with further refinement during ontogeny) early even though they have a relatively prolonged juvenile period. In *Cebus* and *Saimiri*, the development of adult locomotor competency and finer manipulative abilities associated with foraging and diet appears early during ontogeny (also see MacKinnon, this volume). However, other insectivorous New World primates, such as tamarins and marmosets, appear to disassociate locomotor development and fine manipulative skills. Tamarins and marmosets locomote independently at 3 months of age, but are provisioned with insects and fruits by adult caretakers well into the juvenile period (7–9 months) (Garber and Leigh, 1997).

Anatomical research on changes in body mass and limb growth suggests that howlers grow at a relatively faster rate than capuchins (Lumer and Schultz,

1947; Stahl *et al.*, 1968; Fleagle and Samonds, 1975). Behavioral and life history data suggest that *C. capucinus* have a longer juvenile or preadult stage than *Alouatta* (Glander, 1980; Jack and Fedigan, 2004). However, in the present study, juvenile *Alouatta* do not exhibit adult-like locomotor behavior until late in development. Juvenile howlers differ from adults in the frequency of bimanual pull-up, bridging, and leaping behaviors. Several factors could account for the differences observed in leaping and climbing in this study including: (1) age-based differences in diet or body mass; (2) social dominance affecting patterns of habitat utilization and positional behavior; and (3) differences in patterns of the development of fine motor skills. Martin (1996) has argued that infant brain size at birth is constrained by maternal metabolic requirements. One possible advantage of mothers investing heavily in prenatal brain tissue is that it allows young animals to develop motor skills required for adult-like foraging patterns at a relatively early age. However, howler neonatal brains are 55.9% of adult brain mass, while capuchin brains are 36% of adult brain mass (Harvey *et al.*, 1987). In the case of *Alouatta*, mothers may invest more in prenatal brain growth in each infant as a strategy to maximize reproductive output. *Cebus*, on the other hand, despite having a small neonatal brain weight, exhibit a relatively fast rate of postnatal brain growth following birth (Fragaszy *et al.*, 2004). Despite their larger body mass, howlers are reported to have an earlier age at first reproduction and a shorter interbirth interval than *Cebus* (Fedigan and Rose, 1995; Ross, 1991). The consequences of juveniles developing adult-like locomotor and manipulative skills early remain unclear, however, given these differences in positional behavior, it can be hypothesized that compared to *Cebus*, young *Alouatta* have different diets and exploit their environment differently than do adults.

SUMMARY

Adult-like locomotor behavioral patterns were found to develop at an earlier age in *C. capucinus* than in *A. palliata*. By 5 months of age, young *Cebus* resemble adults in the frequency of positional behaviors while young *Alouatta* individuals vary significantly from the adult condition. Young *Alouatta* (6–24 months of age) differed from adult *Alouatta* in leaping, bridging, and climbing. Leaping modes significantly decreased during ontogeny in *Alouatta* while bridging modes increased significantly during ontogeny. *Cebus* was observed leaping between gaps more often (juveniles and adults) than howlers, but did not exhibit significant changes in gap crossing behaviors beyond the infant stage of development. These data suggest that in a comparison of these New World monkeys, interspecific ontogenetic differences in linear growth, body mass, and life history timing did not predictably influence locomotion.

ACKNOWLEDGMENTS

I thank Alejandro Estrada, Paul Garber, Leandra Luecke, and Mary Pavelka for inviting me to participate in this symposium. Sarah Smith, Greg Bridgett, Teague O'Mara, and Chris Schaffer provided help during field research. Funding for this project was provided by National Science Foundation (Doctoral dissertation improvement grant #0228924), American Museum of Natural History, Sigma Xi, and the University of Arizona (College of Social and Behavioral Sciences, Graduate College, Department of Anthropology, and Department of Ecology and Evolutionary Biology). I thank Paul Garber, M. E. Morbeck, Mary Pavelka, and Luke Premo for helpful suggestions. Finally, Alvaro Molina, Renee Molina, Israel Mesen Rubi, Raquel Mesen Rubi, Reinaldo Aguilar, Liz Bezanson, and Warren Bezanson provided logistical support at Estación Biológica La Suerte.

REFERENCES

Adams, D. C. and Anthony, C. D. 1996, Using randomization techniques to analyse behavioural data. *Anim. Behav.* 51:733–738.

Altmann, J. 1974, Observational study of behavior: Sampling methods. *Behaviour* 49:227–265.

Bejder, L., Fletcher, D., and Bräger, S. 1998, A method for testing association patterns of social animals. *Anim. Behav.* 56:719–725.

Bergeson, D. J. 1996, The positional behavior and prehensile tail use of *Alouatta palliata*, *Ateles geoffroyi*, and *Cebus capucinus*. PhD Dissertation, Washington University, St Louis.

Bergeson, D. J. 1998, Patterns of suspensory feeding in *Alouatta palliata*, *Ateles geoffroyi*, and *Cebus capucinus*, in: E. Strasser, J. Fleagle, A. Rosenberger, and A. McHenry, eds., *Primate Locomotion: Recent Advances*, Plenum Press, New York, pp. 45–60.

Bezanson, M. F. 1999, Positional behavior and prehensile-tail use in *Alouatta palliata*. *Am. J. Phys. Anthropol.* 108 (Suppl. 28):92.

Bicca-Marques, J. C. and Calegaro-Marques, C. 1995, Locomotion of black howlers in a habitat with discontinuous canopy. *Folia Primatol.* 64:55–61.

Boinski, S. and Fragaszy, D. M. 1989, The ontogeny of foraging in squirrel monkeys, *Saimiri oerstedii. Anim. Behav.* 37:415–428.

Cant, J. G. H. 1986, Locomotion and feeding postures of spider and howling monkeys: Field study and evolutionary interpretation. *Folia Primatol.* 46:1–14.

Cant, J. G. H. 1992, Positional behavior and body size of arboreal primates: A theoretical framework for field studies and an illustration of its application. *Am. J. Phys. Anthropol.* 88:273–283.

Cant, J. G. H. and Temerin, A. 1984, A conceptual approach to foraging adaptations in primates, in: P. Rodman and J. Cant, eds., *Adaptations for Foraging in Nonhuman Primates: Contributions to an Organismal Biology of Prosimians, Monkeys, and Apes*, Columbia University Press, New York, pp. 304–342.

Dagosto, M. 1994, Testing positional behavior of Malagasy lemurs: A randomization approach. *Am. J. Phys. Anthropol.* 94:189–202.

Dagosto, M. and Gebo, D. L. 1998, Methodological issues in studying positional behavior: Meeting Ripley's challenge, in: E. Strasser, J. Fleagle, A. Rosenberger, and H. McHenry, eds., *Primate Locomotion: Recent Advances*, Plenum, New York, pp. 5–29.

Doran, D. M. 1992, The ontogeny of chimpanzee and pygmy chimpanzee locomotor behavior: A case study of paedomorphism and its behavioral correlates. *J. Hum. Evol.* 23:139–157.

Doran, D. M. 1997, Ontogeny of locomotion in mountain gorillas and chimpanzees. *J. Hum. Evol.* 32:323–344.

Fedigan, L. M. and Rose, L. M. 1995, Interbirth interval variation in three sympatric species of Neotropical monkey. *Am. J. Primatol.* 3:9–24.

Fleagle, J. G. and Mittermeier, R. A. 1980, Locomotor behavior, body size, and comparative ecology of seven Surinam monkeys. *Am. J. Phys. Anthropol.* 52:301–314.

Fleagle, J. G. and Samonds, K. W. 1975, Physical growth of cebus monkeys (*Cebus albifrons*) during the first year of life. *Growth* 39:35–52.

Ford, S. M. 1994, Evolution of sexual dimorphism in body weight of platyrrhines. *Am. J. Primatol.* 34:221–244.

Fragaszy, D. M. and Adams-Curtis, L. 1997, Developmental changes in manipulation in tufted capuchins (*Cebus apella*) from birth through two years and their relation to foraging and weaning. *J. Comp. Psychol.* 111:201–211.

Fragaszy, D. M. and Boinski, S. 1995, Patterns of individual differences in diet choice and efficiency of foraging in wedge-capped capuchins (*Cebus olivaceous*). *J. Comp. Psychol.* 109:339–348.

Fragaszy, D. M., Visalberghi, E., and Fedigan, L. M. 2004, *The Complete Capuchin: The Biology of the Genus Cebus.* Cambridge University Press, Cambridge.

Freese, C. H. and Oppenheimer, J. R. 1981, The capuchin monkeys, genus *Cebus*, in: A. Coimbra-Filho and R. Mittermeier, eds., *Ecology and Behavior of Neotropical Primates*, Rio de Janeiro, Academia Brasileira de Ciências, pp. 331–390.

Garber, P. A. 1984, Use of habitat and positional behavior in a Neotropical primate, *Saguinus oedipus*, in: P. S. Rodman and J. G. H. Cant, eds., *Adaptations for Foraging in Nonhuman Primates: Contributions to an Organismal Biology of Prosimians, Monkeys, and Apes*, Columbia University Press, New York, pp. 112–133.

Garber, P. A. Primate behavior and ecology, in: S. Bearder, C. J. Campbell, A. Fuentes, K. C. MacKinnon, and M. Panger, eds., *Primates in Perspective*, Oxford University Press, Oxford, in press.

Garber, P. A. and Leigh, S. R. 1997, Ontogenetic variation in small-bodied New World primates: Implications for patterns of reproduction and infant care. *Folia Primatol.* 68:1–22.

Garber, P. A. and Rehg, J. A. 1999, The ecological role of the prehensile tail in *Cebus capucinus. Am. J. Phys. Anthropol.* 110:325–339.

Gebo, D. L. 1992, Locomotor and postural behavior in *Alouatta palliata* and *Cebus capucinus. Am. J. Primatol.* 26:277–290.

Gebo, D. L. and Chapman, C. A. 1995, Positional behavior in five sympatric Old World monkeys. *Am. J. Phys. Anthropol.* 97:49–76.

Glander, K. E. 1980, Reproduction and population growth in free-ranging mantled howling monkeys. *Am. J. Phys. Anthropol.* 53:25–36.

Grand, T. 1984, Motion economy within the canopy: Four strategies for mobility, in: P. S. Rodman and J. G. H. Cant, eds., *Adaptations for Foraging in Nonhuman Primates: Contributions to an Organismal Biology of Prosimians, Monkeys, and Apes*, Columbia University Press, New York, pp. 54–72.

Harvey, P. H., Martin, R. D., and Clutton-Brock, T. H. 1987, Life histories in comparative perspective, in: B. B. Smuts, D. L. Cheney, R. M. Seyfarth, R. W. Wrangham, and T. T. Struhsaker, eds., *Primate Societies*, University of Chicago Press, Chicago, pp. 181–196.

Hunt, K. D., Cant, J. G. H., Gebo, D. L., Rose, M. D., Walker, S. E., and Youlatos, D. 1996, Standardized descriptions of primate locomotor and postural modes. *Primates* 37:363–387.

Hurov, J. R. 1991, Rethinking primate locomotion: What can we learn from development? *J. Mot. Behav.* 23:211–218.

Jack, K. M. and Fedigan, L. 2004, Male dispersal patterns in white-faced capuchins, *Cebus capucinus.* Part 1: Patterns and causes of natal emigration. *Anim. Behav.* 67:761–769.

Jungers, W. L. and Fleagle, J. G. 1980, Postnatal growth allometry of the extremities in *Cebus albifrons* and *Cebus apella*: A longitudinal and comparative study. *Am. J. Phys. Anthropol.* 53:471–478.

Lumer, H. and Schultz, A. H. 1947, Relative growth of the limb segments and tail in *Ateles geoffroyi* and *Cebus capucinus*. *Hum. Biol.* 19:53–67.

MacKinnon, K. C. 1995, Age differences in foraging patterns and spatial associations of the white-faced capuchin monkey (*Cebus capucinus*) in Costa Rica. MA Thesis, University of Alberta.

Manly, B. F. J. 1997, *Randomization, Bootstrap, and Monte Carlo Methods in Biology*. Chapman and Hall, New York.

Martin, P. and Bateson, P. 1993, *Measuring Behavior: An Introductory Guide*. Cambridge University Press, New York.

Martin, R. D. 1996, Scaling of the mammalian brain: The maternal energy hypothesis. *News Physiol. Sci.* 11:149–156.

McGraw, W. S. 1996, The positional behavior, support use, and support availability in the Tai Forest, Ivory Coast. *Am. J. Phys. Anthropol.* 100:507–522.

Mendel, F. 1976, Postural and locomotor behavior of *Alouatta palliata* on various substrates. *Folia Primatol.* 26:36–53.

Oppenheimer, J. R. 1968, Behavior and ecology of the white-faced monkey, *Cebus capucinus*, on Barro Colorado Island. PhD Dissertation, University of Illinois, Urbana, IL.

Prost, J. 1965, A definitional system for the classification of primate locomotion. *Am. Anthropol.* 67:1198–1214.

Richard, A. 1970, A comparative study of the activity patterns and behavior of *Alouatta villosa* and *Ateles geoffroyi*. *Folia Primatol.* 12:241–263.

Ripley, S. 1967, The leaping of langurs, a problem in the study of locomotor adaptation. *Am. J. Phys. Anthropol.* 26:149–170.

Ross, C. 1991, Life history pattern in New World primates. *Int. J. Primatol.* 12:481–502.

Sanford, R. L., Paaby, P., Lavall, J. C., and Phillips, E. 1994, Climate, geomorphology, and aquatic systems, in: L. A. Mc Dade, K. S. Bawa, H. A. Hespenheide, and G. S. Hartshorn, eds., *La Selva: Ecology and Natural History of a Neotropical Rainforest*, University of Chicago Press, Chicago, pp. 19–33.

Schön Ybarra, M. A. 1984, Locomotion and postures of red howlers in a deciduous forest-savanna interface. *Am. J. Phys. Anthropol.* 63:65–76.

Schön Ybarra, M. A. and Schön, M. A. 1987, Positional behavior and limb bone adaptations in red howling monkeys. *Folia Primatol.* 49:70–89.

Stahl, W. R., Malinow, M. R., Maruffo, C. A., Locker Pope, B., and Depaoli, R. 1968, Growth and age estimation of howler monkeys. Biology of the howler monkey. *Bibl. Primatol.* 7:59–80.

Stone, A. 2004, Juvenile feeding ecology and life history in a neotropical primate, the squirrel monkey (*Saimiri sciureus*). PhD Thesis, University of Illinois at Urbana-Champaign.

Walker, S. E. 1996, Evolution of positional behavior in the saki/uakaris (Pithecia, Chiropotes, Cacajao), in: M. A. Norconk, A. L. Rosenberger, and P. A. Garber, eds., *Adaptive Radiations of Neotropical Primates*, Plenum Press, New York, pp. 335–367.

Wells, J. P. and Turnquist, J. E. 2001, Ontogeny of locomotion in rhesus macaques (*Macaca mulatta*): II. Postural and locomotor behavior and habitat use in a free-ranging colony. *Am. J. Phys. Anthropol.* 115:80–94.

Youlatos, D. 1993, Passages within a discontinuous canopy: Bridging in the red howler monkey (*Alouatta seniculus*). *Folia Primatol.* 61:144–147.

Youlatos, D. 1998, Seasonal variation in positional behavior of red howling monkeys (*Alouatta seniculus*). *Primates* 39:449–457.

CHAPTER FOURTEEN

Food Choice by Juvenile Capuchin Monkeys (*Cebus capucinus*) in a Tropical Dry Forest

Katherine C. MacKinnon

INTRODUCTION

Capuchin monkeys are highly adaptable in their ability to occupy a wide array of habitat types, and are extremely flexible in their ability to use a range of foods such as insects, fruits, and vertebrate prey (Chapman and Fedigan, 1990; Panger *et al.*, 2002; Fragaszy *et al.*, 2004). Capuchins are described as manipulative and extractive foragers, which enables them to exploit hard-to-obtain and hard-to-process resources including larvae from embedded substrates, hard-shelled fruits and nuts, and fast-moving vertebrate prey such as squirrels, tree rats, birds, and lizards (Fedigan, 1990; Rose, 1997; Panger *et al.*, 2002). Juvenile capuchins are reported to be skilled foragers from a young age (MacKinnon, 1995) and exploit many of the same foods as adults. For example, in a study by Fragaszy and Boinski (1995), it was found that sex was a more powerful

Katherine C. MacKinnon • Department of Sociology and Criminal Justice, and Center for International Studies, Saint Louis University, 210 Fitzgerald Hall, 3500 Lindell Blvd., St. Louis, MO.

New Perspectives in the Study of Mesoamerican Primates: Distribution, Ecology, Behavior, and Conservation, edited by Alejandro Estrada, Paul A. Garber, Mary S. M. Pavelka, and LeAndra Luecke. Springer, New York, 2005.

predictor of variations in foraging activities and foods eaten than was age. In this chapter, I will examine how the diets of small and large juvenile capuchin monkeys vary with seasonal changes in a tropical dry forest. I present juvenile capuchin dietary preferences during dry and wet seasons in northwestern Costa Rica, and compare these with those of adults.

Due to an extended period of immaturity, ontogenetic factors such as physiological constraints, the effects of experience, and differing nutritional requirements play an important role in food choice and acquisition in young primates (Janson and van Schaik, 1993; Altmann, 1998). Four general theories have been proposed to explain differences in ingestion rates, food types chosen, and methods of food processing that may occur between juveniles and adults. Janson and van Schaik (1993) have argued that slow growth in juvenile primates, in particular *Macaca* and *Cebus*, represents an evolutionary strategy to avoid feeding competition with larger, stronger, and more dominant adults. These authors suggest that as adult–juvenile feeding competition increases, juveniles would be forced to the periphery of the group, an area of high predation risk. Growing slowly might decrease the problems of competing with adults for food, as well as lessen the chance of starvation. A second theory concerning the ontogeny of primate feeding patterns suggests that juveniles and adults differ in diet and feeding behavior in response to variable opportunities for learning (Fragaszy and Visalberghi, 1989; Visalberghi and Fragaszy, 1990, 2002). Capuchins forage in a social context, and while there is much debate about whether capuchins can truly imitate (e.g., Visalberghi and Fragaszy, 1990, 2002; Visalberghi and Limongelli, 1996; Custance *et al.*, 1999; Visalberghi and Addessi, 2003), young animals do intensely observe the behavior of others around them. Therefore, varying group compositions might allow differing opportunities for observational learning (Whiten, 1989; Custance *et al.*, 1999) and hence variation in diet. A third model suggests that younger primates lack the strength and dexterity to open hard-to-extract resources (e.g., hard-shelled fruits) or the motor skills required to break open substrates for embedded invertebrates (Gibson, 1986; Fragaszy and Boinski, 1995). In this regard, juveniles and adults might exploit different diets, with adults consuming more foods that present a challenge, and juveniles consuming smaller, softer, or more easily obtained foods. The final model suggests that juveniles and adults might exploit different diets based on differences in nutritional requirements associated with rates of brain and body growth (Altmann, 1998).

Data on juvenile feeding behavior in nonhuman primates are not common. Fragaszy (1986) and Fragaszy and Boinski (1995) studied the feeding behavior of *Cebus olivaceous* and found that dietary profiles of immature male and female capuchins are quite similar, suggesting that sex differences may not become significant until later subadult/adult stages. However, adult diets were characterized by greater amounts of fruits, whereas juvenile diets contained proportionally more plant foods (other than fruits) and insignificantly more animal foods. The most detailed data on the ontogeny of feeding behavior come from Altmann's (1998) study of yellow baboons (*Papio cynocephalus*). He reports that the diets of yearling baboons affected their chances of surviving to reproductive maturity and the longevity of those that survived. For example, juvenile females with poor diets had lower rates of reproductive success as adults than juvenile females with adequate diets. Like capuchins, yellow baboons have extended periods of juvenility, live in highly seasonal environments, feed on a wide variety of foods, and are flexible and opportunistic in their dietary choices (Altmann, 1998).

Capuchin physical development is slow compared to other New World monkey species, and the major life history stages occur later in capuchins when contrasted to similar-sized primates (Robinson and Janson, 1987; Fragaszy, 1990). An adult female *C. capucinus* in the wild first gives birth around age 7, and interbirth intervals average about 2 years (Fedigan and Rose, 1995). Males at age 7–10 years are still considered subadult, but are on the threshold of adult status and engage in sexual mountings with adult females. *C. capucinus* have a gestation length of 157–167 days (Freese and Oppenheimer, 1981), which is average for primates of their weight (Harvey and Clutton-Brock, 1985). The maximum lifespan published for a capuchin in captivity is nearly 55 years (Hakeem *et al.*, 1996). However, lifespan in wild-living capuchins is presumed to be considerably shorter, perhaps by at least half when considering predation, disease, and infections from fight wounds (Fedigan, pers. comm.).

Among primates, capuchins are especially altricial at birth (Fragaszy, 1990). They acquire postural control, prehension, and locomotion later than squirrel monkeys, to which they are closely related, and even later than some Old World monkey species (e.g., baboons and macaques) (Fragaszy, 1990; and see Bezanson, this volume, for a discussion of locomotor development in young capuchins). The reasons for such a prolonged physical dependency are unclear. For example, capuchin and baboon brains are approximately the same size at birth, relative to adult brain size (~40% of adult size). However, baboon motor development is greatly accelerated compared to capuchins (Harvey and Clutton-Brock, 1985). Neonatal brain to body weight ratios are also not adequate correlates to motor ability at birth: capuchin and squirrel monkey ratios are virtually identical in this regard (around 13–14%,), yet squirrel monkeys are far more advanced in motor ability at an early age than capuchins (Fragaszy *et al.*, 1991). However, while possessing comparatively small brains at birth (i.e., 40% of adult size compared to 63% for *Saimiri*), capuchins possess an unusually large brain for their body size as adults (Harvey and Clutton-Brock, 1985). Specifically, they have well-developed cerebellum, neocortex, and dorsal thalamus areas (Bauchot, 1982).

Capuchins have a highly manipulative and extractive foraging strategy and extensive cognitive abilities (Parker and Gibson, 1977; Gibson, 1986, 1990); they may need a longer period of development for brain growth and cognitive functioning associated with learning their foraging and social behavior repertoires. Thus, given their extended period of growth and development, studies of foraging behavior in juvenile capuchins offer insight into the set of factors that influence diet and feeding behavior in other juvenile primates.

In examining the juvenile feeding patterns in white-faced capuchins, the following four specific questions are addressed: (1) What is the dietary pattern of younger and older juveniles? (2) Do younger and older juveniles shift their diet seasonally, and if so, how does this compare to the dietary behavior of adults? (3) Are there food categories exploited by adult capuchins that are not exploited by younger and older juveniles? (4) Which model (juvenile risk aversion, learning opportunities, motor skills, or nutritional needs) best explains the ontogeny of feeding behavior in white-faced capuchins?

METHODS

The Study Site and Study Species

This study took place in the Area de Conservación Guanacaste (ACG), Sector Santa Rosa, located in the Guanacaste province of northwestern Costa Rica (10°50′N, 85°37′W). Sector Santa Rosa is 10,800 ha in size at a mean elevation of 280 m above sea level. Vegetation is a mixture of grassland (abandoned cattle pasture), dry deciduous forest, semi-evergreen forest, fragments of old oak forest, and riparian patches (Janzen, 1983; Chapman, 1987; Fedigan *et al.*,

1996). This Pacific coastal zone region has two distinct seasons: dry and wet. Virtually no rain falls from mid-December to mid-May. During this time, the many deciduous plant species lose their leaves and the land is dry. In the wet season, more fleshy fruits are available, lianas are abundant, and the forest is green and lush. Yearly rainfall at Santa Rosa averages between 1500 and 2000 mm (Janzen, 1983).

Cebus capucinus are the most dimorphic capuchin species (Masterson and Hartwig, 1998), and live in large multi-male, multi-female social groups of 4–30 individuals (Freese and Oppenheimer, 1981; Perry, 1997), with an average group size of 15 (Fragaszy *et al.*, 1990; Fedigan *et al.*, 1996; Fedigan and Jack, 2001; DeGama-Blanchet and Fedigan, this volume). Previous studies indicate that the diet of *C. capucinus* consists of 50–80% fruit, 20–30% animal material (invertebrate and vertebrate), and 10% plant material, with much variation between sites and geographical regions (Chapman and Fedigan, 1990; Janson and Boinski, 1992; Rose, 1994; Panger *et al.*, 2002; Fragaszy *et al.*, 2004).

Data Collection

Data were collected on habituated individuals in two social groups. The individuals analyzed here include 6 small juveniles, 6 large juveniles, and 12 adults. As the animals were used to human observers, observation distances of 5–10 m were common, with distances sometimes as close as 1–3 m. Individuals were identified by size, characteristics of body hair markings, brow hair length, hair cap shape, and facial markings.

The feeding data presented here are part of a larger study on the social behavior of immature capuchins conducted between January 1998 and December 1998. Approximately 236 h of focal animal data were collected on younger juveniles (approximately 1–3 years of age, noticeably smaller, highly active and curious, weaned and thus able to obtain all of their nutritional requirements from the surrounding habitat) and older juveniles (approximately 3–5 years in age, larger than small juveniles but lacking the robust body morphology of fully mature adults). Play is common in both age categories, and sex differences are usually not measurable until subadulthood. An additional 164 h of focal data were collected on adult animals.

Focal animal samples (Altmann, 1974) of 10 min in length were recorded on a handheld Psion computer, using a focal observation behavioral program designed for recording primate behavior (J. Silk, Psion "FOCOBS" program). During the sample period, all behaviors were recorded according to an ethogram designed for this study (MacKinnon, 2002). All feeding behaviors (e.g., solitary eat, eat in proximity, and eat in contact) are combined here, with the majority of data from solitary eat. This was done in order to obtain information about all foods eaten during sampling. The start and stop time of feeding bouts were recorded, and the food type and plant part were noted. Throughout the year, a monthly food list was maintained, and collected plant samples were identified with plant keys and with the help of botanists and ecologists working in the park.

Data Analysis

All the foods eaten are assigned to four plant food categories (soft fruits, hard fruits, seeds, and plants-other) and four animal food categories (caterpillars, invertebrates-embedded, invertebrates-other, and vertebrates); see Table 1 for specific examples of foods in these categories. The total time spent eating items from each food category was divided by the eat time for all categories, to yield a percentage for each category. These data were broken down by younger juvenile, older juvenile, and adult age classes, as well as dry and wet seasons. For data analyses, I used percentage of feeding time, rather than percentage of feeding bouts, for two reasons. First, percentage of feeding time accurately

Table 1. Food categories used in analyses

Plant food	Animal food
Soft fruits: ripe fleshy fruits that require little processing (e.g., *Ficus* sp., *Guettarda* sp.)	Vertebrates: e.g., birds, eggs, coati pups, lizards, rodents
Hard fruits: fruits that have a hard outer casing and require more processing (e.g., *Quercus* acorns, *Acacia* pods, *Inga* pods)	Caterpillars: e.g., Arctiidae, Geometridae, Nymphalidae, Saturnidae, Sphingidae, Tortrisidae
Seeds: foods that capuchins target only for seed material (e.g., *Luehea* spp., *Lasiacis* sp.)	Invertebrates-embedded: small invertebrate or larvae embedded in plant material (excluding caterpillars)
Plants-other: flowers, stems, shoots, leaves, buds	Invertebrates-other: primarily large arthropods (e.g., grasshoppers, katydids, cicadas, tree cockroaches, scorpions) (excluding caterpillars)

represents the proportional amount of time an individual spent consuming a particular food item. Second, I was also recording social behavior during this study, and different social situations were recorded as separate bouts. For instance, if an animal was eating *Ficus* fruits alone, and was then joined by an animal with whom it ate in proximity, that would represent continuous eating of figs in terms of time spent on that food, but two separate bouts: a "solitary eat" bout and a social "eat in proximity" bout.

To test for differences among age classes, Kruskal–Wallis statistical tests were performed using GraphPad Prism, version 4.0a for Macintosh. The confidence parameter was set at $p = 0.05$. If significance was found using a Kruskal–Wallis test, additional paired comparisons were made using two-tailed Mann–Whitney U tests. The correction formula of 0.05 divided by the number of paired comparisons made for that behavior, and was used for the alpha in paired comparisons to control for Type 1 Errors. To test for differences within each age class by season for the various food categories, unpaired t-tests (with Welch's correction) were performed with the confidence parameter set at $p = 0.05$.

RESULTS

The capuchins in this study ate foods from 63 plant species in 34 families (Appendix 1). Over the course of the year, there were no significant differences among age classes in percentage of eat time spent on soft fruits, hard fruits, seeds, and plant-other categories. Soft fruits were consumed for roughly half of mean sample eat time, hard fruits for around 10%, seeds for 7–10%, and plant-other for less than 4% (Table 2).

Throughout the year, the capuchins also ate a wide variety of animal food items, mostly arthropods. There was more variability in time spent eating these foods among the age classes. Invertebrates-embedded were consumed for 13–16% of sample eat time; small juveniles had the highest rates, with statistical significance between small juveniles and adults ($U = 58.0$, $W = 14.0$, $p = 0.0415$). Caterpillars were eaten for 5–13% of total sample eat time with no significance among age classes. Invertebrates-other were consumed for 2–4% of focal eat time; large juveniles had the highest rates, with statistical significance between large juveniles and adults ($U = 59.5$, $W = 12.5$, $p = 0.0312$), and vertebrate prey was eaten for up to 2% of focal eat time, with no significance among age classes (Table 3).

Table 2. Percentage of focal eat time that each age class spent on plant food categories. Mean and standard deviation noted

	Soft fruits	Hard fruits	Seeds	Plants-other
Small juvenile				
Entire year	44.27 ± 6.37	10.88 ± 4.52	7.4 ± 3.41	3.86 ± 3.99
Dry season	58.33 ± 10.21	5.39 ± 4.8	8.56 ± 3.63	0.99 ± 1.16
Wet season	27.28 ± 15.89	16.95 ± 15.3	5.43 ± 8.82	7.2 ± 7.72
Large juvenile				
Entire year	46.52 ± 8.73	10.4 ± 3.65	7.92 ± 5.51	1.32 ± 1.76
Dry season	52.77 ± 14.14	9.49 ± 5.99	10.2 ± 7.05	0.63 ± 0.78
Wet season	35.05 ± 20.04	11.28 ± 13.51	2.58 ± 4.81	2.5 ± 5.27
Adult				
Entire year	51.83 ± 15.55	10.31 ± 9.6	10.57 ± 9.09	1.13 ± 2.24
Dry season	61.02 ± 16.72	6.42 ± 8.8	10.5 ± 10.23	0.21 ± 0.6
Wet season	18.34 ± 20.55	17.96 ± 18.35	3.85 ± 7.51	8.46 ± 20.66

Dry Season

When the eating results are broken down by season, several trends emerge. In terms of plant foods eaten during the dry season, soft fruits comprised over 50% of the mean eat time for all age classes, hard fruits were eaten between 5% and 9%, seeds were consumed between 8% and 10%, and food items in the category plants-other were eaten <1% of dry season eat time by all age classes (Table 2). Within age class categories, small juveniles spent a greater percentage of focal eat time on soft fruits in the dry (58.3%) versus wet (27.3%) season, and the results

Table 3. Percentage of focal eat time that each age class spent on animal food categories. Mean and standard deviation noted

	Caterpillars	Invertebrates (embedded)	Invertebrates (other)	Vertebrates
Small juvenile				
Entire year	13.23 ± 6.44	16.03 ± 3.5	2.02 ± 1.27	0.79 ± 1.01
Dry season	0	21.19 ± 4.53	2.36 ± 1.6	1.42 ± 1.92
Wet season	31.19 ± 16.46	8.97 ± 6.7	1.72 ± 1.53	0
Large juvenile				
Entire year	10.67 ± 3.73	13.99 ± 5.65	4.2 ± 3.33	1.6 ± 2.81
Dry season	0	16.11 ± 9.07	5.49 ± 4.55	2.39 ± 4.44
Wet season	34.69 ± 15.68	8.07 ± 6.98	1.18 ± 1.24	0.21 ± 0.52
Adult				
Entire year	5.43 ± 6.81	13.41 ± 11.17	2.52 ± 2.11	1.97 ± 4.42
Dry season	0	14.7 ± 11.73	2.02 ± 2.04	2.03 ± 4.42
Wet season	22 ± 19.55	23.42 ± 37.18	3.19 ± 4.38	0

were very significant ($t = 4.03$, df $= 8$, $p = 0.0038$). Similarly, adults ate more soft fruits in the dry season (61% of eat time) versus the wet season (18.3%), with very strong statistical significance ($t = 4.89$, df $= 12$, $p = 0.0004$). Large juveniles showed no significance in amount of time eating soft fruits between seasons, but they did spend more time consuming seeds in the dry season (10.2%) versus wet season (2.6%), which approached significance ($t = 2.19$, df $= 8$, $p = 0.06$). No other plant food categories showed significance within age classes.

When examining dry season preferences for animal foods, several interesting results appear. The amount of eat time small juveniles spent on invertebrates-embedded during the rainless months (21.2%) compared to the wet season (9%) was very significant ($t = 3.70$, df $= 8$, $p = 0.006$). For large juveniles and adults, the variation in eat time for invertebrates-embedded was not significant. Large juveniles approached significance in the amount of time they spent eating invertebrates-other during the dry season months (5.5%) versus wet season months (1.1%; $t = 2.24$, df $= 5$, $p = 0.07$); the small juveniles and adults did not. Vertebrate prey items were consumed more during the dry season versus wet season, but with no significance among age classes. Large juveniles devoted the highest percentage of eat time to vertebrate prey in the dry season (2.4%), followed by adults (2%), and then small juveniles (1.4%).

Wet Season

During the wet season, all age classes spent the majority of their eating time on soft fruits (18–35%), followed by hard fruits (11–18%). Within each age class, however, there were several trends. Each age class spent more of their eating time on hard fruits and plant-other in the wet than in the dry season months, and less time on soft fruits and seeds in the wet versus dry months (Table 2). None of the results approached significance.

For animal foods, caterpillars accounted for 22–34% of the mean eat time for all age classes, and invertebrates-embedded were consumed between 8% and 23% of eat time across age categories (Table 3). Interestingly, adults spent a greater fraction of their eat time on invertebrates-embedded in the wet season (23.4%) than in the dry season (14.7%), reversing the trend found in the juveniles (e.g., 21.2% dry, 9.0% wet for small juveniles). The large standard deviation associated with the adult wet time (37.2%) makes this trend inconsequential, however. Vertebrate prey was virtually never eaten during focal sampling time in this season.

DISCUSSION

Juveniles are very capable foragers from a relatively young age (i.e., 1–2 years old), and both age classes (from 1 to 4+ years) are able to find, process, and eat the wide range of food types typical of the adult capuchin diet. The juvenile dietary pattern follows closely that of adult animals, with a focus on soft fruits and invertebrate prey. Small juveniles, large juveniles, and adults shift their diets seasonally, with notable emphases on soft fruits, embedded invertebrates, and seeds during the dry season, and soft fruits and caterpillars during the rainy months (Tables 2 and 3). During this study, there were no food categories exploited only by adults.

Of the four models considered in this study, not all make predictions that can be addressed by these data. As a consequence of their slower metabolic growth, and in order to reduce feeding competition with adults, Janson and van Schaik (1993) predict that juvenile monkeys will prefer more protein-rich and easily digested foods. In the present study, I cannot quantify how much food juveniles are actually consuming, but I do know the amount of time spent eating. If the assumption is made that food consumed is proportional to time spent eating, the data seem to support the prediction. Specifically, small juveniles had the highest rates of eating caterpillars and invertebrates-embedded throughout the year, which is consistent with this model's prediction for a young, omnivorous primate.

The learning model predicts that juvenile primates will have different foraging behaviors and rates of eating than adults because they have not had as much time to learn complex foraging skills, either by trial and error or observational learning (Custance *et al.*, 1999). I cannot fully address the predictions made by this model, as I did not investigate novel feeding opportunities where learning presumably takes place. However, exploration of hidden resources likely contributes to acquiring foraging proficiency. In an earlier study, I documented how the frequency rates of the foraging behavior "explore" showed a propensity to decrease from infant to adult (MacKinnon, 1995). Infants in that study made more mistakes while foraging (e.g., they had higher rates of grabbing and missing insects, and lower rates of eating compared to older animals), suggesting that efficient foraging abilities do not become fully developed until older age class stages. An accumulation of various foraging skills is most likely acquired through inconsistent trial and error, as well as observational learning, and practice (Whiten, 1989; Visalberghi and Fragaszy, 1990, 2002; Custance *et al.*, 1999).

The motor skills model predicts that younger individuals will be unable to exploit difficult-to-process foods because of a lack of strength and dexterity compared to older animals. My observations seem to support this prediction: small juveniles ate vertebrate prey at a rate of less than half of either large juveniles or adults. Capuchins are opportunistic vertebrate predators, and happen upon various animals as they search for insects, larvae, and fruits. Many of these animals have strong beaks and claws, and a 2-year-old capuchin seems to be no match for a tree squirrel or parakeet (pers. obs.). Larger juveniles and adults are better able to subdue such large, biting, squirming prey (for a discussion of vertebrate predation in this species, see Fedigan, 1990). Thus, young capuchins' food choice may be constrained by the size and strength required to handle and process food items (Gibson, 1986; Fragaszy and Boinski, 1995).

Finally, it is not possible to compare Altmann's (1998) nutrition model with these data. Altmann presented quantitative nutritional data for a number of baboons of a similar age (weanlings) over the course of a year. The observations and generalizations made in that study are not applicable here. Specifically, Altmann does not address diet variation across broad age classes, my data cannot be extended to a detailed analysis of the nutritional intake of the study animals, and no assessment of how the observed diets relate to a putative "ideal diet" can be made.

Compared to other New World monkeys (e.g., *Saimiri*), capuchins have extended life history stages including a long period of juvenility. Is such a lengthy period of immaturity necessary for the development of species-appropriate foraging skills? It appears, no. Rather, the complex manipulative foraging skills employed by capuchins to exploit certain food groups are present during the small juvenile stage, and are refined as the young capuchin grows to adulthood.

Small and large juvenile capuchins eat a wide variety of foods in the dry and wet seasons of a tropical dry forest. Overall, the dietary profiles of small and large juveniles are remarkably similar, as are the profiles of all juveniles and adults. Of the four models discussed, the juvenile risk aversion and motor ability models are best supported with the data presented here, where feeding differences occur between age classes.

DIRECTIONS FOR FUTURE RESEARCH

Small sample sizes are one of the unfortunate realities of field-based primate studies, and the present study is no exception. Data on more juveniles and larger sample sizes on feeding rates across age classes would allow trends (seasonal and age) to be better distinguished and stronger tests versus existing models to be performed. I suspect the broad food categories used herein (e.g., soft fruits, invertebrates-embedded) obscure much of the feeding variability that is present in the different age classes.

In order to thoroughly address questions of ontogeny in the wild, individual capuchin monkeys need to be observed as they mature through many stages of development; the current study, while a year in length, gives "snapshot" perspectives on the feeding behavior of small and large juveniles, not longitudinal skill acquisition by individual animals. Finally, I examine duration and rates of eating behavior in capuchins but do not evaluate their success rates or efficiency. For instance, a small juvenile may spend a given amount of time eating invertebrates-embedded, but the number of insects or larvae consumed might be considerably less than for adults, as the juvenile is acquiring this complex foraging skill. Further research on the actual rates of ingestion of specific food types (e.g., those that are embedded in difficult to process substrates) is needed in order to differentiate effective feeding from less efficient trial and error skill acquisition. The juvenile stage of development in capuchins—far from being just an intermediate stage between infant and adult—offers rich possibilities for the examination of the effects of extended life history stages on behavioral variables in wild populations.

SUMMARY

Dietary preferences of white-faced capuchins monkeys (*C. capucinus*) in northwestern Costa Rica were examined across age classes and between two distinct seasons over the course of 1 year. The findings show that the dietary profiles of small juveniles, large juveniles, and adults are quite similar, suggesting that juveniles are efficient foragers from a young age. The predictions of several ontogenetic models are presented, and the data lend limited support to the juvenile risk aversion and motor ability models.

ACKNOWLEDGMENTS

I would like to thank the editors of this volume, Alejandro Estrada, Paul Garber, Mary Pavelka, and LeAndra Luecke, for the invitation to participate. Alejandro Estrada, Paul Garber, and Mary Pavelka provided patience and many helpful suggestions on earlier versions of this chapter. I also thank Phyllis Dolhinow,

Linda Fedigan, and Kathy Jack for assistance throughout this project, and the research and scientific community of the Area de Conservación Guanacaste, especially Róger Blanco Segura, for support during three field seasons in the 1990s. For assistance with plant and insect identifications, I thank INBio staff and Alejandro Masis. Many thanks go to Matthew Wyczalkowski for immeasurable help with database organization. Funding for fieldwork was provided by the National Science Foundation (Grant SBR-9732926), and a University of California at Berkeley Social Sciences and Humanities Research Grant.

REFERENCES

Altmann, J. 1974, Observational study of behavior: Sampling methods. *Behaviour* 49:227–265.

Altmann, S. A. 1998, *Foraging for Survival: Yearling Baboons in Africa*. University of Chicago Press, Chicago.

Bauchot, R. 1982, Brain organization and taxonomic relationships in Insectivora and Primates, in: E. Armstrong and D. Falk, eds., *Primate Brain Evolution*, Plenum Press, New York, pp. 163–175.

Chapman, C. A. 1987, Flexibility in diets of three species of Costa Rican primates. *Folia Primatol.* 49:90–105.

Chapman, C. A. and Fedigan, L. M. 1990, Dietary differences between neighboring *Cebus capucinus* groups: Local traditions, food availability or responses to food profitability? *Folia Primatol.* 54:177–186.

Custance, D., Whiten, A., and Fredman, A. 1999, Social learning of an artificial fruit task in capuchin monkeys (*Cebus apella*). *J. Comp. Psychol.* 113:13–23.

Fedigan, L. M. 1990, Vertebrate predation in *Cebus capucinus*: Meat-eating in a neotropical monkey. *Folia Primatol.* 54:196–205.

Fedigan, L. M. and Jack, K. 2001, Neotropical primates in a regenerating Costa Rican dry forest: A comparison of howler and capuchin population patterns. *Int. J. Primatol.* 22(5):689–713.

Fedigan, L. M. and Rose, L. M. 1995, Interbirth interval variation in three sympatric species of neotropical monkey. *Am. J. Primatol.* 37(1):9–24.

Fedigan, L. M., Rose, L. M., and Avila, R. M. 1996, See how they grow: Tracking capuchin monkey (*Cebus capucinus*) populations in a regenerating Costa Rican dry forest, in: M. A. Norconk, A. L. Rosenberger, and P. A. Garber eds., *Adaptive Radiations Of Neotropical Primates*, Plenum Press, New York.

Fragaszy, D. M. 1986, Time budgets and foraging behavior in wedge-capped capuchins (*Cebus olivaceus*): Age and sex differences, in: D. M. Taub and F. A. King, eds., *Current*

Perspectives in Primate Social Dynamics, Van Nostrand Reinhold Co., New York, pp. 159–174.

Fragaszy, D. M. 1990, Early behavioral development in capuchins (*Cebus*). *Folia Primatol.* 54:119–128.

Fragaszy, D. M., Baer, J., and Adams-Curtis, L. 1991, Behavioral development and maternal care in tufted capuchins (*Cebus capucinus*) and squirrel monkeys (*Saimiri sciureus*) from birth through seven months. *Develop. Psychobiol.* 24(6):375–393.

Fragaszy, D. M. and Boinski, S. 1995, Patterns of individual diet choice and efficiency of foraging in wedge-capped capuchin monkeys (*Cebus olivaceus*). *J. Comp. Psychol.* 109(4):339–348.

Fragaszy, D. M. and Visalberghi, E. 1989, Social influences on the acquisition and use of tools in tufted capuchin monkeys (*Cebus apella*). *J. Comp. Psychol.* 103:159–170.

Fragaszy, D. M., Visalberghi, E., and Fedigan, L. M. 2004, *The Complete Capuchin: The Biology of the Genus Cebus.* Cambridge University Press, New York.

Fragaszy, D. M., Visalberghi, E., and Robinson, J. G. 1990, Variability and adaptability in the genus *Cebus. Folia Primatol.* 54:114–118.

Freese, C. H. and Oppenheimer, J. R. 1981, The capuchin monkeys, genus *Cebus*, in: A. F. Coimbra-Filho and R. A. Mittermeier, eds., *Ecology and Behavior of Neotropical Primates, Vol. I*, Academia Brasiliera de Ciencias, Rio de Janeiro, pp. 331–400.

Gibson, K. R. 1986, Cognition, brain size and the extraction of embedded food resources, in: J. G. Else and P. C. Lee, eds., *Primate Ontogeny, Cognition, and Social Behaviour*, Cambridge University Press, Cambridge, pp. 39–103.

Gibson, K. R. 1990, Tool use, imitation, and deception in a captive cebus monkey, in: S. T. Parker, K. R. Gibson, eds., "Language" and Intelligence in Monkeys and Apes: Comparative Developmental Perspectives, Cambridge University Press, New York, pp. 205–218.

Hakeem, A., Sandoval, R. G., Jones, M., and Allman, J. 1996, Brain and life span in primates, in: J. E. Birren and K. W. Schaie, eds., *Handbook of the Psychology of Aging, 4th Edn.*, Academic Press, San Diego, pp. 78–104.

Harvey, P. and Clutton-Brock, T. 1985, Life history variation in Primates. *Evolution* 39:559–581.

Janson, C. H. and Boinski, S. 1992, Morphological and behavioral adaptations for foraging in generalist primates: The case of the Cebines. *Am. J. Primatol.* 88: 483–498.

Janson, C. H. and van Schaik, C. P. 1993, Ecological risk aversion in juvenile primates: Slow and steady wins the race, in: M. E. Pereira and L. A. Fairbanks, eds., *Juvenile Primates: Life History, Development, and Behavior*, Oxford University Press, New York, pp. 57–74.

Janzen, D. H. 1983, *Costa Rican Natural History.* University of Chicago Press, Chicago.

MacKinnon, K. C. 1995, Age differences in foraging patterns and spatial associations of the white-faced capuchin monkey (*Cebus capucinus*) in Costa Rica. MA Thesis, University of Alberta.

MacKinnon, K. C. 2002, Social development of wild white-faced capuchin monkeys (*Cebus capucinus)* in Costa Rica: An examination of social interactions between immatures and adult males. PhD Dissertation, University of California at Berkeley.

Masterson, T. J. and Hartwig, W. C. 1998, Degrees of sexual dimorphism in *Cebus* and other New World monkeys. *Am. J. Phys. Anthropol.* 107(3):243–256.

Panger, M., Perry, S., Rose, L. M., Gros-Louis, J., Vogel, E., MacKinnon, K. C., and Baker, M. 2002, Cross-site differences in the foraging behavior of white-faced capuchins (*Cebus capucinus*). *Am. J. Phys. Anthropol.* 119:52–66.

Parker, S. and Gibson, K. 1977, Object manipulation, tool use, and sensorimotor intelligence as feeding adaptations in *Cebus* monkeys and great apes. *J. Human Evol.* 6:623–641.

Perry, S. 1997, Male-female social relationships in wild white-faced capuchins (*Cebus capucinus*). *Behaviour* 134:477–510.

Robinson, J. G. and Janson, C. H. 1987, Capuchins, squirrel monkeys, and Atelines: Socioecological convergence with old world primates, in: B. B. Smuts, D. L. Cheney, R. M. Seyfarth, R. W. Wrangham, and T. T. Struhsaker, eds., *Primate Societies,* University of Chicago Press, Chicago, pp. 69–82.

Rose, L. M. 1994, Sex differences in diet and foraging behavior in white-faced capuchins (*Cebus capucinus*). *Int. J. Primatol.* 15(1):95–114

Rose, L. M. 1997, Vertebrate predation and food-sharing in *Cebus* and *Pan. Int. J. Primatol.* 18(5):727–765.

Visalberghi, E. and Addessi, E. 2003, Food for thought: Social learning about food in feeding capuchin monkeys, in: D. M. Fragaszy and S. Perry, eds., *The Biology of Traditions*, Cambridge University Press, Cambridge, pp. 187–212.

Visalberghi, E. and Fragaszy, D. M. 1990, Do monkeys ape?, in: S. T. Parker and K. R. Gibson, eds., *"Language" and Intelligence in Monkeys and Apes: Comparative Developmental Perspectives*, Cambridge University Press, Cambridge, pp. 247–273.

Visalberghi, E. and Fragaszy, D. M. 2002, Do monkeys ape? Ten years after, in: K. Dautenhahn and C. L. Nehaniv, eds., *Imitation in Animals and Artifacts*, MIT Press, Cambridge, MA, pp. 471–479.

Visalberghi, E. and Limongelli, L. 1996, Action and understanding: Tool use revisited through the mind of capuchin monkeys, in: A. Russon, K. Bard, and S. Parker, eds., *Reaching into Thought: The Minds of the Great Apes*, Cambridge University Press, Cambridge, pp. 57–79.

Whiten, A. 1989, Transmission mechanisms in primate cultural evolution. *Trends Ecol. Evol.* 4(3):61–62.

APPENDIX 1. *Cebus capucinus* PLANT FOODS AND PARTS EATEN, 1998

Family	Genus	Species	Flowers	Fruits/Seeds	Branches/Thorns (invertebrates)	Old pods (invertebrates)	Leaves	Pith/Shoots	Other
Anacardiaceae	*Spondias*	*mombin*		×					
		purpurea		×					
Annonaceae	*Annona*	*purpurea*[a]		×					
		reticulata[a]		×				×	
Apocynaceae	*Stemmadenia*	*obovata*		×					
Araliaceae	*Sciododendron*	*excelsum*		×					×
Asclepiadaceae	*Asclepias*	*curassavica*		×					
Bignonaceae	*Tabebuia*	*ochracea*[a]		×					×
	Pithecodenium	*crucigerum*[a]		×		×			
Bixaceae	*Cochlospermum*	*vitifolium*		×	×(ants)			×	
Boraginaceae	*Cordia*	*alliodora*[a]		×	×(ants)				
		panamensis		×					
Bromeliaceae	*Bromelia*	*pinguin*		×				×	
	Tillandesia	*circinnata*		×				×	
Burseraceae	*Bursera*	*simarouba*		×	×(larvae)			×	
Cecropiaceae	*Cecropia*	*peltata*		×					
Combretaceae	*Combretum*	*farinosum*[a]	×	×					
Elaeocarpaceae	*Muntingia*	*calabura*		×					
	Sloanea	*terniflora*		×					
Euphorbiaceae	*Margaritaria*	*nobilis*[a]		×					
Fagaceae	*Quercus*	*oleoides*		×					
Flacourtiaceae	*Casearia*	*sylvestrius*		×					
	Prockia	*cruces*[a]		×					
	Zuelania	*guidonia*		×					
Gramineae	*Lasiacis*	sp.		×					
Leguminosae	*Acacia*	*collinsii*		×	×(larvae)				
	Albizia	*adinocephala*[a]		×					×
	Enterolobium	*cyclocarpum*[a]		×		×			
	Gliricidia	*sepium*		×	×(ants)				
	Inga	*vera*		×					
	Mucuna	*pruriens*[a]		×					
	Pithecellobium	*saman*[a]		×					

Malpighiaceae	*Byrsonima*	*crassifolia*	×			
Malvaceae	*Malvaviscus*	*arboreus*[a]	×			
Meliaceae	*Trichilia*	*cuneata*	×			
Moraceae	*Chlorophora*	*tinctoria*	×			
	Ficus	spp.	×			
Myrtaceae	*Psidium*	*guajava*[a]	×			
		guincense[a]	×			
Passifloraceae	*Passiflora*	sp.	×			
Piperaceae	*Piper*	*auritum*	×(fur-rub)		×(fur-rub)	
Rhamnaceae	*Karwinskia*	*calderoni*	×			
Rubiaceae	*Alibertia*	*edulis*[a]	×			
	Calycophyllum	*candissimum*[a]	×			
	Chomelia	*spinosa*[a]	×			
	Genipa	*americana*	×			
	Guettarda	*macrosperma*	×			
	Hamelia	*mateus*[a]	×			
	Psychotria	*horizontalis*	×			
	Randia	*echinocarpa*	×			
		subchordata	×			
Sapindaceae	*Allophylus*	*occidentalis*	×			
	Dipterodendron	*costaricene*[a]	×			
	Paullinia	*cururu*	×			
Sapotaceae	*Manilkara*	*zapota*	×			
Simaroubaceae	*Simarouba*	*glauca*	×			
Sterculiaceae	*Guazuma*	*ulmifolia*[a]	×	×		×
Theophrastaceae	*Jacquinia*	*pungens*	×			
Tiliaceae	*Abeiba*	*tibourbou*[a]	×	×		
	Luehea	*candida*	×			
		speciosa	×			
	Muntingia	*calabura*	×			
Vitaceae	*Cissus*	*rhombifolia*[a]	×			

[a] Not recorded for juveniles during focal sampling.

CHAPTER FIFTEEN

Why Be Alpha Male? Dominance and Reproductive Success in Wild White-Faced Capuchins (*Cebus capucinus*)

Katharine M. Jack and Linda M. Fedigan

INTRODUCTION

Most social mammals residing in multimale–multifemale groups display some sort of dominance hierarchy, although the stability and determinants of these hierarchies vary across species and according to sex. In species where females are philopatric, female dominance is usually based on kinship, whereas male dominance is determined by the outcome of male–male competition (Preuschoft and

Katharine M. Jack • Department of Anthropology, Tulane University, 1021 Audubon Street, New Orleans, LA 70118. **Linda Fedigan** • Department of Anthropology, University of Calgary, 2500 University Dr. NW, Calgary, Alberta, Canada 2N 1N4.

New Perspectives in the Study of Mesoamerican Primates: Distribution, Ecology, Behavior, and Conservation, edited by Alejandro Estrada, Paul A. Garber, Mary S. M. Pavelka, and LeAndra Luecke. Springer, New York, 2005.

Paul, 2000; Walters and Seyfarth, 1987). Sexual selection theory predicts that males who win intrasexual competition will receive reproductive benefits and this prediction has led to the long-standing assumption that male dominance rank is positively correlated to reproductive success (Andersson, 1994). However, the distribution of reproduction is extremely variable in animal societies (Keller and Reeve, 1994), and it should be noted that in many species mating success is not synonymous with reproductive success (e.g. Inoue *et al.*, 1993).

The correlation between male dominance rank and reproductive success has a particularly long history of debate in primate studies (e.g. Fedigan, 1983; Cowlishaw and Dunbar, 1991; De Ruiter and van Hooff, 1993; Ellis, 1995). Although this issue has been more thoroughly investigated in the Order Primates than in any other taxon (see Berard, 1999 for review), the correlation between male dominance and reproductive success is less than straightforward, with results ranging from no correlation to a significant positive or negative correlation depending on the species, seasonality, and/or housing conditions (e.g. Altmann *et al.*, 1996; Paul, 1997). Here, we examine the relationships between male dominance rank and reproductive success in two groups of wild white-faced capuchins (*Cebus capucinus*) residing in Santa Rosa National Park, Costa Rica, between 1993 and 2000.

Social System of White-Faced Capuchins

White-faced capuchins are medium-sized, Neotropical primates (in our study population males weigh approximately 3.3 kg and females 2.3 kg; see Fedigan and Rose, 1995) and range throughout Latin America from Honduras through the northwest coast of Ecuador. In general, capuchins (*Cebus*) more closely resemble Old World monkeys than do other Neotropical genera in that *Cebus* species reside in groups comprised of multiple related females, immigrant males, and their immature offspring. However, unlike Old World monkeys, capuchin groups are composed of nearly equal ratios of adult males and females (Robinson and Janson, 1987). White-faced capuchin groups are comprised of approximately 17 individuals with, on average, four adult males and five adult females (Fedigan and Jack, 2001). This species is moderately sexually dimorphic, with males being 25–35% larger than females (Fedigan, 1993). Female white-faced capuchins give birth approximately every 27 months (Fedigan, 2003) and, although they engage in non-conceptive matings throughout the year (Manson *et al.*, 1997), they display a birth peak between January and April (Fedigan *et al.*,

1996). Males emigrate from their natal group at around 4 years of age and they continue to change groups throughout their lives approximately every 4 years (Jack and Fedigan, 2004a,b).

Within groups, both males and females form linear dominance hierarchies; however, they are determined through very different mechanisms. Female dominance is related to matrilineal kinship and the maintenance of coalition partners through reciprocal grooming and proximity (Perry, 1995; Fragaszy *et al.*, 2004), whereas male dominance appears to be largely determined by the outcome of intrasexual competition (Perry, 1998a,b; Fragaszy *et al.*, 2004), although overt aggression among co-resident males is rare (Jack, 2001b). Male dominance is relatively unstable over time due to the frequent dispersal of group males (i.e. males emigrating from or immigrating into the groups; see Jack and Fedigan, 2004b) or, less commonly, through rank reversals within groups (Perry, 1998a, pers. obs.).

Why be Alpha Male?

We have often asked the question, "Why be alpha male?" As is the case in most primate species, alpha male white-faced capuchins work harder than other group members in that they spend more time engaged in vigilance, expend greater effort in deterring predators and extra-group males, and they are the most active participants during inter-group conflicts (Rose and Fedigan, 1995). What are the benefits of all these efforts? Enhanced mating and reproductive success has long been considered the major benefit of high status (reviewed in Berard, 1999). However, our observations of white-faced capuchins over the past two decades indicate that although alpha males may obtain a slightly greater proportion of copulations (see Rose, 1998), they are by no means exclusively selected as mates by females (Fedigan, 2003; Fragaszy *et al.*, 2004). Indeed the mating system of white-faced capuchins appears to be very egalitarian; subordinate males, including sub-adults, will mate in full view of alpha males and they make no effort to hide their mating activities. Copulations in this species are very conspicuous, involving a coordinated dance display performed by the male and the female, and accompanied by specific vocalizations and facial expressions (see Manson *et al.*, 1997 for a complete description). Thus far, we have no concrete evidence of overt mating competition occurring among co-resident males (Carnegie, unpublished data; Jack, 2003). To date, the reproductive system of this species has not been investigated and it remains to be determined if the egalitarian mating system that we have observed among co-resident males equates to shared reproduction within the group.

Although initially developed to explain differences in the reproductive output of females in cooperative and highly social groups, models of reproductive skew and concession theory have also been used to determine the optimal amount of reproduction a dominant needs to concede to a subordinate in order to keep him/her in the group and peacefully cooperating (e.g. Vehrencamp, 1979; Emlen, 1982; Reeve and Ratnieks, 1993; Reeve and Emlen, 2000). High skew societies are those in which reproduction is dominated by one or a few breeders, whereas low skew societies are those in which reproduction is more equitably distributed among group members. Concession theory predicts that the differences in the reproductive output of dominants and subordinates will be lower when the presence of subordinates in a group increases the fitness of dominants (Clutton-Brock, 1998). That is, when subordinates provide fitness benefits to dominant males (i.e. cooperation in resource and/or mate defense), dominants will concede a portion of reproduction to subordinates as a "staying incentive" in order to keep them cooperating in the group.

The presence of subordinate males does impose some costs to dominants in white-faced capuchin groups (e.g. increased social vigilance: Jack, 2001a; Perry, 1998b; increased foraging competition: Rose and Fedigan, 1995). However, in this species, male cooperation is necessary to enter groups, which is most often achieved through aggressive takeovers (Fedigan and Jack, 2004), and to retain membership in these groups (Rose and Fedigan, 1995; Perry, 1998b; Jack, 2001b; Jack and Fedigan, 2004b). In all cases of successful group takeovers, invading males are of superior physical strength, or they form coalitions that out-number resident males (Fedigan and Jack, 2004). Given the necessity of male cooperation in this species, it is not surprising to find that male–male relationships within groups are very tolerant, with low levels of intragroup aggression (Jack, 2003; Perry, 1998b). We have also recorded high levels of affiliative interactions among some resident males and that the maintenance of these male–male bonds can persist through multiple emigrations (Jack, 2003; see also Jack and Fedigan, 2004b). We therefore interpret the relaxed mating system of white-faced capuchins as a means by which alpha males can maintain the cooperation of co-resident males.

This study examines male reproductive success in wild white-faced capuchins to determine if high dominance rank confers a reproductive advantage. We also address the issue of reproductive skew in this species, by examining whether or not alpha males provide their co-resident males with staying incentives in terms of reproductive opportunities.

METHODS

Study Site

Data presented here are based on two of our long-term study groups of wild white-faced capuchins residing in the Santa Rosa Sector of the Area de Conservacion Guanacaste in Costa Rica. Formerly known as Santa Rosa National Park (SRNP), the sector lies 35 km northwest of Liberia and approximately 30 km south of the Nicaraguan border, in the Guanacaste Province. SRNP is comprised of approximately 108 km^2 of dry deciduous forest and reclaimed pasture in varying stages of regeneration, and ranges from sea level to 300 m in altitude (see Fedigan *et al.*, 1996; Fedigan and Jack, 2001 for additional site details).

Study Groups

LF began studying the capuchins of SRNP in 1982 (Fedigan, 1986; Fedigan *et al.*, 1985; Chapman *et al.*, 1988) and research has been on-going since that time. Although numerous groups have been studied over the years, the data presented here focus on our two long-term study groups, Cerco de Piedra (CP) and Los Valles (LV). Data collected on these groups have included, but are not limited to, the recording of group demographics and dominance relationships (see Jack and Fedigan, 2004a,b; Fedigan and Jack, 2004, for additional details on long-term monitoring of these two groups). Here, we address the issue of male dominance and reproductive success by examining the paternity of infants born into the LV group over a 6.25-year period (November 1993 to January 2000) and infants born into the CP group over a 6.5-year period (October 1993 to April 2000).

Individual group members were identified by natural markings such as peak shape, scars, and missing or broken digits and according to their age-sex class. Our long-term observations of this species demonstrate that male white-faced capuchins do not reach full adult body size until they are 10 years of age (see Jack and Fedigan, 2004a,b), whereas sexual maturity occurs at approximately 8 years of age (Freese and Oppenheimer, 1981). In general, juvenile males are not sexually active within our groups; and since we began our intensive observations in 1985, we have only once observed a single juvenile male (an immigrant male aged 6 or 7 years) copulate with group females (see Jack and Fedigan, 2004a). Therefore, in this study, we include only those males classified as adult (≥10 years) or subadult (7–10 years), all of whom were immigrants into the study groups. Male dominance rank was monitored continuously throughout the study period and determined by the direction of agonistic signals, supplantation, and approach/retreat interactions. Across all study years, multiple adult and/or subadult males (≥2, maximum six) resided in each of the study groups and all males could be easily ranked in a linear hierarchy.

DNA Sample Collection and Analysis

We collected hair and/or fecal samples from all individuals present in the two study groups between 1997 and 2000. Additionally, we were able to collect samples from all former resident males, by following them after their transfer into other social groups. Fecal samples were collected immediately upon defecation and stored in vials containing 95% ethanol (see Gerloff *et al.*, 1995). Hair samples were collected using a modified blow darting technique (using a blunt dart with duct tape) and stored in paper in a dry location. Whenever possible, both hair and multiple fecal samples were collected for each individual. DNA extractions, using QIAGEN kits appropriate to sample types, and genotyping (PCR) were performed by Dr. David Paetkau, Wildlife Genetics International (unpublished data). Paternity exclusions detailed here are based on the use of three dinucleotide microsatellite markers with previously demonstrated utility in New World primates. One marker PEPL4 was developed for *Lagothrix* (Escobar-Parámo, 2000), while the remaining two markers (D3S1210 and D8S165) are human derived loci that have previously amplified well for *Saimiri boliviensis* (Witte and Rogers, 1999). Twenty-three additional markers were tested but were either found to be monomorphic in the Santa Rosa capuchins or the PCR was illegible (see Table 1 for details on additional markers tested). The three markers utilized here had either 3, 4, or 5 alleles, and DNA amplification (PCR) was independently repeated for each locus a minimum of two times, to combat the problems of false genotyping often associated with PCR products taken from non-invasive samples such as we use here (see Gerloff *et al.*, 1995).

Table 1. Microsatellite markers tested in two groups of *Cebus capucinus*

Marker	Repeat length	Amplification?	Legible?	Variable?
PEPL4	2	Yes	Yes	Yes; 5 alleles
D8S165	2	Yes	Yes	Yes; 3 alleles
D3S1210	2	Yes	Yes	Yes; 4 alleles
D6S260	2	Yes	Yes	2 alleles, 1 rare
D14S51	2	Yes	Yes	2 alleles, 1 rare
Ap6	2	Yes	Yes	Monomorphic
PEPC40	2	Yes	Yes	Monomorphic
PEPC59	2	Yes	Yes	Monomorphic
PEPC8	2	Yes	Yes	Monomorphic
PEPC3	2	Yes	No	3 alleles?
D3S1229	2	Yes	No	Yes?
D5S117	2	Yes	Marginal	Monomorphic?
Ap20	2	No		
Sw21F	2	No		
Sw65B	2	No		
CYP19	4	Yes	Yes	2 alleles, 1 rare
D21S1443	4	Yes	Yes	Monomorphic
THO1	4	Yes	Marginal	2 alleles
D1S518	4	Yes	Marginal	2 or 3 alleles?
D14S118	4	Yes	No	Monomorphic?
D8S588	4	Yes	No	3 alleles?
D9S746	4	Yes	No	Unknown
D12S1025	4	No		
D4S1628	4	No		
D8S373	4	No		
VWF-TNR	4	No		

Paternity Determination

By comparing the genotypes of infants and mothers, we were able to deduce paternal genotypes. We then examined the genotypes of all non-natal males, both adult and subadult, present in the group when the infant was conceived. Conception dates were determined by counting back 164 days from the birth date of each infant (see Fedigan and Rose, 1995; Robinson and Janson, 1987), although all males present in the month before and after the possible conception dates were also investigated for possession of the paternal alleles. Each male who did not possess the paternal alleles was excluded as a possible sire of the offspring. A male was considered the likely sire of an infant only if all other males could be excluded and the particular male in question possessed the paternal genotype at all three loci (e.g. Borries *et al.*, 1999; Soltis *et al.*, 2001).

Although four males in our sample resided first in one study group and then in the other, they are considered here as distinct males within each group and their reproductive success (and dominance rank) within each of the groups is treated independently.

RESULTS

Los Valles Group

Between November 1993 and January 2000, nine immigrant males (adults and subadults) resided in LV. Six of these males were adults (BU, MO, NO, DI, PI, and SI) and three were subadults (SP, LE, and TR). Several of the males present in the group occupied more than one rank during their tenure, and each is considered according to his rank at the particular time that a conception was estimated to occur. Therefore, throughout the study period, the nine immigrant males occupied ranks as follows: three alpha males (BU, NO, and DI) and nine subordinates (BU, MO, LE, SP, DI, PI, SI, NO, and TR; see Table 2). A total of 15 infants were born into the group during the study period. Three of these infants disappeared from the group prior to being sampled; two disappeared together, accompanied by their mothers and one subadult male (group fission was suspected); while the third infant disappeared at 6 months of age and was presumed dead. These three infants are excluded from our analyses. Of the 12 remaining infants in our LV sample, three of the paternity exclusions (HE, SA, and CH) were incomplete as we were unable to type the infant's mothers (they died or disappeared prior to sampling). Finally, we could not exclude multiple possible sires from two additional LV infants, LZ and SA, who were siblings.

Table 2 lists these 12 infants born into LV during the study period, as well as adult and subadult males present (possible sires listed in descending order of rank), and the identity of males not excluded as possible sires. Of the seven infants for whom only one male could not be excluded, the alpha male was the sole non-excluded male for six (85.7%), while the beta male could not be excluded for the one remaining infant (14.3%; Table 2). In the case where the beta male was the probable sire of the infant, he was the only other adult male present in the group at the time of conception. In addition, this male (SI) had been the alpha male of the neighboring BH group where he and the infant's mother had resided prior to transferring together into LV 2 years earlier.

Paternity exclusions were incomplete for 5 of the 12 infants in our sample (i.e. multiple males could not be excluded as possible sires). The alpha male was included among the possible sires in four of these cases (infants LZ, ST, SA,

Table 2. Paternity exclusions for infants born into study groups

Los Valles: November 1993 to January 2000

Infant	Mother	Conception date	Males present and rank[a]	Males not excluded
LZ	BL	20-Nov-1993	BU, MO	BU, MO
HE	BO[c]	22-Jul-1994	BU, MO, SP[b], LE[b]	MO and LE[b]
AL	KL	27-Aug-1995	BU, SP[b], LE[b]	BU
ST	CA[c]	13-Aug-1995	BU, SP[b], LE[b]	BU, SP[b], LE[b]
SA	BL	20-Sep-1995	BU, SP[b], LE[b]	BU, SP[b]
CH	BO[c]	23-Mar-1996	BU, SP[b], LE[b]	BU, SP[b]
PP	FE	21-Oct-1997	NO, DI, BU, PI, SI, TR[b]	NO
MA	KL	21-Nov-1997	DI, NO, PI, SI, TR[b]	DI
SO	DL	27-Nov-1998	DI, SI, PI	DI
CY	BL	21-Feb-1999	DI, SI	DI
DJ	FE	22-Jun-1999	DI, SI	SI
Y2K	KL	22-Jul-1999	DI, SI	DI

Cerco de Piedra: October 1993 to April 2000

Name	Mother	Concept	Males present	Males not excluded
NY	LI	30-Jul-1993	NO, PI, DI	NO
PU	TU[c]	22-Apr-1993	NO, PI, DI	NO
RA	LI	5-Dec-1995	NO, PI, DI	NO
TI	SE	5-Dec-1995	NO, PI, DI	NO
SI2	LI	24-Feb-1998	NO, TR[b]	NO
ZA	SE	23-Aug-1998	NO, TR[b]	NO
BA	LI	19-Nov-1999	NO, TR[b]	NO

All infant names listed in bold indicate incomplete paternity exclusions.
[a] All males are listed in descending order of their rank at time of infant's conception.
[b] Subadult male.
[c] Mother not typed.

and CH in Table 1). The last case of unsuccessful exclusion (see infant HE in Table 2) involved a beta male and one subadult male, indicating that this infant was sired by a subordinate male.

In summary, of the 15 infants born into the LV study group, three were untyped, paternity exclusions were incomplete for five and complete for the seven remaining infants. However, if we consider the infants sired by males according to rank-class (alpha or subordinate), which enables us to include the infant HE as we were able to exclude the alpha as a possible sire, six of eight (75%) were sired by alphas, and two of eight (25%; including HE) were sired by subordinate males.

Cerco de Piedra Group

Four immigrant males (one alpha and three subordinates) resided in the CP group between October 1993 and April 2000. The same adult male (NO) was alpha throughout this entire period, with the exception of 9 months during which time he took up residence as alpha male in the LV group. He followed his two subordinate males (PI and DI) who had aggressively taken over the neighboring LV group 1 month prior. During his 9 month absence, no new males entered the CP group, which remained without a resident male. This enabled NO to return to CP after being ousted from LV by his two former subordinates, at which time he returned to his position of alpha male accompanied by a subadult male from LV (TR). Note that these two males, and others who resided in both of our study groups (see below), are treated as different individuals in each group (e.g. NO/CP and NO/LV).

Nine infants were born in CP during the study period (Table 2). Samples were not obtained from two of these infants, as one died of apparently natural causes at 3 months of age and the other disappeared at 26 months with his mother, another adult female and an older brother (group fission suspected). Paternity exclusions were performed for the remaining seven infants; and in every case only the alpha male, NO, was the sole male that could not be excluded as a possible sire.

Dominance and Reproductive Success (Combined Group Data)

Over the course of our study, four males resided in each of the LV and CP groups. In addition, several males in LV occupied a different rank at varying points throughout the study. Given that our interest here is not in individual reproductive success, but rather reproductive success according to male rank, each of these males is considered according to his rank at the estimated time of each conception (see Curie-Cohen *et al.*, 1983 for similar treatment of male rank changes in rhesus macaques). For example, Table 2 shows that DI was beta male in LV when PP was conceived in October 1997, and the alpha male when the remainder of the LV infants were conceived. Such rank change can occur due to reversals within the group (i.e. a low ranking male moving up, which only rarely occurs), through the immigration of new males into the group, or through the emigration of resident males out of the group (see Fedigan and Jack, 2004; Jack and Fedigan, 2004b). Given the frequent fluctuation of rank among subordinate males, and so as not to artificially inflate our sample size, we have opted to consider each male as either alpha or subordinate rather than by his linear rank. Calculated in this way, there was a total of 16 "ranked males" residing in our groups during the study period, and each is assigned a status of either alpha (n = 4) or subordinate (n = 12) according to his particular rank-class at the estimated time of conception. At no time during the course of our study was there only one male (i.e. the alpha male) present in the group when a conception was estimated to have taken place. We follow Smith (1981) and Curie-Cohen (1983) and assume that each male has an equal opportunity of siring each of the infants whose paternity was not determined. In this way, we were able to include incomplete paternity exclusions, and assign each male a reproductive success score (RS, Equation (1)) for each rank category he occupied in the group.

$$\text{RS} = \text{Infants sired} + \frac{\text{Possible infants sired}}{\text{Total no. of males not excluded}} \qquad (1)$$

We also calculated an adjusted RS score (RSA, Equation (2)), which takes into account the total number of infants conceived within the group while each male was at a particular rank.

$$\text{RSA} = \frac{\text{RS}}{\text{No. of infants conceived during tenure}} \qquad (2)$$

This score represents the total proportion of infants that were born into the group that the male was likely to have sired. The closer this number is to 1, the higher that male's reproductive success. The RS and RSA for each male at each rank-class he occupied are given in Table 3, as are the minimum and maximum number of infants that each male sired. Although we have previously demonstrated that alpha and subordinate males do not differ significantly from one another in terms of tenure lengths within our study groups (Jack and Fedigan, 2004b), this adjusted reproductive success score enables us to control any bias that may be present in the number of offspring that each male had the opportunity to sire.

Of the 19 infants that were genotyped, alpha males ($n = 4$) sired a minimum of 13 (complete exclusions only) and a maximum of 17 infants (including the four incomplete exclusions; 68.4–89.5%), whereas subordinates ($n = 12$) were possible sires for a minimum of two and a maximum of six infants (10.5–31.6%; Table 4). Within our two groups, both RS and RSA were significantly correlated with male dominance rank-class over the course of the study period (RS:

Table 3. Male dominance rank and reproductive success

	Dominance rank(s)	No. of Infants born/typed[a]	Min. and Max[b] No. of infants sired	Reproductive success (RS[c])	Reproductive success—adjusted (RSA[d])
BU (LV)	1	6	1–5	2.84	0.473
BU (LV)	2	1	0	0.00	0.00
MO (LV)	2	1	0–2	1.00	0.50
SP (LV)	3 and 2	5	0–3	1.33	0.266
LE (LV)	3 and 4	5	0–2	0.83	0.166
NO (LV)	1	1	1	1.00	1.00
DI (LV)	2	1	0	0.00	0.00
DI (LV)	1	5	4	4.00	0.80
NO (LV)	2	1	0	0.00	0.00
TR (LV)	5 and 6	5	0	0.00	0.00
PI (LV)	3	3	0	0.00	0.00
SI (LV)	2, 4, and 5	6	1	1.00	0.167
NO (CP)	1	7	7	7.00	1.00
DI (CP)	2	4	0	0.00	0.00
PI (CP)	3	4	0	0.00	0.00
TR (CP)	2	3	0	0.00	0.00
Alpha total (mean)			13–17	14.84	Mean = 0.82 ($n = 4$)
Subordinate total (mean)			2[e]–6	4.16	Mean = 0.06 ($n = 12$)

[a] Excludes five infants that were not genotyped.
[b] Minimum represents the number of offspring assigned to that male, while maximum is the minimum plus all additional infants for which that male could not be excluded as a possible sire.
[c] $\text{RS} = \text{Infants sired} + \dfrac{\text{Possible infants sired}}{\text{Total no. of males not excluded}}$.
[d] $\text{RSA} = \dfrac{\text{RS}}{\text{No. of infants conceived during tenure}}$.
[e] Includes one case where two subordinate males were the only group males not excluded as a possible sire.

Table 4. Possible reproductive success according to male rank status (alpha versus subordinate)

Group	No. infants born	Total no. sampled[a]	Likely no. of infants sired (single non-excluded male)		Possible additional infants sired (multiple non-excluded male)	
			Alpha	Subordinate[b]	Alpha	Subordinate
Los Valles	15	12	6	2	4	4
Cerco de Piedra	9	7	7	0	0	0
Total	24	19	13	2	4	4

[a] Excludes those infants that died before sampling.
[b] Includes one case where two subordinate males were the only group males not excluded as a possible sire.

$r_s = 0.738$, $n = 16$, $p = 0.001$; RSA: $r_s = .770$, $n = 16$, $p < 0.001$), with alphas siring significantly more offspring than subordinates.

DISCUSSION

The data presented here cover a combined total of 12.75 years (1993–2000), during which time 4 alpha males and 12 subordinate males resided in our two study groups, and a total of 24 infants were born. Due to deaths or disappearances of 5 infants prior to sampling, only 19 of those infants were included in this study. Our examination of male reproductive success in these two groups of white-faced capuchins found a significant positive correlation between male rank-class (alpha or subordinate) and reproductive success, with alpha males siring most (68.4–89.4%) infants born into our study groups. These data show that being the top-ranked male within a group has definite reproductive benefits and that, although dominant males are providing subordinates with mating opportunities (see Rose, 1998), reproductive opportunities (i.e. paternity) are not truly being shared. Paul (2002) correctly points out that even if a female engages in multiple matings, as is the case with female white-faced capuchins, this does not necessarily mean that matings are random across female reproductive states (also see Carnegie *et al.*, this volume). Our finding that alpha males are fathering the majority of infants within our study groups may indicate that females are choosing to mate with alpha males during their conceptive periods or that fertile females are being monopolized by alpha males. Although we have not observed active mate guarding within our groups, it is possible that a more subtle form of competition is occurring, and subordinate males may avoid mating with females when dominants are showing interest in them. In a later study of the same groups, Carnegie (2004; Carnegie *et al.*, this volume) found that alpha males were more likely to mate with cycling, periovulatory females, whereas subordinate males were more likely to mate with pregnant females. Whatever the explanation, it is obvious that subordinates are not being rewarded with reproductive opportunities, by either group females or alpha males, in exchange for their cooperation in group defense.

Given the benefits that subordinates provide to dominants in terms of group defense from the possibility of takeovers by outsiders, we predicted that this species would display low reproductive skew (i.e. paternity would be shared by group males). In contrast to our expectations, and despite their very egalitarian mating system, white-faced capuchins exhibit high reproductive skew (i.e. paternity is dominated by one or a few breeders). According to concession theory (e.g. Reeve and Ratnieks, 1993; Reeve *et al.*, 1998), high reproductive skew is predicted if (1) there are ecological constraints on dispersal (e.g. habitat saturation) or (2) dominants and subordinates are related (in which case both alphas and subordinates would receive inclusive fitness benefits). Our long-term studies of the white-faced capuchin population within SRNP have shown that the population has increased significantly in the past two decades; however, it has done so through increased group size rather than through an increase in the number of groups (Fedigan and Jack, 2001). There appears to be strong ecological constraints on the number of groups that the habitat can support, because during the 6-month dry season these monkeys become central-place foragers, focusing their activities around locally available water sources from which they drink on a daily basis. These water sources are very limited and the distribution of groups throughout the park is dictated by the location of these water sources and the number (and size) of groups sharing them. In addition, sex ratios at birth are skewed towards males (Fedigan, 2003), and adult sex ratios within groups of white-faced capuchins are nearly one to one (Fedigan and Jack, 2001); both factors potentially leading to increased male–male competition. Such ecological and demographic factors, either individually or combined, may necessitate that, after joining a group, males must bide their time at a low rank and tolerate low reproductive success. If this is the case, the "staying incentive" that dominants are providing their subordinates may merely be their tolerance of these males within the group.

Johnstone and Cant (1999) recently suggested that in some species, where dispersal patterns are largely influenced by ecological constraints, such as those described here, subordinates may keep their mating and reproduction to a minimum so as not to risk eviction from the group by dominants. However, we have yet to see a dominant male harass a subordinate male engaged in copulation or forcibly evict a subordinate co-resident male from the group. Indeed, when subordinates are away from the group, alpha males go to great lengths to locate them. In such cases, alpha males will emit lost-calls (specialized long-distance vocalizations used to locate individuals who are lost or away from the group). When subordinates are reunited with the group, the alpha male often initiates a reunion display to welcome them back and ease social tension (see Fedigan and Jack, 2004). An analysis of our long-term data on male dispersal patterns showed that subordinates are more likely to disperse from the group voluntarily (i.e. they are not aggressively evicted by dominant males), perhaps in response to an alpha's unwillingness to yield sufficient reproductive incentives (Jack and Fedigan, 2004c). These factors (alphas actively trying to keep subordinates in the group and subordinates voluntarily emigrating) provide strong evidence against the suggestion that subordinates keep their mating to a minimum to avoid eviction, and that ecological constraints on dispersal cause the high reproductive skew that we see in white-faced capuchins.

High-skew societies are also thought to occur when dominants and subordinates are related. White-faced capuchins are a male dispersed species, which generally implies that co-resident males are unrelated. However, in the absence of male philopatry, kinship among males can be maintained through parallel dispersal (Van Hooff, 2000). Parallel dispersal occurs when male siblings emigrate together, or when males preferentially disperse into groups containing familiar, previously dispersed, males. Male white-faced capuchins engage in parallel dispersal at very high rates, and this pattern of coordinated emigration and immigration remains high even in adulthood (Jack and Fedigan, 2004a,b). Moore (1992) suggests that natal individuals dispersing together and subsequently joining the same group, as has been observed among white-faced capuchins, may result in a level of relatedness within the new group that is comparable to that found in groups made up of philopatric individuals. Indeed, we have seen several cohorts of males reside in consecutive groups together, and we have also observed males reuniting with previously familiar males after more than 5 years of separation. Given these observations, it is possible that white-faced capuchins display high reproductive skew within groups because co-resident males are related. That is, even when they do not directly sire infants, subordinate males may achieve inclusive fitness benefits via the enhanced reproductive success of the related alpha males with whom they have cooperated. However, we await additional kinship analysis as further support for this interpretation.

SUMMARY

The relationship between male dominance rank and reproductive success has been a long-debated topic in primate behavior. While most early studies looked at mating success as a proxy measure of reproductive success, recent advances in using small quantities of DNA obtained from hair and feces have enabled paternity testing even in a field setting. In this study, we examine the relationship between male dominance rank and reproductive success in two groups of wild white-faced capuchin monkeys (*Cebus capucinus*) residing in Santa Rosa National Park between October 1993 and January 2000. A total of four alpha males and 12 subordinates resided within the two groups during the study period and 25 infants were born. Of these infants, only 19 were genotyped due to the deaths or disappearances of six infants prior to sampling. Paternity was determined using DNA extracted from non-invasively obtained hair and fecal samples and amplified using PCR. Using this method, we were able to perform complete exclusions on 15 of the 19 infants genotyped. Our analysis revealed that alpha males in the two study groups sired significantly more offspring than did subordinates; alphas sired a minimum of 13 infants and a maximum of 17, while subordinates sired between two and six of the infants born in our groups. This multimale–multifemale species displays an extremely egalitarian mating system and overt mate guarding by dominant males has not been observed. However, the data presented here indicate that there is a definite advantage to being an alpha male in this species. This finding may explain the high rate of male secondary dispersal and, in particular, the voluntary dispersal of subordinate males observed in our long-term study groups.

ACKNOWLEDGMENTS

We are grateful to the National Park Service of Costa Rica for allowing us to work in SRNP from 1983 to 1988 and to the administrators of the Area de Conservacion Guanacaste (especially Roger Blanco Segura, the Research Director) for permission to continue research in the park to the present day. We thank the many people who contributed to the demographic and dominance data presented in this chapter, especially Rodrigo Morera Avila, Lisa Rose, Katherine MacKinnon, Craig Lamarsh, Dale Morris, Sasha Gilmore and Sarah Carnegie. We are also grateful to Sara Phillips for editorial assistance. Katharine Jack gratefully acknowledges the financial support from the Natural Sciences and Engineering Research Council of Canada (NSERCC), the Royal Anthropological Institute, Sigma-Xi, the Faculty of Graduate Studies and Research/Department of Anthropology at the University of Alberta, the Alberta Heritage Scholarship Fund, the Killam Foundation, and the National Geographic Society. Linda Fedigan's research is funded by the Canada Research Chairs Program and by an on-going operating grant from NSERCC (#A7723).

REFERENCES

Altmann, J., Alberts, S., Haines, S. A., Dubach, J., Muruthi, P., Coole, T., Geffen, E., Cheesman, D. J., Mututua, R. S., Saiyalel, S. N., Wayne, R. K., Lacy R. C., and Bruford, M. W. 1996, Behavior predicts genetic structure in a wild primate group. *Proc. Natl. Acad. Sci.* 93:5797–5801.

Andersson, M. 1994, *Sexual Selection.* Princeton University Press, Princeton, New Jersey.

Berard, J. 1999, A four-year study of the association between male dominance rank, residency status, and reproductive activity in rhesus macaques (*Macaca mulatta*). *Primates* 40:159–175.

Borries, C., Launhardt, K., Epplen, C., Epplen, J. T., and Winkler, P. 1999, DNA analysis support the hypothesis that infanticide in adaptive in langur monkeys. *Proc. R. Soc. Lond. B* 266:901–904.

Carnegie, S. D. 2004, The relationship between ovarian hormones and behavior in white-faced capuchins, *Cebus capucinus.* MA Thesis, University of Alberta, Edmonton, Canada.

Chapman, C., Fedigan, L., and Fedigan, L. 1988, A comparison of transect methods of estimating population densities of Costa Rican primates. *Brenesia* 30:67–80.

Clutton-Brock, T. 1998, Reproductive skew, concessions and limited control. *Tree* 13:288–291.

Cowlishaw, G. and R. Dunbar. 1991, Dominance rank and mating success in male primates. *Anim. Behav.* 41:1045–1056.

Curie-Cohen, M., Yoshihara, D., Luttrell, L., Beneforado, K., MacCluer, J. W., and Stone., W. 1983, The effects of dominance on mating behavior and paternity in a captive troop of rhesus monkeys (*Macaca mulatta*). *Am. J. Primatol.* 5:127–138.

De Ruiter, J. R. and van Hoof, J. A. R. A. M. 1993, Male dominance rank and reproductive success in primate groups. *Primates* 34:513–523.

Ellis, L. 1995, Dominance and reproductive success among non-human animals: A cross-species comparison. *Ethol. Sociobiol.* 16:257–333.

Emlen, S. T. 1982, The evolution of helping II: The role of behavioral conflict. *Am. Nat.* 119:40–53.

Escobar-Parámo, P. 2000, Microsatellite primers for the wild brown capuchin monkey *Cebus apella. Mol. Ecol.* 9:107–108.

Fedigan, L. 1983, Dominance and reproductive success in primates. *Yearb. Phys. Anthropol.* 26:91–129.

Fedigan, L. 1986, Demographic trends in the *Alouatta palliata* and *Cebus capucinus* populations of Santa Rosa National Park, Costa Rica, in: J. G. Else and P. C. Lee, eds., *Primate Ecology and Conservation*, Cambridge University Press, New York, pp. 285–293

Fedigan, L. 1993, Sex differences and intersexual relations in adult white-faced capuchins (*Cebus capucinus*). *Int. J. Primatol.* 14:853–877.

Fedigan, L. 2003, Impact of male takeovers on infant deaths, births and conceptions in *Cebus capucinus* at Santa Rosa, Costa Rica. *Int. J. Primatol.* 24:723–741.

Fedigan, L., Fedigan, L., and Chapman, C. 1985, A census of *Alouatta palliata* and *Cebus capucinus* monkeys in Santa Rosa National Park, Costa Rica. *Brenesia* 23: 309–322.

Fedigan, L. and Jack, K. 2001, Neotropical primates in a regenerating Costa Rican dry forest: A comparison of howler and capuchin population patterns. *Int. J. Primatol.* 22:689–713.

Fedigan L. and Jack, K. 2004. The demographic and reproductive context of male replacements in Cebus capucinus: How, when and why do male white-faced capuchins take over groups? *Behaviour* 141:755–775

Fedigan, L. and Rose, L. M. 1995, Interbirth interval variation in three sympatric species of Neotropical monkey. *Am. J. Primatol.* 3:9–24.

Fedigan, L., Rose, L., and Avila, R. 1996, See how they grow: Tracking capuchin monkey (*Cebus capucinus*) populations in a regenerating Costa Rican dry forest, in: M. A. Norconk, A. L. Rosenberger, and P. A. Garber, eds., *Adaptive Radiations of Neotropical Primates*, Plenum Press, New York, pp. 289–307.

Fragaszy, D., Vesalberghi, E., and Fedigan, L. 2004, *The Complete Capuchin: The Biology of the Genus Cebus.* Cambridge University Press, Cambridge.

Freese, C. H. and Oppenheimer, J. R. 1981, The capuchin monkeys, genus *Cebus*, in: A. F. Coimbra-Filho and R. A. Mittermeier, eds., *Ecology and Behavior of the Neotropical Primates*, Academica Brasileira de Ciencias, Rio de Janeiro, pp. 331–390.

Gerloff, U., Schloetterer, C., Rassmann, K., Rambold, I., Hohmann, G., Fruth, B., and Tautz, D. 1995, Amplification of hypervariable simple sequence repeats (microsatellites) from excremental DNA of wild living bonobos (*Pan paniscus*). *Mol. Ecol.* 4:515–518.

Inoue, M., Mitsunaga, F., Nozaki, M., Ohsawa, H., Takenaka, A., Sugiyama, Y., Shimuzu, K., and Takenaka, O. 1993, Male dominance rank and reproductive success in an enclosed group of Japanese macaques: With special reference to post-conception mating. *Primates* 34:503–511.

Jack, K. 2001a, The effect of male emigration on the vigilance behavior of coresident males in white-faced capuchins (*Cebus capucinus*). *Int. J. Primatol.* 22:715–732.

Jack, K. 2001b, Life history patterns of male white-faced capuchins: Male bonding and the evolution of multimale groups. PhD Thesis, University of Alberta, Edmonton, Canada.

Jack, K. 2003, Affiliative relationships among male white-faced capuchins (*Cebus capucinus*): Evidence of male bonding in a female bonded species. *Folia Primatol.* 74:1–16.

Jack, K. and Fedigan, L. M. 2004a, Male dispersal patterns in white-faced capuchins (*Cebus capucinus*) Part 1: Patterns and causes of natal emigration. *Anim. Behav.* 67(4):761–769.

Jack, K. and Fedigan, L. M. 2004b, Male dispersal patterns in white-faced capuchins (*Cebus capucinus*) Part 2: Patterns and causes of secondary dispersal. *Anim. Behav.* 67(4):771–782.

Jack, K. and Fedigan, L. M. 2004c, How are male dispersal patterns, dominance rank and reproductive success related in wild white-faced capuchins (*Cebus capucinus*) in Santa Rosa National Park, Costa Rica? *Am. J. Primatol.* 62:88.

Johnstone, R. A. and Cant, M. A. 1999, Reproductive skew and the threat of eviction: A new perspective. *Proc. R. Soc. Lond. B* 266:275–279.

Keller, L. and Reeve, H. 1994, Partitioning of reproduction in animal societies. *Trends Ecol. Evol.* 9:98–102.

Manson, J. H., Perry, S., and Parrish, A. R. 1997, Nonconceptive sexual behavior in bonobos and capuchins. *Int. J. Primatol.* 18:767–786.

Moore, J. J. 1992, Dispersal, nepotism, and primate social behavior. *Int. J. Primatol.* 13:361–378.

Paul, A. 1997, Breeding seasonality affects the association between dominance and reproductive success in non-human primates. *Folia Primatol.* 68:344–349.

Paul, A. 2002, Sexual selection and mate choice. *Int. J. Primatol.* 23:877–904.

Perry, S. 1995, Social relationships in wild white-faced capuchin monkeys (*Cebus capucinus*). PhD Thesis, University of Michigan, Ann Arbor, Michigan.

Perry, S. 1998a, A case report of a male rank reversal in a group of wild white-faced capuchins (*Cebus capucinus*). *Primates* 39:51–70.

Perry, S. 1998b, Male–male social relationships in wild white-faced capuchins, *Cebus capucinus. Behaviour* 135:139–172.

Preuschoft, S. and Paul, A. 2000, Dominance, egalitarianism, and stalemate: An experimental approach to male–male competition in Barbary macaques, in: P. M. Kappeler, ed., *Primate Males: Causes and Consequences of Variation in Group Composition*, Cambridge University Press, Cambridge, pp. 205–216.

Reeve, H. K. and Emlen, S. T. 2000, Reproductive skew and group size: An N-person staying incentive model. *Int. Soc. Behav. Ecol.* 11:640–647.

Reeve, H. K., Emlen, S. T., and Keller, L. 1998, Reproductive sharing in animal societies: Reproductive incentives or incomplete control by dominant breeders? *Behav. Ecol.* 9:267–278.

Reeve, H. K. and Ratnieks, F. L. W. 1993, Queen-queen conflict in polygynous societies: Mutual tolerance and reproductive skew, in: L. Keller, ed., *Queen Number and Sociality in Insects*, Oxford University Press, Oxford, pp. 45–85.

Robinson, J. G. and Janson, C. H. 1987, Capuchins, squirrel monkeys, and atelines: Sociological convergence with Old World primates, in: B. B. Smuts, D. L. Cheney, R. M.

Seyfarth, R. W. Wrangham, and T. T. Struhsaker, eds., *Primate Societies*, University of Chicago Press, Chicago, pp. 69–82.

Rose, L. M. 1998, Behavioral ecology of white-faced *capuchins* (*Cebus capucinus*) in Costa Rica. PhD Dissertation, Washington University-St. Louis MI, University Microfilms International.

Rose, L. M. and Fedigan, L. M. 1995, Vigilance in white-faced capuchins, *Cebus capucinus*, in Costa Rica. *Anim. Behav.* 49:63–70.

Smith, D. G. 1981, The association between rank and reproductive success of male rhesus monkeys. *Am. J. Primatol.* 1:83–90.

Soltis, J., Thomsen, R., and Takenaka, O. 2001, The interaction of male and female reproductive strategies and paternity in wild Japanese macaques, *Macaca fuscata*. *Anim. Behav.* 62:485–494.

Van Hooff, J. A. R. A. M. 2000, Relationships among non-human primate males: A deductive framework, in: P. M. Kappeler, ed., *Primate Males: Causes and Consequences of Variation in Group Composition*, Cambridge University Press, Cambridge, pp.183–191.

Vehrencamp, S. L. 1979, The roles of individual, kin, and group selection in the evolution of sociality, in: P. Marler and J. G. Vandenberg, eds., *Handbook of Behavioral Neurobiology: Social Behavior and Communication*, Plenum Press, New York, pp. 351–394.

Walters, J. R. and Seyfarth, R. M. 1987, Conflict and cooperation, in: B. B. Smuts, D. L. Cheney, R. M. Seyfarth, R. W. Wrangham, and T. T. Struhsaker, eds., *Primate Societies*, University of Chicago Press, Chicago, pp. 306–317.

Witte, S. and Rogers, J. 1999, Microsatellite polymorphisms in Bolivian squirrel monkeys (*Saimiri boliviensis*). *Am. J. Primatol.* 47:75–84.

CHAPTER SIXTEEN

Post-conceptive Mating in White-Faced Capuchins, *Cebus capucinus*: Hormonal and Sociosexual Patterns of Cycling, Noncycling, and Pregnant Females

Sarah D. Carnegie, Linda M. Fedigan, and Toni E. Ziegler

Sarah D. Carnegie • Department of Anthropology, University of Alberta, Edmonton, Alberta, Canada. **Linda M. Fedigan** • Department of Anthropology, University of Calgary, Calgary, Alberta, Canada. **Toni E. Ziegler** • National Primate Research Center, University of Wisconsin, Madison, Wisconsin.

New Perspectives in the Study of Mesoamerican Primates: Distribution, Ecology, Behavior, and Conservation, edited by Alejandro Estrada, Paul A. Garber, Mary S. M. Pavelka, and LeAndra Luecke. Springer, New York, 2005.

INTRODUCTION

In many species of nonhuman primates, sociosexual behaviors are known to vary across a female's reproductive cycles and thus numerous studies have focused on the relationship between behavior and hormonal fluctuations during the ovarian cycle (e.g., *Macaca fuscata*: Enomoto *et al.*, 1979; *Papio ursinus*: Saayman, 1970; *Erythrocebus patas*: Loy, 1981; *Cebus apella*: Carosi *et al.*, 1999; *Brachyteles arachnoides*: Strier and Ziegler, 1997). Baum (1983) and Beach (1976) have argued that ovarian hormones (i.e., estradiol and progesterone) correlate with aspects of sociosexual behavior such that high levels of estrogen are associated with an increase in proceptive behaviors and female attractiveness, and high levels of progesterone are associated with reduced female proceptivity and female attractiveness. The recent development of non-invasive fecal collection and hormone extraction techniques that can be successfully used in the field have resulted in a growing number of studies that relate hormones to behavior in free ranging primates (*Propithecus verreauxi*: Brockman *et al.*, 1995; *Brachyteles arachnoides*: Strier and Ziegler, 1997; *Macaca nemestrina*: Risler *et al.*, 1987; *Papio cynocephalus*: Wasser *et al.*, 1991).

After reaching sexual maturity, a female primate repeatedly experiences three different reproductive states: ovarian cycling, pregnancy, and anovulation. The period of time when the female experiences fluctuating ovarian hormone levels (i.e., estrogen and progesterone) is known as the "cycling" state. It is during this time that ovulation occurs and a female is able to conceive an infant. The second state, pregnancy, is when the female is carrying a developing fetus. During this period, ovarian hormones remain at elevated levels and help to maintain the fetus throughout gestation. The third state, "anovulation", occurs after parturition in the early phase of lactation. One physiological response of the female primate's body to lactation is the suppression of ovarian function. During anovulation, the hormone fluctuations normally experienced during the "cycling" state do not occur and therefore ovulation and conception cannot occur (Dixson, 1998). Late in lactation, or after the infant is weaned, normal ovarian cycling resumes and the female can conceive another infant. For the purposes of this study, we will refer to females who are not pregnant and not experiencing hormonal cycles as "noncycling." Only a few primate studies have compared the behavioral changes across the different reproductive states (e.g., *Cercocebus torquatus atys*: Gordon *et al.*, 1991; Gust, 1994; *Pan troglodytes*: Wallis, 1982). Our study uses hormone profiles, created from the analysis of fecal ovarian steroids, to determine the reproductive status of wild female white-faced capuchins, and then examines sexual and affiliative behaviors that are exhibited across the three reproductive states. We have three objectives: (1) to profile the ovarian steroid pattern of female capuchins in different reproductive states, (2) to determine the behavioral indicators from which one could reliably infer the reproductive status of wild female capuchins in the absence of hormonal data, and (3) to explore the reproductive strategies used by wild female capuchins.

Hrdy (1974, 1979) suggested that mating during nonconceptive periods (i.e., when females are either pregnant or anovulatory due to lactation) is a female behavioral counter-strategy to reduce the threat of infanticide by invading males (also see Agrell *et al.*, 1998). Mating during pregnancy (also referred to as "post-conceptive" mating) has been reported in a number of primate species who also experience infanticide (*Cercocebus torquatus atys*: Gordon *et al.*, 1991; *Papio hamadryas*: Zinner and Deschner, 2000; *Pan troglodytes*: Wallis, 1982; also *Cebus capucinus*: Manson *et al.*, 1997). Females of some of these species even go so far as to exhibit "false conceptive signals" that confuse males as to their true reproductive state. For instance, female sooty mangabeys (*Cercocebus torquatus atys*) and hamadryas baboons (*Papio hamadryas*) have been found to display post-conceptive swellings that mimic the swellings experienced during their conceptive periods. However, it was the subordinate sooty mangabey males who mated with the post-conceptive females, while the alpha males mated only with females displaying true conceptive swellings. Thus, the authors inferred that these alpha males were able to discriminate between conceptive and post-conceptive swellings, but they were unable to explain what cues this discrimination was based upon, and why subordinate males chose to mate with pregnant females (Gust, 1994).

Manson *et al.* (1997) examined nonconceptive (pregnant or lactating) mating behavior in white-faced capuchins and found that the pregnant females mated more frequently than the potentially conceptive (cycling) females. In their study, reproductive state was inferred from behavioral observations rather than determined from hormones. The authors concluded that post-conceptive mating in this species functions to confuse paternity among males and reduce the risk of infanticide. To support this theory, Fedigan (2003) and Fedigan and Jack (2004) found that the immigration of new males into white-faced capuchin groups usually occurs through aggressive group takeovers and that these takeovers coincide with the deaths and disappearances of infants.

To the observer, white-faced capuchin females appear inconspicuous in displaying any behavioral or morphological cues that indicate their conceptive phase (i.e., they seem to conceal their ovulation), but adult male capuchin behavior suggests they may be able to recognize this phase in cycling females (Carnegie *et al.*, in press). However, for post-conceptive mating to act as a successful counter-strategy to infanticide, males should not be able to detect the ovulatory phase, and they also should not associate the timing of mating with the timing of births (Zinner and Deschner, 2000). If post-conceptive mating in female white-faced capuchins is a reproductive-strategy, we would expect to see both cycling and pregnant females responding positively to male sexual solicitations by mating with them.

METHODS

Study Site and Species

Our study took place in Santa Rosa National Park (SNRP), Costa Rica (the original sector of the Area de Conservacion Guanacaste, ACG), between the months of January and June, 2002. The park encompasses about 108 km^2 of dry, deciduous forest which experiences a distinct dry season (mid-December to mid-May) and wet season (late May to early December; average annual rainfall is 1473 mm; Fedigan and Jack, 2001). Data for this project were collected during the dry season because most trees lose their leaves and the increased visibility makes it easier to follow individuals and collect fecal samples. The white-faced capuchins (*C. capucinus*) in SRNP live sympatrically with two other species of nonhuman primates; *Alouatta palliata* (mantled howling monkey) and *Ateles geoffroyi* (black handed spider monkey).

White-faced capuchins live in multimale/multifemale social groups that consist of related females, immigrant males, and immature offspring. White-faced capuchins engage in sexual behavior (copulations and solicitations) throughout the year but we have found a birth peak between January and April (Fedigan, 2003). Females become reproductively mature (i.e., start ovarian cycling) around the age of 5 years and the first age of first birth in this population is between 6 and 7 years. Females do not show any conspicuous signs of ovulation (i.e., no morphological changes and overt behavioral changes; Carnegie *et al.*, in press). There is a discernable linear hierarchy within a group of capuchins and

Table 1. Group, age, rank, reproductive state, and history for each female in 2002

Female	Group	Age[a]	Rank	Reproductive state	Parity	Age of last infant[b]
Limp	CP	24	Alpha	Cycling	Multiparous	12
Kathy Lee	LV	13	Alpha	Cycling	Multiparous	12
Blanquita	LV	20	Subordinate	Cycling	Multiparous	9
Timone	CP	6	Subordinate	Cycling	Nulliparous	0
Nyla	CP	8	High rank	Noncycle	Multiparous	12
Seria	CP	13	Subordinate	Noncycle	Multiparous	7
Fiesty	LV	20	Subordinate	Noncycle	Multiparous	9
Pumba	CP	8	Subordinate	Pregnant	Multiparous	24
Dos Leches	LV	11	High rank	Pregnant	Multiparous	12[c]
Salsa	LV	6	Subordinate	Pregnant	Nulliparous[d]	0

[a] Age of female in years.
[b] Age of infant in months from parturition to January 2002.
[c] This female lost her last infant (disappeared) less than 1 month after it was born resulting in a shorter interbirth interval.
[d] This female gave birth to her first infant during the study (May 2002).

within the sexes (Perry, 1997). One male is usually dominant over the others but not to the point of exclusive access to females, as is commonly seen in *C. apella* or *C. olivaceus* (Janson, 1984; Robinson, 1988; Fragaszy *et al.*, 2004).

The subjects for this study consisted of 10 wild adult female white-faced capuchins that were part of two habituated groups; Cerco de Piedra and Los Valles (5 per group). Females ranged in age from 6 to ~24 years. Additionally, there were six adult males: two alpha males and four subordinate males. Individuals were identified by natural markings such as broken digits, scars, hair coloring, and the brow and peak shape. This population of capuchins has been under study by LMF's research team since 1983. Capuchins are female bonded (Fragaszy *et al.*, 2004), and kinship relations among females are known for the last 15–20 years. Table 1 lists information on the group, rank, and reproductive history for each of the 10 females.

Behavioral and Fecal Data Collection and Analysis

Between January and June, 2002, we collected 443 h of focal animal data (ranging from 39 to 47 h per female) between 6 am and 6 pm. Focal sessions lasted for 15 min each, during which time all behaviors and interactions were recorded continuously (Altmann, 1974). We used hand-held computers (PSION

Workabout MX) and entered data into a software program entitled "Behavior", which was designed by Syscan International Inc. (Montreal, Quebec) for our use. We employed an exhaustive ethogram developed over the years by previous *C. capucinus* researchers to identify and code behaviors.

We collected fecal samples from each focal female on every day that behavioral collection occurred. A minimum of two to three samples per week per female were enough to assess ovarian patterns in females but additional samples were collected in case females could not be located in subsequent days (Hodges and Heistermann, 2003). After the pregnant females gave birth, we collected only one sample per week per female which was sufficient to monitor post-partum ovarian function. In total, we collected 600 samples from the 10 subject females. Fecal collection and initial extraction of the steroids in the field followed the techniques described in Strier and Ziegler (1997) and Strier *et al.* (2003). The samples were refrigerated until transported to the National Primate Research Center (NPRC, University of Wisconsin) in Madison for the laboratory analysis.

We conducted the hormone assays at the NPRC (University of Wisconsin) during July and August 2002. Before these analyses, the third author validated 10 previously collected fecal samples in January 2001 for estradiol and progesterone and the enzyme-immunoassays and radio-immunoassays that were used to assess those hormones, respectively. Before assays were performed, it was necessary to perform solvolysis to break conjugated steroids into unconjugated forms. This procedure has been previously described in Ziegler *et al.* (1996) and Strier *et al.* (1999).

We used the results of the fecal assays to create hormone profiles for each adult female and from the profiles we were able to categorize females as being in one of the three reproductive states: cycling, noncycling, or pregnancy. We identified noncycling females by nonfluctuating and sustained baseline levels of estradiol and progesterone. The gestational period of pregnant females was identified by elevated levels of both steroid hormones, which drop drastically to baseline levels after parturition. Cycling females were identified by fluctuating levels of estradiol and progesterone which are representative of the periovulatory and non-ovulatory phases of the ovarian cycle (Dixson, 1998; Carnegie *et al.*, submitted).

For unknown reasons, four of the females whom we classified as "cycling" actually stopped cycling mid-way through the study (e.g., they did not become pregnant). Therefore, for this analysis, we used only behavioral data collected from them while they were cycling (cycling: 55.25 h). We analyzed behavioral data collected from the pregnant females ($N = 3$) only while they were pregnant and not after they gave birth (68.75 h). For consistency, we analyzed behavioral data collected from the noncycling females ($N = 3$) between January 15 and April 30 (68.5 h). We choose the later date because most of the cycling females had stopped cycling at this time and there was only one pregnant female still to give birth (her parturition date was May 13).

The behavioral variables analyzed for reproductive state variation were: urine washing (rubbing urine into the hands and feet), copulations and courtship displays (sexual behavior), and the hypothesized behavioral indicators of attractivity, proceptivity, and receptivity. For proceptivity indicators, we used grooming (frequency of groom solicits and frequency and duration of grooming bouts) *directed* by the subject female to adult males. We considered the same behaviors as indicators of attractivity when they were *received* from adult males. We measured receptivity as a percentage of male courtship displays to which females responded positively by presenting for mounting and facilitating copulations.

We compared each of the behavioral frequencies across each category of females (noncycling, cycling, and pregnant) by calculating a mean rate (frequency per hour) for each of the subject females for each reproductive state. We determined mean grooming duration by tallying the total amount of time females spent in grooming bouts and calculating the proportion of that time they spent grooming (or spent being groomed by) adult males.

We used Hinde's Index (Hinde and Atkinson, 1970) to decide which member of the male–female dyad was responsible for maintaining proximity. Hinde's Index is the proportion of all of the dyad's *approaches* directed by the subject female, minus the proportion of all of the dyad's *leaves* directed by the female. A negative index indicates that the male was responsible for maintaining proximity; a positive index suggests that the female was responsible. *Approaches* to within 3 m and *leaves* beyond 3 m were used in this analysis.

We used Kruskall–Wallis one-way analysis of variance (nonparametric test) to compare behavioral frequencies among reproductive states (Siegel and Castellan, 1988; Zar, 1999). All frequencies showing a significant difference were further analyzed using a multiple comparison test to determine where the difference existed among the three categories of females (Siegel and Castellan, 1988).

RESULTS

Hormone Validations and Profiles

Mean steroid recoveries were within acceptable values for progesterone (90.7%) and estradiol (64.5%; recoveries should normally be greater than 75%, but E2 recoveries were consistent over all eight assays so this value was accepted). Pooled samples compared for accuracy and parallelism to the standard curve were high for both progesterone (P) and estradiol (E2) assays (>100% accuracy; accepted values are between 80% and 120%; slopes did not differ statistically). Additionally, mean intra-assay coefficient of variations (CV) were 6.89% for P and 4.73% for E2 (acceptable values are less than 10%). Inter-assay CV values were 8.56% for P and 8.20% for E2 (acceptable values are less than 20%).

The hormone profiles created from the fecal analysis clearly revealed that three females were noncycling for the duration of the study, three females were pregnant at the beginning of the study (and subsequently gave birth during the study), and four females displayed regular cycling for a period of time and then stopped mid-way through the study period without becoming pregnant (cycling females). Figures 1a, 1b, and 1c illustrate representative hormone profiles for noncycling, cycling, and pregnant females, respectively.

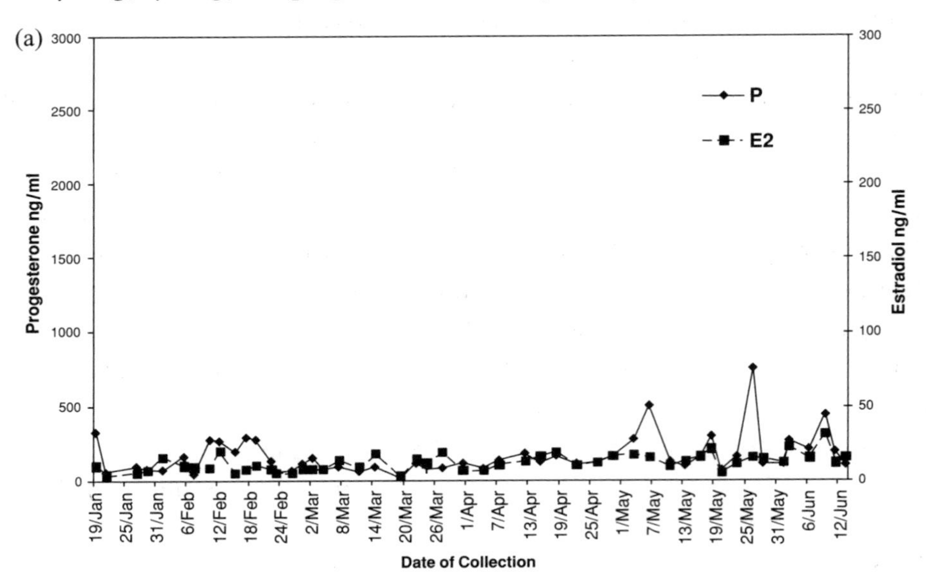

Figure 1. (a) Representative hormone profile for one noncycling female white-faced capuchin. Estradiol (E2) and progesterone (P) remained near baseline levels for the entire duration of the study; (*Continued*)

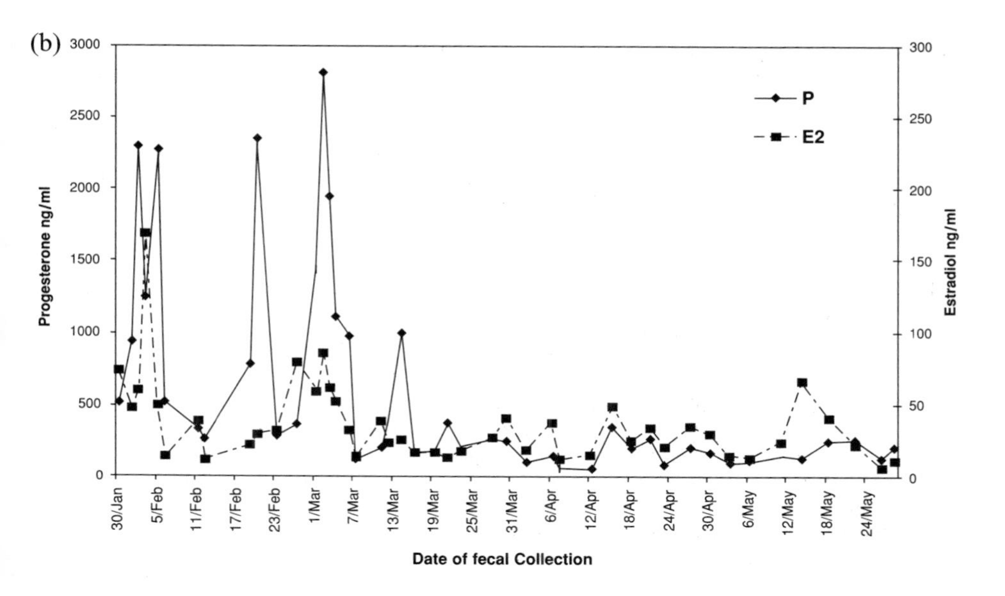

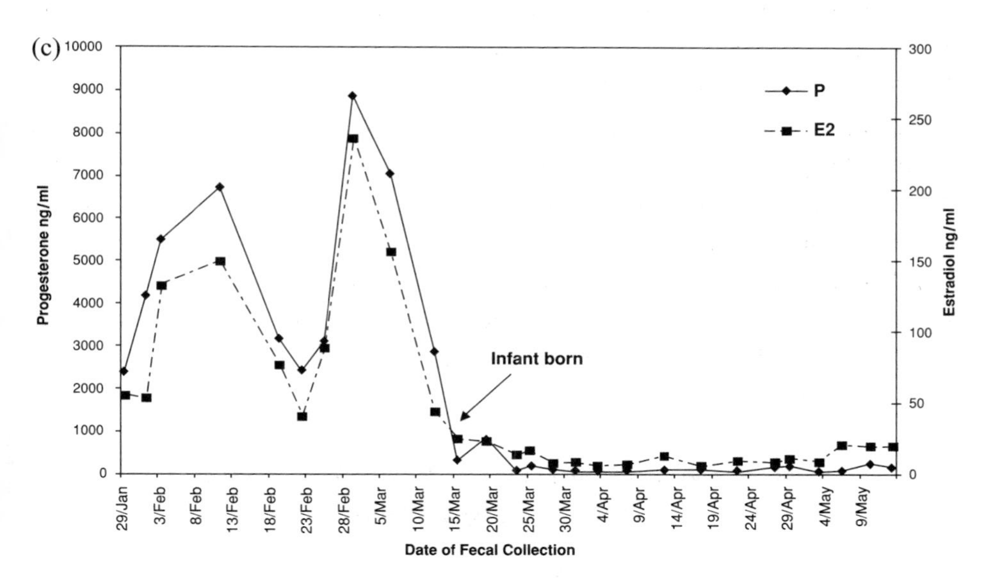

Figure 1. (*Continued*) (b) representative hormone profile for one cycling female white-faced capuchin. This female appeared to have already started cycling when observations commenced and displayed two cycles before she stopped in early March; and (c) representative hormone profile for one pregnant female white-faced capuchin. Pregnancy is represented by elevated progesterone (P) and estradiol (E2) which drop to baseline levels 1–2 days after parturition. The date of parturition is labeled (March 14).

Table 2. Summary of the ovarian hormone patterns for the cycling females

Female	Group	Length of complete cycle 1/ cycle 2 (days)[a]	Length of total cycling period (days)[b]	Date of cycling cessation	Length of noncycling period (days)[c]	Length of time since last infant born (months)[d]
TI	CP	25/–	47	March 19	86	0
LI	CP	22/–	41	April 12	62	12
KL	LV	26/–	48	March 16	73	12
BL	LV	14/13	38	April 18	40	9

[a] A complete cycle was calculated from one progesterone surge to the next.
[b] The length of the cycling period was calculated from the first day of the P surge to the end of the cycling when *P* levels decreased and did not rise again.
[c] The length of the noncycling period was calculated from the date of cycling cessation to the date the last sample was collected.
[d] This population of capuchins have a 2.5-year interbirth interval (Fedigan, 2003); gestation lengths are approximately 5.5 months (Nagle and Denari, 1982).

Of the four cycling females, three did not show cycling until 14, 41, and 43 days after their first collected sample. Our observations on the fourth female appeared to start at the beginning of her luteal phase. This phase is characterized by elevated levels of E2 and P that are produced by the *corpus luteum*, which is formed from the follicle after ovulation. The onset of cycling was indicated by the abrupt rise in the concentrations of progesterone and estradiol from near baseline levels. Regular cycling for all females ceased when both steroid hormones dropped to near baseline and did not elevate again to the previous (ovulatory) levels. We calculated between one and three regular cycles for each female during observations that extended over an average of 43.5 days (range: 40–86 days). Table 2 summarizes the ovarian hormone patterns for each of the cycling females.

Behavioral Variation Among Noncycling, Cycling, and Pregnant Females

Sexual Behavior

We found significant differences in the rates of copulations and courtship displays across the three reproductive states (copulations: $\chi^2 = 7.52$, df $= 2$, and $p = 0.023$; courtship displays: $\chi^2 = 7.891$, df $= 2$, and $p = 0.019$). Multiple comparison tests revealed that pregnant females copulated and received more courtship displays from all adult males than did cycling females (Figure 2).

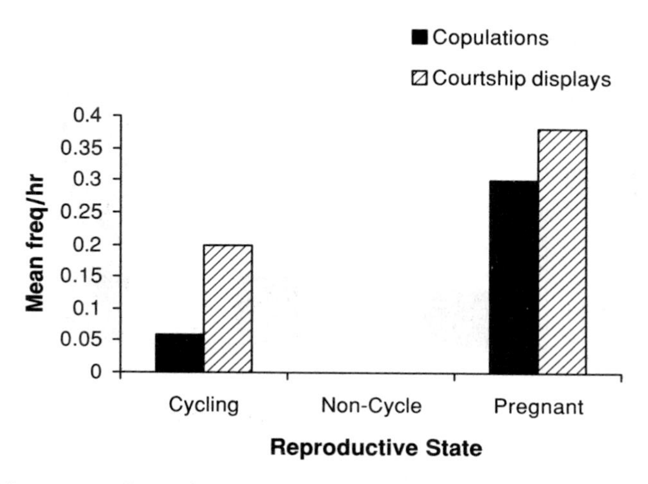

Figure 2. Mean rate of copulations with adult males and the mean rate of courtship displays (sexual solicitations) that cycling, noncycling, and pregnant females received from adult males.

Noncycling females were never observed to copulate with, or receive a courtship display from, an adult male.

In total, there were 22 copulations observed and 19 of these involved pregnant females mating with subordinate males. There were 32 courtship displays observed and 21 of these involved pregnant females—19 of which came from subordinate males. Additionally, pregnant females responded positively to 100% of the courtship displays they received from subordinate males (19 displays) by presenting for and facilitating copulation, whereas they never responded positively to any of the alpha males' solicitations (three displays). Furthermore, one and two days, respectively, *after* giving birth, two of the "pregnant" females copulated once each with a subordinate male.

The remaining three copulations occurred between cycling females and adult males two of which were with alpha males. Cycling females received the remaining 11 courtship displays and 8 of these were from the alpha males (73%). Cycling females responded positively to only 3 of the 11 courtship displays directed to them (27%).

Proceptivity

The rate of *grooming bouts directed to subordinate males* by subject females varied significantly across states ($\chi^2 = 6.85$, df $= 2$, and $p = 0.033$). Multiple

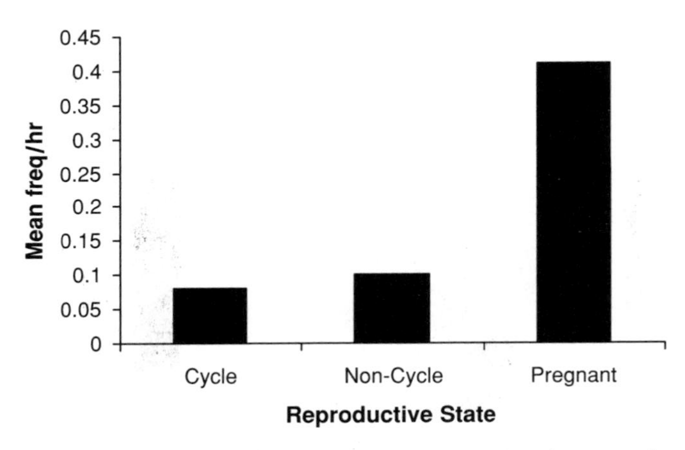

Figure 3. Mean rate of grooming bouts directed to subordinate males by cycling, noncycling, and pregnant females.

comparison tests revealed that pregnant females directed grooming bouts to subordinate males at a significantly higher rate (0.41/h) than did cycling (0.08/h) and noncycling females (0.10/h) (Figure 3).

Attractivity

The rates of *grooming bouts* and *groom solicits received from alpha males* by subject females varied significantly across reproductive states (grooming bouts; $\chi^2 = 7.63$, df $= 2$, $p = 0.02$; groom solicits: $\chi^2 = 6.82$, df $= 2$, $p = 0.032$). Multiple comparison tests show that cycling females received grooming bouts from alpha males at a significantly higher rate (0.13/h) than did either the pregnant (0.02/h) or the noncycling females (none) (Figure 4a). Cycling females received solicitations to be groomed from alpha males at a higher rate (0.22/h) than did the noncycling females (0.01/h) but the rate was comparable to pregnant females (0.15/h) (Figure 4b).

Proximity

Hinde's index was calculated to determine which sex was responsible for maintaining proximity in the male–female dyads (Table 3). In the "cycling female" category, only 10 out of a possible 12 dyads could be calculated as 1 male was never within proximity to subject females. All four dyads involving the alpha males were negative, which implies the alpha males were responsible for

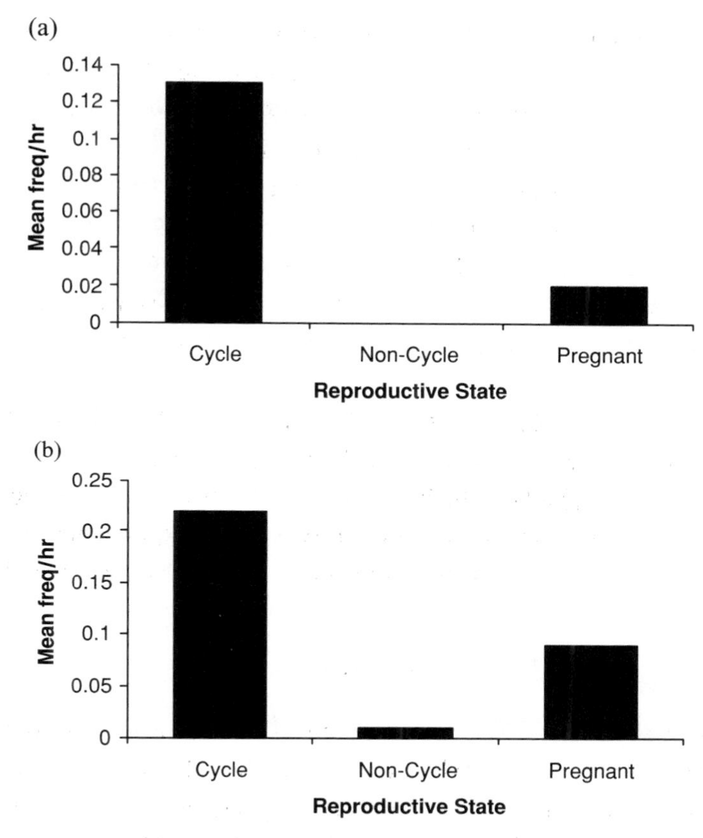

Figure 4. (a) Mean rate of grooming bouts that cycling, noncycling, and pregnant females received from alpha males and (b) mean rate of groom solicits that cycling, noncycling, and pregnant females received from alpha males.

Table 3. Percentage of dyads for which either the male or the female maintained proximity in each reproductive state. Values calculated using Hinde's Index

	Cycling	Noncycling	Pregnant
Male maintained			
Alpha	100	0	67
Subordinate	67	100	80
Female maintained			
Alpha	0	100	33
Subordinate	0	0	0
Neutral	33	0	20

maintaining proximity 100% of the time with cycling females. Four of the six dyads involving the subordinate males were negative (67%) and the two remaining dyads were neutral (33%), which implies that there was no difference between the sexes in who maintained the proximity.

In the pregnant female category, 8 out of a possible 10 dyads could be calculated. Two of the three dyads involving the alpha males were negative (67%) and one was positive (33%), which implies the pregnant female was responsible for maintaining proximity in that dyad. Of the five dyads involving the subordinate males, four were negative (80%), and the remaining dyad was neutral (20%).

Among noncycling females, 8 out of the possible 10 dyads could be calculated. In contrast to the other females, all three of the dyads involving the alpha males were positive (100%) which implies that the noncycling females were entirely responsible for maintaining proximity with the alpha males. In contrast, all five of the dyads involving the subordinate males were negative (100%).

Urine Washing

We found a significant difference in the rate of urine washing between cycling, noncycling, and pregnant females ($\chi^2 = 6.56$, df $= 2$, and $p = 0.03$). Cycling females performed urine washes at a significantly higher rate (3.12/h) compared to both pregnant (1.54/h) and noncycling females (1.67/h) (Figure 5).

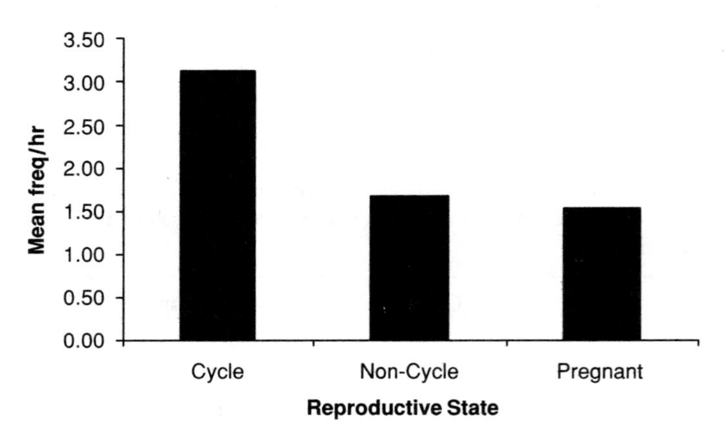

Figure 5. Mean rate of urine washing displayed by cycling, noncycling, and pregnant females.

DISCUSSION

Fecal Hormone Analysis and Interpretation of Hormone Profiles

The results from the hormone assays revealed that the collection and extraction techniques used in the field and the laboratory were successful methods for studying ovarian steroids in white-faced capuchin feces. Noncycling, cycling, and pregnant females were easily recognized from the hormone profiles created from the fecal analysis.

The three noncycling females were identified by their sustained baseline levels of estradiol and progesterone that lasted throughout the study. We could confirm that the pregnant females were in fact pregnant by their hormone profiles (characteristic elevated levels of ovarian hormones during gestation, which dropped to baseline after parturition), their expanding bellies, and the fact that they all gave birth during the study. We were not able to calculate gestation lengths as all three females were already pregnant when we began in January 2002. However, we assume that white-faced capuchins have gestation lengths similar to *C. apella*, of 22–23 weeks (Nagle and Denari, 1982), therefore, these females probably conceived between October and December of 2001.

The four cycling females stopped cycling between March and April (2002) after each experiencing one to three cycles, even though pregnancy had not been achieved. After the cycles stopped, hormone levels stayed low and fluctuated slightly but never reached or sustained baseline levels compared to the noncycling females (see Figure 1b). It is unclear why these four females stopped cycling and did not get pregnant. The following year (2003), all four of these females gave birth between the months of February and May; therefore, they must have started to cycle again and conceived between September and December, 2002 (assuming a 5.5-month gestation period; Nagle and Denari, 1982). Further studies on this cessation and resumption of cycling are presently being conducted.

Behavioral Indicators of Reproductive State

Our original prediction that cycling and pregnant females would be involved in sexual and social behaviors at higher rates than noncycling females is supported by our findings in this study. In fact, one of our most interesting findings is that pregnant females mated more often than any of the other females and they did so almost always with subordinate males. Furthermore, the cycling females mated only with the alpha males, and did not mate with any subordinate males, even though they were solicited by them. Pregnant females received significantly more sexual solicitations from subordinate males and were over three times more likely to respond positively to solicitations from males compared to the cycling females. Moreover, alpha males directed courtship displays toward cycling females at three times the rate they directed them to pregnant females.

We also found that proceptive and attractive behaviors involved cycling and pregnant females at much higher rates compared to the noncycling females. Specifically, the rate of grooming bouts that pregnant females directed toward subordinate males was much higher than grooming directed by cycling or noncycling females. Additionally, cycling females received groom solicits and grooming bouts from alpha males at the highest rate. We found that alpha males maintained proximity to cycling females, and subordinate males maintained proximity to pregnant females more than to any other category of female (i.e., cycling or noncycling). In contrast, we also found that noncycling females were responsible for maintaining proximity to the alpha males in all cases.

One exception to our predictions was that noncycling females maintained a close proximity to the alpha males (suggesting proceptivity). Among the three noncycling females, only one was very high ranking and the other two were low ranking and often remained on the outskirts of the group; therefore, it is possible that the two females were trying to maintain a close proximity to the alpha males in an effort to associate with high ranking individuals. By staying close to the alpha male, and therefore more central within the group, they might have had better access to resources or benefit from greater group protection in case of an inter-group encounter or predator attack.

Reproductive Strategies in White-Faced Capuchins

In light of our findings on sexual behavior in this species, two questions are raised concerning male and female mating strategies: (1) why do subordinate males court pregnant females more often than they do cycling females? and (2) why did pregnant females mate with subordinate males as often as they did, and not with the alpha males, even after they were solicited by the latter?

One explanation for mating behavior between pregnant females and subordinate males may be related to the elevated levels of stimulating estrogen that pregnant females produce during gestation; in that high estrogen levels may influence males to be more attracted to pregnant females and females to be more proceptive to males (Beach, 1976; Baum, 1983). However, if this were the case, we would expect the alpha males to be just as attracted to the pregnant females as were the subordinate males.

Gordon *et al.* (1991) and Gust (1994) found that alpha male sooty mangabeys (*Cercocebus torquatus atys*) discriminate between a female's fertile and nonfertile or post-conceptive swellings. The researchers found that alpha males only mated with females during their fertile swellings and never mated with pregnant females, whereas subordinate males mated with pregnant females during their nonfertile swelling phases. They found no difference in the concentrations of estradiol or progesterone between the two types of swellings, but the concentration of lutenizing hormone (LH) was significantly higher in the fertile females. The implication is that alpha males may be sensitive to the olfactory cues of the differing LH concentrations between the two groups of females, which helps them to discriminate between a conceptive and nonconceptive female. There is some evidence to support the notion that LH stimulates sexual behavior in male rats and male primates; however, it is inconclusive and more research is still needed (see Dixson, 1998).

Urine washing occurs in many primate species including squirrel monkeys, woolly spider monkeys, and all species of capuchins (Boinski, 1992; Milton, 1985; Robinson, 1979). Urine washing has been suggested to function as an olfactory indicator of reproductive status (Boinski, 1992; Perret, 1992), and it may act as a non-affiliative proceptive behavior. Often, we see male white-faced capuchins smelling and licking branches where a female has performed a urine wash. Previously (Carnegie *et al.*, in press), we demonstrated that urine washing was not related to ovarian phase (i.e., periovulatory versus non-ovulatory), and in the present study, we found that urine washing occurred at significantly higher rates among the cycling females compared to the noncycling and pregnant females. Therefore, in this species, olfaction may play a role in communicating reproductive state rather than ovarian phase. The use of olfaction by anthropoid primates is not as well studied as it is in prosimians (Schilling, 1979) and further studies need to be done to understand how olfaction is used in cebids and other anthropoids.

It is unclear why subordinate males would sexually solicit pregnant females when there is no obvious reproductive advantage to this behavior. Gust (1994) could not conclude that hormonal differences in female sooty mangabeys accounted for the behavioral differences observed between the alphas and subordinate males. Nevertheless, if alpha males are able to discriminate between LH concentrations in cycling and pregnant females, it is more than likely that subordinate males can do the same. Therefore, it is possible that from the subordinate males' perspective, post-conceptive mating has little or nothing to do with increasing reproductive fitness but instead functions to build social bonds and increase sexual familiarity and/or reduce inter-sexual tension (Gust, 1994; Manson *et al.*, 1997).

As to our second question, pregnant females may choose to mate with subordinate males in order to confuse the male as to the paternity of the soon-to-be-born infant (Hrdy, 1974, 1979). Infanticide does occur in this species and it is often carried out by immigrating and/or lower ranked males after a rank reversal (Fedigan, 2003). It is argued that in forming a sociosexual bond with the mother, subordinate males are less likely to harm a newborn and instead will stay in the group to help protect the infant from predators and invading males (Hrdy, 1974; Agrell *et al.*, 1998; Fedigan, 2003).

Previously, we have found that these capuchin females mate more often with the alpha males during their conceptive phase (two out of three copulations), and paternity studies have shown that the alpha males sire almost all of the infants within their groups (Jack and Fedigan, this volume; Jack and Fedigan, 2003). Therefore, it can be assumed that the pregnant females have already mated and conceived infants with the alpha male during their regular ovarian cycling period. Since lower ranked (and immigrating) males may pose a high risk to a new infant in terms of infanticide, by being receptive to the male's solicitations during gestation and even within days of parturition, the pregnant female may be forming a positive bond with a male, thereby reducing the likelihood of aggressive encounters with him once the infant is born.

Forming and maintaining strong social bonds could also be facilitated through grooming. Our finding that the pregnant females were grooming the subordinate males more often than the alpha males supports the theory that the pregnant females are trying to form positive bonds with males that could potentially harm their infant in the event of a rank reversal. Moreover, formation of these bonds may also facilitate a protective relationship and, in the case of an extra group male takeover, these subordinate males may help protect the female and her new infant. Additionally, the close proximity that is required for grooming means that the female is accessible to that male for mating.

However, for these strategies to be successful, males must not be able to associate the timing of mating with the timing of births (Zinner and Deschner, 2000).

SUMMARY

In many primate species, sociosexual behaviors are known to vary throughout a female's reproductive cycle, yet few studies have focused on changes in behavior across reproductive states. We examined sexual and affiliative behaviors that are exhibited across three reproductive states: cycling, noncycling, and pregnancy, in wild white-faced capuchins. To reliably determine reproductive state, we analyzed fecal steroids to create hormone profiles for each of the subject females. Our objectives were to determine the behavioral indicators from which one could reliably infer the reproductive status of wild female capuchins in the absence of hormonal data, and in doing so, to explore the reproductive strategies used by female capuchins.

Our hormonal analysis was successful and revealed some interesting aspects of female capuchin reproduction. We found that the cycling females stopped cycling mid-way through the study but did not become pregnant. We are uncertain as to why they stopped cycling and further research into this finding is presently underway. We found that cycling and pregnant females copulated and received more courtship displays from adult males than did noncycling females. Sexual behaviors were more commonly seen between pregnant females and subordinate males, and between cycling females and alpha males. Pregnant females displayed proceptive behaviors at higher rates toward subordinate males, and cycling females received attractivity indicators at higher rates from alpha males. Noncycling females were rarely, if at all, involved in sexual, proceptive, or attractive behavioral indicators. However, they were responsible for maintaining close proximity to the alpha males 100% of the time. Finally, we found that urine washing occurred at higher rates among the cycling females compared to the other females.

We conclude that the high rate of sexual behavior and proceptivity between the pregnant females and the subordinate males may be associated with a female-strategy to prevent infanticide. By mating with lower ranked males and by forming positive bonds with them, pregnant females confuse males as to the paternity of the new infant and encourage them to be protective in the eventuality of an aggressive male take-over.

ACKNOWLEDGMENTS

We would like to thank the editors for inviting us to contribute to this book: A. Estrada, P. Garber, M. Pavelka, and L. Luecke. We are grateful to the Costa Rican National Park Service for allowing us to work in SRNP from January to June, 2002, and to the administrators of the ACG for permission to continue to work in the park to present day (especially, R. Blanco Seguro). Thanks to G. McCabe for her assistance in the field, to D. Wittwer and S. Jacoris for their help with the laboratory analysis, and to K. Jack and J. Addicott for statistical advice and comments on drafts of the manuscript. This research was funded by operating grant A7723 from the Natural Sciences and Engineering Research Council of Canada and by the Canada Research Chairs Program (LMF), and by grants from American Society of Primatologists, Sigma-Xi, and FGSR/Dept of Anthropology, University of Alberta Research Fund (SDC), and by base grant RR000167 of the National Primate Research Center (TEZ).

REFERENCES

Agrell, J., Wolff, J., and Ylonen, H. 1998, Counter-strategies to infanticide in mammals: Costs and consequences. *Oikos* 83:507–517.

Altmann, J. 1974, Observational study of behavior: Sampling methods. *Behavior* 49–50:227–265.

Baum, J. M. 1983, Hormonal modulation of sexuality in female primates. *BioScience* 33:578–582.

Beach, F. A. 1976, Sexual attractivity, proceptivity and receptivity in female mammals. *Horm. Behav.* 7:105–138.

Boinski, S. 1992, Olfactory communication among Costa Rican squirrel monkeys: A field study. *Folia Primatol.* 59:127–136.

Brockman, D. K., Whitten, P. L., Russell, E., Richard, A. F., and Izard, M. K. 1995, Application of fecal steroid techniques to the reproductive endocrinology of female Verreaux's sifaka (*Propithecus verreauxi*). *Am. J. Primatol.* 36:313–325.

Carnegie, S. D., Fedigan, L. M., and Ziegler, T. E. In press, Behavioral indicators of ovarian phase in white-faced capuchins, *Cebus capucinus*. *Am J. Primatol.* 67.

Carosi, M., Heistermann, M., and Visalberghi, E. 1999, Display of proceptive behaviors in relation to urinary and fecal progestin levels over the ovarian cycle in female tufted capuchin monkeys. *Horm. Behav.* 36:252–265.

Dixson, A. 1998, *Primate Sexuality: Comparative Studies of Prosimians, Monkeys, Apes and Human Beings.* Oxford University Press, New York.

Enomoto, T., Seiki, K., and Haruki, Y. 1979, On the correlation between sexual behavior and ovarian hormone level during the menstrual cycle in captive Japanese monkeys. *Primates* 20:563–570.

Fedigan, L. M. 2003, Impact of male take-overs on infant deaths, births and conceptions in *Cebus capucinus* at Santa Rosa, Costa Rica. *Int. J. Primatol.* 24:723–741.

Fedigan, L. M. and Jack, K. 2001, Neotropical primates in a regenerating Costa Rican dry forest: A comparison of howler and capuchin population patterns. *Int. J. Primatol.* 22:689–713.

Fedigan, L. M. and Jack, K. M. 2004. The demographic and reproductive context of male replacements in *Cebus capucinus. Behaviour* 141:755–775.

Fragaszy, D., Visalberghi, E., and Fedigan, L. M. 2004, *The Complete Capuchin: Biology of the Genus Cebus.* Cambridge University Press, Cambridge.

Gordon, T. P., Gust, D. A., Busse, C. D., and Wilson, M. E. 1991, Hormones and sexual behavior associated with postconception perineal swelling in the sooty mangabey (*Cercocebus torquatus atys*). *Int. J. Primatol.* 12:585–597.

Gust, D. A. 1994, Alpha-male sooty mangabeys differentiate between females' fertile and their postconception maximal swellings. *Int. J. Primatol.* 15:289–301.

Hinde, R. A. and Atkinson, S. 1970, Assessing the roles of social partners in maintaining mutual proximity as exemplified by mother infant relations in rhesus monkeys. *Anim. Behav.* 18:169–176.

Hodges, J. K. and Heistermann, M. 2003, Field endocrinology: Monitoring hormonal changes in free-ranging primates, in: J. M. Setchell and D. J. Curtis, eds.,*Field and Laboratory Methods in Primatology: A Practical Guide*, Cambridge University Press, Cambridge, pp. 282–294.

Hrdy, S. B. 1974, Male-male competition and infanticide among the langurs (*Presbytis entellus*) of Abu, Rajasthan. *Folia Primatol.* 22:19–58.

Hrdy, S. B. 1979, Infanticide among animals: A review, classification, and examination of the implications for the reproductive strategies of females. *Ethol. Sociobiol.* 1:13–40.

Jack, K. and Fedigan, L. M. 2003, Male dominance and reproductive success in white-faced capuchins (*Cebus capucinus*). *Am. J. Phys. Anthropol.* 36(Suppl):121–122.

Janson, C. H. 1984, Female choice and mating system of the brown capuchin monkey *Cebus apella* (Primates: Cebidae). *Z. Tierpsychol.* 65:177–200.

Loy, J. D. 1981, The reproduction and heterosexual behaviors of adult patas monkeys in captivity. *Anim. Behav.* 29:714–726.

Manson, J. H., Perry, S., and Parish, A. R. 1997, Nonconceptive sexual behavior in bonobos and capuchins. *Int. J. Primatol.* 18:767–786.

Milton, K. 1985, Urine washing behavior in the woolly spider monkey (*Brachyteles arachnoides*). *Z. Tierpsychol.* 67:154–160.

Nagle, C. A. and Denari, J. H. 1982, The reproductive biology of capuchin monkeys, in: P. J. S. Olney, ed., *International Zoo Yearbook*, vol. 22, Dorset Press, Dorchester, pp. 143–150.

Perret, M. 1992, Environmental and social determinants of sexual function in the male lesser mouse lemur (*Microcebus murinus*). *Folia Primatol.* 59:1–25.

Perry, S. 1997, Male–female social relationships in wild white-faced capuchins, (*Cebus capucinus*). *Behaviour* 134:477–510.

Risler, L., Wasser, S. K., and Sackett, G. P. 1987, Measurement of excreted steroids in *Macaca nemestrina.Am. J. Primatol.* 12:91–100.

Robinson. J. G. 1979, Correlates of urine washing in the wedge-capped capuchin *Cebus nigrivittatus*, in: J. F. Eisenberg, ed., *Vertebrate Ecology in the Northern Tropics*, Smithsonian Institution Press, Washington, D.C., pp. 137–143.

Robinson, J. G. 1988, Group size in wedge-capped capuchin monkeys, *Cebus olivaceus*, and the reproductive success of males and females. *Behav. Ecol. Sociobiol.* 23: 187–197.

Saayman, G. S. 1970, The menstral cycles and sexual behavior in a troop of free-ranging chacma baboons (*Papio ursinus*). *Folia Primatol.* 12:81–110.

Schilling, A. 1979, Olfactory communication in prosimians, in: G. S. Doyle and R. D. Martin, eds., *The Study of Prosimian Behavior*, Academic Press, New York, pp. 461–542.

Siegel, S. and Castellan N. J. Jr. 1988, *Non Parametric Statistics for the Behavioral Sciences*, 2nd Edn. McGraw-Hill Inc., Mexico.

Strier, K. B., Lynch, W. J., and Ziegler, T. E. 2003, Hormonal changes during the mating and conception seasons of wild northern muriquis (*Brachyteles arachnoides hypoxanthus*). *Am. J. Primatol.* 61:85–99.

Strier, K. B. and Ziegler, T. E. 1997, Behavioral and endocrine characteristics of the reproductive cycle in wild muriqui monkeys, *Brachyteles arachnoides. Am. J. Primatol.* 42:299–310.

Strier, K. B., Ziegler, T. E., and Wittwer, D. 1999, Seasonal and social correlates of fecal testosterone and cortisol levels in wild male muriquis (*Brachyteles arachnoides*). *Horm. Behav.* 35:125–134.

Wallis, J. 1982, Sexual behavior of captive chimpanzees (*Pan troglodytes*): Pregnant versus cycling females. *Am. J. Primatol.* 3:77–88.

Wasser, S. K., Monfort, S. L., and Wildt, D. E. 1991, Rapid extraction of faecal steroids for measuring reproductive cyclicity and early pregnancy in free-ranging yellow baboons (*Papio cynocephalus cynocephalus*). *J. Reprod. Fertil.* 92:415–423.

Zar, J. 1999, *Biostatistical Analysis*, 4th Edn. Prentice Hall, New Jersey.

Ziegler, T. E., Scheffler, G., Wittwer, D. J., Schultz-Darken, N., Snowdon, C. T., and Abbott, D. H. 1996, Metabolism of reproductive steroids during the ovarian cycle in

two species of Callitrichids, *Saguinus oedipus* and *Callithrix jacchus*, and estimation of the ovulatory period from fecal steroids. *Biol. Reprod.* 54:91–99.
Zinner, D. and Deschner, T. 2000, Sexual swellings in female hamadryas baboons after male take-overs: "Deceptive" swellings as a possible female counter strategy against infanticide. *Am. J. Primatol.* 52:157–168.

PART FOUR

Conservation and Management Policies

Introduction: Conservation and Management Policies

P. A. Garber, A. Estrada, and M. S. M. Pavelka

Given current rates of deforestation (440,000 ha per year, see chapter 1), increased use of land for cattle ranching and agriculture, and human population growth approaching a rate of 3% per year, informed and immediate conservation policies are needed to protect habitats and primate populations throughout Mesoamerica. To be effective, such policies must be based on reliable field data, take advantage of recent advances in new methods and technologies in data analysis, consider the needs of local communities, be supported by a national campaign to promote conservation education, and present economic incentives that support ecologically responsible land use. One such initiative is the Mesoamerican Biological Corridor project. Here, Mesoamerican countries have formally agreed to implement a

P. A. Garber • Department of Anthropology, University of Illinois at Urbana-Champaign, 109 Davenport Hall, Urbana, IL 61801, USA. **A. Estrada** • Estación Biológica Los Tuxtlas, Instituto de Biología, Universidad Nacional Autónoma de México, Apartado Postal 176, San Andrés Tuxtla, Veracruz, México. **M. S. M. Pavelka** • Department of Anthropology, University of Calgary, 2500 University Drive NW, Calgary, Alberta, Canada T2N 1N4.

New Perspectives in the Study of Mesoamerican Primates: Distribution, Ecology, Behavior, and Conservation, edited by Alejandro Estrada, Paul A. Garber, Mary S. M. Pavelka, and LeAndra Luecke. Springer, New York, 2005.

network of interconnected protected forest corridors linking fragmented habitats and animal populations within a plan of sustainable economic development. International projects are critical for developing large-scale conservation plans. However, effective long-term conservation efforts also must focus on the local level in order to develop management policies that fit within the cultural framework of individual communities. Chapters in this section combine empirical data, new methodological tools, and models derived from landscape ecology, population biology, and community conservation to develop effective management strategies.

In chapter 17, Milton reviews demographic data collected over the past four decades on the size and growth rate of the spider monkey population on Barro Colorado Island, Panama. Young spider monkeys were introduced onto the island in the early to mid-1960s. The founding genetic population was composed of one juvenile male and four juvenile females. The current population is 28 animals. Despite an absence of predators, human hunting, and large-bodied competitors, the number of spider monkeys on the island has increased very slowly. Milton cautions that even in protected and productive rainforest habitats, reintroduction programs involving primates with a long interbirth interval and a long juvenile period require several decades of sustained population growth before such programs can be considered successful.

Estrada and colleagues offer a model of agroecosystem sustainability in which select forms of agriculture (shade coffee, cacao, and cardamom; monocultures such as mangos, bananas, and palms) and patterns of human land use (planting of live fences) are compatible with the long-term persistence of some primate populations. The presence of forest fragments and other types of arboreal agroecosystems, including linear strips of man-made vegetation in heterogeneous fragmented landscapes "may provide temporary habitat, may function as stepping stones, and may increase area of vegetation and availability of potential resources, among other benefits, for isolated segments of populations of a broad spectrum of animal species, including primates." On the basis of data collected in Mexico, Guatemala, and Costa Rica, these authors report that spider monkeys, capuchins, howlers, and squirrel monkeys exploit resources in agrosystems and use the available forest canopy to travel between forest patches. Although some agroforests were not suitable for primates or contained populations that were clearly in decline or considered as pests, given continued pressure from humans to alter natural landscapes, the maintenance of ecologically diverse fragmented landscapes may enable primate populations to survive in a commensurate relationship with human populations.

In their chapter on primate populations in the protected forests of Maya archeological sites, Estrada and colleagues assess the demography and conservation status of black howlers (*Alouatta pigra*) and spider monkeys (*Ateles geoffroyi*) in southern Mexico and Guatemala. Their surveys indicate that mean population densities, mean group size, and the ratio of immature to adult females of both primate species in forests surrounding protected Maya sites were more similar to values in areas of extensive forest than in areas of fragmented forests. Efforts by the Mexican and Guatemalan governments to promote cultural patrimony through tourism generate critical revenue for the local and national economy and provide important conservation benefits by maintaining plant and animal diversity and enhancing public awareness of environmental issues. The continued protection of large forested areas surrounding Maya archeological sites represents an important conservation tool in northern Mesoamerica.

The results of a population survey of black howlers, mantled howlers, and spider monkeys in Mexico's Yucatan peninsula are presented in the chapter by Serio-Silva and colleagues. Although much of the Yucatan remains forested (estimated 93,942.39 km^2 of forest totaling 63.9% of the peninsula's total surface area), human impact has resulted in forest fragmentation with few remaining sites of semi-evergreen forest larger than 1,000 km^2 in area. Mantled howlers (*Alouatta palliata*) were thought to be absent from the region; however, a small number of groups were sighted on private and unprotected land in the southern State of Campeche. Troop size and composition of black howler and spider monkey populations in the Yucatan were found to be similar to those reported in other parts of their range. Approximately 51% of the primate sites surveyed had no legal protection. These authors argue that "because the Yucatan peninsula still contains large tracks of forested habitats, this region must be considered among the highest priority conservation regions in Mesoamerica."

Mandujano and colleagues examine questions of behavioral plasticity or the ability of primates to respond to forest fragmentation and habitat loss. Using metapopulation and ecological network models, these authors found that, on the basis of data collected from 92 forest patches ranging in size from 0.5 to 76 ha, the amount of suitable habitat (patch size and presence of large canopy trees) had a greater impact on mantled howler monkey (*Alouatta palliata*) population viability than did the degree to which forest patches were connected. These authors recommend that in addition to restoring and connecting habitats, conservation efforts focusing on larger and less disturbed forest patches offer the greatest opportunity to sustain howler monkey populations.

The chapter by Alexander *et al.* demonstrates the value of using landsat imagery, remote sensing, and graphic information system technologies to describe, measure, and analyze processes of habitat change and their effect on the structure of primate populations. These authors describe specific changes in vegetation cover, habitat fragmentation, plant species diversity, and black howler monkey populations density resulting from the devastation caused by Hurricane Iris after it struck the southern coast of Belize. Their results indicate a decrease in howler survivorship and group stability with increased habitat fragmentation, reduced patch size, and decreased habitat diversity. Combining the use of new technologies with theories of island biogeography, Alexander and colleagues provide conservation biologists with a strong framework for documenting how processes of habitat destruction and habitat recovery affect ecosystem stability and animal populations.

Together, these chapters provide a series of models or examples of research strategies and analytical tools that are critical for assessing relationships between habitat fragmentation, population viability, and ecosystem management, and the immediate priorities and effectiveness of primate conservation policies. Using approaches that have proven successful in regions of Mesoamerica such as Mexico, Guatemala, Belize, and Costa Rica offers hope that increased commitment at the local and national levels will aid in reversing decades of habitat loss and protect wild primate populations throughout the region.

CHAPTER SEVENTEEN

Growth of a Reintroduced Spider Monkey (*Ateles geoffroyi*) Population on Barro Colorado Island, Panama

Katharine Milton and Mariah E. Hopkins

INTRODUCTION

The release of a number of young spider monkeys (*Ateles geoffroyi*) onto Barro Colorado Island (BCI) Panama, in the early 1960s provides a unique opportunity to examine the growth of a small founder population reintroduced into a protected reserve of documented size. Increasingly, as human intervention continues to subdivide remaining large blocks of tropical forest into smaller units, the resulting fragments can be viewed as "islands," surrounded by human-modified landscapes. What rate of population growth can be predicted for spider monkeys and similar species reintroduced into suitable forest fragments? As

Katharine Milton and Mariah E. Hopkins • Department of Environmental Science, Policy and Management, Division Insect Biology, University of California, Berkeley, CA 94720.

New Perspectives in the Study of Mesoamerican Primates: Distribution, Ecology, Behavior, and Conservation, edited by Alejandro Estrada, Paul A. Garber, Mary S. M. Pavelka, and LeAndra Luecke. Springer, New York, 2005.

will be shown, the slow expansion of this population over a period of >44 years points to the need for caution in assuming that reintroductions, even under the most fortuitous conditions, will necessarily result in the rapid repopulation of a given area. On a brighter note, though only four reproductively viable individuals survived the reintroduction, to date, the expanding population appears normal in all respects.

BACKGROUND

Barro Colorado Island is a 1600 ha protected nature reserve located in Lake Gatun, the principal water supply for the Panama Canal. The damming of the Chagras River in 1912 to create Lake Gatun flooded adjacent mainland areas. Only higher peaks and plateaus ultimately remained above water, creating islands. At present, BCI, the largest island in the lake, is densely covered in lowland tropical forest; some areas consist of old second growth, while others are made up of undisturbed primary forest estimated to be 500 or more years in age (Hubbell and Foster, 1990). Detailed descriptions of the climate and physical characteristics of BCI as well as its flora and fauna can be found in the literature (e.g. Leigh *et al.*, 1982; Milton *et al.*, 2005).

Five primate species occur naturally in this area of Panama. Founder populations of three of these species (i.e. *Alouatta palliata*, *Saguinus geoffroyi*, *Cebus capucinus*) and possibly four (*Aotus zonalis*) were trapped on what became BCI by the rising lake water and their descendents can be found on the island to this day. Though black-handed spider monkeys (*A. geoffroyi*) are also native to this region of Panama, by 1912, they had been entirely extirpated from the area by hunters and none remained to populate BCI.

Initial Reintroduction

Beginning in December 1959 and continuing intermittently until mid-1966, the then-Director of the Smithsonian Tropical Research Institute (STRI) in Panama, the late Martin Monihan, began to release immature A. *geoffroyi* onto BCI. It was hoped that the young monkeys would survive and lead to the re-establishment of a free-ranging spider monkey population.

At the time of the reintroduction effort, it was common to find young monkeys of various species routinely offered for sale in the public market in Panama City. The usual way to secure a young monkey for the pet trade is to shoot the mother and remove the clinging, dependent young infant from her body. Thus, it is highly probable that all of the immature spider monkeys purchased for release onto BCI were tiny dependent infants at the time of their initial capture. Some likely were kept as pets for shorter or longer periods of time before being offered for sale and brought to BCI. Given the usual banana-dominated diet offered to captive monkeys in Panama, most young monkeys likely were not in optimal physical condition at the time of purchase.

The exact number of spider monkeys released onto BCI is not known (Dare, 1974). Conversations with the BCI caretaker for the young monkeys, the late Bonifacio de Leon, indicate that 18 or more monkeys were ultimately released. Their estimated ages ranged from perhaps 1 year to 3 or 4 years (Dare, 1974). Some were released near the laboratory clearing on the northeastern side of BCI while others were released at more distant locales (B. de Leon, pers. comm.). All were offered fresh fruits and other foods each day on feeding platforms constructed in each release area as the monkeys were viewed as far too young and/or inexperienced to forage successfully in the BCI forest. This provisioning was gradually ended as the young monkeys matured and became more self-sufficient at finding wild foods.

Even with nutritional supplementation, most monkeys did not survive the reintroduction. Data indicate that by the mid-1960s, only five (one male and four females) of the released monkeys were still alive (Eisenberg and Kuhen, 1966; Richard, 1970). The single male survivor, Chombo, was estimated to be 3–4 years old at the time of his release on BCI in late December 1959 (Dare, 1974). Three or all four of the surviving females were released onto BCI in 1960; at the time of their release, three were estimated to be one to one and one-half years old and the fourth was estimated to be 3 years old (Dare, 1974). There is mention of a fifth female possibly released on BCI between 1965 and 1966 at an estimated age of 3 years (Dare, 1974). This fifth female may have died shortly thereafter or one of the original 1960-introduced females may have died as various observers on BCI in the mid-to-late 1960s consistently report only five adults (one male and four females; Eisenberg and Kuhen, 1966; Richard, 1970). One of the four surviving females (KH) appears to have been barren. Though Dare (1974) mentions a possible birth by this female in 1969, in KM's experience (from March 1974 to this female's death at some point in 1989), this female was never noted with an infant.

An accurate record of the BCI spider monkey population in terms of the timing and number of births, the identity of the mother in each case and the timing and number of deaths can never be provided since no one collected the necessary continuous long-term data. In spite of this, as shown in Table 1,

Table 1. Censuses of BCI spider monkey population, 1964–2003

Year	Adult M	Adult F	Juv M	Juv F	Total Juv	Infant M	Infant F	Total infant	Total adult	Total immatures	Total	Obser.
1964	0	0	1	4	5	0	0	0	0	5	5	E & K
1966	1	4	0	0	0	1	2	3	5	3	8	E & K, Dare
1968	1	4	0	0	0	2	2	4	5	4	9	Richard
1970	1	4	1	1	2	2	0	2	5	8	13	Dare
1972	3	5	4	0	4	2	2	4	8	8	16	Dare
1974	5	4	1	2	3	1	1	2	9	5	14	KM
1978	4	4	1	2	3	2	0	2	8	5	13	KM
1979	4	4	1	2	3	3	1	4	8	7	15	KM, Glanz
1980	3	5	2	1	3	1	1	2	8	5	13	KM
1981	3	5	2	1	3	1	2	3	8	6	14	KM
1983	3	6	2	2	4	1–2	0–1	2	9	6	15	KM
1984	4	6	1	0	1	1	2	3	10	4	14	KM
1985	4	4	1	1	2	2	3	5	8	7	15	KM
1986	3	5	1	3	4	1	3	4	8	8	16	KM
1987	3	5	1	3	4	1	3	4	8	8	16	KM
1988	3	5	1	3	4	1	2	3	8	7	15	KM
1989	4	5	1	2	3	1	2	3	9	6	15	Ahumada
1990	3	5	1	2	3	1	2	3	8	6	14	KM
1991	5	5	2	3	5	2	1	3	10	8	18	KM
1995	6	4	1	5	6	1	2	3	10	9	19	KM
1997	5	6	2	7	9	0	1	1	11	10	21	Campbell
1998	4	7	1	6	7	0	2	2	11	9	20	Campbell
1999	4	8	1	6	7	5	3	8	12	15	27	Campbell
2001	4	10			7			6	14	13	27	KM
2002	3	9			8			7	12	15	27	KM
2003	4	10			7			7	14	14	28	KM

sufficient observational records exist such that a reasonable overview of the growth of this founder population can in fact be provided.

GROWTH OF THE POPULATION

Reproductive maturity for wild female spider monkeys on BCI is placed at $\geq$6.5 years (Milton, 1981). Therefore, 3–5 years had to pass for any of the surviving founder females to reach reproductive maturity. The first mention of spider monkey infants born on BCI comes from Eisenberg and Kuhen (1966), who noted the presence of three infant spider monkeys in the spring of 1966. There is strong general consensus that these were the first spider monkey infants born on BCI.

As Chombo was the only male to survive the reintroduction, it is certain that he fathered *all* infants born between 1959 and 1971 (by which time at least one F1 male could have reached reproductive maturity). As the dominant male, Chombo may have continued to father most or all infants born on BCI until his death in 1978, at an estimated age of 22 years.

After the three confirmed births in 1966, the birth record becomes blurred until around 1972. Data suggest that some 9–11 infants may have been born between 1966 and 1972 (Dare, 1974). Though various "spider monkey births" are noted in the BCI record books over this period, these record books were casual accounts of events noted by island visitors and cannot be relied on in terms of accuracy. Given the 28–36 month interbirth interval characteristic of wild spider monkeys (Milton, 1981; Symington, 1988), some of these "birth" records are likely redundant. Young spider monkeys are deceptively neonatal such that human observers not familiar with their growth trajectory invariably greatly underestimate their actual age.

In 1972, a USA graduate student, Ron Dare, came to BCI to carry out a study of spider monkey behavior for his doctoral dissertation. He created a photographic record for each individual, finding a total of 13 spider monkeys at the start (March 1972) of his study and 15 at the conclusion (December 1972). During his study, one adult F1 male died (born in 1966) and three infants (one male and two females) were born (Dare, 1974).

In March 1974, KM came to BCI to carry out fieldwork on the howler monkey (*A. palliata*) population and began to note down occasional information on the spider monkeys as well; in 1978–1979, she collected feeding and ranging data on the spider monkey population (Milton, 1993). By late 1974, there was a total of 14 spider monkeys on BCI—five clearly recognizable as the original founders and nine F1 offspring. These latter included four subadult to adult males, two juvenile females, one juvenile male and two infants, a male and a female, both born in 1974. A female infant was born in 1975, and two male infants were born in 1977. Another male infant as well as the first F2 generation infant, a female, were born in 1979. If all of these individuals had lived, this should have given a population of ≥19 spider monkeys on BCI by late 1979. However, during this same period, the founder male and one juvenile male had died, one founder female had to be sacrificed for a rabies test and her infant daughter was later sent to a US zoo. A December 1979 census showed a total of only 15 individuals (Milton, pers.obs.; see also Glanz, 1982).

After 1979, KM no longer lived on BCI but tried to census the BCI spider monkey population whenever she visited the island—generally at least once in a year. Though often able to count what appeared to be the total spider monkey population, by the mid-1980s, she was unable to distinguish younger members as individuals and her counts surely missed some births and deaths.

In 1988, a Colombian undergraduate student, Jorge Ahumada, came to BCI to study grooming behavior in *Ateles.* Ahumada provided accurate identification for each spider monkey present in the population during his stay on the island and KM attempted to link her identifications of older individuals with his. The same 15 individuals were alive at both the start (October 1988) and conclusion of his study (January 1989; Ahumada, 1992). On KM's visits to BCI, she continued to try and census the spider monkey population whenever the opportunity presented itself.

In fall 1997, a graduate student from the USA, Christina Campbell, came to BCI to carry out a dissertation study of reproductive behavior in spider monkeys. Like Ahumada, she described each member of the population. By 1997, too much time had passed for KM to be able to link her and Ahumada's identification records to Campbell's 1997 identity data (Campbell, 2000). Campbell found a total of 21 spider monkeys on BCI at the start of her study (August 1997) and 20 at the conclusion (August 1998). There were various changes in group composition during her study—three individuals died (one adult male, one juvenile male, and one new infant of undetermined sex) and two infants were born (both female).

Thus, overall, there are a series of accurate "touchstone" censuses of all spider monkeys on BCI, each described and known as an individual, for the particular years in which detailed observations were being carried out by one or another spider monkey researcher. Then, there are a number of opportunistic censuses that often lack individual identification. A summary of all available census data is presented in Table 1.

Pattern of Growth

Census data over the period 1960–2003 show that once the population grew to a size of around 14–16 animals in the early 1970s, it then hit what might be described as a long plateau in terms of any consistent population growth for almost three decades (Figure 1). The population was not in stasis during this period—data indicate the birth of new individuals and the disappearance and presumed death of others. For example, in 1984, one of the three remaining founder females (Freckles) died at an estimated age of 24 years; in 1987 another founder female (Blackie) died at an estimated age of 27 years. Both females continued to produce offspring until they disappeared. The barren female (KH), the only remaining member of the original cohort, died in 1989 at an estimated age of 26 years. There were always infants and juveniles present in the population. But it seems certain that between 1972 when Dare left BCI and 1998 when Campbell left BCI, a period of almost 26 years, the population had increased by only three individuals (Table 1 and Figure 1).

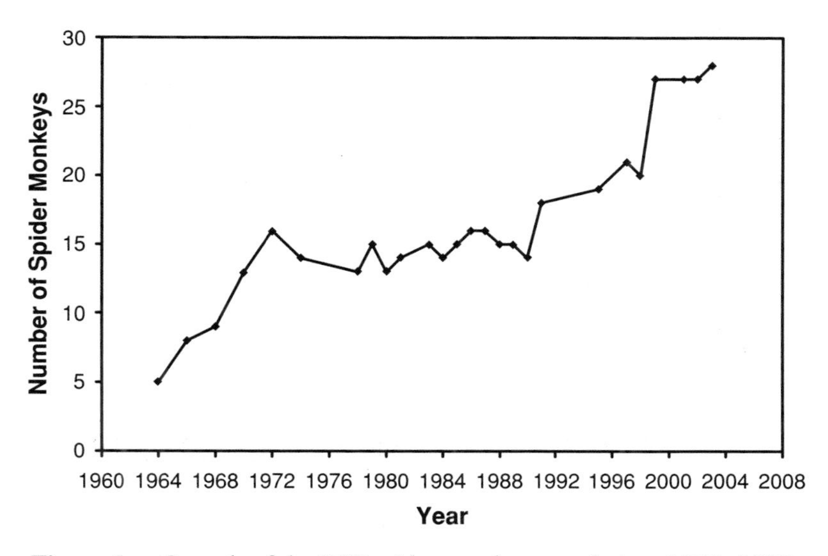

Figure 1. Growth of the BCI spider monkey population, 1964–2003.

The nature of exponential growth is well appreciated. Founder populations tend to grow slowly initially and then, if they persist, there is typically a sharp upward growth trajectory. It would seem that by the new millennium, the BCI spider monkey population had finally turned the corner and begun to accelerate notably in terms of size. In 1997, Campbell counted 21 spider monkeys; but by 1999, there was a total of 27 as six infants were born in 1999 (Campbell, 2000). KM counted 27 spider monkeys in 2001 and 2002 and 28 in 2003. Therefore, over the approximately 44-year period (1960–2003) since the initial reintroduction effort, overall population size has increased almost six-fold (from an initial size of five individuals to a current size of 28), an increase of around 4% per year.

Population Model

An age-based population growth model was generated using the program MATLAB 7 (MathWorks, 2004) in order to approximate the observed growth curve of the BCI spider monkey population (Figure 2). The model begins with one male and three breeding females, and assumes that the fourth female had no offspring. Each female is allowed to give birth every 36 months and age at sexual maturity for females is placed at 7 years. The sex ratio of male and female

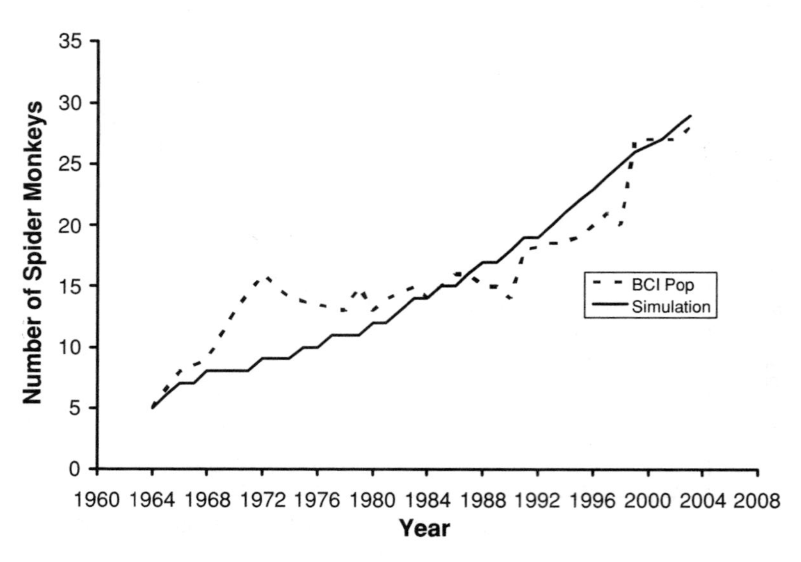

Figure 2. Comparison of simulated and observed growth for BCI spider monkey population.

births is assumed to be equal. These reproductive parameters were derived from estimated pedigrees for this population.

The population growth rate, as well as the numbers of individuals in each age/sex class was then simulated for 40 years according to different survivorship values. Simulated population growth was most sensitive to varying levels of adult female survivorship. Even with all other parameters set at 100% survivorship, adult female survivorship from year to year has to be above 90% in order for the population to reach current levels. The best-fit curve which approximates observed population trends on BCI contains modest mortality values (i.e. large survivorship rates) for all age-sex classes except for juvenile males approaching adulthood. Adult females have a survivorship probability between 98% and 99% per year. Male and female infants were given a fairly large survivorship probability of 75% per year. Male and female juveniles were given a survivorship probability of 90% per year, with one exception: the probability of a male juvenile spider monkey transitioning to an adult male spider monkey was lowered to 50% per year. These values yielded simulated population demographics that are very similar to those present on BCI today (2003): 29 individuals total, 4 males, 11 females, and 14 juveniles and infants (Figure 3).

However, while this simulation can illustrate the average reproductive rates and survivorship rates for various age/sex classes within the spider monkey

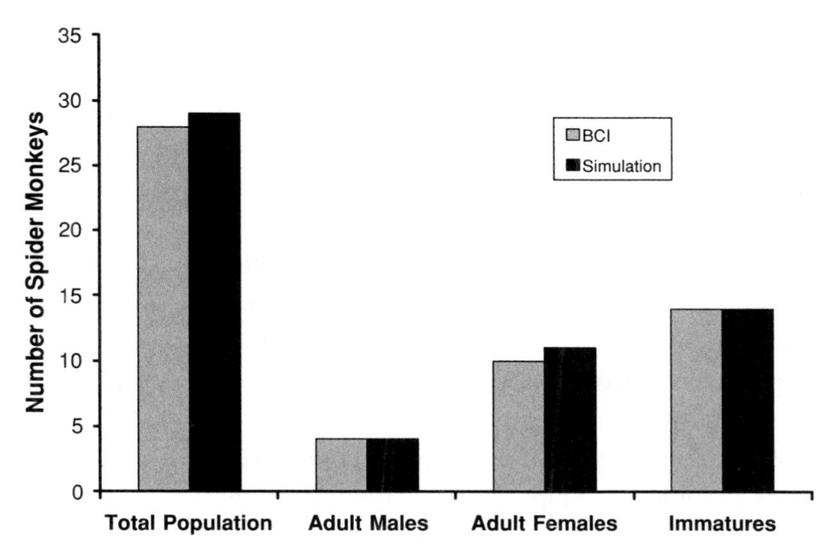

Figure 3. Comparison of simulated and observed BCI spider monkey population demographics in 2003.

population over the last 40 years, it cannot approximate the variation (i.e. population increases and decreases over short periods of time) that occurs between the starting and ending points (see Figure 2). This variation could result from a number of factors, including biased sex ratios, census errors, or human intervention (such as the removal of a breeding female and her daughter from the population in the 1970s). It is of interest to note that if 100% survivorship is assumed for all age/sex classes, given the BCI reproductive parameters, the BCI population should have grown to $\geq$170 individuals by now. In fact, it is less than 30 individuals (Table 1).

NATAL SEX RATIO

Some data suggest that spider monkeys may be able to bias the sex of their infants (Symington, 1987). In Peru, a 5-year birth record showed that significantly more female infants (32) were born than male infants (12) (Symington, 1987). It was suggested that at this site, lower-ranking females tended to produce daughters almost exclusively, while higher-ranking females biased their investment somewhat less strongly toward sons (Symington, 1987). At this Peruvian site, over the 5-year period, 21 out of 21 infants produced by low-ranking females were female while high-ranking females produced 12 male and 11 female infants (Symington, 1987). In contrast to the Peruvian situation, birth records for BCI (which are not complete but which should not show observer bias) suggest a natal sex ratio of approximately 1:1. However, it is the case that over the first reproductive decade, a disproportionate number of the F1 generation on BCI were male. Dare's (1974) data show that, in 1973, there were seven F1 male spider monkeys alive on BCI and only two F1 females—these in addition to the five original founders. The lack of maturing young females certainly contributed to the slow initial growth of the population. Also, as mentioned above, in late 1975, one founder female had to be sacrificed for a rabies test and her infant daughter was later shipped to a zoo in the USA. The human-mediated loss of these two females likewise had a negative impact on initial population growth.

At present, there are an unprecedented number of adult female spider monkeys on BCI (Table 1). Campbell (2000) noted that, of the six 1999 births, five were female and only one was male. Perhaps there is finally a sufficient cohort of reproductively active female spider monkeys on BCI to have high-ranking and low-ranking female status. Further observation of the natal sex ratio over the next decade or so will help to test this possibility.

ADULT SEX RATIO

Though overall data suggest a natal sex ratio of approximately 1:1 for the BCI spider monkey population, it is certain that, initially, male infants outnumbered female infants. Available data suggest that most of the initial F1 males survived in the 1970s. Both sexes can live upto >20 years of age (Milton, 1981). Yet, there never seems to have been a time when more than six and typically three or four adult males were present in the population together. Even today with some 10 adult females in the population, there are only 4 adult males (M:F adult sex ratio = 1:2.5).

It is the case that there tend to be more adult females than males in spider monkey groups. Klein and Klein (1977) reported that the two study groups of *Ateles belzebuth* they worked with at La Macarena, Colombia, were composed of 5 adult males and 12 adult females (1:2.4) and 3 adult males and 11 adult females (1:3.5), respectively. Decades later, Shimooka (2003) likewise reported 5 adult males and 10–11 adult females in his study group of *A. belzebuth* at La Macarena. Symington's two *Ateles paniscus* communities in Peru showed the same pattern—the East community had 7 adult males and 13 adult females, while the Lake community had 6 adult males and 11 adult females—an overall total of 13 adult males and 24 adult females and an adult sex ratio of approximately 1:2 (Symington, 1986, 1987, 1988). Similar female-biased adult sex ratios have been reported for spider monkey groups at various other sites (Campbell, 2000). The BCI data, which consistently show no more than three to six fully adult males present at any given time, indicate that the normative number of adult males per group for *Ateles* spp. (i.e. 3–6 males per group), is somehow maintained regardless of total group size or the number of adult females in it (Table 1).

CAN BCI SUPPORT MORE THAN ONE GROUP?

Barro Colorado Island seems sufficiently large and productive to support more than a single spider monkey group, though this is only conjecture. In the late 1970s, KM estimated that some members of the BCI spider monkey population, which then consisted of some 14–15 individuals, might range over an area as large as 800 ha (Milton, 1981). In 1998, Campbell (2000) estimated that that the 20–21 BCI spider monkeys she observed were using an area of some 960 ha. At present, there are some 28 spider monkeys in total on BCI yielding a density of 0.017 spider monkeys per hectare. Symington's (1987) two groups in Peru ranged over home range areas estimated at only 150–250 ha (375 ha for both communities together, 52 monkeys total for both communities yielding a density of 0.138 spider monkey per ha). Campbell (2000) noted that the density of spider monkeys on BCI was much lower than that recorded for various other sites.

As BCI is an island, there is a large edge effect as well as several narrow peninsulas. Island topography may reduce suitable habitat such that only a single spider monkey group can be supported due to the need of its members to range over a large portion of the 1600 ha island to encounter sufficient high quality food throughout the year. During the late rainy–early dry season on BCI, spider monkeys appear nutritionally stressed (Milton, 1981; unpublished data), suggesting that sufficient suitable food may at times be scarce.

It will be interesting to follow the growth trajectory of the population now that it has reached a size commensurate with many spider monkey groups (communities) elsewhere. If the adult females on BCI continue to produce new infants approximately every 3 years, with approximately 10 adult females now present, all else being equal, a considerable increase in spider monkey numbers seems possible over the next decade.

BOTTLENECK EFFECT AND ESTIMATED HETEROZYGOSITY

As there was only a single founder male, his Y chromosome is and always has been the only Y chromosome available on BCI. For the first 10 or so years after reintroduction, the founder male, therefore, must have fathered all infants, male and female. This means that at sexual maturity, the first few female members of the F1 generation could mate only with their father (who died in 1978) or a half- or full sib. However, even though the BCI spider monkey population can be said to have passed through a decided founder bottleneck, contrary to what might seem inevitable, this does not mean that average heterozygosity fell to dangerously low levels (Nei *et al.*, 1975).

There are no data to suggest that any of the breeding founders were related. Therefore, one unrelated male and three (or initially, possibly four) unrelated females provided the initial gene pool. In the wild, one generally finds three to six adult males per group, but because males of this genus are philopatric, all males of a given group may be close kin. The most dominant male may also have the highest reproductive success. The initial BCI situation with only a single male clearly lacks the genetic scope provided by four or five adult males, even if all are close kin. But as most or all males in mainland groups may be related and one male in particular may sire a high percentage of infants at any one time, over the short term, the BCI situation actually may not be as peculiar as it first appears. Wild spider monkey groups elsewhere typically have more than three to four reproductively active adult females but, similar to the initial BCI situation, these females likely are not related as in the genus *Ateles*, juvenile females disperse from their natal groups (Symington, 1987). However, unlike mainland groups, there was no way for new genetic material to enter the BCI population except through mutations as no gene flow was possible.

What level of heterozygosity may have been lost in the founder event? Inbreeding and loss of genetic diversity are an inevitable consequence of small population size (Wayne *et al.*, 1991; Bouzat *et al.*, 1998). Although the BCI spider monkey population began with one male and four females, the effective population size of the founder population was likely reduced by two factors: (1) one of the females is presumed to be barren at the time of introduction, and (2) all females had to mate with the same male (i.e. the assumption of random mating is somewhat violated by a biased sex ratio). Once these two factors are taken into account, the effective population size of the BCI founders becomes three individuals:

$$N_e = \frac{4\,N_{ef}\,N_{em}}{(N_{ef} + N_{em})} \tag{1}$$

where

N_e = Effective population size
N_{ef} = No. females
N_{em} = No. males

This unequal sex ratio of breeding males and females has continued throughout the subsequent generations of spider monkey populations, and all of the calculations involved in the subsequent assessments of inbreeding adjust for this factor.

Although the reduction of a population to an effective size of three individuals is a drastic decrease in population size, the immediate impact of this bottleneck on population genetic diversity (as measured by average heterozygosity) is relatively small. Estimates indicate that the BCI spider monkey population immediately after placement on BCI likely retained approximately 83.3% of the heterozygosity present within a mainland randomly mating population:

$$\frac{H_1}{H_0} = 1 - \left(\frac{1}{2\,N_e}\right) \tag{2}$$

where

H_1 = Heterozygosity immediately after the bottleneck
H_0 = Heterozygosity before bottleneck
N_e = effective population size

However, sustained reductions in population size after a bottleneck, such as that which has occurred on BCI, can substantially reduce average heterozygosity and allele frequencies within a population over time (Nei *et al.*, 1975; Frankham *et al.*, 2002). In the small closed population of spider monkeys on BCI, inbreeding is unavoidable as all individuals become related by descent. A common measure of this effect is the inbreeding coefficient (F_t) which measures the probability that an individual in a given generation (t) will receive identical alleles from both parents due to common descent:

$$F_t = 1 - \prod_{i=1}^{t}\left[1 - \frac{1}{2\,N_{ei}}\right] \tag{3}$$

where

F_t = Inbreeding coefficient at generation t
N_{ei} = Effective population size in the ith generation

All equations above are taked from Frankham *et al.*, 2002.

Due to an inability to construct a complete pedigree for the BCI spider monkey population, it is impossible to calculate an exact inbreeding coefficient in the absence of genetic data. However, from the partial pedigree reconstructed between 1966 and 2003, we can garner approximate estimations of the effective population size in each generation, thereby allowing the calculation of a conservative approximation of the inbreeding coefficient for the current generation. For example, from census data, we can confirm that there were at least nine breeding F1 adults present in the population at various times. If we assume that each successive generation yields at least as many individuals as the founding population (a conservative assumption given that the F1 generation was at least three times the size of the founding population), the youngest generation of breeding spider monkeys on BCI would have an estimated inbreeding coefficient of approximately 0.338. This value indicates that approximately 66.2% of the average heterozygosity of a mainland population remains on BCI, given the population growth trend over the past 40 years ($H_1/H_0 = 1 - F_t$; Frankham *et al.*, 2002). Yet, given the properties of Equation (3), as the effective population size on BCI continues to increase, the corresponding decrease in average heterozygosity becomes negligible. For example, if the effective population size of the next generation is 14 individuals (the current number of breeding adults in the population), the corresponding decrease in average heterozygosity would only be approximately 4%.

NOISE

Two other features of spider monkey population dynamics on BCI require mention, i.e. racetrack monkeys and small island monkeys.

Racetrack Monkeys

In April 1991, STRI permitted the release of five spider monkeys, apparently all *A. geoffroyi*, onto BCI. These monkeys were released at the end of Armour trail—an area of old growth forest on the extreme south-western side of the island, largely removed from the normal ranging circumference of the resident spider monkey population. The five racetrack monkeys were of various ages and apparently consisted of one male and four females. They had been captive animals, exhibited in a cage at the public racetrack in Panama City for the amusement of its patrons. For some unknown reason, possibly to enhance genetic diversity of the resident spider monkey population, STRI decided to permit the release of these racetrack monkeys onto BCI. To our knowledge, no study was carried out to determine the parasite loads, possible health problems or other features of the racetrack monkeys—features that should have been evaluated prior to permitting the release of these monkeys into a protected nature reserve—one that now had a viable spider monkey population of its own as well as viable populations of three or four other primate species. At least one juvenile female from the racetrack cohort had no teeth—so it is not clear how she was expected to survive her new freedom in the forest.

Data suggest that none of the racetrack spider monkeys survived. KM and several colleagues, a veterinarian and two researchers from the Primate Center at U.C. Davis, arrived on BCI the day after the release and spent several days working with howler monkeys in the release area. After their departure, KM remained on BCI for several more weeks, often working in the Armour release area. During that entire period, she noted nothing unusual in the resident spider monkey population nor did she hear any outbursts of spider monkey vocalizations in the forest that might suggest any newcomers had met the residents. The amount of noise that a group of excited spider monkeys can make is extreme and such noise carries clearly over a long distance in the forest (KM, pers. obs.). We hypothesize that the racetrack monkeys starved due to their inexperience in finding wild foods. They were probably also in poor physical condition at the time of release. KM later recovered the skulls of two spider monkeys in the release area (including the toothless juvenile) but the fate of the other three individuals is not known.

Small Island Monkeys

Occasionally, there are reports of black-handed spider monkeys sighted on one or another of the smaller islands adjacent to BCI. It has been speculated that people with boats may at times release unwanted pet monkeys onto smaller islands in Lake Gatun, which could be the source of some or all of these individuals. In June 2004, we and various colleagues saw an adult male spider monkey and a female spider monkey, both apparently in excellent condition, on Orchid Island, a small forested island directly adjacent to BCI. These two individuals may have swum the short distance from BCI to Orchid Island. The factor or factors that would induce spider monkeys to enter the caiman-infested lake and swim to Orchid Island are not known. The forest on Orchid Island is far less diverse than that on BCI (KM, pers. obs.). Both of the Orchid Island animals appeared to be young adults; they were not at all tame, suggesting that they were not former pets or, if so, did not retain fond memories of past human interactions. Perhaps these two individuals voluntarily emigrated from BCI in search of new habitat.

Campbell (2000) reported that one female spider monkey, known to her as an individual and initially seen on a small island adjacent to BCI, was later seen on BCI with the BCI spider monkey group, suggesting that some spider monkeys may swim between BCI and nearby islands. This remains to be confirmed as, to date, there have been no observations of spider monkeys swimming in Lake Gatun.

Future genetic studies may indicate whether any of the racetrack monkeys were incorporated into the resident population on BCI and also help to resolve the possible origins of spider monkeys occasionally sighted on nearby islands. If the racetrack monkeys perished, all individuals on BCI should be the descendents of the few initial founders. Conversely, if one or more of the racetrack monkeys survived, they may have provided new genetic input to the resident population. Genetic study will also provide an actual inbreeding coefficient that can be compared with the above estimate as well as with those of mainland spider monkey populations elsewhere. Such data should be of interest to conservation biologists for it will help to answer questions related to minimum size and composition thresholds for similar reintroduction efforts.

SUMMARY

The present-day BCI spider monkey population on Barro Colorado Island, Republic of Panama, results from the reintroduction of $\geq$19 young spider monkeys onto the island in the early 1960s. One male and four females survived the reintroduction though one of these females apparently was barren. No data suggest that any of the five founder individuals were related. The first three F1 infants were born in early 1966. By late 1974, there was a total of 14 spider monkeys on BCI—five clearly recognizable as the original founders and nine F1 offspring. These latter included four subadult to adult males, two juvenile females, one juvenile male and two infants, a male and a female, both born in 1974. Thus, a preponderance of the initial surviving members of the F1 generation was male. Over the next 25 or so years, little change was noted in the size of the BCI spider monkey population. Though births and deaths were recorded, the overall size of the population remained approximately the same (14–16 individuals). Only in the late 1990s did the BCI spider monkey population appear to turn the corner and begin to accelerate notably in terms of size. In 1997, there was an estimated total of 20 spider monkeys on BCI and by 1999, this number had increased to 27 (Campbell 2000). The composition and size of the present-day BCI spider monkey population now approximates those of various other wild spider monkey populations elsewhere.

It may seem surprising that a population originating from such a small number of founders could survive and grow successfully over time as this one has done without manifesting any inbreeding or behavioral problems. To date, however, the picture emerging from BCI suggests that, given suitable protected habitat and nutritional supplementation in the early stages of the founder event, even a small number of unrelated young animals as socially and ecologically complex as spider monkeys can reach successful reproductive adulthood without adult role models or further human intervention and then continue on successfully as a viable breeding population into the indefinite future.

ACKNOWLEDGMENTS

We thank the Smithsonian Tropical Research Institute for use of their facilities on BCI. We thank George Roderick for drawing our attention to Nei *et al.* (1975).

REFERENCES

Ahumada, J. A. 1992, Grooming behavior of spider monkeys (*Ateles geoffroyi*) on Barro Colorado Island, Panama. *Int. J. Primatol.* 13:33–49.

Bouzat, J. L., Lewin, H. A., and Paige, K. N. 1998, The ghost of genetic diversity past: Historical DNA analysis of the greater prairie chicken. *Am. Nat.* 152:1–6.

Campbell, C. 2000, The reproductive biology of black-handed spider monkeys. Unpublished Doctoral Dissertation, Department of Anthropology, University of California, Berkeley.

Dare, R. J. 1974, The social behavior and ecology of spider monkeys, *Atleles geoffroyi*, on Barro Colorado Island. Unpublished Doctoral Dissertation, Department of Anthropology, University of Oregon.

Eisenberg, J. F. and Kuhen, R. E. 1966, The behavior of *Ateles geoffroyi* and related species. *Smiths. Miscel. Coll.* 151(8), Pub. #4683. Washington, D.C.

Frankham, R., Ballou, J. D., and Briscoe, D. A. 2002, *Introduction to Conservation Genetics.* Cambridge University Press, Cambridge.

Glanz, W.E. 1982, The terrestrial mammal fauna of Barro Colorado Island: Census and long-term changes, in: E. G. Leigh, A. S. Rand, and D. M. Windsor, eds., *The Ecology of a Tropical Forest,* Smithsonian Press, Washington, D.C., pp. 455–468.

Hubbell, S. P. and Foster, R. B. 1990, Structure, dynamics and equilibrium status of old-growth forest on Baro Colorado Island, in: A. H. Gentry, ed., *Four Neotropical Rainforests,* Yale University Press, New Haven, C.T., pp. 522–541.

Klein, L. L. and Klein, D. B. 1977, Feeding behaviour of the Colombian spider monkey, in: T. H. Clutton-Brock, ed., *Primate Ecology*, Academic Press, New York, pp. 153–182.

Leigh, E. G., Rand, A. S., and Windsor, D. M. 1982, *The Ecology of a Tropical Forest.* Smithsonian Press, Washington, D.C.

MathWorks. 2004. Matlab: the Language fo Technical Computing. Version 7.0.1, Release 14. MathWorks, Natick, Massachusetts, USA.

Milton, K. 1981, Estimates of reproductive parameters for free-ranging *Ateles geoffroyi*. *Primates* 22:754–759.

Milton, K. 1993, Diet and social organization of a free-ranging spider monkey population: The development of species-typical behaviors in the absence of adults, in: M. E. Peirera and L. A. Fairbanks, eds., *Juvenile Primates: Life History, Development and Behavior*, Oxford University Press, New York, pp. 173–181.

Milton, K., Giacalone, J., Wright, S. J., and Stockmayer, G. 2005, Do populations fluctuations of neotropical mammals reflect fruit production estimates? The evidence from Barro Colorado Island, in: L. Dew and J. P. Boubli, eds., *Tropical Fruits and Frugivores: The Search for Strong Interactors*, Kluwer Publishing, Dordrecht, The Netherlands, pp. 5–35.

Nei, M., Maruyama, T., and Chakraborty R. 1975, The bottleneck effect and genetic variability in populations. *Evolution* 29:1–10.

Richard, A. 1970, A comparative study of the activity patterns and behavior of *Alouatta villosa* and *Ateles geoffroyi*. *Folia Primatol.* 12:241–263.

Shimooka, Y. 2003, Seasonal variation in association patterns of wild spider monkeys (*Ateles belzebuth belzebuth*) at La Macarena, Colombia. *Primates* 44:83–90.

Symington, M. J. 1986, Ecological determinants of fission-fusion sociality in *Ateles* and *Pan*, in: J. G. Else and P. C. Lee, eds., *Primate Ecology and Conservation*, Cambridge University Press, Cambridge, pp. 181–190.

Symington, M. 1988, Demography, ranging patterns and activity budgets of black spider monkeys (*Ateles paniscus chamek*) in the Manu National Park, Peru. *Am. J. Primatol.* 15:45–67.

Symington, M. McF. 1987, Sex ratio and maternal rank in wild spider monkeys: When daughters disperse. *Behav. Ecol. Sociobiol.* 20:421–425.

Wayne, R. K., Lehman N., Girman D., Gogan P. J. P., Gilbert D. A., Hansen K., Peterson R. O., Seal U. S., Eisenhawker A., Mech L. D., and Krumenaker R. J. 1991, Conservation genetics of the endangered Isle Royale gray wolf. *Conserv. Biol.* 5:41–51.

CHAPTER EIGHTEEN

Primates in Agroecosystems: Conservation Value of Some Agricultural Practices in Mesoamerican Landscapes

Alejandro Estrada, Joel Saenz, Celia Harvey, Eduardo Naranjo, David Muñoz, and Marleny Rosales-Meda

INTRODUCTION

While there is a general perception that agricultural activities are the principal threat to biodiversity in the tropics (Donald, 2004; Henle *et al.*, 2004a,b), recent assessments suggest that some agroecosystems in fragmented landscapes may favor the persistence of diverse assemblages of animal species

Alejandro Estrada • Estación de Biología Los Tuxtlas, Instituto de Biología, Universidad Nacional Autónoma de México, Mexico. **Joel Saenz** • Programa Regional de Manejo de Vida Silvestre, Universidad Nacional Autonoma (UNA), Costa Rica. **Celia Harvey** • Departmento de Agricultura y Agroforesteria, CATIE, Costa Rica. **Eduardo Naranjo and David Muñoz** • Colegio de la Frontera Sur, San Cristóbal de las Casas, Chiapas, Mexico. **Marleny Rosales-Meda** • Universidad de San Carlos, Guatemala.

New Perspectives in the Study of Mesoamerican Primates: Distribution, Ecology, Behavior, and Conservation, edited by Alejandro Estrada, Paul A. Garber, Mary S. M. Pavelka, and LeAndra Luecke. Springer, New York, 2005.

(Vandermeer, 2003; Schroth *et al.*, 2004). Agroecosystems, covering more than one-quarter of the global land area, or almost 5 billion hectares, are ecosystems in which people have deliberately selected crop plants and livestock animals to replace the natural flora and fauna (Altieri, 2004). There are highly simplified agroecosystems (e.g., pasturelands, intensive cereal cropping, and monocultures), and there are also others that support high biodiversity in the form of polycultures and/or agroforestry patterns (Pimentel *et al.*, 1992; Moguel and Toledo, 1999; Vandermeer, 2003; Henle *et al.*, 2004a,b; Melbourne *et al.*, 2004; Schroth *et al.*, 2004).

Recent evidence suggests that some agroecosystems may be important in sustaining vertebrate biodiversity in human modified tropical landscapes, as they may provide temporary habitat, may function as stepping stones, and may increase area of vegetation and availability of potential resources, among other benefits, for isolated segments of populations of a broad spectrum of animal species (Estrada *et al.*, 1993, 1994; Villaseñor and Hutto, 1995; Rice and Greenberg, 2000; Estrada and Coates-Estrada, 2000, 2002a,b; Perfecto and Armbrecht, 2003; Daily *et al.*, 2003; Greenberg, 2004; Harvey *et al.*, 2004). In the case of primates, there are a few reports indicating presence of primates in agroecosystems. For example, cabruca cacao in Brazil has attracted attention because of its ability to harbor primates such as the golden-headed lion tamarin (*Leontopithecus chrysonelis*), an endangered species (Rice and Greenberg, 2000). Similarly in Gulung Palung National Park in Kalimantan, primates such as leaf monkeys (*Presbytis rubicunda*) and gibbons (*Hylobates agilis*) are found in agroforests (Salafsky, 1993). Michon and de Foresta (1995) report the presence of seven primate species: macaques, leaf monkeys, gibbons, and siamangs (*Hylobates syndactylus*) in rubber (*Hevea brasiliensis*) and dammar (*Shorea javanica*) agroforests, and five species in durian (*Durio zibethinus*) agroforests in Sumatra and noted that their density was similar to that in primary forest. In Costa Rica, howler monkeys (*Alouatta palliata*) have been found in shaded coffee plantations (Somarriba *et al.*, 2004), and in Los Tuxtlas, Mexico, howler (*A. palliata*) and spider (*Ateles geoffroyi*) monkeys have been observed to be present in forest-shaded cacao and coffee plantations (Estrada *et al.*, 1994; Estrada and Coates-Estrada, 1996).

Extending from southern tropical Mexico to the Colombian border of Panama, Mesoamerica harbors the northernmost representatives of the primate order in the American continent. Primate species diversity is represented by 22 taxa. These belong to three families (Callithrichidae, Cebidae, and Atelidae), six genera (*Sanguinus, Aotus, Alouatta, Ateles, Saimiri,* and *Cebus*), and eight major species (see Rylands *et al.*, this volume; Rodriguez-Luna *et al.*, 1996; Rowe, 1996; Nowak, 1999). Major proximate threats to primate habitats in the region are agricultural activities aimed at building up pasturelands for cattle-raising and at expanding crop-land to raise food crops. General deforestation rate in the region is exceedingly high, estimated at 440,000 ha per year (Sader *et al.*, 1999; FAO, 2000), and current estimates (see Estrada *et al.*, this volume) indicate that about 70% of the original forest cover present in the region has been lost as a result of human activity.

While extensive pastureland for cattle grazing dominates fragmented landscapes in Mesoamerica, many of these also harbor various types of arboreal and non arboreal agroecosystems such as forest-shaded and tree-shaded (trees planted by man) coffee (*Coffea arabica*), cacao (*Theobroma cacao*), and cardamom (*Eletteria cardamomum*; Zingiberaceae), unshaded arboreal crops (e.g., allspice, *Pimienta dioica*, citrus, *Citrus* spp.), and non arboreal cultivars such as bananas, *Musa* spp., and corn, among others (FAO, 1999, 2001). Many of these landscapes also harbor thousands of meters of linear strips of live fences (single or double rows of trees) planted by local people to delimit pastures and agricultural lots (Harvey *et al.*, 2004). In Central American landscapes dominated by cattle production, live fences occur between 49% and 89% of all farms, with a mean density of 0.14 km of live fence per ha of farmland (Harvey *et al.*, submitted). These land-use practices have resulted in varied and highly heterogeneous landscapes in which natural, semi-natural, and introduced patches of vegetation coexist.

In this paper, we explore the value of some agricultural practices for the persistence of primate population in human-modified landscapes in Mesoamerica. We present results from surveys of primate populations in agroecosystems in fragmented landscapes in Los Tuxtlas, Mexico, in Lachuá, Guatemala, and in three localities (Central Pacific, Cañas, and Rio Frio) in Costa Rica. Data were used to determine the types of agroecosystems in which primate populations are present and the species involved, to assess how primate population parameters such as density, group size, and immature to adult female ratios vary among agroecosystems and with respect to those of populations of the same species in extensive and in isolated forest remnants. We also examined data from an ongoing study of the feeding ecology of howler monkeys (*A. palliata*) living in a cacao plantation in the lowlands of the Tabasco, Mexico, with the aim of assessing how primates sustain themselves in such agroecosystems.

METHODS

Los Tuxtlas, Veracruz, Mexico

In the region of Los Tuxtlas, in southern Veracruz, Mexico, we focused our investigation on a 300 km^2 fragmented landscape by the Gulf of Mexico coast (95°00′W, 18°25′N; mean annual precipitation 4900 mm; altitudinal gradient sea level to 550 m.a.s.l.) (Figure 1). In this landscape, forest fragments coexist in a mosaic of vegetation consisting of pastures lands (the dominant vegetation) with interdispersed seasonal non arboreal (corn, jalapeño chili pepper, beans, tobacco, and bananas) and perennial arboreal (cacao, coffee, oranges, and allspice—*P. dioica*, Myrtaceae) crops. Both cacao and coffee, and mixed crops of these two plants, are grown under the shade of rain forest arboreal vegetation or less commonly under the shade of coconut palms and other plants, or under the shade of banana shrubs and planted trees. In this landscape, pasturelands harbor extensive networks of live fences, which consist of live posts of *Bursera simaruba* (Burseraceae) and *Gliricidia sepium* (Fabaceae), among other species, planted by the local inhabitants to hold barbed wire to delimit boundaries of properties.

Presence or absence of howler (*A. palliata*) and spider monkeys (*A. geoffroyi*) was investigated in 132 agricultural sites representing 12 types of agroecosystems: cacao shaded by rain forest vegetation (10 sites), cacao shaded by legume trees (6 sites), and cacao shaded by coconut and other trees and banana (5 sites), coffee shaded by forest vegetation (10 sites), mixed cacao/coffee under the shade of forest vegetation (8 sites), citrus (10 sites), allspice (*P. dioica*) (10 sites), mango (*Mangifera indica*; 8 sites), bananas (10 sites), mixed mango/citrus/banana (5 sites), and young live fences (DBH of posts <30 cm; 25 sites of 5 km in length each) and old (DBH >30 cm; 25 sites of 5 km in length each) present in this landscape (tree species: *B. simaruba*, *G. sepium*, *Ficus* spp.). Data on population parameters were gathered following standardized sampling protocols (National Research Council, 1992; Wilson *et al.*, 1996) in some of the forest-shaded cacao, coffee, and mixed cacao/coffee plantations in which primates were present. Population data were also collected for primates found in forest fragments in the same countryside, and in a natural

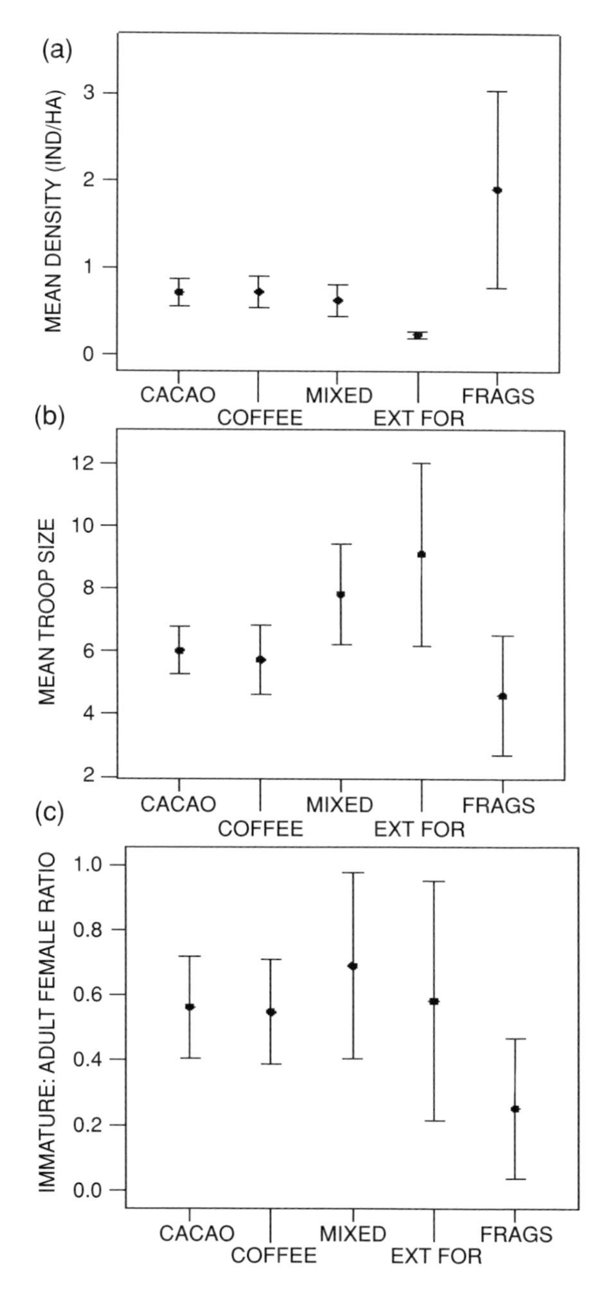

Figure 1. Mean (±SD) population density (a), mean troop size (b), and mean immature to adult female ratio (c), in populations of mantled howler monkeys (*Alouatta palliata*) residing in shaded (rain forest vegetation) cacao, coffee and mixed (cacao and coffee) agroecosystems, in extensive forests (>100 km^2; "EXT FOR") and in forest fragments (<10 km^2; "FRAGS") in Los Tuxtlas, Mexico.

protected area (15,000 ha) abutting the landscape. Interviews with the owners of the farms provided general information on whether the presence of the primates in the plantations had a neutral (no damage) or negative (damage) impact upon the cultivated fruit trees (e.g., cacao, coffee, bananas, and citrus).

Lachuá, Guatemala

In Guatemala, surveys of black howler monkeys (*Alouatta pigra*) were conducted, during 2002–2003 in cardamom plantations (*E. cardamomum*; Zingiberaceae) growing both under the shade of rain forest vegetation and in forest fragments not being used for agricultural purposes in a 230 km^2 fragmented landscape in the ecoregion of Lachuá (15°59′N, 90°36′W; mean annual precipitation 2252 ± 328 mm; altitudinal gradient 150–300 m.a.s.l.). Mean values for the demographic parameters of interest were compared with the overall means of eight populations of *A. pigra* existing in extensive forest tracts (>1000 ha; see Van Belle, this volume).

Costa Rica

Squirrel, howler, and capuchin monkeys were surveyed in agroecosystems present in fragmented landscapes in three distinct geographic locations. Population parameters for these primates were estimated in these habitats and in forest fragments. In the Central Pacific region of Costa Rica, one study area was a 540 km^2 landscape where the national park Manuel Antonio is located (9°59′N, 84°5′W and 9°43′N, 84°12′W; mean annual precipitation 3860 mm; altitudinal gradient sea level to 500 m.a.s.l.). In this landscape, forest fragments are surrounded by pastures and interdispersed with patches of second growth are forestry (*Tectona grandis* and *Gmelina arborea*) and African palm plantations (*Eliaeis guinneensis*), and other arboreal monocultures (e.g., bananas and mangos). Squirrel (*Saimiri oerstedii citrinellus*) and capuchin (*Cebus capucinus*) monkeys were found in some of these plantations.

Another study area was located in Cañas, province of Guanacaste (10°22′N and 85°08′W; annual precipitation varies from 1000 to 2500 mm; altitude 100–250 m.a.s.l.). The study site was a 100 km^2 fragmented landscape (original forest cover was tropical dry forest) which along with pastures also harbored patches of forestry (*Albizia saman* and *T. grandis*) and fruit plantations (e.g., banana—*Musa* spp., mango—*M. indica*, guayaba—*Psidum guajava*, guaba—*Inga* spp., among others), as well as networks of live fences (primarily live posts of *Bursera simarouba* (54.2% of the trees), *Pachiraquinat* (27.6% of the trees, *Ficus* spp. (3.8% of the trees), *G. sepium* (1.9% of trees), and *Tabebuia rosea* (1.9%), among others). The landscape is dominated by cattle production, with pastures covering 48.4% of the landscape and the remaining forest patches and riparian forests covering 23.3% of the landscape; the remaining area is under crop production (principally sugarcane) or small forest plantations. Howler monkeys (*A. palliata*) were present in this landscape and were observed opportunistically during an in-depth study of other taxa (birds, bats, dung beetles, and butterflies in this area).

A third study area consisted of another 100 km^2 fragmented landscape located in Rio Frio, Saraquipí, province of Heredia (10°22′N and 83°54′W; mean annual precipitation 3962 mm; altitude 80–250 m.a.s.l.) whose original vegetation was tropical wet forest. Pastures dominated the landscape (accounting for 45% of the landscape), but also present were forest fragments, riparian forests, fruit-tree groves (mainly *Citrus* sp.) palmito plantations, live fences (principally *Erythrina costarricensis* and *G. sepium*), and forestry plantations. Forest fragments and riparian forests together account for 20.7% of the landscape, but most of these are small (<10 ha).

Comalcalco, Tabasco, Mexico

The feeding ecology of a small population of mantled howler monkeys (*A. palliata*) was investigated for a 9-month-long period in 2003, in a 12 ha cacao plantation located in Comalcalco (18°26′N, 93°32′W; mean annual precipitation 2700 mm, altitude 10 m.a.s.l.), Tabasco, Mexico. Cacao trees in this plantation were mainly shaded by trees of *Pithecelobium saman* (Fabaceae) and *G. sepium* (Fabaceae) planted by the owners of the plantation about 50 years earlier. Individuals of another 28 tree species (e.g., *Ficus* spp., *M. indica*, *Cedrela odorata*, etc.) providing shade to the cacao trees were present in the plantation and these became established by planting by humans or via seed dispersal by birds, bats, and/or the primates that exist there. Observations of the feeding behavior of the howler monkey troop were conducted following standardized procedures (see Estrada *et al.*, 1999; Garcia del Valle *et al.*, 2003; Fuentes *et al.*, 2003; Muñoz, 2004 for details).

RESULTS

Los Tuxtlas, Mexico

In Los Tuxtlas, the presence of primates was detected in 8 of the 12 types of agroecosystems investigated. In these, we counted 184 monkeys, of which 73% were howler monkeys and 27% were spider monkeys. Howler and spider monkeys were present in 38% and in 16% of the agroecosystem sites we surveyed ($N = 82$; excluding live fences), respectively (Table 1). Both howler and spider monkeys were detected in the shaded cacao, coffee, and mixed plantations, in the mango/citrus/banana plantations and in old live fences. In addition, howler monkeys were detected in cacao plantations shaded by coconut palms or shaded by other trees and bananas, and in banana plantations (Table 1). No primates were found residing in the citrus, allspice, and mango groves we surveyed, but in a few instances we observed some individuals moving through sections of these groves that were adjacent to the forest where they resided. Presence of howler and of spider monkeys was observed only in old live fences (Table 1). According to the information provided to us by the owners of the farms, the howler and spider monkeys cause no damage to the cacao and coffee plants. However, crop damage was reported in the banana, mango, citrus, and allspice plantations (Table 1).

Comparison of Demographic Traits

A. palliata

Mean howler monkey population densities in the forest-shaded cacao (0.71 ± 0.15 individuals/ha), coffee (0.71 ± 0.18 individuals/ha), and mixed cacao/coffee (0.64 ± 0.21 individuals/ha) plantations, were significantly higher than in extensive forests ($0.23 + 0.05$ individuals/ha) (t-test, $p < 0.01$ in all cases; Figure 1), but significantly lower than in forest fragments (1.9 ± 1.13 individuals/ha; t-test, $p < 0.01$ in all cases; Figure 1a). Mean troop size was higher in the howler populations living in extensive forests (9.1 ± 2.93 individuals; $N = 20$ troops) than in cacao (6.0 ± 0.75 individuals; $N = 8$ troops) and coffee (6.0 ± 0.82 individuals; $N = 7$ troops) agroecosystems (t-test, $p < 0.01$). This mean value did not differ with respect to that in mixed cacao/coffee plantations (8.3 ± 8.68 individuals; $N = 6$ troops) (Figure 1), but

Table 1. Agroecosystems surveyed for presence of primates in Los Tuxtlas, Mexico. Impact (0) = neutral, (M) = minor, (−) = negative (e.g., crop/plant damage)

Agroecosystem	Condition	Number of sites surveyed	*Alouatta palliata* sites present	*Ateles geoffroyi* sites present	Impact
Los Tuxtlas, Mexico					
Cacao	Forest shade	10	6	4	(0)
Cacao	Legume trees shade	6	3	1	(0)
Coffee	Forest shade	10	6	2	(0)
Mixed (cacao/coffee)	Forest shade	8	6	3	(0)
Cacao	Coconut/banana shade	5	3	0	(0)
Citrus	Not shaded	10	0	0	(M)
Allspice	Not shaded	10	0	0	(M)
Mango	Not shaded	8	0	0	(M)
Mango/citrus/bananas	Not shaded	5	3	1	(−)
Bananas	Not shaded	10	4	2	(−)
Total		82	31	13	
Percentage of sites with presence			38	16	
Young live fences (mean DBH <25 cm)		25	0	0	(0)
Old live fences (mean DBH >25 cm)		25	12	3	(0)
Total sites with presence		132	43	16	
Total agroecosystems with presence			8	7	

mean howler monkey troop size was significantly higher in the agroecosystems and in the extensive forest than in the fragments (4.6 ± 1.92 individuals; $N=$ 37 troops; t-test, $p < 0.01$ in all cases) (Figure 1b). Mean immature to adult female ratios were 0.56 ± 0.15 in cacao, 0.55 ± 0.16 in coffee, and 0.58 ± 0.24 in mixed cacao/coffee. While these values did not differ from that in extensive forests (0.52 ± 0.37; Figure 1c), they were significantly higher (t-test, $p < 0.01$ in all cases) than the mean value in forest fragments (0.25 ± 0.21) (Figure 1).

A. geoffroyi

Spider monkeys live in small temporary subgroups of unstable composition, which are part of larger groups or communities. Because of the fusion–fission nature of their social organization, it is rare to see all members of the community in the same location, suggesting that it is not easy to make generalizations on density and/or subgroup size for this primate species (Kinzey, 1997). Bearing this in mind, the mean values we present next for population density, subgroup size, and immature to adult female ratios are gross estimates. Mean population density estimates for spider monkeys in agroecosystems (0.36 ± 0.35 individuals/ha in cacao, 0.45 ± 0.07 individuals/ha in coffee, and 0.68 ± 0.02 individuals/ha in mixed cacao/coffee) did not differ from those in extensive forests (0.37 ± 0.28 individuals/ha), but were significantly higher than those in forest fragments (0.04 ± 0.03 individuals/ha; t-test, $p < 0.01$ in all cases; Figure 2a). Mean spider monkey subgroup size (5.3 ± 1.52 individuals in cacao, $N = 6$ subgroups; 5.0 ± 1.41 individuals in coffee, $N = 5$ subgroups; 6.6 ± 1.52 individuals in mixed cacao/coffee, $N = 5$ subgroups; 6.0 ± 1.54 individuals in extensive forests, $N = 30$ subgroups; 5.0 ± 2.62 individuals in fragmented forests, $N = 10$ subgroups) did not differ statistically among habitats, and the only noticeable feature was the large variations in spider monkey mean subgroup size in the forest fragments compared to the smaller variation found in agroecosystems and extensive forests (Figure 2b). Mean immature to adult female ratios (0.72 ± 0.25 in cacao, 0.67 ± 0.29 in coffee, and 0.72 ± 0.25 in mixed cacao/coffee) did not differ from the mean value in extensive forests (0.88 ± 0.27), but they were significantly higher (t-test, $p < 0.05$ in all cases) than in forest fragments (0.19 ± 0.07) (Figure 2c).

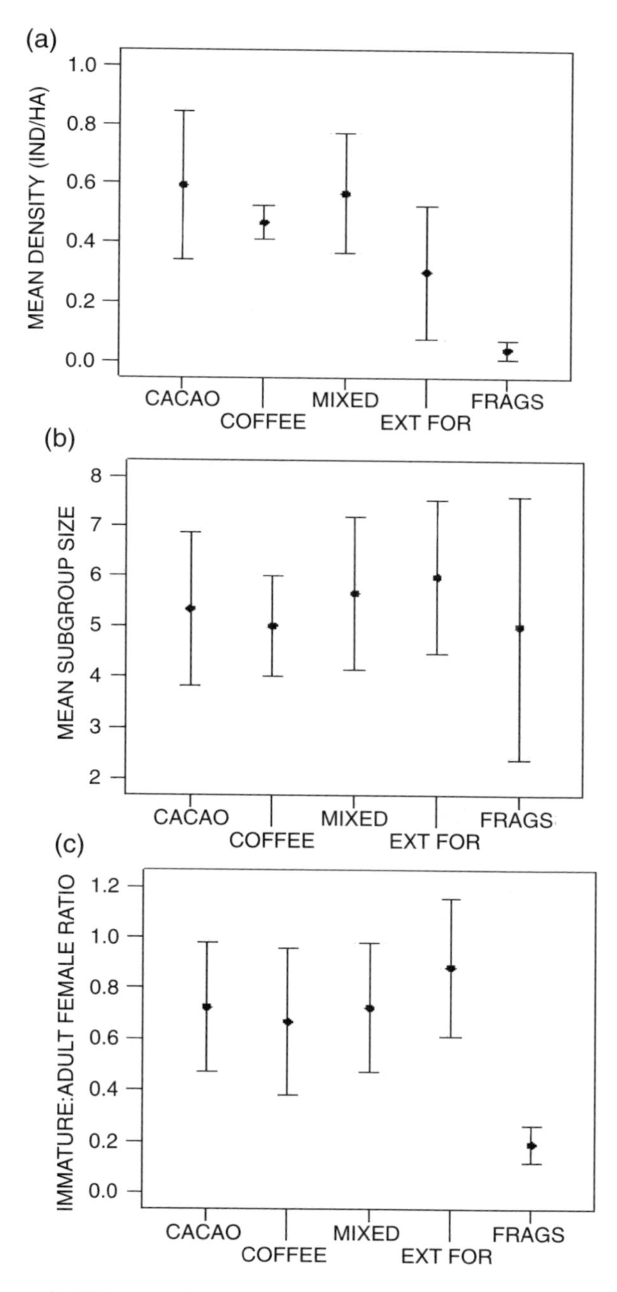

Figure 2. Mean (±SD) population density (a), mean troop size (b), and mean immature to adult female ratio (c), in populations of spider monkeys (*Ateles geoffroyi*) residing in shaded (rain forest vegetation) cacao, coffee and mixed (cacao and coffee) agroecosystems, in extensive forests (>100 km^2; "EXT FOR") and in forest fragments (<10 km^2; "FRAGS") in Los Tuxtlas, Mexico.

Lachuá, Guatemala

A. pigra

In Lachuá, Guatemala, howler monkey presence was recorded in forest-shaded cardamom and in coffee agroecosystems, but demographic data were collected only in the former habitat. Mean population density for black howler monkeys in cardamom agroecosystems was significantly smaller (0.59 ± 0.86 individuals/ha) than in forest fragments (2.48 ± 3.76 individuals/ha) (t-test, $p < 0.01$), but did not differ from mean values in extensive forests (0.21 ± 0.10 individuals/ha) (Figure 3a). Mean troop size in the cardamom plantations (5.36 ± 1.75 individuals; $N = 11$ troops) did not differ from mean values in forest fragments (5.00 ± 1.73 individuals; $N = 9$ troops), but it was significantly smaller (t-test, $p = 0.01$) than in extensive forests (6.54 ± 1.20 individuals; $N = 120$ troops) (Figure 3b). On average, mean immature to adult female ratios were higher in the cardamom (1.08 ± 0.55) and extensive forests (1.09 ± 1.40) than in forest fragments (0.85 ± 0.65), but the values in the cardamom plantations did not differ statistically from those in extensive and in fragmented forests (Figure 3c).

Costa Rica

In the fragmented landscape of the Central Pacific region, surveys showed the presence of squirrel and capuchin monkeys in fruiting-tree groves and in African palm plantations. Estimated mean densities for squirrel monkeys in the former habitat were 0.42 ± 0.14 individuals/ha, while in the latter was 0.14 ± 0.08 individuals/ha. These values did not differ from those in forest fragments (0.35 ± 0.24 individuals/ha) (Kruskal–Wallis test, $p = 0.17$) (Figure 4a). Mean troop size in the fruit and palm plantations was 19.0 ± 1.41 and 22.6 ± 9.87 individuals, respectively. These values were within the range of those in forest fragments (29.14 ± 13.9 individuals; Kruskal–Wallis test, $p = 0.35$) (Figure 4b). Mean immature to adult female ratios were 0.18 ± 0.03 in the fruit and 0.14 ± 0.05 in the palm plantation. These values did not differ from those in forest fragments (0.16 ± 0.06; Kruskal–Wallis test, $p = 0.44$) (Figure 4c).

In the case of capuchin monkeys, mean density was $0.63 + 0.11$ individuals/ha in the fruit and 0.10 ± 0.01 in the palm plantations, but values did not differ from those in forest fragments (0.31 ± 0.26) (Figure 5a). Mean troop size in the fruit and palm plantations was 12.5 ± 2.12 and 7.6 ± 1.53 individuals, respectively. The values in the fruit plantations fell within the range of those

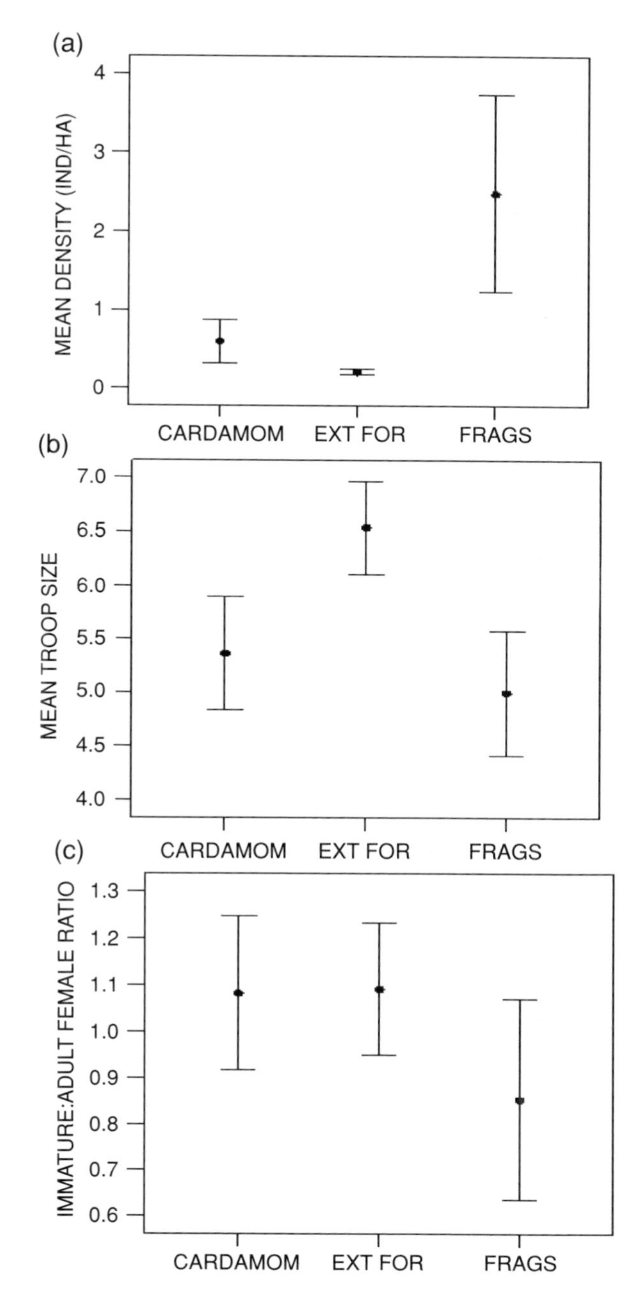

Figure 3. Mean (±SD) population density (a), mean troop size (b), and mean immature to adult female ratio (c), in populations of black howler monkeys (*Alouatta pigra*) living in cardamom agroecosystems shaded by rain forest vegetation, in extensive forests (>100 km^2; "EXT FOR") and in forest fragments (<10 km^2; "FRAGS") in Lachuá, Guatemala.

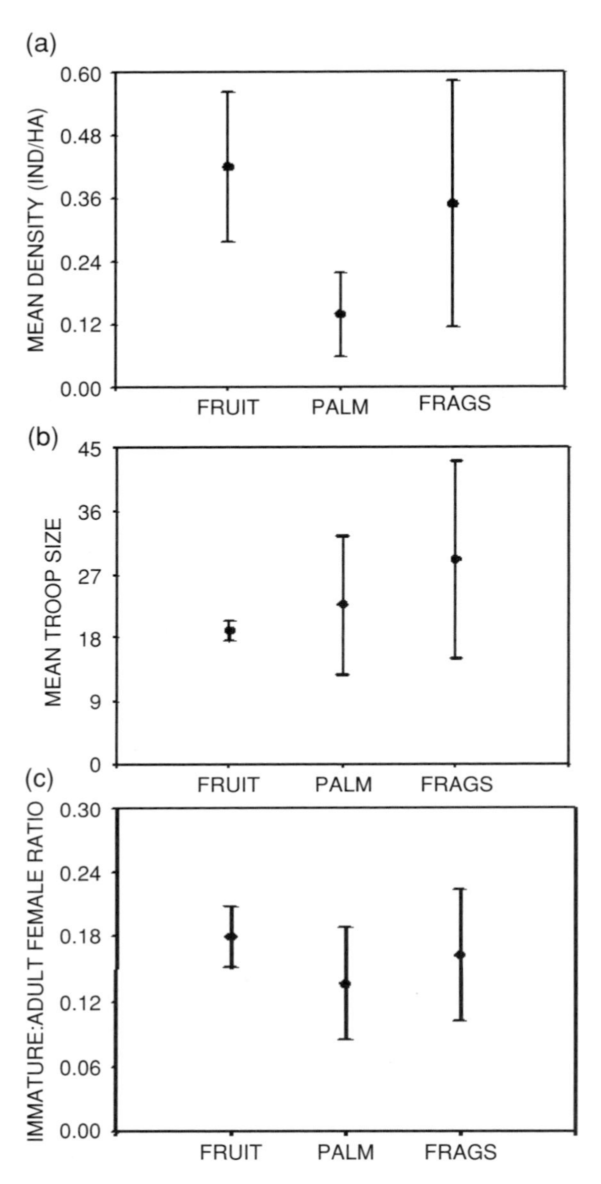

Figure 4. Mean (±SD) population density (a), mean troop size (b), and mean immature to adult female ratio (c), in populations of squirrel monkeys (*Saimiri oerstedii*) in fruit (FRUIT) and palm (PALM) agroecosystems and in forest fragments (FRAGS <10 km^2) in Central Pacific Costa Rica.

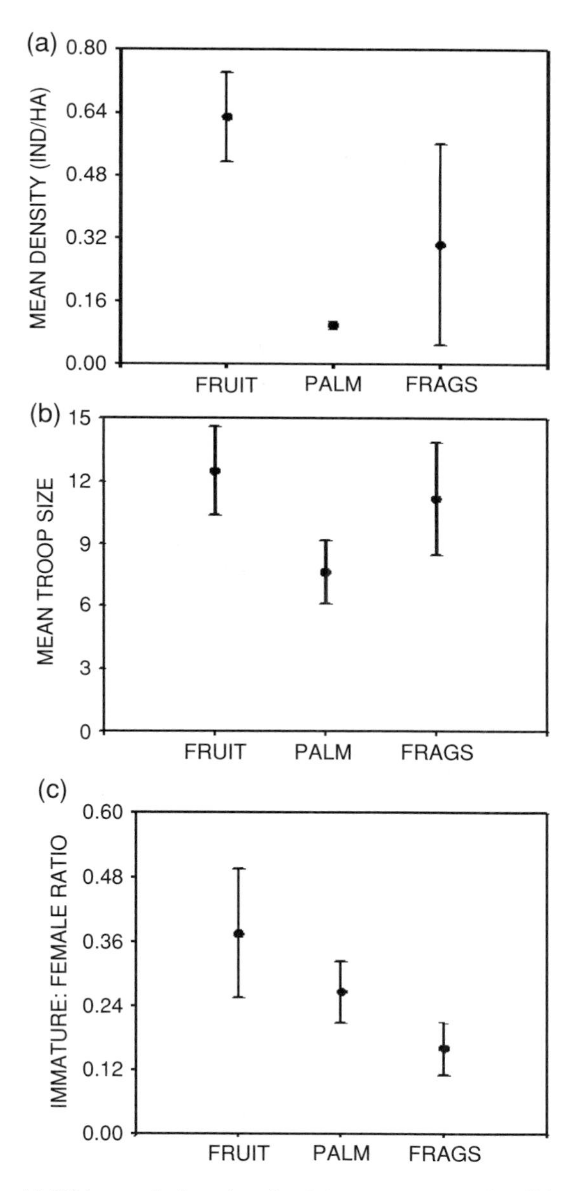

Figure 5. Mean (±SD) population density (a), mean troop size (b), and mean immature to adult female ratio (c), in populations of capuchin monkeys (*Cebus capucinus*) in fruit (FRUIT) and palm (PALM) agroecosystems and in forest fragments (FRAGS <10 km^2) in Central Pacific Costa Rica.

in forest fragments (11.1 ± 2.70 individuals; Kruskal–Wallis test, $p = 0.7$), but those in the palm plantations were significantly lower ($p < 0.05$) (Figure 5b). Mean immature to adult female ratios were 0.38 ± 0.12 in the fruit and 0.27 ± 0.06 in the palm plantation. In the forest fragments, mean values were 0.16 ± 0.05. The three habitats differed significantly in this measure (Kruskal–Wallis test, $p = 0.005$) (Figure 5c).

In Cañas, howler monkeys were found not only in forest fragments, but also in fruiting tree groves, forestry plantations, and in live fences. Mean density values in the first two agroecosystems were 0.55 ± 0.07 and 0.39 ± 0.16 individuals/ha, respectively, whereas in forest fragments the mean density was 0.48 ± 0.14 individuals/ha; habitats did not differ in this measure (Kruskal–Wallis test, $p = 0.09$) (Figure 6a). Mean troop size was 5.5 ± 0.71 and 8.5 ± 0.71 individuals at the fruit and at the forestry plantations, respectively. In the forest fragments, mean troop size was 7.8 ± 2.1 individuals. These values did not differ among habitats (Kruskal–Wallis test, $p = 0.21$) (Figure 6b). Mean immature to adult female ratios were 0.37 ± 0.05 and 0.41 ± 0.04 in the fruit and forestry plantation, respectively, and these were higher than in forest fragments (0.28 ± 0.08; Kruskal–Wallis test, $p = 0.03$) (Figure 6c). In Rio Frio, howler monkeys were observed in live fences and forest fragments; capuchins, on the other hand, were only found in forest fragments.

Howler Monkey Feeding Ecology in a Cacao Plantation, Comalcalco, Mexico

The howler population living in the 12-ha cacao plantation consisted of a single troop of 24 individuals (5 adult males, 11 adult females, 6 juveniles, and 2 infants; estimated population density was 2.0 individuals/ha). This troop (part of a larger howler monkey population once existing in the area when it was forested) has been living in the cacao plantation for as long as the plantation has been in existence (about 50 years; owners, pers. comm.). The 9-month-long investigation of the feeding ecology of the howler monkey troop in the cacao plantation revealed that howler monkeys did not use the *T. cacao* leaves, fruit or flowers as food. Instead they concentrated their foraging on the leaves, fruits, and flowers of 16 plant species (11 plant families), that together with other tree species, provided the shade to the cacao trees. Thirteen of the plant species used by the howlers as source of food were trees (nine botanical families), and the others were a liana and two epiphytes (Table 2). Three tree species, *Ficus*

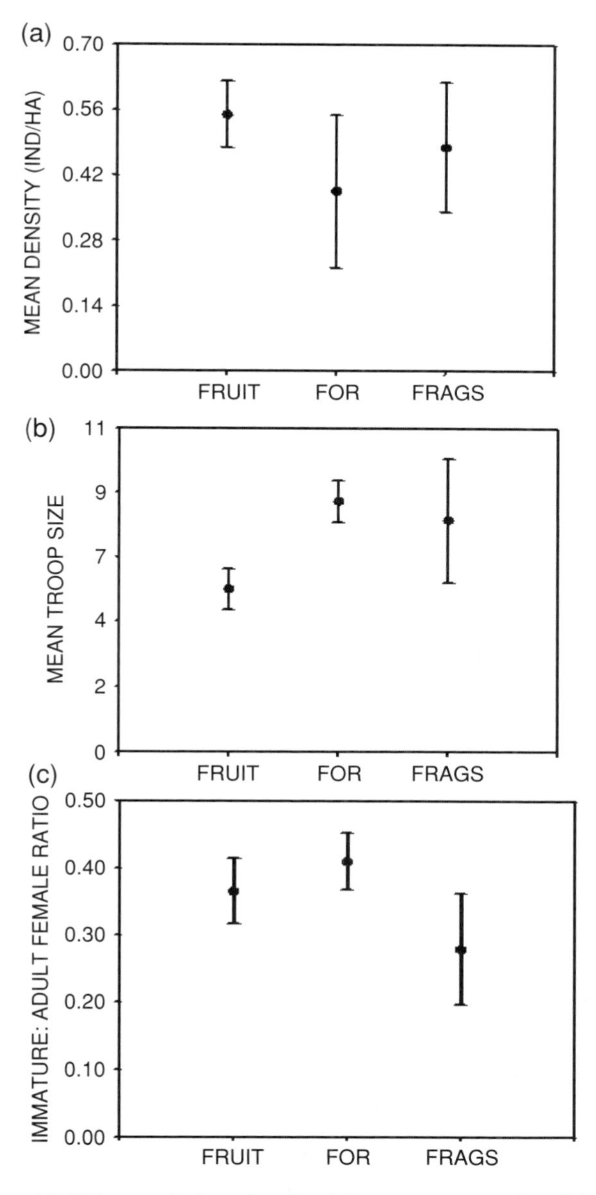

Figure 6. Mean (±SD) population density (a), mean troop size (b), and mean immature to adult female ratio (c), in populations of howler monkeys (*Alouatta palliata*) in fruit (FRUIT) and forestry (FOR) plantations and in forest fragments (FRAGS <10 km^2) in Cañas, Costa Rica.

Table 2. Plant species used as a source of food by a troop of howler monkeys ($N = 24$) living in a cacao plantation in Comalcalco, Tabasco, Mexico. Species with no code in parenthesis are trees. (E) = epiphyte, (V) = vine. Species are ranked by percent of feeding time

Species	Family	Trees used	Months used	Trees in site	Percent of feeding time
Ficus cotinifolia	Moraceae	22	9	36	41.6
Pithecellobium saman	Fabaceae	41	9	99	15.6
Gliricidia sepium	Fabaceae	30	9	103	12.7
Ficus sp.	Moraceae	4	6	6	8.7
Ficus obtusifolia	Moraceae	11	8	18	7.1
Spondias mombin	Anacardiaceae	6	3	13	5.3
Diphysa robinioides	Fabaceae	13	8	35	3.5
Manilkara zapota	Sapotaceae	1	2	1	2.1
Mangifira indica	Anacardiaceae	2	3	7	0.9
Busera simaruba	Burseraceae	1	1	6	0.7
Eritrina americana	Fabaceae	6	3	55	0.5
Cecropia obtusifolia	Cecropiaceae	1	1	14	0.4
Selenicereus sp. (E)	Cactaceae	1	4	–	0.4
Terminalia amazonia	Combretaceae	1	2	3	0.3
Paullinia pinata (V)	Sapindaceae	1	1	–	0.2
Syngonium podophyllum (E)	Araceae	1	1	–	0.03

cotinifolia, *P. saman*, and *G. sepium*, accounted for 70% of total feeding time recorded, and three additional tree species contributed to 21%; the rest of the tree species accounted for another 8% (Table 2).

The number of species used per month as a source of food by the howler monkeys in the cacao plantation ranged from 5 to 11 (mean 8.0 ± 1.63), and Sorensen's index of species overlap between adjacent months ranged from 0.57 to 0.84 (mean 0.76 ± 0.09) (Figure 7). Consumption of young leaves (50.7%; range 23.0–69.9% of feeding time per month) and of mature fruits (29.1%; range 11.9–63.6% of feeding time per month) predominated in the howlers' diet (Figure 8). Three important correlations were detected in patterns of resource use. First, use of tree species was positively associated to their relative abundance in the plantation. Second, percent of feeding time per species was found positively associated to the number of months species were used as a source of food by the howlers. Third, the number of plant parts used per species was found to be positively associated to percent of feeding time per species (Figure 9).

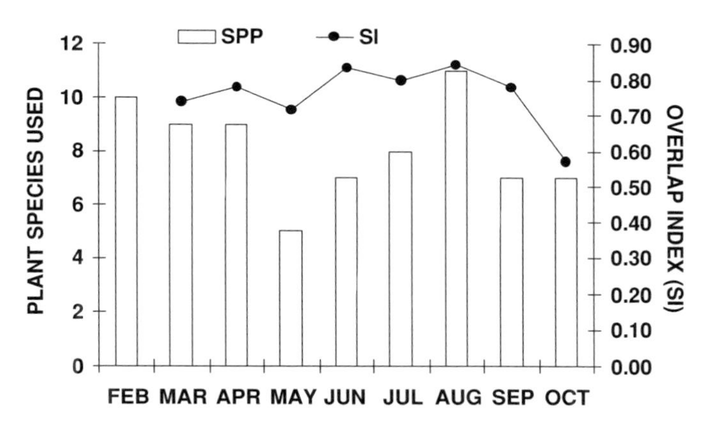

Figure 7. Monthly variation in use of plant species by a howler monkey troop living in a cacao plantation shaded by planted trees in Comalcalco, Tabasco, Mexico. Also shown are the values of Sorensen's index of species overlap between adjacent months (a value of 0 = no overlap, a value of 1 = 100% overlap).

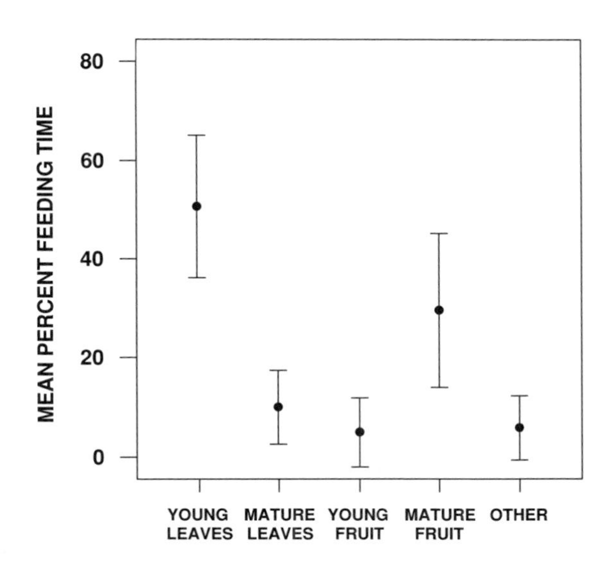

Figure 8. Mean monthly percent of feeding time spent by the howlers consuming the plant parts that were their major source of food in the cacao plantation in Comalcalco, Tabasco, Mexico.

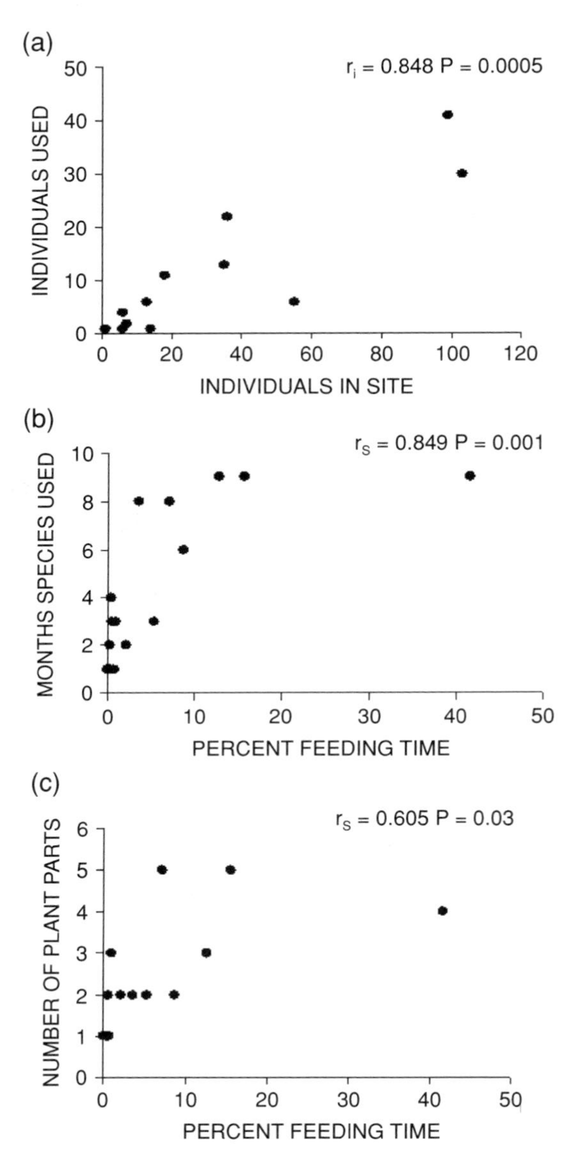

Figure 9. Associations between number of trees of species used by the howler monkeys as a source of food present in a cacao plantation and number of trees actually used by the monkeys for this purpose (a), number of month each species in the howlers' diet was used and percent of feeding time (b), and number of plant parts used (young leaves, mature leaves, young fruit, mature fruit, and others) per species and percent of feeding time (c).

DISCUSSION

General Aspects

The results of our study showed that, in fragmented Mesoamerican landscapes, primates use some agroecosystems as habitat for permanent and/or temporary residency. Some of these habitats seem also to facilitate the movements of primates in the fragmented landscape. For example, howler, spider, squirrel, and capuchin monkeys have been observed moving from forest patch to forest patch by making use of arboreal agroecosystems in the landscape, such as shaded coffee and cacao agroecosystems. Further, mature live fence trees (>25 cm DBH) with wide, intact canopies may be used by heavy primates such as howler and spider monkeys for the same purpose, while younger live fences may support smaller squirrel and capuchin monkeys, thus avoiding movement on the ground and through open areas. Live fences may also offer food to primates that visit these habitats. For example, howler, spider, squirrel, and capuchin monkeys have been observed consuming the leaves and fruits of *B. simaruba* and of *Ficus* spp., *G. sepium*, *Spondias* spp., *Cordia* spp., which in many localities in Mesoamerican are some of the most important tree species with which local people build live fences (Harvey *et al.*, submitted). These species have also been reported as top ranking tree species in the leaf and fruit diet of some of these primates (Milton, 1980; Estrada, 1984; Ramos-Fernández and Ayala-Orozco, 2003).

Data also showed that not all agroecosystems may be suitable for primate visitation and/or residency. For example, our surveys indicated that primates did not permanently or temporarily reside in citrus, allspice, and mango groves and only occasionally visited banana plantations. Usually, these plantations were bordering the forest patch in which the monkeys resided or they traveled to them by moving along a strip of forest or of old live fences, returning to their forest patch shortly afterwards. Several factors may mitigate against visitation and or residency by primates in these habitats. The wide inter-row space between the cultivated plants and their sparse vegetation mean lack of suitable structures for arboreal locomotion by large monkeys such as howler and spiders. Extreme climatic conditions in this habitats, as well as greater exposure to potential predators, including humans and dogs, may deter primates from visiting or establishing temporary or permanent residency in these agroecosystems. In the case of live fences, the narrow width of live fences (the average width of live fences is generally <8 m; Harvey *et al.*, submitted) means that these elements are only able to serve as passageways, rather than as true habitats.

In contrast, in agroecosystems such as cacao, coffee, mixed cacao/coffee, and cardamom, growing under the shade of rain forest trees, and in forestry plantations, the complexity of the mid and upper canopy, enhanced by the numerous epiphytes, vines, lianas, and other climbing plants, present on the trees, offer many potential food resources, shelter, resting sites, and cover from potential dangers for howler, spider, squirrel, and capuchin monkeys making use of these habitats (Estrada and Coates-Estrada, 1996; Greenberg, 2004). The presence in these agroecosystems of tall (>20 m) rain forest trees of plant families such as Moraceae (e.g., *Ficus* spp., *Poulsenia armata*, *Brosimum alicastrum*), Fabaceae (e.g., *Pterocarpus rorhii*, *Lonchocarpus guatemalensis*), Sapotaceae (e.g., *Manilkara zapota*, *Pouteria campechiana*), Anacardiaceae (*Spondias radlkoferi*), Lauraceae (e.g., *Nectandra ambigens*, *Ocotea* spp.), and Annonaceae (e.g., *Rollinia jimenezii*), among others, means the existence of a contiguous canopy cover for these arboreal primates. Further, some of these tree species are also known to be an important source of leaves and of fruit for the monkeys (Estrada, 1984; Estrada *et al.*, 1999; Ramos-Fernández and Ayala-Orozco, 2003). Data from Costa Rica suggest that squirrel and capuchin monkeys may also reside in large (>100 ha) African palm plantations because these contain small patches of other trees where the monkeys find shelter and a relatively high abundance of potential food represented by the palm fruit (they feed on the sugary pulp encasing the seed) and by insects found in the palm fronds and trunk.

Although there was some variability in the demographic parameters examined within and between populations of the primate species present in the agroecosystems investigated in Los Tuxtlas, population density, mean troop size, and immature to adult female ratios of these populations more closely resembled those in extensive forest tracts than in fragmented landscapes. While high population densities for *A. palliata* and *A. pigra* in forest fragments forests are suggestive of saturation of remnants (Estrada and Coates-Estrada, 1996; see Van Belle, this volume), in spider monkeys, the lower densities detected in forest fragments than in agroecosystems and extensive forests, may be the result of hunting and low fruit availability.

In Los Tuxtlas, the smaller mean group sizes and lower immature to adult female ratio of howler and of spider monkeys in forest fragments than in extensive forests and in agroecosystems suggest lower reproductive potential. Both howler and spider monkeys are subjected to hunting, illegal pet trade, and continued habitat degradation in small forest fragments, pressures with a higher impact upon spider monkey populations (Duarte and Estrada, 2003). In contrast, hunting of monkeys is practically non existent in the cacao and coffee plantations where howler and spider monkeys exist.

In general, population parameters such as mean group size and immature to adult female ratios of primates in agroecosystems were higher than in forest fragments and in some cases approached those in extensive forests. This suggests that primate populations living in the agroecosystems sustain reproductive potential. Permanent and semi-permanent residency in agroecosystems and use of these as stop-overs are the feature of the various ways in which primates use these habitats in the landscapes investigated. Howler monkeys have been observed to reside in several cacao, coffee, and cacao/coffee plantations in Los Tuxtlas for >15 years, but spider monkeys residency in these habitats is less permanent, 3, 6, and 12 months, after which they have moved to nearby patches of forest vegetation or to other shaded plantations. Here, the patchy nature of the resources preferred by spider monkeys (e.g., mature fruit) may exert important constraints upon the length of time they can reside in cacao or in coffee plantation, as these usually constitute small (4–15 ha) units of vegetation in the landscape. In contrast, howler monkeys can persist for several decades in these habitats, as our study in Comalcalco has shown, by exploiting the leaves, fruit, and flowers of individuals of major tree species (*F. cotinifolia, P. saman,* and *G. sepium*) providing shade to the cacao plants, trees which they consistently seek out in the plantation.

Impact of Primates in Agroecosystems

Long-term observations of primates in the cacao, coffee, and cacao/coffee agroecosystems and interviews with the farmers in Los Tuxtlas, indicate that the monkeys residing in these habitats do not feed on the economically important fruits; instead they concentrate their feeding on leaves and fruits of the tall rain forest trees providing the shade for the plantation. In Los Tuxtlas as well as in other sites in Mesoamerica, farmers may tolerate a certain amount of damage to fruit crops such as citrus, bananas, allspice, among others, especially when these crops constitute a minor source of income in their subsistence. However, when the plantations are a key source of income and excessive damage is produced by the monkeys, humans respond by shooting the monkeys or by aggressively chasing them away. Such reactions may deter primates from residing or regularly visiting these habitats.

Notwithstanding the above, the presence and activities of primates may be beneficial to the plantations. For example, the feeding activities of howlers at these habitats may favor primary productivity by accelerating the flow of nutrients and the conversion of matter and energy (Estrada and Coates-Estrada, 1993). The ingestion of fruits may favor the dispersal of seeds of species that are their sources of fruit, contributing to the persistence of trees of these species in the plantations (Estrada and Coates-Estrada, 1991).

Monkey defecation may also result in important additions and dispersal of nutrients to the soil of the plantation. It has been reported that the waste excretion of howlers tend to be very nutrient rich (Milton *et al.*, 1980; Nagy and Milton, 1979), producing dung that contains 1.8–2.1% N and 0.3–0.4% P (based on dry mass measurements; Milton *et al.*, 1980). In contrast, concentrations of nutrients in leaf litter are ~1% N and 0.04% P for tropical moist forests (Vitousek and Sanford, 1986, cited in Feeley, 2004). Using these data, Feeley (2004) reports that total soil nitrogen concentration under the trees in which howlers defecate in Venezuelan forests was 1.6–1.7 times greater than in control sites (test plots in surrounding soil), and that phosphorus concentration was 3.8–6 times greater under their resting or resting/feeding trees than in the surrounding soil, probably enriching the soil and nutrient uptake of these trees (Feeley, 2004). In the agroecosystems in which primates reside, this may benefit not only the trees use for resting and/or feeding, but also the cacao, coffee, cardamom, and other cultivated plants growing directly under these.

Conservation Implications

In spite of the preliminary nature of our investigation, it is evident that certain types of agroecosystems in Mesoamerican fragmented landscapes have an important potential in favoring the persistence of primate populations. These agroecosystems may be used as stepping-stones when primates move through the landscape or as foraging habitats or as habitats for temporary or permanent residency. Our surveys in Los Tuxtlas (Mexico), Lachuá (Guatemala), and in the three landscapes in Costa Rica, showed that 15 types of economically important agroecosystems are used by the Mesoamerican primate species investigated (Table 3). Seven of these are shaded either by rain forest vegetation or by arboreal vegetation planted by man, and monkeys were found temporarily

Table 3. Summary of agroecosystems in which primate populations were found in Mexico (Los Tuxtlas), Guatemala (Lachuá), and Costa Rica (Central Pacific, Cañas, and Rio Frío). The asterisk indicates the habitats in which monkeys were found permanently or temporarily residing. The other habitats are used as foraging stop-overs or as stepping stones when moving in the fragmented landscape

Agroecosystem	Condition	*Alouatta palliata*	*Alouatta pigra*	*Ateles geoffroyi*	*Saimiri oerstedii*	*Cebus capucinus*
Cacao*	Forest shade	×		×		
Cacao*	Legume trees shade	×				
Coffee*	Forest shade	×	×	×		
Mixed (cacao/coffee)*	Forest shade	×		×		
Cacao*	Coconut banana shade/	×				
Cardamom*	Forest shade		×			
Forestry plantations*	Shaded	×			×	×
Citrus	Not shaded	×				
Allspice	Not shaded	×				
Mango	Not shaded	×		×	×	×
Mango/citrus/ bananas	Not shaded	×		×		
Bananas	Not shaded	×		×		
Fruit-tree groves	Not shaded	×		×	×	×
African Palm*	Not shaded				×	×
Live fences	Not shaded	×	×	×	×	×

or permanently residing in some of these. The others are basically used as stop-over habitats to forage or to move from one patch of vegetation to another (Table 3). The presence of extensive networks of live fences in many parts of Mesoamerican modified landscapes seems to enhance connectivity among isolated social units existing in native and anthropogenic patches of vegetation, and monkey may also find food resources in these linear habitats.

Habitat loss and fragmentation reduces the availability of adequate habitats and the effective size of primate populations, and results in isolation of remnant populations which are subjected to stochastic demographic events that put them at risk (Chapman and Ribic, 2002; Henle *et al.*, 2004a,b) (Figure 10a). But tolerance of species to habitat loss and fragmentation may be related to an ability to traverse open areas to reach other forest fragments or other vegetation types and use resources within the matrix (see Mandujano *et al.*, this volume; Law *et al.*, 1999; Schulze *et al.*, 2000). Such tolerance may be enhanced by the presence of patches of agroforests and of other arboreal agroecosystems (*sensu*

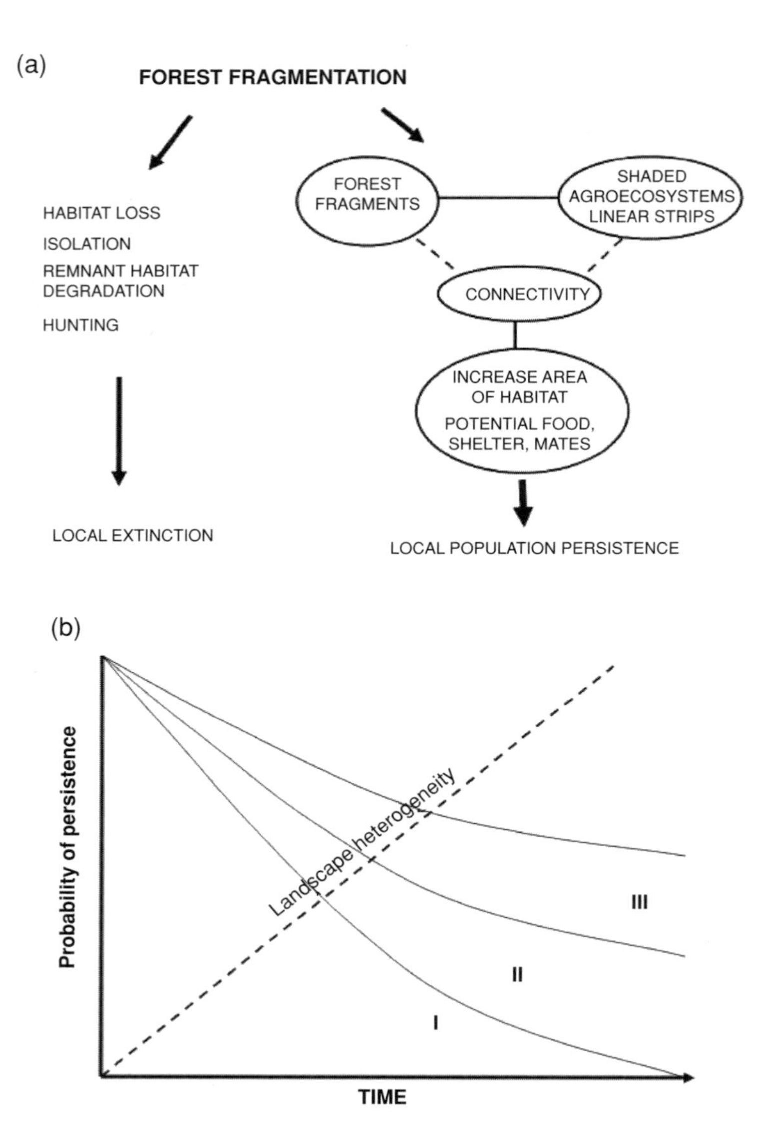

Figure 10. (a) Paths in landscape changes resulting from habitat fragmentation by human activity. The route on the left leads to extensive habitat loss and fragmentation/isolation of remnant primate populations, with a rapid decline toward extinction. In this scenario, the landscape is mainly dominated by pasturelands. The route on the right, consists of landuse patterns in which forest fragments are sourrounded by a heterogenous matrix consisting of pastures and different types of agroecosystems. Such conditions may allow primate populations to persist in the human modified landscape. (b) Three scenarios of land management with varying effects upon primate population persistence. (I) A few forest fragments and pasturelands (low landscape heterogeneity), (II) forest fragments, a few arboreal agroecosystems and networks of live fences, (III) forest fragments and a complex and diverse assemblage of arboreal agroecosystems and agroforests, and complex networks of linear strips of vegetation (high landscape heterogeneity).

Schroth *et al.*, 2004) in intermediate positions (Figure 10a). Heterogeneity of the landscape, involving various types of arboreal agroecosystems, including live fences, may be an important general feature of some landscapes favoring dispersal and possibly connectivity between isolated segments of primate populations (see Mandujano *et al.*, this volume; Laurance *et al.*, 2002; Harvey *et al.*, 2004). The presence of agroecosystems in fragmented landscapes may represent, for remnant primate fauna, increased area of vegetation available, increased diversity of resources and habitats potentially available, increased stop-over points in the matrix, and possibly reduced isolation of populations in forest fragments. The net medium and long-term effect of the interaction among these factors may be persistence of primate populations/species in the landscape (Figure 10a).

Depending on the complexity of the landscape, we could contemplate three conservations scenarios. In one, continuous forest is reduced to a collection of a few forest fragments, with primate populations undergoing fragmentation and isolation and rapid decline in population sizes (I, Figure 10b). In a second scenario, the landscape may contain forest fragments and patches of one or two (e.g., forest-shaded cacao and coffee) types of agroecosystems, including live fences. Under this scenario (II, Figure 10b), the enhanced structural and functional connectivity may allow primate populations to persist for a longer time than in the first scenario. A third, and more complex scenario, may be one in which the fragmented landscape is highly heterogeneous. Here, in addition to forest fragments, the landscape has more patches of more types of arboreal shaded agroecosystems (e.g., cacao, coffee, cacao/coffee, and cardamom) and of sun-loving arboreal plantations (e.g., citrus, allspice, etc.) located at distances not far from one another and from forest patches. Interdispersed in the landscape are also linear strips of forest vegetation along rivers and streams and a complex network of live fences that interconnect the various patches of forest and man-made arboreal vegetation in the landscape. Under this scenario, the likelihood of primate population persistence (assuming no other pressures) may be at its highest (III, Figure 10b).

Caveats to Consider

Adjacency and/or proximity of agroforests and of other types of arboreal agroecosystems to forest fragments, as well as the presence of networks of live fences may favor dispersal of primates in fragmented landscapes. However, we also need to consider to what extent such movements may place individuals and/or groups in a perilous matrix where they are more exposed to the elements, to humans, to dogs, and to other dangers. Moreover, in fragmented landscapes forest fragments and agroecosystems may also act as ecological sinks and traps for primate populations (Kristan, 2003; Murphy and Lovett-Doust, 2004; Laurance and Vasconcelos, 2004). For example, our records for Los Tuxtlas showed that not all plantations surveyed were occupied by howler and/or spider monkeys. Thus, howlers and spider monkeys were absent in 62% and in 84% of the plantation sites surveyed, respectively. This suggests that in many cases resources may not be sufficient to support primate populations, structural connectivity of landscape vegetation units may be insufficient to facilitate dispersal or that people or other stochastic events may have eradicated the primates from these habitats. It is also not clear about the threshold level of landscape tree cover below which primates would be lost from an agroecosystem. Is there an overall level of tree cover and connectivity that must be maintained or is a threshold level of forest cover that is more important? Another thing that is often not known is the degree to which primates in agroecosystems are actively moving to other forested areas and depend on these other areas.

Changes in regional and world market demands may result in the disappearance of or in change in the local and regional distribution of agroforests and of other agroecosystems where primate populations can exist. For example, the current trend to switch from forest-shade coffee to sun-loving coffee in many Mesoamerican countries may mean an important loss of habitats where primate populations could persist (Perfecto and Armbrecht, 2003). Similarly, the trend to expand cultivation of sun-loving coffee at the expense of areas dedicated to the cultivation of forest-shaded cacao has similar consequences (Rice and Greenberg, 2000; FAO, 2004). In many areas of Mesoamerica, forest-shaded cacao and cacao agroforestry systems have been abandoned due to disease problems, and converted to other land uses (such as pastures, banana or plantain production), which have lower value for biodiversity conservation.

In conclusion, our investigation suggests that further research is needed to document the value of certain types of agroecosystems for the persistence of primate populations in fragmented landscapes in Mesoamerica and also to work with farmers to seek ways in which agricultural landscapes can be managed sustainably for both productive and conservation goals. Such research needs to assess how the primate species present in such landscapes respond to the presence of different types of agroecosystems and to their spatial configuration, to determine threshold levels of tree and forest cover within agroecosystems for primate habitation, to determine the medium and long-term conservation value of specific agroecosystems, and to evaluate how primate populations can be managed in those cases where they may become agricultural pests. Such research is of relevance in light of the proposed Mesoamerican Biological Corridor project (UNDP, 1999; World Bank, 2004; http://www.biomeso.net/), in which a major objective is to sustain local biodiversity and diminish isolation of animal and plant populations in natural protected areas. To achieve this, the MBC project contemplates the sustainable use of fragmented landscapes in intermediate locations among natural protected areas in the region.

SUMMARY

While there is a general perception that agricultural activities are the principal threat to primate biodiversity in the tropics, empirical evidence was presented in this paper to investigate the value of certain types of agroecosystems for sustaining primate populations in fragmented landscapes in Mesoamerica. Presence of primates was investigated in Los Tuxtlas, Mexico, in Lachuá, Guatemala, and in three landscapes in Costa Rica. We also compared the similarity in population parameters (density, group size, and immature to adult female ratios) of five primate species (*A. palliata*, *A. pigra*, *A. geoffroyi*, *S. oerstedii*, and *C. capucinus*) living in agroecosystems with those of the same species living in extensive and/or in fragmented forests. Primates were found in 15 agroecosystems. Some species were found residing in shaded agroecosystems (e.g., cacao, coffee), but not in unshaded plantations (e.g., citrus, allspice), which were used as foraging or stop-over habitats. For howler and spider monkeys in Mexico, mean values of primate demographic parameters in agroecosystems more closely resembled those in extensive than in fragmented forests. Those for squirrel and capuchin monkeys fell within the range of populations in forest fragments. Farmers reported crop damage by primates in banana, mango, citrus, and allspice plantations, but responses toward the monkeys' activities ranged from tolerance to expulsion. No damage was reported by howler and spider monkeys to the shaded cacao, coffee, and cardamom plants or in forestry plantations. Some primate species can persist in cacao plantation by exploiting the leaves and fruits of tree species providing shade for the cultivated plants, while others can do so by visiting various agroecosystems on a regular basis. Our study suggests that certain types of agroecosystems, specifically those grown under the shade of forest or of planted trees, favor the persistence of primate populations in fragmented landscapes. At these habitats, the presence and feeding activities of primates may benefit the plantations by accelerating primary productivity, by dispersing the seeds of their fruit sources, and by adding important amounts of nutrients, via their defecation, to the soil of the plantation.

ACKNOWLEDGMENTS

We are grateful to the Cleveland Zoo Scott Neotropical Fund and to Universidad Nacional Autónoma de México for support, to the government of Guatemala for permission to conduct the primate surveys in Lachuá. We thank the support from the EU and from Universidad Nacional. Heredia and the Tropical Agricultural Research and Higher Education Center (CATIE) of Costa Rica. We also acknowledge the support from CONACYT (Mexico) and from ECOSUR-San Cristobal de las Casas (Mexico). Research by J. Saenz and C. Harvey in Cañas and Rio Frio in Costa Rica was partially supported by the FRAGMENT project, funded by the EU (ICA4-CT-2001-10099). Finally, we are grateful to Paul Garber for insightful comments on earlier versions of the manuscript, and to the WRPRC for library support.

REFERENCES

Altieri, M. A. 2004, Globally Important Indigenous Agricultural Heritage Systems (GIAHS): Extent, significance, and implications for development. http://www.fao.org/ag/agl/agll/giahs/documents/backgroundpapers_altieri.doc

Chapman, E. W. and Ribic, C. A. 2002, The impact of buffer strips and stream-side grazing on small mammals in southwestern Wisconsin. *Agric. Ecosyst. Environ.* 88:45–59.

Daily, G., Ceballos, G., Pacheco, J., Suzan, G., and Sanchez-Azofeifa, A. 2003, Country side biogeography of neotropical mammals: Conservation opportunities in agricultural landscapes in Costa Rica. *Conserv. Biol.* 17:1815–1826.

Donald, P. F. 2004, Biodiversity impacts of some agricultural commodity production systems. *Conserv. Biol.* 18:17–37.

Duarte, A. and Estrada, A. 2003, Primates as pets in Mexico City: An assessment of species involved, source of origin and general aspects of treatment. *Am. J. Primatol.* 61:53–60.

Estrada, A. 1984, Resource use by howler monkeys in the rain forest of Los Tuxtlas, Veracruz, Mexico. *Int. J. Primatol.* 5:105–131.

Estrada, A. and Coates-Estrada, R. 1991, Howling monkeys (*Alouatta palliata*), dung beetles (Scarabaeidae) and seed dispersal: Ecological interactions in the tropical rain forest of Los Tuxtlas, Veracruz, Mexico. *J. Trop. Ecol.* 7:459–474.

Estrada, A. and Coates-Estrada, R. 1993, Aspects of ecological impact of howling monkeys (*Alouatta palliata*) on their habitat: A review, in: A. Estrada, E. Rodriguez Luna, R. Lopez-Wilchis, and R. Coates-Estrada, eds., *Avances en: Estudios Primatologicos en Mexico I. Asociacion Mexicana de Primatologia*, A.C. y Patronatto Pro-Universidad Veracruzana, A.C. Xalapa, Veracruz, Mexico, pp. 87–117.

Estrada, A. and Coates-Estrada, R. 1996, Tropical rain forest fragmentation and wild populations of primates at Los Tuxtlas. *Int. J. Primatol.* 5:759–783.

Estrada, A. and Coates-Estrada, R. 2000, Bird species richness in vegetation fences and in strips of residual rain forest vegetation at Los Tuxtlas, Mexico. *Biodiv. Conserv.* 9:1399–1416.

Estrada, A. and Coates-Estrada, R. 2002a, Bats in continuous forest, forest fragments and in an agricultural mosaic habitat-island at Los Tuxtlas, Mexico. *Biol. Conserv.* 2:237–245.

Estrada, A. and Coates-Estrada, R. 2002b, Dung beetles in continuous forest, forest fragments and in an agricultural mosaic habitat-island at Los Tuxtlas, Mexico. *Biodiv. Conserv.* 11:1903–1918.

Estrada, A., Coates-Estrada, R., and Meritt, D. Jr. 1993, Bat species richness and abundance in tropical rain forest fragments and in agricultural habitats at Los Tuxtlas, Mexico. *Ecography* 16:309–318.

Estrada, A., Coates-Estrada, R., and Meritt, D. Jr. 1994, Non flying mammals and landscape changes in the tropical rain forest region of Los Tuxtlas, Mexico. *Ecography* 17:229–241.

Estrada, A., Juan Solano, S., Ortíz Martínez, T., and Coates-Estrada, R. 1999, Feeding and general activity patterns of a howler monkey (*Alouatta palliata*) troop living in a forest fragment at Los Tuxtlas, Mexico. *Am. J. Primatol.* 48:167–183.

FAO 1999, *State of the World's Forest.* FAO, Rome, Italy.

FAO 2000, State of Forestry in the Region-2000. Latin American and Caribbean Forestry Commission FAO. Forestry series no. 15. FAO http://www.rlc.fao.org

FAO 2001, Food Security (www.document). URL, http:/www.faor.org/biodiversity

FAO 2004, FAO Statistical Data Bases. http://faostat.fao.org/

Feeley, K. 2004, The role of clumped defecation in the spatial distribution of soil nutrients and the availability of nutrients for plant uptake. *J. Trop. Ecol.* 20:1–4.

Fuentes, E., Estrada, A., Franco, B., Magaña, M., Decena, Y., Muñoz, D., and García, Y. 2003, Reporte Preliminar Sobre el Uso de Recursos Alimenticios por Una Tropa de Monos Aulladores, *Alouatta palliata*, en El Parque La Venta, Tabasco, México. *Neotrop. Primates* 11:24–29.

García del Valle, Y., Muñoz, D., Magaña-Alejandro, M., Estrada, A., and Franco, B. 2003, Selectividad del Alimento en una tropa de monos aulladores *Alouatta palliata* en el Parque Yumká, Tabasco, México. *Univ. Cien. (Mexico)* 37:34–42.

Greenberg, R. 2004, Biodiversity in the cacao agroecosystem: Shade management and landscape considerations. *Migratory Bird Center Website.* Smithsonian Institution, Washington, D.C.

Harvey, C., Tucker, N., and Estrada, A. 2004, Can live fences, isolated trees and windbreaks help conserve biodiversity within fragmented tropical landscapes? in: G. Schroth, G. Fonseca, C. Gascon, H. Vasconcelos, A. M. Izac, and C. Harvey, eds., *Agroforestry and Conservation of Biodiversity in Tropical Landscapes.* Island Press Inc., New York, pp. 261–289.

Harvey, C. A., Villanueva, C., Villacis, J., Chacon, M., Munoz, D., Lopez, M., Ibrahim, M., Gomez, R., Taylor, R., Martinez, J., Navas, A., Saenz, J., Sanchez, D., Medina, A., Vilchez, S., Hernandez, B., Perez, A., Ruiz, F., Lopez, F., Lang, I., and Sinclair, F. L. Contribution of live fences to the productivity and ecological integrity of agricultural landscapes in Central America. *Agriculture, Ecosystems and Environment*, submitted.

Henle, K., Davoes, K. F., Kleyer, M., Margules, C., and Settele, J. 2004a, Predictors of species sensitivity to fragmentation. *Biodiv. Conserv.* 13:207–251.

Henle, K., Lindemayer, D. B., Margules, C. R., Saunders, D. A., and Wissel, C. 2004b, Species survival in fragmented landscapes: Where are we now? *Biodiv. Conserv.* 13:1–8.

Kinzey, W. G. 1997, Ateles, in: W. G. Kinzey, ed., *New World Primates: Ecology, Evolution and Behavior*, Aldine de Gruyter, New York, pp. 192–199.

Kristan, B. K. III. 2003, The role of habitat selection behavior in population dynamics: Source-sink systems and ecological traps. *Oikos* 103:457–468.

Laurance, W. F., Lovejoy, T. E., Vasconcelos, H. L., Bruna, E. M., Dirham, R. K., and Stoufer, P. C. 2002, Ecosystem decay of Amazonian forest fragments: A 22 year investigation. *Conserv. Biol.* 16:605–618.

Laurance, W. F. and Vasconcelos, H. L. 2004, Ecological effects of habitat fragmentation in the tropics, in: G. Schroth, G. Fonseca, C. Gascon, H. Vasconcelos, A. M. Izac, and C. Harvey, eds., *Agroforestry and Conservation of Biodiversity in Tropical Landscapes*, Island Press Inc., New York, pp. 33–49.

Law, B. S., Anderson, J., and Chidel, M. 1999, Bat communities in a fragmented landscape on the south-west slopes of New South Wales, Australia. *Biol. Conserv.* 88:333–345.

Melbourne, B., Davies, K., Margules, C. R., Lindenmayer, D. B., Saunders, D. A., Wissel, C., and Henle, K. 2004, Species survival in fragmented landscapes: Where to from here? *Biodiv. Conserv.* 13:275–284.

Michon, G. and de Foresta, H. 1995, The Indonesian agro-forest model, in: P. Halladay and D. A. Gimour, eds., *Conserving Biodiversity Outside Protected areas: The role of traditional agroecosystems*, IUCN, Gland, Switzerland, pp. 90–106.

Milton, K. 1980. The Foraging Strategy of Howler Monkeys: *A Study in Primate Economics*. Columbia University Press, New York.

Milton, K., Van Soest, P. J. and Robertson, J. B. 1980, Digestive efficiencies of wild howler monkeys. *Physiol. Zool.* 4:402–409.

Moguel, P. and Toledo, V. M. 1999, Biodiversity conservation in traditional coffee systems of Mexico. *Conserv. Biol.* 13:11–21.

Muñoz, D. 2004, Uso de Recursos Alimentarios por Monos Aulladores (*Alouatta palliata*) en un Cacaotal del Estado de Tabasco, México. MSc Thesis, Colegio de la Frontera Sur, Chiapas, Mexico.

Murphy, H. T. and Lovett-Doust, J. 2004, Context and connectivity in plant populations and landscape mosaics: Does the matrix matter? *Oikos* 105:3–14.

Nagy, K. and Milton, K. 1979, Aspects of dietary quality, nutrient assimilation and water balance in wild howler monkeys (*Alouatta palliata*). *Oecologia* 39:249–258.

National Research Council. 1992, *Techniques for the Study of Primate Population Ecology*. National Academy Press, Washington, D.C.

Nowak, R. M. 1999, *Walker's Primates of the World*. The John Hopkins University Press, Baltimore.

Perfecto, I. and Armbrecht, I. 2003, The coffee agroecosystem in the Neotropics: Combining ecological and economic goals, in: J. H. Vandermeer, ed., *Tropical Agroecosystems*, CRC Press, New York, pp. 159–194.

Pimentel, D., Stachow, D. A., Takacs, H. W., Brubaker, A. R., Dumas, J. J., Meaney, J. A. S., O'Neil, D., Onsi, E. and Corzilius, D. B. 1992, Conserving biological diversity in agricultural/forestry systems. *Bioscience* 5:354–362.

Ramos-Fernández, G. and Ayala-Orozco, B. 2003, Population size and habitat use of spider monkeys in Punta Laguna, Mexico, in: L. K. Marsh, ed., *Primates in Fragments: Ecology and Conservation*, Kluwer/Plenum Press, New York, pp. 191–209.

Rice, R. A. and Greenberg, R. 2000, Cacao cultivation and the conservation of biological diversity. *Ambio* 3:167–176.

Rodriguez-Luna, E., Cortez-Ortiz, L., Mittermeier, R. A., Rylands, A. B., Wong-Reyes, G., Carrillo, E., Matamoros, Y., Núñez, F., and Motta-Gill, J. 1996, Hacia un plan de accion para los primates Mesoamericanos. *Neotrop. Primates* 4(Suppl.):112–133.

Rosales-Meda, M. 2003, Abundancia, distribución y composición de tropas del mono aullador negro (*Alouatta pigra*) en diferentes remanentes de bosque en la eco región, Lachuá. BSc Thesis, Universidad de San Carlos, Guatemala.

Rowe, N. 1996, *A Pictorial Guide to the Living Primates*. Pogonias Press, Charsletown, RI.

Sader, S. S., Hayes, D. J., Irwin, D. E., and Saatchi, S. S. 1999, Preliminary forest cover change estimates for Central America (1990's), with reference to the proposed Mesoamerican Biological Corridor. NASA Jet Propulsion Lab, http://www.ghcc.msfc.nasa.gov/corredor/

Salafsky, N. 1993, Mammalian use of a buffer zone agroforestry system bordering Ganung Palung National Park, west Kalimantan, Indonesia. *Conserv. Biol.* 7:928–933.

Schroth, G., Fonseca, G., Gascon, C., Vasconcelos, H., Izac, A. M., and Harvey, C., eds. 2004, *Agroforestry and Conservation of Biodiversity in Tropical Landscapes.* Island Press Inc., New York.

Schulze, M. D., Seavy, N. E. and Whitacre, D. F. 2000, A comparison of phyllostomid bat assemblages in undisturbed Neotropical forest and in forest fragments of a slash-and burn farming mosaic in Petén, Guatemala. *Biotropica* 32:174–184.

Somarriba, E., Harvey, C. A., Samper, M., Anthony, F., Gonzales, J., Slaver, Ch., and Rice, R. A. 2004, Biodiversity conservation in Neotropical coffee (*Coffea Arabica*) plantations, in: G. Schroth, G. Fonseca, C. Gascon, H. Vasconcelos, A. M. Izac, and C. Harvey, eds., *Agroforestry and Conservation of Biodiversity in Tropical Landscapes*, Island Press Inc., New York, pp. 198–226.

UNDP. 1999, Establishment of a Programme for the Consolidation of the Mesoamerican Biological Corridor Project document (RLA/97/G31). United Nations Development Programme. Global Environment Facility. Project of the Governments of Belize, Costa Rica, El Salvador, Guatemala, Honduras, Mexico, Nicaragua, Panama. L:\bd\regional\mesoamerica\MBCprodoc

Vandermeer, J. H. 2003, *Tropical Agroecosystems.* CRC Press, Boca Raton, FL.

Villaseñor, J. F. and Hutto, R. L. 1995, The importance of agricultural areas for the conservation of neotropical migratory landbirds in Western Mexico, in: M. Wilson, and S. Sader, eds., *Conservation of Neotropical Migratory Birds in Mexico*, Maine Agricultural and Forest Experiment Station. Miscellaneous Publications 727, pp. 59–80.

Wilson, D. E., Cole, F. R., Nichols, J. D., Rudran, R., and Foster, M., eds. 1996, *Measuring and Monitoring Biological Diversity: Standard Methods for Mammals (Biological Diversity Handbook Series)*. Smithsonian Institution Press, Washington, D.C.

World Bank. 2004, http://www.worldbank.org/data/

CHAPTER NINETEEN

Primate Populations in the Protected Forests of Maya Archaeological Sites in Southern Mexico and Guatemala

Alejandro Estrada, Sarie Van Belle, LeAndra Luecke, and Marleney Rosales

INTRODUCTION

The tropical forests of southern Mexico harbor the northernmost Neotropical primates. These are represented by four taxa (*Alouatta palliata*, *Alouatta pigra*, *Ateles geoffroyi vellerosus*, and *Ateles geoffroyi yucatanensis*), two species of howler monkeys, one of which, *A. pigra*, is endemic to the region of Mesoamerica shared by Mexico, Belize, and Guatemala (Horwich and Johnson,

Alejandro Estrada • Estación de Biología Los Tuxtlas, Instituto de Biología, Universidad Nacional Autónoma de México, Mexico. **Sarie Van Belle** • Department of Zoology, University of Wisconsin-Madison, 250 N Mills Street, Madison, WI 53706. **LeAndra Luecke** • Department of Anthropology, Washington University, St. Louis, Missouri, USA. **Marleney Rosales** • Universidad de San Carlos, Guatemala.

New Perspectives in the Study of Mesoamerican Primates: Distribution, Ecology, Behavior, and Conservation, edited by Alejandro Estrada, Paul A. Garber, Mary S. M. Pavelka, and LeAndra Luecke. Springer, New York, 2005.

1986; Rylands *et al.*, 1995; Watts and Rico-Gray, 1987). The other howler monkey species, *A. palliata*, is found in all of Mesoamerica, and in northern Colombia, and Ecuador (Rowe, 1996). The other two taxa are two subspecies of spider monkeys, *A. g. vellerosus* and *A. g. yucatanensis*. The former subspecies is present in most of southern Mexico, while the latter is endemic to the Yucatan peninsula (Watts and Rico-Gray, 1987; Rowe, 1996).

In Mexico, *A. palliata* has been intensively studied at the northernmost limit of its geographic range in Los Tuxtlas, Veracruz. Much is known about extant populations and about some aspects of their ecology, behavior, and conservation in the region (Estrada, 1982; Estrada and Coates-Estrada, 1996; Estrada *et al.*, 2001; Rodríguez-Toledo *et al.*, 2003). About 80% of the geographic distribution of *A. pigra* is found in Mexico, but most information available for this howler monkey species is derived mainly from only a few sites in Belize (Horwich and Johnson, 1986; Ostro *et al.*, 1999; Pavelka *et al.*, 2003) and from Tikal in Guatemala (Coelho *et al.*, 1976; Schlichte, 1978). More recently, published information on populations of *A. pigra* has become available for other sites in southern Mexico (Gonzales-Kirchener, 1998; Estrada *et al.*, 2002a,b, 2004).

In the case of the spider monkeys, information is available for populations of *A. g. vellerosus* in Los Tuxtlas (Estrada and Coates-Estrada, 1996; Rodríguez-Toledo *et al.*, 2003; González-Zamora, 2003), but for *A. g. yucatanensis*, information is particularly scanty and derived from studies of social behavior in a fragmented landscape in the Punta Laguna reserve in the northeast of the Yucatan peninsula (Ramos-Fernández and Ayala-Orozco, 2003), and from population surveys in extensive forests in other parts of the peninsula (Cant, 1978; Gonzales-Kirchener, 1999; Estrada *et al.*, 2004; Barrueta *et al.*, 2003).

The scarcity of information for the primate taxa present in southeast Mexico and the rapid fragmentation and conversion of their natural habitat to pasture lands and agricultural fields at a rate of −1.1% (UNEP, 2004; CCAD, 2003), coupled with pressures from illegal hunting and the trafficking of infants as pets (Duarte and Estrada, 2003), makes the task of conservation of extant primate populations particularly problematic. Adequate conservation assessments, at regional and local scales of howler and spider monkeys in southeast Mexico require demographic data on populations existing in various types of protected forests and in human-modified landscapes. While some initial progress has been made in this direction (Estrada *et al.*, 2001, 2002a,b, 2004), more information

Figure 1. Maya region in southern Mexico, Guatemala, Belize, El Salvador, and Honduras. Gray areas are Natural Protected Areas in southeast Mexico, in northern Guatemala, and in Belize. The approximate location of Maya archeological sites investigated are shown as black triangles. Numbers refer to the identity of the site as in Table 1.

is needed on land-use patterns that put populations at risk or that favor the persistence of such populations.

The Maya civilization developed in the area of Mesoamerica shared by Mexico, Belize, Guatemala, El Salvador, and Honduras. It has an antiquity close to 3000 years, reached its peak between 200 and 700 AD, but was a civilization long gone by the time the Spaniards arrived to the area (Gómez-Pompa *et al.*, 2003) (Figure 1). The Maya built huge cities such as Tikal, Palenque, and Chichen-Itza, among others, harboring tens of thousands of inhabitants and had sophisticated numerical and calendar systems, as well as writing (Flannery, 1982). They lived in the lowlands, an area dominated by various types of tropical rain forest vegetation and archeologists estimate that at its peak, population density in urban areas exceeded 1000 people/km^2, ranged from 500 to 800 people/km^2 for peri-urban areas and approached 200 people/km^2 for large rural areas (Turner *et al.*, 2003). To feed the population, the Maya used extensive techno-managerial strategies to manage landscape mosaics composed of various simple crops and of complex agroecosystems, in which the use of forest fruiting trees and of other forest plants were particularly important (Barnhart, 2001; Allen and Rincón, 2003; Turner *et al.*, 2003). For some still debated reasons, Mayan cities were "suddenly" abandoned between 700 and 1200 AD, and the area depopulated rapidly (Turner *et al.*, 2003; Gómez-Pompa, 2003; Allen and Rincón, 2003).

The evidence of this great civilization can be seen today in the hundreds of Maya archeological sites scattered in southern Mexico, Belize, Guatemala, Honduras, and El Salvador in which the tips of monumental buildings rise above the forest canopy. In Mexico, Maya archeological sites and culture are an important part of the cultural patrimony of the nation, and the sites attract public attention and are visited by large numbers of national and foreign visitors every year. Because of their national and world-wide cultural value, the Mexican government has developed an extensive program of work and protection for the Maya archeological sites. This includes not only the area where major Mayan ruins are concentrated, but also several square kilometers of land around them (INAH, 2004). Archeological activity in many Maya sites is usually localized in space and time. This and the use of sonar and satellite technologies to map the ancient structures causes negligible disturbance to the buried building and to the surrounding vegetation (Barnhardt, 2001). While some Maya sites occur as small isolated forest units in the modern landscape, others are embedded in larger forested areas that are part of a system of natural protected areas in southern Mexico, and thus benefit also from the protection afforded to these areas by the Mexican government.

The protected forests surrounding Maya archeological sites not only safeguard the ancient ruins, but they also protect important source populations of local biodiversity. Primate populations exist in many of these forests, but until now no systematic documentation of these populations was available (see Estrada *et al.*, 2002a,b, 2004). In this paper, we present results from the first phase of a series of primate population surveys aimed at assessing the conservation value of protected forests surrounding Maya archeological sites for primate populations in southern Mexico. Because of their close vicinity to Mexico and because they occur within the same eco-region, we also present information on primate surveys in three Maya sites located in Guatemala.

METHODS

During several field sessions conducted in each year between 2001 and 2003, we explored the forests surrounding 19 Maya archeological sites found within the geographic range documented for *Alouatta* and *Ateles* (Figure 1). The areas of forest protected by each site ranged from 200 to 2700 ha (mean 484.2 ± 316.6 ha; median 500 ha) (Table 1). Six sites were embedded in larger (>30,000 ha) protected forested areas (four in Mexico and two in Guatemala) (Table 1). Tall evergreen rain forest (annual rainfall >3500 mm) was the dominant vegetation in seven of the Maya sites explored and semi-evergreen rain forest (annual rainfall >2000 and <3000 mm) was predominant in 12 sites (Table 1).

A total of 400 field days were completed surveying the sites. The majority of the sites were visited 2–3 times during the 3-year period. To locate *Alouatta* and *Ateles*, we triangulated at dawn (04:00–08:00 h) from the top of the tallest Maya temples short and long distance vocalizations produced by howler monkeys. On a few occasions, vocalizations by spider monkeys were recorded, but *Ateles* was mostly encountered during ground surveys. Sightings of monkeys from the same vantage points also were recorded. Once triangulation was completed, we carried out ground surveys following standardized sampling protocols (National Research Council, 1992; Wilson *et al.*, 1996) to locate primate groups and to obtain specific counts of individuals of each sex and age (see Estrada *et al.*, 2002a,b, 2004 for details of procedures). Interviews with local guards and other staff working at the site and with Maya indigenous inhabitants living in the area provided additional information on the presence and history of the primate populations at the sites.

RESULTS

Primate Populations

Primate populations were detected in 90% of the Maya sites surveyed (Table 1). In the westernmost site of the Maya area, Comalcalco, we found a small population of *A. palliata* inhabiting the surrounding forest. This forest is a mixture

Table 1. Demographic parameters for populations of *Alouatta* and of *Ateles* found in the Maya sites investigated. Tikal, Uaxactum, and Lachúa are found in Guatemala. El Tormento, Palenque, Yaxchilán, Jbptomor, Tikal, and Uaxactum are sites embedded in larger forests that are part of the system of Natural Protected Areas in Mexico and in Guatemala. NP—no primates present; NA—data not available, but primates present. Imm:F—immature to adult female ratio. Site numbers refer to map locations in Figure 1

	Site	Area (ha) sampled	Density (*Alouatta* individuals/km^2)	Mean troop size of *Alouatta*	Density (*Ateles* individuals/km^2)	Mean group size	*Alouatta* Imm:F	*Ateles* Imm:F	*Alouatta* sex ratio	*Ateles* sex ratio
1	Comalcalco	200	11.0	22.0 ± 0.0	NP	NP	0.66	NP	1.60	NP
2	Palenque[a,b]	600	23.0	7.0 ± 2.8	NP	NP	1.48	NP	1.14	NP
3	Lacanjá[b]	500	6.0	6.2 ± 1.3	2.0	3.1 ± 2.1	0.35	0.30	0.60	0.40
4	Bonampák[b]	500	4.0	6.5 ± 1.2	2.0	3.0 ± 1.2	0.34	0.35	1.06	0.45
5	Lachúa[b]	200	6.8	5.3 ± 1.7	8.0	6.4 ± 3.2	0.97	0.72	0.79	0.86
6	Tikal[a,b]	500	17.8	8.7 ± 2.2	56.4	4.7 ± 2.6	1.23	0.71	1.40	1.61
7	Uaxactum[a,b]	500	10.7	6.5 ± 1.3	12.0	6.0 ± 1.7	1.22	0.70	1.30	1.00
8	Yaxchilán[a,b]	100	12.8	6.6 ± 2.1	17.2	5.6 ± 3.0	0.97	0.83	1.08	0.83
9	Balancan	100	NP	NP	NP	NP	NP	NP	NP	NP
10	Tormento[a]	500	12.7	6.7 ± 2.7	6.4	4.6 ± 1.2	0.94	0.46	1.63	1.50
11	Calakmul[a,b]	500	15.2	7.5 ± 1.9	17.0	7.7 ± 3.8	1.22	0.74	1.25	2.07
12	Xpujil	200	9.0	6.5 ± 1.0	NP	NP	0.45	NP	0.45	NP
13	Hormiguero	300	NP	NP	NA	NA	NA	NA	NA	NA
14	Rio Bec	400	NA	NA	NA	NA	NA	NA	NA	NA
15	Kohunlich	400	4.5	6.5 ± 1.2	3.0	3.0 ± 0.8	0.50	0.40	0.60	0.50
16	Sian Ka'an	500	NA	NA	NA	NA	NA	NA	NA	NA
17	Xcaret	200	6.0	5.5 ± 0.7	NP	NP	0.70	NP	1.04	NP
18	Coba	300	NP	NP	NP	NP	NP	NP	NP	NP
19	Jbptomor	500	NP	NP	10	4.5 ± 2.3	NP	0.7	NP	1.50

[a] Sites embedded in larger protected areas. Palenque National Park Palenque 20 km^2; Calakmul 7000 km^2; El Tormento 14 km^2; Yaxchilán 27 km^2; Tikal and Uaxactum 3000 km^2. Area for other sites is total area of protected forest.

[b] Tall evergreen rain forest (annual rainfall >3500 mm); rest of sites semi-evergreen rain forest (annual rainfall >2000 and <3000 mm).

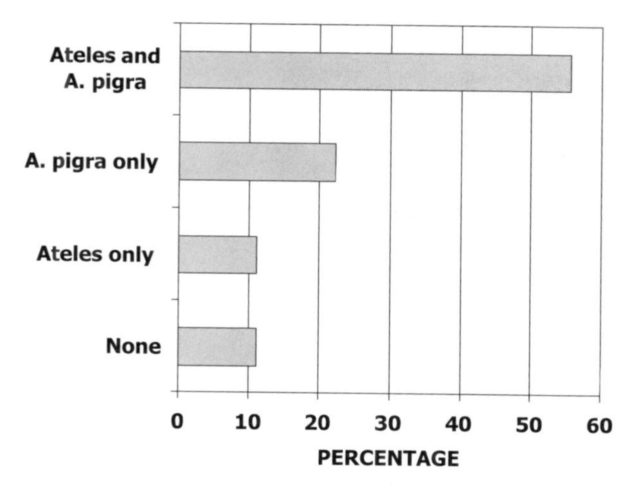

Figure 2. Proportion of Maya sites in the core of the Maya region in which populations of *Alouatta pigra* and of *Ateles geoffroyi yucatanensis* were found.

of secondary growth and primary forest, with remnants of once cultivated cacao trees. Adjacent to it is a fully operational cacao plantation. The howlers' range encompassed both the differently aged forest surrounding the site and the forest providing shade for the cacao trees in the plantation. Population density for *A. palliata* at the site was estimated at 11.0 individuals/km^2.

At the other sites ($N = 18$) the breakdown by taxa was as follows: 61% of the sites had populations of *A. geoffroyi* and of *A. pigra*, 17% had only *A. pigra*, 11% had only *Ateles*, and 11% had no primates (Figure 2). Estimated population densities for *A. pigra* varied from 3.5 individuals/km^2 (Lacanjá) to 23 individuals/km^2 (Palenque). Overall mean density was 10.8 ± 5.7 individuals/km^2; median 9.85 individuals/km^2 (Figure 3; Table 1). Estimated population densities for *A. geoffroyi* varied from 2.0 individuals/km^2 (Bonampák) to 56 individuals/km^2 (Tikal). Overall mean density was 13.4 ± 16.1 individuals/km^2; median 9.0 individuals/km^2 (Figure 3; Table 1). For both *A. pigra* and *A. geoffroyi* population, density was not associated with area encompassed by the protected forest ($r_s = 0.140$, $p = 0.643$ and $r_s = 0.078$, $p = 0.811$, respectively).

Mean troop size in *A. pigra* was 6.6 ± 0.87 individuals, and mean subgroup size in *A. geoffroyi* was 4.8 ± 1.5 individuals. Adult female to immature ratio for *A. pigra* ranged from 0.34 (Bonampák) to 1.48 (Palenque) (overall mean 0.84 ± 0.37). In *Ateles*, this ratio ranged from 0.30 (Lacanjá) to 0.83 (Yaxchilán) (overall mean 0.59 ± 0.19, Table 1). Adult sex ratios in *A. pigra*

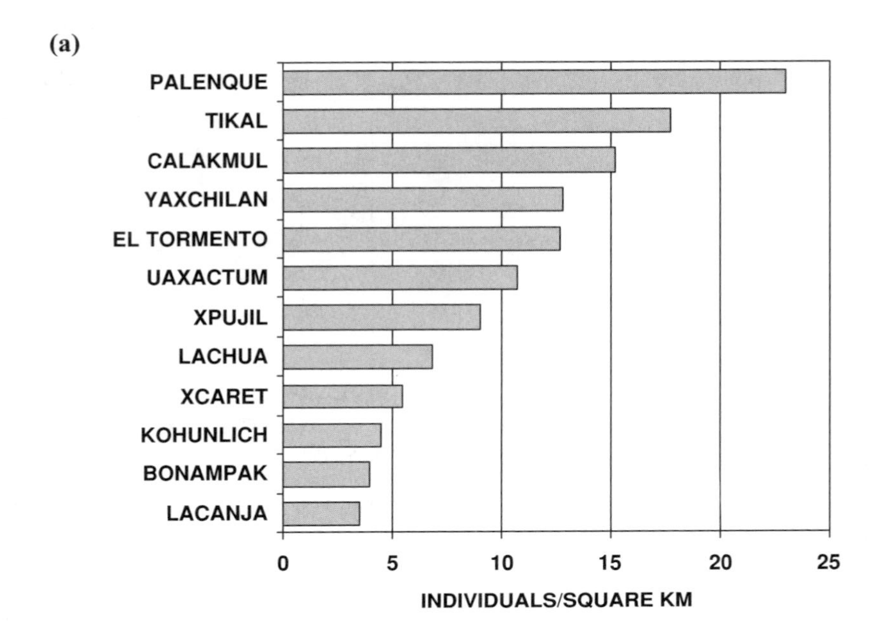

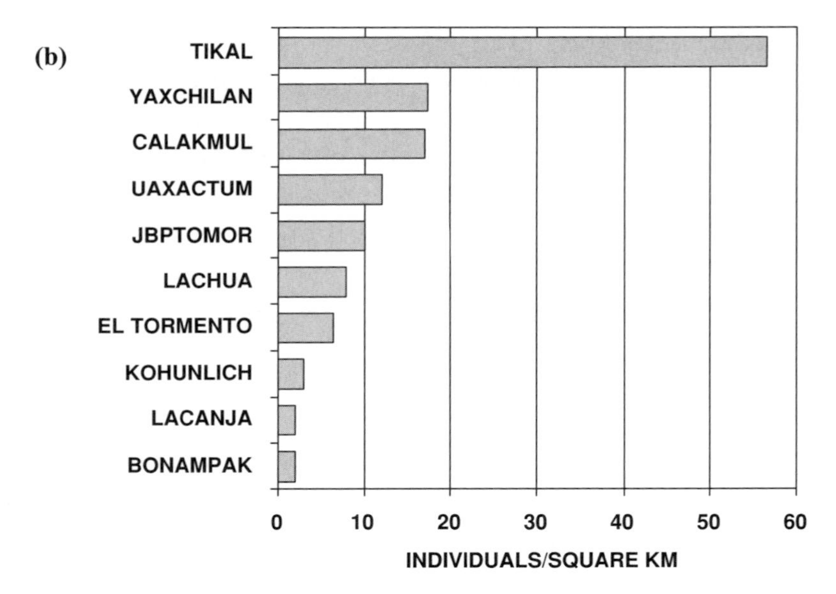

Figure 3. Estimated densities for (a) *Alouatta pigra* and (b) *Ateles geoffroyi yucatanensis* at the Maya sites investigated. Tikal, Uaxactum, and Lachúa are sites located in Guatemala.

ranged from 1:0.45 to 1:1.63 (overall mean 1:1.07 ± 1:0.38). In *Ateles*, this ratio ranged from 1:0.40 to 1:2.07 (overall mean 1:1.07 ± 1:0.57, Table 1).

DISCUSSION

Our study showed that the protected forests surrounding many archeological sites in the Maya region harbor populations of *Alouatta* and of *Ateles*. While there is much variability in the sizes of the areas of the protected forests around Maya sites, in their degree of connectedness to larger tracts of forests or to fragmented landscapes, and in whether the sites are embedded in natural protected areas or not, the conservation value of these forests for primate populations is significant.

This is particularly important considering that deforestation rates in southern Mexico for the period 1990–2000 are reported to be about −1.1% per year (FAO, 2000; World Resources Institute, 2004). Currently only about 28% of the total land area in southeast Mexico (this includes the Mexican states of Chiapas, Tabasco, Campeche, Yucatan, and Quintana Roo) is covered with forests. According to FAO statistics, forest cover in the region has been decreasing steadily, while pastures and crop land areas have been expanding for the period 1960–2000 (Figure 4) (FAO, 2001, 2004; World Resources Institute, 2004). Current human population in the region is estimated at about 8.0 million people and density is *ca.* 31 people/km^2 (FAO, 2004; World Resources Institute, 2004). Growth projections show a steady increase in human population from 2004 to 2030 (Figure 5) (FAO, 2004). This will result in

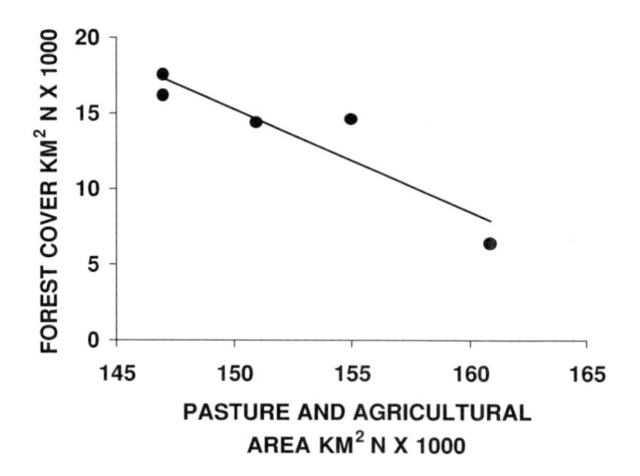

Figure 4. Relationship between increase in pasture and in agricultural land area and decrease in forest cover in 10-year intervals for the period 1960–2000 in southeast Mexico (source of data FAO, 2004). Trend line is shown only for illustrative purposes.

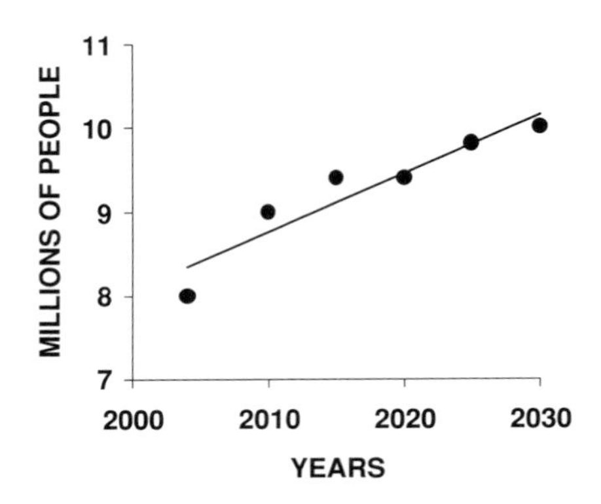

Figure 5. Population growth projection in 5-year intervals for southern Mexico based on FAO statistics (source of data FAO, 2004). Trend line is shown only for illustrative purposes.

greater demand for land, and increased food production to satisfy local and global market demands that will exert pressures upon natural resources resulting in extensive losses of primate habitat.

Concerned with the need to preserve the regional biodiversity, Mexico established a system of 29 natural protected areas (NPA) in the Maya region. These encompass some 3.8 million ha (UNDP, 1999; CCAD/UNEP/GEF. 2002). However, several of these protected areas preserve ecosystems that may be unsuitable for primate habitation (e.g., low deciduous forest, mangroves, coastal sand dunes, and marine ecosystems), others do protect areas harboring tall evergreen and semi-evergreen rain forest vegetation suitable for the existence of primates (Estrada *et al*., 2004; Van Belle and Estrada, this volume). Because of surrounding deforestation, many of these reserves are becoming virtual islands of native vegetation in human-modified landscapes (UNEP, 2004). In some of these, conservation efforts may not be effective due to illegal extraction of wood, poaching, and squatters (WWF, 2002; CCAD, 2003). Hunting and the illegal pet trade in surrounding areas and at the periphery of some of these reserves exert further pressures on remnant primate populations (Duarte and Estrada, 2003). In this context, the value of protected forests surrounding Maya sites to primate conservation merit consideration for the following reasons: most of the forested areas surrounding the ruins are protected—such protection is enforced 24 ha day, deforestation activities are non existent, and as our surveys have shown, many of these forests preserve populations of the four primate taxa that occur in southern Mexico, and in particular of the two primate taxa endemic to the Maya region.

In the majority of the Maya sites, we noted a high proportion of the primate populations concentrated in the forest area encircling the site. Usually, such area ranged from 1 to 5 km^2. Although troops of howlers and subgroups of spider monkeys ranged in various parts of the protected forest, they tended to make regular forays into the areas where the largest concentration of vestiges of Maya buildings were found. In those sites embedded in larger natural protected areas, line transect surveys of ≥ 3 and ≤ 5 km in length radiating away from the core area (4–5 km^2) of the Maya sites, showed a much lower encounter rate for both spider and howler monkeys than found in line transects within the area occupied by the ancient Maya city. For example, whereas in the core area of the Maya site of Calakmul (400 ha), howler troops were detected at a rate of 0.16 ± 0.09 troops/km surveyed, in the trails radiating 3–5 km away from the edges of this area, detection rate of troops was significantly lower (0.09 ± 0.01 troops/km) (A. Estrada, unpublished data).

The higher concentration of primates by the Maya archeological remains could be the result of the abundance in these areas of tree species of the genus *Brosimum*, *Ficus*, and *Poulsenia* (Moraceae), *Manilkara*, and *Pouteria* (Sapotaceae), *Spondias* (Anacardiaceae), and *Bursera* (Burseraceae: Lundel, 1937; Rojas, 2000; Valle, 2000). These and other tree species are known to have been used intensively in agroforestry practices by the ancient Maya (Puleston, 1982; Gómez-Pompa *et al.*, 2003). Botanical surveys show a non random skewed distribution and concentration of fruit-producing trees of these species toward Maya settlements or toward their immediate vicinity (Flannery, 1982; Puleston, 1982; Barnhart, 2001; Gómez-Pompa *et al.*, 2003). Interestingly, these species of trees also have been documented as important components in the diet of howler and spider monkeys in the region (Coelho *et al.*, 1976; Cant, 1990; Silver *et al.*, 1998; see Rivera and Calmé, this volume). Thus, it seems that agroforestry management practiced by the Maya thousands of years ago, favor primate populations at these sites today.

Conservation of Primate Populations

A comparison with extensive and fragmented forests showed that mean population densities for both primates in the protected forests of Maya archeological sites more closely approached average values in extensive forest tracts, than those in fragmented landscapes. Similarly, mean group size and immature to adult female ratio for *A. pigra* and *A. geoffroyi* in Maya sites fell within the range of variation for populations found in extensive forest tracts (Table 2).

Table 2. Comparison of demographic parameters of populations of black howler monkeys (*Alouatta pigra*) and of spider monkeys (*Ateles geoffroyi*) in Maya sites, in extensive forest tracts (>2000 ha), and in fragmented landscapes

	Density (individuals/km²)	Mean group size	Mean immature to adult female ratio
Alouatta pigra			
Maya sites	10.8 ± 5.70	6.6 ± 0.87	0.84 ± 0.37
Extensive forests ($N = 3$)[a]	28.5 ± 14.31	5.4 ± 0.20	0.77 + 0.39
Fragmented landscapes ($N = 3$)[b]	116.5 ± 8.33	5.5 ± 0.36	1.18 ± 0.26
Ateles geoffroyi			
Maya sites	13.4 ± 16.1	4.8 ± 1.50	0.59 ± 0.19
Extensive forests ($N = 3$)[c]	14.0 ± 13.22	3.9 ± 0.54	0.79 ± 0.17
Fragmented landscapes ($N = 2$)[d]	58.6 ± 38.7	5.4 ± 3.6	0.43 ± 0.21

[a] Biosphere Reserve Montes Azules and Reforma Community Reserve in Chiapas, and Lachúa Reserve in Guatemala (see Van Belle, this volume).
[b] Community Baboon Sanctuary (see Van Belle, this volume), Belize, Palenque, Mexico (Estrada *et al.*, 2002a,b), Lachúa, Guatemala (Rosales-Meda, 2003).
[c] Muchukux and Najil Tucha in Quintana Roo (Gonzales-Kirchener, 1999), Los Tuxtlas, Veracruz (Estrada, unpublished data).
[d] Los Tuxtlas, Mexico (three fragmented landscapes) (Estrada and Coates-Estrada, 1996; Estrada, unpublished data; González-Zamora, 2003).

This suggests that primate populations in the protected forests of Maya archeological sites may have enough habitat and resources to adjust their populations in a fashion similar to populations found in extensive forests, whereas populations of the same species in forest fragments are at a saturated state with respect to the area of habitat available.

Decades of efforts by the Mexican government in protecting the forested land surrounding Maya archeological sites has resulted in the protection of howler and spider monkey populations inhabiting these forests, attesting to the value of these landscape units as foci of conservation for primate populations in southeast Mexico. In contrast, adjacent landscapes have been impacted heavily by human activity, primate populations have become locally extinct and remnant populations live in small isolated forest fragments at very high densities (see Estrada *et al.*, 2002b for an example).

Conservation Awareness and the Maya Archeological Sites

Maya archeological sites preserve the cultural patrimony of the localities and regions in which they occur in southeast Mexico, and they are an important attraction for all local, national, and foreign tourists. The flux of yearly tourist visitors to some of these sites is staggering. Statistics from the Mexican National

Institute of Anthropology (INAH, 2004) indicate, for example, that between 2002 and 2003 about 6 million people visited the Maya archeological sites in the region. Interestingly, the majority (*ca.* 75%) were nationals of Mexico. Such a high volume of tourism visiting the Maya sites has an important economic impact at local and regional levels. By visiting these sites, tourists also benefit from dissemination of information about the Mayan culture and their environment.

Because of their enormous cultural appeal to the public, Maya archeological sites are a unique resource for dissemination of information regarding not only the ancient culture, but also the biological richness of the tropical rain forest ecosystem that surrounds the ruins. Such processes of information exchange enhance conservation awareness among local inhabitants and the general public. The large economic input derived from the high volume of visitors to the sites, directly and indirectly benefit the localities and regions where these are found. This places an important premium on local and national governments to safeguard these sites and the surrounding natural environment from any disturbance, with the net synergistic result of conserving, along with the rest of the biota, the primate populations existing within it (Figure 6).

In conclusion, the initial phase of our study suggests that the protected forests surrounding many Maya archeological sites are important foci of conservation for populations of primate taxa existing in northern Mesoamerica. Further research should be directed to gathering comparable information

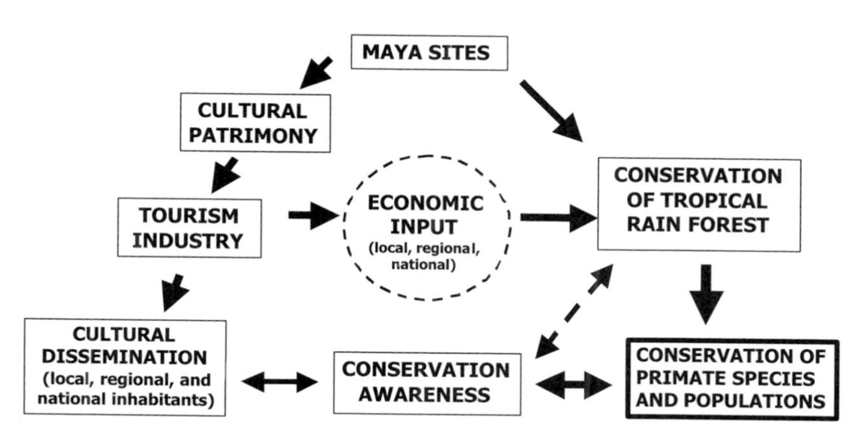

Figure 6. Importance of Maya sites for the conservation of primate populations in southern Mexico. Protection of the cultural patrimony attracts large number of tourists, with an important economic impact at local, regional and national levels. Dissemination of information about the Maya coupled with the protection of tropical rain forests surrounding the sites enhances conservation awareness among visitors to the Maya sites.

on primate populations in additional sites in Mexico, Guatemala, Belize, El Salvador, and Honduras. Such studies will enable us to develop a database from which to evaluate stability and variability in population parameters for *Alouatta* and *Ateles* and to assess to what extent the forests surrounding Maya archeological sites represent a network of viable primate reserves in the Maya region of Mesoamerica.

SUMMARY

Habitat destruction and fragmentation resulting from human activity impinge on the viability of existing primate populations in the lowlands of southeast Mexico, specifically in the Maya region. While habitat conservation occurs in the few natural protected areas and national parks in the region, weak supervision in many results in illegal extraction of woods, poaching, and squatters. Other protected forests in the Maya region are those surrounding Maya archeological sites. Because of the national and world-wide cultural importance, the Mexican government has an extensive program of protection of Maya archeological sites that involves not only the area where the ancient ruins are located, but also several square kilometers of land around them. Primate populations exist in many of these forests, but until now information about the species involved and about their demographic features was not available. In this paper, we report the results of surveys of primate populations conducted between 2001 and 2003 in the forests surrounding 19 Maya sites (17 in Mexico and 2 in Guatemala). Primate populations were discovered in 90% of the sites surveyed. In Comalcalco, the westernmost site of the Maya region, we found a population of *A. palliata* inhabiting the surrounding forest and for which density was estimated at 11.0 individuals/km^2. At the other sites ($N = 18$) the breakdown by taxa was as follows: 61% of the sites had populations of *A. geoffroyi* and of *A. pigra*, 17% had only *A. pigra*, 11% had only *Ateles*, and 11% had no primates (Figure 2). Estimated population densities for *A. pigra* varied from 3.5 individuals/km^2 (Lacanjá) to 23 individuals/km^2 (Palenque). Overall mean density was 10.8 ± 5.7 individuals/km^2; median 9.85 individuals/km^2. Estimated population densities for *A. geoffroyi* varied from 2.0 individuals/km^2 (Bonampák) to 56 individuals/km^2 (Tikal). Overall mean density was 13.4 ± 16.1 individuals/km^2; median 9.0 individuals/km^2. Average population parameters (density, troop size and immature to adult female ratios) for primate populations living in the protected forests of Maya sites more closely approximated average values of populations living in extensive than in fragmented forests. This suggests that primate populations in the protected forests of Maya sites may have enough habitat and resources to adjust their populations in a fashion similar to populations found in extensive forests, whereas populations of the same species in forest fragments are at a saturated state with respect to the area of habitat available. Our study suggests that the protected forests surrounding many Maya archeological sites are important foci of conservation for populations of primate taxa existing in northern Mesoamerica.

ACKNOWLEDGMENTS

We are grateful to the Cleveland Zoo Scott Neotropical Fund, the American Society of Primatologists, the Primate Conservation Inc., and the Universidad Nacional Autónoma de México for support to sustain the Maya Sites and Primate Conservation project. We thank the Mexican Institute of Anthropology (INAH) for permission to work at all the Maya sites explored in Mexico and we are grateful to the government of Guatemala for permission to conduct the primate surveys in Tikal and Uaxactum and in Lachúa, and for providing invaluable logistical support. LL acknowledges support from the School of Liberal Arts at University of Texas. SVB acknowledges the support of the Flemish Government of Belgium, Secretaria de Relaciones Exteriores de México and of Universidad Nacional Autónoma de México. MR thanks Universidad Nacional Autónoma de México and Universidad de San Carlos, Guatemala, for support. We thank Lucía Castellanos, Adrián Mendoza, Yasminda García, and David Muñoz for assistance in the field, and we like to extend a special thanks to the many local Maya indigenous people who assisted us while doing field work in southeast Mexico and in Guatemala. We thank the WRPRC for library support.

REFERENCES

Allen, F. M. and Rincón, E. 2003, The changing global environment and the lowland Maya: Past patterns and current dynamics, in: A. Gómez-Pompa, M. F. Allen, S. L. Fedick, and J. J. Jiménez-Osornio, eds., *The Lowland Maya Area. Three Millenia at the Human-Wildland Interface*, Food Products Press, New York, pp. 13–30.

Barnhart, E. L. 2001, The Palenque Mapping Project: Settlement and urbanism at an ancient Maya city. PhD Dissertation, The University of Texas at Austin.

Barrueta, T., Estrada, A., Pozo, C., and Calmé, S. 2003, Reconocimiento Demográfico de *Alouatta pigra* y *Ateles geoffroyi* en la Reserva El Tormento, Campeche, México. *Neotrop. Primates* 11:165–169.

Cant, J. G. H. 1978, Population survey of the spider monkey (*Ateles geoffroyi*) at Tikal, Guatemala. *Primates* 19:525–535.

Cant, J. G. H. 1990, Feeding ecology of spider monkeys, *Ateles geoffroyi* at Tikal, Guatemala. *Hum. Evol.* 5:269–281.

CCAD. 2003, Memorias. Primer Congreso Mesoamericano de Areas Protegidas. CCAD (Consejo Centro Americano de Areas Protegidas). Managua, Nicaragua. Document available through http://www.biomeso.net/

CCAD/UNEP/GEF. 2002, El corredor biológico mesoamericano México. Proyecto Para La Consolidación del Corredor Biológico Mesoamericano. CCAD Comisión Centroamericana de Ambiente y Desarrollo. UNEP United Nations Environmental Program, GEF UN Global Environmental Facility.

Coelho, A. M. Jr., Coelho, L., Bramblett, C., Bramblett, S., and Quick, L. B. 1976, Ecology, population characteristics, and sympatric associations in primates: A socio-bioenergetic analysis of howler and spider monkeys in Tikal, Guatemala. *Yearb. Phys. Anthropol.* 20:96–135.

Duarte, A. and Estrada, A. 2003, Primates as pets in Mexico City: An assessment of species involved, source of origin and general aspects of treatment. *Am. J. Primatol.* 61:53–60.

Estrada, A. 1982, Survey and census of howler monkeys (*Alouatta palliata*) in the rain forest of Los Tuxtlas, Veracruz, Mexico. *Am. J. Primatol.* 2:363–372.

Estrada, A., Castellanos, L., García, Y., Franco, B., Muñoz, D., Ibarra, A., Rivera, A., Fuentes, E., and Jimenez, C. 2002a, Survey of the black howler monkey, *Alouatta pigra*, population at the Mayan site of Palenque, Chiapas, Mexico. *Primates* 44:51–58.

Estrada, A. and Coates-Estrada, R. 1996, Tropical rain forest fragmentation and wild populations of primates at Los Tuxtlas. *Int. J. Primatol.* 5:759–783.

Estrada, A., García, Y., Muñoz, D., and Franco, B. 2001, Survey of the population of howler monkeys (*Alouatta palliata*) at Yumká Park in Tabasco, Mexico. *Neotrop. Primates* 9:12–14.

Estrada, A., Luecke, L., Van Belle, S., Barrueta, E., and Rosales-Meda, M. 2004, Survey of black howler (*Alouatta pigra*) and spider (*Ateles geoffroyi*) monkeys in the Mayan sites of Calakmul and Yaxchilán, Mexico and Tikal, Guatemala. *Primates* 45:33–39.

Estrada, A., Mendoza, A., Castellanos, L., Pacheco, R., Van Belle, S., García, Y., and Muñoz, Z. 2002b, Population of the black howler monkey (*Alouatta pigra*) in a fragmented landscape in Palenque, Chiapas, Mexico. *Am. J. Primatol.* 58:45–55.

FAO. 2000, *State of Forestry in the Region-2000.* Latin American and Caribbean Forestry Commission FAO. Forestry series no. 15. http://www.rlc.fao.org

FAO. 2001, Food Security (www.document). http:/www.faor.org/biodiversity

FAO. 2004, FAO Statistical Data Bases. http://faostat.fao.org/

Flannery, K. V. ed. 1982, *Maya Subsistence. Studies in Memory of Dennis E. Puleston.* Academic Press Inc., New York.

Gómez-Pompa, A. 2003, Research challenges for the lowland Maya area: An introduction, in: A. Gómez-Pompa, M. F. Allen, S. L. Fedick, and J. J. Jiménez-Osornio, eds., *The Lowland Maya Area. Three Millenia at the Human-Wildland Interface*, Food Products Press, New York, pp. 3–9.

Gómez-Pompa, A., Allen, M. F., Fedick, S. L., and Jiménez-Osornio, J. J. eds. 2003, *The Lowland Maya Area. Three Millenia at the Human-Wildland Interface.* Food Products Press, New York.

Gonzales-Kirchener, J. P. 1998, Group size and population density of the black howler monkey (*Alouatta pigra*) in Muchukux forest, Quintana Roo, Mexico. *Folia Primatol.* 69:260–265.

Gonzales-Kirchener, J. P. 1999, Habitat use, population density and subgrouping pattern of the Yucatan spider monkey (*Ateles geoffroyi yucatanensis*) in Quintana Roo, Mexico. *Folia Primatol.* 70:55–60.

González-Zamora, A. 2003, *Uso de Fragmentos por Mono Araña (Ateles geoffroyi vellerosus) en Fragmentos del Sur de Los Tuxtlas, Ver.* Tesis de Maestría (Manejo de Fauna Silvestre), Instituto de Ecología A. C., Xalapa, Ver.

Horwich, R. H. and Johnson, E. W. 1986, Geographic distribution of the black howler monkey (*Alouatta pigra*) in Central America. *Primates* 27:53–62.

INAH. 2004, http://www.inah.gob.mx/index_.html

Lundel, C. L. 1937, *The Vegetation of Peten.* Carnegie Institution of Washington. Pub. No. 478. Washington, D.C.

National Research Council. 1992, *Techniques for the Study of Primate Population Ecology.* National Academy Press, Washington, D.C.

Ostro, L. E. T., Silver, S. C., Koontz, F. W., Young, T. P., and Horwich, R. H. 1999, Ranging behavior of translocated and established groups of black howler monkeys *Alouatta pigra* in Belize, Central America. *Biol. Conserv.* 87:181–190.

Pavelka, M. S. M., Brusselers, O. T., Nowak, D., and Behie, A. M. 2003, Population reduction and social disorganization in *Alouatta pigra* following a hurricane. *Int. J. Primatol.* 24:1037–1055.

Puleston, D. E. 1982, The role of ramón in Maya subsistence, in: K. V. Flannery, ed., *Maya Subsistence. Studies in Memory of Dennis E. Puleston*, Academic Press Inc., New York, pp. 353–366.

Ramos-Fernández, G. and Ayala-Orozco, B. 2003, Population size and habitat use of spider monkeys in Punta Laguna, Mexico, in: L. K. Marsh, ed., *Primates in Fragments: Ecology and Conservation*, Kluwer/Plenum Press, New York, pp. 191–209.

Rodríguez-Toledo, E. M., Mandujano, S., and Garcia-Ordunia, F. 2003, Relationships between forest fragments and howler monkeys (*Alouatta palliata mexicana*) in southern Veracruz, Mexico, in: L. K. Marsh, ed., *Primates in Fragments; Ecology and Conservation*, Kluwer Press, New York, pp. 79–97.

Rojas, S. 2000, La reserva de la biosfera de Calakmul. *Arqueol. Mex.* 7:46–51.

Rosales Meda, M. M. 2003, Abundancia, distribucion y composicion de tropas del mono aullador negro (*Alouatta pigra*) en diferentes remanentes de bosque en la ecoregion Lachu´a. B. Sc. Thesis Universidad de San Carlos de Guatemala, Guatemala.

Rowe, N. 1996, *A Pictorial Guide to the Primates*. Pogonian Press, Rhode Island.

Rylands, A., Mittermeier, R. A., and Rodríguez-Luna, E. 1995, A species list for the New World primates (Platyrrhini): Distribution by country, endemism, and conservation status according to the Mace-Land system. *Neotrop. Primates* 3(Suppl.):114–164.

Schlichte, H. 1978, A preliminary report on the habitat utilization of a group of howler monkeys (*Alouatta villosa pigra*) in the national Park of Tikal, Guatemala, in: G. G. Montgomery, ed., *The Ecology of Arboreal Folivores*, Smithsonian Institution Press, Washington, D.C., pp. 551–561.

Silver, S. C., Ostro, L. E. T., Yeager, C. P., and Horwich, R. 1998, Feeding ecology of the black howler monkey (*Alouatta pigra*) in northern Belize. *Am. J. Primatol.* 45:263–279.

Turner II, B. L., Klepeis, P., and Schneider, L. C. 2003, Three millennia in the southern Yucatan peninsula: Implications for occupancy, use and carrying capacity, in: A. Gómez-Pompa, M. F. Allen, S. L. Fedick, and J. J. Jiménez-Osornio, eds., *The Lowland Maya Area. Three Millenia at the Human-Wildland Interface*, Food Products Press, New York, pp. 361–388.

UNDP. 1999, Establishment of a Programme for the Consolidation of the Mesoamerican Biological Corridor Project Document (RLA/97/G31). Global Environment Facility. Project of the Governments of Belize, Costa Rica, El Salvador, Guatemala, Honduras, Mexico, Nicaragua, Panama. L:\bd\regional\mesoamerica\MBCprodoc.

UNEP. 2004, United Nations Environmental Program. www.rolac.unep.mx/recnat/esp/CBM/

Valle, J. A. 2000, Analisis Estructural de una Hectarea de Selva Alta Perennifolia en el Monumento Natural Yaxchilán (Chiapas), México. BSc Thesis, University of Mexico (UNAM).

Watts, E. and Rico-Gray, V. 1987, Los primates de la peninsula de Yucatán, México: Estudio preliminar sobre su distribución actual y estado de conservación. *Biótica* 12:57–66.

Wilson, D. E., Cole, F. R., Nichols, J. D., Rudran, R., and Foster, M. eds. 1996, *Measuring and Monitoring Biological Diversity: Standard Methods for Mammals (Biological Diversity Handbook Series)*. Smithsonian Institution Press, Washington, D.C.

World Bank. 2004, http://www.worldbank.org/data/

World Resources Institute. 2004, http://earthtrends.wri.org/

WWF. 2002, Living Planet Report 2002. WWF, UNDP, WCMC. World Wide Fund for Nature. Gland, Switzerland. http://www.panda.org/

CHAPTER TWENTY

Mapping Primate Populations in the Yucatan Peninsula, Mexico: A First Assessment

Juan Carlos Serio-Silva, Víctor Rico-Gray, and Gabriel Ramos-Fernández

INTRODUCTION

The Yucatan Peninsula: Pioneer Research

The Yucatan Peninsula occupies an important place in Mexican geography and was the indigenous homeland of the Maya, one of the most significant pre-Hispanic societies in the New World (Taube, 2003). Mayan groups inhabiting the Mexican portion of the Yucatan Peninsula (states of Campeche, Yucatan, and Quintana Roo) participated in a complex network of cultural, political, and economic activities, and developed land use patterns that contributed to the conservation of vast extensions of the natural landscape (Shaker, 1999).

Juan Carlos Serio-Silva • Departamento de Biodiversidad y Ecología Animal, Instituto de Ecología, A.C., Xalapa, Veracruz, México. **Víctor Rico-Gray** • Departamento de Ecología Funcional, Instituto de Ecología, A.C., Xalapa, Veracruz, México. **Gabriel Ramos-Fernández** • Pronatura Península de Yucatán, Mérida, Yucatán, México.

New Perspectives in the Study of Mesoamerican Primates: Distribution, Ecology, Behavior, and Conservation, edited by Alejandro Estrada, Paul A. Garber, Mary S. M. Pavelka, and LeAndra Luecke. Springer, New York, 2005.

The Distribution of Primates on the Mexican Side of the Yucatan Peninsula

Three primate species are indigenous to Mexico: mantled howlers (*Alouatta palliata mexicana*), black howlers (*Alouatta pigra*), and Geoffroy's spider monkey (two subspecies *Ateles geoffroyi vellerosus* and *Ateles geoffroyi yucatanensis*). Only two of these taxa, *A. pigra* and *A. g. yucatanensis* are currently found in the Yucatan Peninsula (however, see below for confirmed sightings of *A. palliata* in the Yucatan). The black howler monkey exhibits a geographic distribution that includes Belize, Guatemala, and Mexico. *A. pigra* is the only *Alouatta* species present in the Yucatan Peninsula (Smith, 1970; Horwich and Johnson, 1986; Watts and Rico-Gray, 1987). Spider monkey populations (*A. g. yucatanensis*) coexist with *A. pigra* in several localities in this area (Watts and Rico-Gray, 1987); however, habitat destruction, hunting, and the pet trade put these populations at risk (Estrada *et al.*, 2004).

Major Land Use Patterns and Impact on Native Vegetation in the Yucatan Peninsula

Some 50 years ago, approximately 86,000 km^2 of the Yucatan Peninsula were covered with semievergreen forest. At present, however, very few sites currently exist with semievergreen forest fragments larger than 1000 km^2 and deforestation continues at a rate of 8000 km^2 per year (Challenger, 1998). It is clear that habitat destruction is the most significant threat to the survival of primates in the Yucatan Peninsula (Ramos Fernández and Ayala-Orozco, 2003). In some areas of each state, most of the natural vegetation has been modified or destroyed by slash-and-burn agriculture, cattle ranching, and accidental fires caused by slash and burn agriculture (Challenger, 1998). We are facing an important moment in which knowledge of the demography and distribution of primate species in the Yucatan Peninsula is critical to developing effective conservation and management policies.

Early Research on Primate Distribution in Yucatan Peninsula

Despite the need for conservation efforts in the Yucatan, little is known about the Peninsula's natural resources, including its wild primate populations. The first studies of population demography and distribution were conducted by

Watts *et al.* (1986) and Watts and Rico-Gray (1987). These researchers visited 18 forested Yucatan sites and confirmed the presence of *Ateles* and *Alouatta* at only eight of these sites. These authors concluded that habitat destruction, hunting and pet capture were the major factors affecting the presence of primates at these sites. At this same time, Horwich and Johnson (1986) published a report on the distribution and vegetation characteristics of forests inhabited by *A. pigra* in southeastern Mexico, including the Yucatan Peninsula. However, these authors acknowledged that much of their data came from indirect sources, rather than confirmed sightings, and thus should be viewed with great caution. Lara and Jorgenson (1998) also surveyed wild primates in the state of Quintana Roo. They conducted field observations aimed at understanding the relationship between the presence of particular vegetation types and the conservation status of howler and spider monkeys in this region.

Recent Research on Aspects of Ecology and Behavior

More recently, studies of Yucatan's primates have focused on questions of behavior and ecology. For example, Gonzalez-Kirchner (1998, 1999) examined group size, habitat use, and population density in *A. pigra* and *A. g. yucatanensis* in Muchukux, Quintana Roo. Navarro-Fernandez (2000) working in the state of Campeche developed a protocol for using local people to collect data on the location and density of *A. pigra* and *A. g. yucatanensis.* In an attempt to address questions concerning primate conservation and health, Bonilla-Moheno (2002) examined the effects of habitat disturbance and the presence of endoparasites on *A. pigra* and *A. g. yucatanensis* populations in the state of Quintana Roo. She found that the density and diversity of endoparasites in both primate species were greater in disturbed habitats. Similarly, Rangel-Negrín (2003) initiated a study of fecal cortisol levels in populations of *A. g. yucatanensis* inhabiting intact and altered habitats in Quintana Roo, México. Cortisol levels are an indicator of stress and may be a sensitive measure of the health of individuals in a natural population. The results of this study indicate that spider monkeys living in intact forest showed lower cortisol levels than individuals living in altered habitats or monkeys reared as pets or housed in zoos.

Primate population surveys also have been conducted in the protected forest of a reserve of the Mexican Forestry agency in El Tormento, Campeche. Barrueta *et al.* (2003) report the existence of a population of *A. pigra* coexisting with a smaller population of *A. geoffroyi*. In the same site, a 10-month-long study of the foraging ecology of *A. pigra* yielded information on seasonal use of plant species, foraging patterns, and dietary preferences. An additional study examined foraging patterns and habitat preferences of groups of *A. pigra* existing in the continuous forest of the Calakmul Biospere Reserve and in adjacent fragmented landscapes in southern Campeche (see Rivera and Calme, this volume).

Ramos-Fernández and Ayala-Orozco (2003) examined the behavior of spider monkeys in Punta Laguna, Yucatan. This study addressed questions concerning patterns of habitat utilization in two groups of *A. g. yucatanensis* using GIS Technology. Finally, Estrada *et al.* (2004) initiated a series of population demography and group size studies of spider and howler monkeys inhabiting forests in proximity to Mayan archaeological sites, including regions of Campeche. These authors report that the protected forests surrounding Mayan sites contain sustainable populations of *A. pigra* and *A. geoffroyi*, and that studies of these populations should represent an important foci for conservation and management policies in Mesoamérica (see Estrada *et al.*, this volume).

Despite these important studies, information on the presence and conservation status of howler and spider monkey populations across a larger geographic region of the Yucatan Peninsula are lacking. Hence, in this paper we present the results of an area-wide survey that provides information on current locations of *A. pigra*, *A. palliata*, and *A. geoffroyi* populations in the Yucatan Peninsula. In addition, these surveys assessed the legal protection status of the habitats/sites that contained primate populations. We use this information to present a general assessment of the conservation status of primate populations and their habitat on the Yucatan Peninsula.

METHODS

Recognition of Wild Monkey Populations

Fieldwork was conducted during a period of 28 months (January 2000–April 2002). Surveys were conducted for approximately 12 days (13 ± 2 days) every 2 months. In order to census primate populations in areas within each state of the Peninsula (Campeche, Quintana Roo, and Yucatan), we conducted six surveys during the dry season (February–May) and eight surveys during the wet season (June–January). Sampling sites were selected based on data obtained from first published reports (see above), maps, letters, and unpublished documents (personal files of the late Dr. Elizabeth S. Watts). The initial objective was to visit areas that had previously been censused, with the aim of corroborating and updating information on the presence/absence of monkey populations. In addition to these 20 localities, we selected a large number of new localities based on cartographic information of known distribution, location of suitable habitat types, and whether areas represented protected or unprotected sites. Overall, we visited 78 potential primate localities (10 localities were visited on more than one occasion either because they contained large forested areas [national reserves or in the southern border near Belize] or because of a particular interest in the habitat conditions in areas with primate populations). Site by site data on primate populations and habitat characteristics may be available from JCSS upon request.

Primate Surveys: Sampling and Identification of Habitat Characteristics

Three people generally worked together collecting data, including a field guide from a nearby town. Once the research team was formed, surveys were conducted following the transect method proposed by Struhsaker (1981). The maximum length of transects walked in this study was 5–10 km. Fieldwork began between 06:00 and 07:00 h and ended around 17:00–18:00 h, weather permitting; intense rain was an impediment, sometimes limiting visibility in deep forest. This varied along with observation conditions from site to site, as in severely altered areas, monkeys could be followed until the late afternoon, while in well-preserved parts, it became harder to find them after 17:00 h. In most of the sites visited, surveys of three to five transects were completed. At the largest sites, the number of transects walked was 8–10. Transects were traversed at a rate of 1–1.5 km/h, depending on the condition of the forest path; brief observation stops were made to listen for sounds and detect visual clues (feces, consumed fruit, broken branches, movement in the canopy, among others) that might indicate the presence of monkey troops. Special care was taken not to count the same group twice; this was avoided through radio communication between observers when an individual or group was detected. On this basis, the total number of troops (howler monkeys) or subgroups (spider monkeys) during the sampling period was recorded. We estimated the overall abundance of primate species as the number of troops/subgroups sighted per kilometer. During the entire study, we constructed 107 transects which covered a distance of 353.6 km in Campeche ($n = 36$ localities), 58 transects covering a distance of

293.4 km in Quintana Roo ($n = 25$ localities), and 5 transects covering 69 km in Yucatan ($n = 5$ localities). Finally, direct contact (visual or auditory cues) with a troop of primates was considered a "verified", sighting, while information provided by local inhabitants was scored as "reported".

Characteristics of the Troops and Subgroups Located

For each howler monkey troop or spider monkey subgroup located on a transect, a record was made of the place where it was observed and a consecutive letter of the alphabet was used to indicate the number of populations found for each species (Table 1). The following data also were recorded: species, group size, sex–age composition; time and date of sighting, length of observation, transect position, habitat type, conservation status of the forest (altered/preserved: see Serio-Silva and Rico-Gray, 2002), and legal protection status (CONANP, 2004); distance covered from transect tip, perpendicular distance from the transect to the geometric center of the group, and vertical position of the group in accordance with forest strata (National Research Council, 1992).

Geographic Characterization of Potential Available Habitat

On a map, the georeferenced points of each locality and state where the monkeys were sighted were marked and each vegetation type recorded (Flores and Espejel, 1994); their legal protection status also was noted. On the basis of geographic location, using a Global Positioning System (GPS; Garmin GPS 12, Kansas, USA) of areas where monkeys were found, calculations were made of the potential area available for use as natural habitat on the Peninsula. For this purpose, *landsat 5 TM Imagery* (SYPR, 2000) images were processed through Geographic Information Systems.

RESULTS

A total of 78 localities in the three states that comprise the Peninsula were visited, 66 of which contained (verified or reported) wild primates (Table 1, Figure 1). The number of localities visited per state was 36, 25, and 5 for Campeche, Quintana Roo, and Yucatan, respectively. In Campeche, the most common vegetation type surveyed (18 sites on these localities, 50.0%) was

Table 1. List of locations with the presence of wild primates on the Yucatan Peninsula, Mexico

No.	Site name	Vegetation type	Primate species	Troop/subgroups per site	State	Latitude N	Longitude W	Protection status
1	Dzibalchen	MSEF	A.p (V)	(a–b)	Campeche	19°29″	89°44″	Protected
2	Escarcega region	MSEF	A.p (V), A.g.y (V)	(a–n), (a–b)	Campeche	18°36″	90°40″	Not protected
3	Zaragoza CL	MSEF	A.p (V)	(a–f)	Campeche	18°28″	91°11″	Not protected
4	Pejelagarto CL	MSEF	A.p (V)	(a–d)	Campeche	18°11″	90°01″	Not protected
5	Cristalina CL	MSEF	A.p (V)	(a)	Campeche	18°46″	90°55″	Not protected
6	El Desengaňo CL	TSEF	A.p (V)	(a–c)	Campeche	17°52″	90°29″	Not protected
7	El Naranjo CL	TEF	A.p (V)	(a–c)	Campeche	18°05″	91°06″	Not protected
8	El Ramonal CL	TSEF	A.p (V), A.g.y (V)	(a–c), (a–b)	Campeche	17°50″	90°38″	Not protected
9	El Zapote CL	TEF	A.p (V), A.g.y (V)	(a–e), (a–b)	Campeche	18°09″	91°37″	Not protected
10	Conhuas CL	MSEF	A.p (V), A.g.y (V)	(a–d), (a–c)	Campeche	18°42″	89°57″	Not protected
11	Calakmul	TEF	A.p (V), A.g.y (V)	(a–c), (a–b)	Campeche	18°08″	89°35″	Protected
12	Arroyo Negro	TEF	A.p (V), A.g.y (V)	(a–f), (a–c)	Campeche	17°50″	89°11″	Protected
13	Manuel Rejon/Unidad Militar	TEF	A.p (V), A.g.y (V)	(a–e), (a–c)	Campeche	17°56″	89°11″	Protected
14	Narciso Mendoza CL	MSEF	A.p (V)	(a–c)	Campeche	18°20″	89°23″	Protected
15	El Manantial CL	MSEF	A.p (V)	(a–b)	Campeche	18°25″	89°22″	Protected
16	La Victoria CL	MSEF	A.p (V)	(a)	Campeche	18°22″	89°22″	Protected
17	Kankabchen	MSEF	A.p (V)	(a)	Campeche	19°42″	88°56″	Not protected
18	Dos Lagunas	MSEF	A.p (V)	(a)	Campeche	18°48″	89°18″	Not protected
19	Near Hopelchen/Dzibalchen	MSDF	A.p (V)	(a)	Campeche	19°36″	89°51″	Protected
20	Nunkini	MDF	A.g.y (V)	(a)	Campeche	20°24″	90°08″	Protected
21	El Remate	MANG	A.g.y (V)	(a)	Campeche	20°33″	90°23″	Protected
22	Nuevo Coahuila CL	TSEF	A.p (V)	(a–d)	Campeche	17°53″	90°44″	Not protected
23	El Sacrificio	TEF	A.p (V)	(a–e)	Campeche	18°05″	91°36″	Not protected
24	Nuevo Becal	MSEF	A.p (V)	(a–b)	Campeche	18°34″	89°30″	Protected

(*Continued*)

Table 1. *(Continued)*

No.	Site name	Vegetation type	Primate species	Troop/ subgroups per site	State	Latitude N	Longitude W	Protection status
25	Central Chiclera Villahermosa	TEF	A.p (V)	(a–c)	Campeche	17°55″	89°41″	Protected
26	La Esperanza CL	MSEF	A.p (V)	(a–b)	Campeche	18°19″	90°11″	Protected
27	10 km before Constitución	MSEF	A.p (V)	(a)	Campeche	18°34″	90°14″	Protected
28	Los Alacranes	TEF	A.p (V)	(a–c)	Campeche	17°58″	89°12″	Protected
29	Miguel Colorado	MSEF	A.p (V)	(a–b)	Campeche	18°46″	90°40″	Not protected
30	10 km near Chompoton	MSDF	A.p (V)	(a)	Campeche	19°25″	90°43″	Not protected
31	Calkini region	MANG	A.p (Re), A.g.y (V)	(a) (a)	Campeche	20°23″	90°03″	Protected
32	Petenes northern Campeche	MANG	A.p (Re), A.g.y (V)	(a) (a–b)	Campeche	20°23″	90°22″	Protected
33	Tenabo-Hanpolol	LDF	A.p (V), A.g.y (V)	(a) (a)	Campeche	20°00″	90°19″	Protected
34	Conquista Campesina	MSEF	A.p (V), A.p.m (V)	(a–b) (a–c)	Campeche	18°11″	91°17″	Not protected
35	El Alamo Ranch	MSEF	A.p (V), A.p.m (V)	(a–h) (a–b)	Campeche	18°48″	90°54″	Not protected
36	El Suspiro	MSEF	A.p (V), A.g.y (V)	(a–b) (a)	Campeche	18°27″	91°16″	Not protected
37	El Eden	MSDF	A.g.y (V)	(a)	Quintana Roo	21°10″	87°04″	Protected
38	Puerto Morelos	MSEF	A.g.y (V)	(a)	Quintana Roo	20° 50″	86° 54″	Protected
39	Pacchen	MSEF	A.p (V), A.g.y (V)	(a–b) (a–b)	Quintana Roo	20°44″	87°32″	Protected
40	Rancho Chacmuchuc	MSEF	A.p (V)	(a)	Quintana Roo	21°17″	86°52″	Protected
41	Playacar Tourist Complex	MSEF	A.p.m (V)	(a)	Quintana Roo	20°36″	87°05″	Protected
42	"Dos Ojos"/"Dos Aguas" area	MSEF	A.p (V), A.g.y (V)	(a) (a)	Quintana Roo	20°20″	87°24″	Protected
43	Carillo Puerto area	MSEF	A.p (V), A.g.y (V)	(a–h) (a)	Quintana Roo	19°34″	88°02″	Not protected
44	Petcacab area	MSEF	A.p (V), A.g.y (V)	(a–e) (a–f)	Quintana Roo	19°17″	88°13″	Protected

45	Centro Integral de Aprovechamiento de Vida Silvestre (Bacalar)	TSEF	A.p (V), A.g.y (V)	(a–b) (a)	Quintana Roo	18°48″	88°19″	Protected
46	Area de Bacalar—Xul-ha	TSEF	A.p (V), A.g.y (V)	(a) (a)	Quintana Roo	18°35″	88°27″	Not protected
47	Ejido Tres Garantias	TSEF	A.p (V), A.g.y (V)	(a–g) (a–h)	Quintana Roo	18°11″	89°05″	Protected
48	Area de “La Camiseta”	TSEF	A.p (V), A.g.y (V)	(a–d) (a–d)	Quintana Roo	18°07″	89°00″	Not protected
49	El Diez	TSEF	A.p (V), A.g.y (V)	(a) (a)	Quintana Roo	21°01″	87°17″	Not protected
50	Laguna Madera	MSEF	A.p (V), A.g.y (V)	(a–b) (a)	Quintana Roo	20°48″	87°38″	Not protected
51	Chunyaxche	MSEF	A.p (V), A.g.y (V)	(a–d) (a–c)	Quintana Roo	19°57″	87°37″	Protected
52	Sian Ka’an	MSDF	A.p (V), A.g.y (Re)	(a) (a)	Quintana Roo	19°33″	87°44″	Protected
53	Rancho “X” km. 92	MSEF	A.p (V)	(a)	Quintana Roo	19°28″	87°58″	Not protected
54	Bosque Andres Q. Roo	TSEF	A.p (V)	(a)	Quintana Roo	19°22″	88°02″	Not protected
55	Bosque cercano Bacalar	TSEF	A.p (V), A.g.y (Re)	(a) (a)	Quintana Roo	18°46″	88°33″	Not protected
56	Tomas Garrido	TSEF	A.p (V), A.g.y (V)	(a–c) (a–b)	Quintana Roo	18°01″	89°04″	Not protected
57	Estero Franco	TSEF	A.p (V)	(a–e)	Quintana Roo	17°56″	88°52″	Not protected
58	Dos Bocas	MSEF	A.p (V), A.g.y (Re)	(a) (a)	Quintana Roo	17°55″	88°52″	Not protected
59	U Yumil Ceh’	MSEF	A.g.y (V)	(a)	Quintana Roo	18°42″	87°45″	Protected
60	Cobá	MSEF	A.g.y (Re)	(a)	Quintana Roo	20°30″	87°41″	Protected
61	Chunhuhub	MSEF	A.g.y (Re)	(a)	Quintana Roo	19°35″	88°42″	Not protected
62	Tizimin-Panaba	MSDF	A.g.y (V)	(a)	Yucatan	21°14″	88°33″	Not protected
63	Las Coloradas	LDF	A.g.y (V)	(a)	Yucatan	21°35″	88°02″	Not protected
64	Colonia Yucatan	MSDF	A.g.y (V)	(a)	Yucatan	21°13″	87°49″	Not protected
65	Punta Laguna	MSEF	A.p (V), A.g.y (V)	(a) (a–b)	Yucatan	20°38″	87°37″	Protected
66	El Cuyo	LDF	A.g.y (V)	(a)	Yucatan	21°30″	87°43″	Not protected

Codes: CL = Commonland, A.p = *Alouatta pigra*, A.g.y. = *Ateles geoffroyi yucatanensis*, and A.p.m. = *Alouatta palliata mexicana*; (V) = verified and (Re) = reported people; Protected and not protected is related to Mexican legal status of each sampled areas; Letters a–g represent different howler troops and spider monkey subgroups located at each site. Vegetation type: LDF = low, deciduous forest; MSDF = medium, semideciduous forest; MSEF = medium, semievergreen forest, TEF = tall, evergreen forest, TSEF = tall, semievergreen forest, MDF = medium deciduous forest, and MANG = mangrove (Flores and Espejel-Carvajal, 1994).

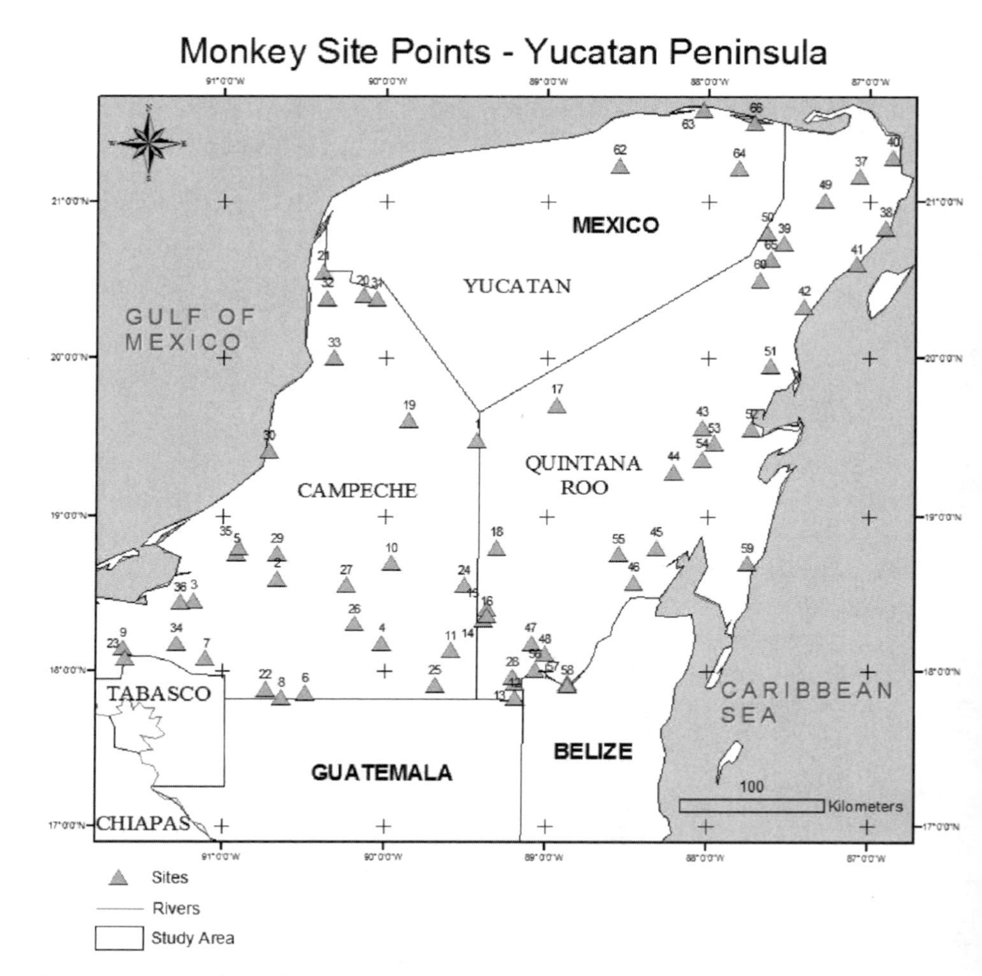

Figure 1. Localities for spider monkeys (*Ateles geoffroyi yucatanensis*), black howler monkeys (*Alouatta pigra*), and Mexican mantled howler monkeys (*Alouatta palliata mexicana*) in the Yucatan Peninsula, Mexico during the 2000–2002 surveys.

medium-height semievergreen forest (MSEF). This same vegetation type also was the most common surveyed in Quintana Roo (14 sites on these localities, 56%). In Yucatan, the most common vegetation was medium-height semideciduous forest (MSDF) and low-height deciduous forest (LDF) (two sites each).

We found the greatest density of *A. pigra* troops inhabiting MSEF forests in Campeche (mean = 5.6 ± 0.16 per km) and Quintana Roo (mean = 2.28 ± 0.18 per km). We found the greatest number of *A. g. yucatanensis* subgroups in MSEF forests (mean = 0.49 ± 0.47 per km) and tall evergreen forest (TEF) forests (mean = 0.87 ± 0.11 per km) forests in Campeche. In Quintana Roo, the greatest number of spider monkey subgroups was found in MSEF vegetation (mean = 1.67 ± 0.10) (Table 2).

Campeche was the state with the greatest abundance of monkey populations ($n = 36$, 54.5%), followed by Quintana Roo ($n = 30$, 37.8%) and Yucatan ($n = 5$, 7.5%). Of all monkey sites, 24 (36.4%) included only *A. pigra*, 15 (22.7%) only *A. g. yucatanensis*, and 25 (37.9%) had both species. Sites with both howler and spider monkeys present were found in southeastern Campeche. We identified two sites (3.0%) that contained coexisting sympatric populations of *A. pigra* and *A. p. mexicana*.

A total of 158 *A. pigra* troops, 5 *A. p. mexicana* troops, and 70 subgroups of *A. g. yucatanensis* were recorded in our survey transects (170 transects, totaling 733.5 km). The probability of finding an *A. pigra* troop on the Yucatan Peninsula was 0.21 troops/km; for *A. g. yucatanensis* subgroups, it was 0.095 subgroups/km; and for *A. p. mexicana*, it was 0.0068 troops/km. The mean number of individuals per *A. pigra* troop and *A. g. yucatanensis* subgroups for our entire sample was 5.7 ± 1.8 and 11.4 ± 6.7, respectively; however, these data varied for each state by sex–age composition and particularly by vegetation type.

The state of Campeche had an average of 5.5 ± 1.8 *A. pigra* individuals per troop, with three solitary individuals also sighted. For *A. g. yucatanensis* the average was 8.9 ± 4.3 individuals per subgroup. In the state of Quintana Roo, the average was 6.4 ± 1.5 individuals per *A. pigra* troop, while it was 12.4 ± 7.1 individuals per *A. g. yucatanensis* subgroup. Finally, for Yucatan, the only *A. pigra* group located consisted of six individuals, while for *A. g. yucatanensis* the average number of individuals per subgroup was 14.7 ± 10.2. For *A. p. mexicana* troops located in various parts of the Peninsula (Campeche and Quintana Roo states), the average number of individuals per troop was 8.8 ± 1.9.

The adult sex-ratio of all *A. pigra* troops was 1:1.59 (male to females). For juveniles this ratio was 1:0.67. The ratio of adult females to immatures was 1:0.84. For *A. g. yucatanensis*, the male–female sex ratio was 1:1.50 for adults and 1:1.22 for juveniles. The ratio of adult females to immature was 1:0.87. The sex-ratios (male to females, juveniles and adult females to immatures) were relatively consistent for *A. pigra* and *A. geoffroyi* across the Peninsula. The adult sex ratio of all three primate species observed is presented in Table 3.

Table 2. Summary of primate troops/subgroups and sites surveyed including vegetation type for each state in the Yucatan Peninsula

State	Number of sites	Vegetation types	Transect length (km)	No. of *A. pigra* troops/km	No. of *A. g. yucatanensis* subgroups/km	No. of *A. palliata* troops/km	Protected status per site and length (km) surveyed in each condition
Campeche	18	MSEF	171	5.6	0.49	0.46	18 protected (159 km)
	8	TEF	83	3.25	0.87	–	18 not protected (195 km)
	3	TSEF	36	0.91	0.15	–	
	2	MANG	35	0.15	0.34	–	
	1	MSDF	13	0.32	–	–	
	1	MDF	4	–	0.23	–	
		LDF	12	0.86	0.09	–	
	Total 36		Total = 354 km				
Quintana Roo	14	MSEF	165	2.28	1.67	0.66	13 protected (188.8 km)
	8	TSEF	105	1.61	0.96	–	12 not protected (104.6 km)
	3	MSDF	23	0.48	0.66	–	
	Total 25		Total = 293 km				
Yucatan	2	MSDF	23.0	–	0.14	–	4 not protected (51 km)
	2	LDF	34.0	–	0.17	–	
	1	MSEF	12.0	0.05	0.11	–	
	Total 5		Total = 69 km				
Total	Total 66		716 km				

Codes: Vegetation type: LDF = low, deciduous forest; MSDF = medium, semideciduous forest; MSEF = medium, semievergreen forest; TEF = tall, evergreen forest; TSEF = tall, semievergreen forest; MDF = medium, deciduous forest; MANG = mangrove (Flores and Espejel, 1994).

Table 3. Sex ratio for howler monkeys (*Alouatta pigra* and *Alouatta palliata*) and subgroups of spider monkeys (*Ateles geoffroyi yucatanensis*) during surveys in Yucatan Peninsula, Mexico

	Alouatta pigra	*Ateles geoffroyi yucatanensis*	*Alouatta palliata mexicana*
Adult sex ratio (male to female)	1:1.59	1:1.50	1:1.72
Juveniles	1:0.67	1:1.22	1:0.26
Adult female—immatures	1:0.84	1:0.87	1:0.84

Sites with Sympatry of *A. palliata* and *A. pigra*

Finally, it is of extreme importance to note that we observed sympatric population of *A. palliata* and *A. pigra* in the "El Alamo Ranch" (Locality # 35) and the "Conquista Campesina" commonland (Locality # 32). These appear to be the only areas whether these two species co-occur. It remains unclear if the range of both howler species traditionally overlapped (see Ford's chapter in this volume) or whether this represents a recent event due to habitat change and forest fragmentation.

Distribution of Vegetation Types in the Yucatan Peninsula

Based on the *landsat* satellite images and Mexican government cartography, we estimated that the potential forested habitat available for primate conservation in the Yucatan Peninsula is 93,942.39 km^2. This amounts to 63.9% of the Peninsula's total surface area and encompasses the entire region examined in our surveys (Figure 1). The distribution of vegetation types in this area is 1332.55 km^2(1.3%) of low semievergreen forest, 4712.01 km^2(5.0%) of MSDF, 8376.93 km^2(8.9%) of LDF, 14,071.23 km^2(15.0%) of TEF, and 65,449.67 km^2(69.7%) of MSEF. Although there continues to remain a substantial area of habitat suitable to primates in the Yucatan, of the total number of sites sampled, 34 (51.5%) were located in unprotected areas, with only 32 sites (48.5%) legally protected (CONANP, 2004). Clearly, primates in these unprotected areas remain vulnerable to human-induced habitat fragmentation associated with agriculture and cattle ranching. Some authors have suggested that the remaining forests of southern Mexico are being impacted by human activity at a variable but high rate. For example, while the overall deforestation rate for the period 1990–2000 for southern Mexico, including the Yucatan Peninsula, has been estimated at −1.1%, in some areas of the Peninsula annual rates of deforestation are 7.7% (Estrada, 2004).

DISCUSSION

Our results indicate that although populations of *A. pigra* and *A. g. yucatanensis* are found throughout the Yucatan Peninsula, approximately half of the sites we censused have no legal or protected status. For example, we found more howler monkey populations in nonprotected areas ($N = 95$) than in protected areas ($N = 52$). In addition, we encountered populations of howler monkeys (*A. pigra*, 0.21 troops/km) more frequently than spider monkeys (0.095 subgroups/km). In the case of *A. pigra*, the mean troop size (5.7 ± 1.8) was similar to that reported by Estrada *et al.* (2004) for this species at other sites in the Peninsula (Calakmul, Campeche 7.5 ± 2.3 individuals) and in other southern sites—Yaxchilan, 6.6 ± 2.1 individuals (Estrada *et al.*, 2002b) and Palenque, Chiapas, 7.0 ± 2.8 individuals (Estrada *et al.*, 2002a). In the Yucatan Peninsula, *A. pigra* populations had an adult male–adult female sex ratio that was higher than that reported by Estrada *et al.* (2002a) for sites in Palenque but similar to those found in Belize (Ostro *et al.*, 1999) and Guatemala (Bolin, 1981). Factors such as forest patch size, forest patch productivity, landscape fragmentation, opportunities for migration, and the presence of corridors between forest patches are likely to play an important role in individual survivorship and the adult sex ratios of primate groups.

In the case of spider monkeys, we found more subgroups in protected areas ($N = 30$ areas than in nonprotected areas ($N = 23$). It is likely that protected areas contain a higher incidence of mature fruit trees and larger or more contiguous tracks of forest. Both of these the factors are critical to spider monkeys that are highly frugivorous and typically exploit home ranges of several hundred hectares. The spider monkey subgroups we observed were considerably larger (11.4–14.7 individuals/subgroup) than subgroups of this subspecies reported at other sites by Gonzalez-Kirchner (1999, 3.8–4.5 individuals/subgroup) and Estrada *et al.* (2002a, 7.7 ± 3.8). Whether this reflects a higher population density in response to a larger resource base or the temporary coalescing of individuals in response to forest fragmentation remains unclear.

Vegetation types have clear effects on the presence and persistence of wild primates in the Yucatan Peninsula. Groups of the three primate species were more likely to be found in MSEF (Table 4). Some vegetation types may be more adversely affected by human impact than others as a result of a particular land use patterns and of high human population densities in their vicinity. MSEF is the most extensive vegetation type in the Yucatan Peninsula and while it seems to be especially important for the persistence of primate populations, it is an ecosystem that produces millions of dollars in internationally traded goods annually, including timber, ornamental palms, latexes, spices, oils and botanical elements (Conservation International, 2000). As a result, this important habitat for primates may be endangered in the near future. However, concerned with the need to preserve this ecosystem, local governments and nongovernmental organizations (NGO) are working toward improved coordination to preserve and manage areas encompassed by MSEF in the Yucatan Peninsula (Conservation International, 2000).

We identified an important association between habitat type (MSEF, tall semievergreen forest (TESF), and TEF) and the mean number of primate troops and subgroups. In the case of black howler monkeys, troop size was the greatest in MSEF. In the case of spider monkeys, the largest subgroups were found in a variety of forest types including TEF, tall, TESF, and low-height deciduous forest (LDF) (Table 4).

During our investigation we confirmed the existence of an area in the state of Campeche where populations of *A. pigra* and *A. p. mexicana* are sympatric. On "El Alamo Ranch" (Locality # 35) and the "Conquista Campesina" commonland (Locality # 32), we documented two and three *A. p. mexicana* troops, respectively, interacting at mean distances of 100–300 m from *A. pigra* troops. The *A. p. mexicana* troops exhibited characteristics that are typical of the species in other regions (e.g., Los Tuxtlas, Veracruz, Mexico), such as dark brown fur color and a yellow-reddish mantle. However, adult males howls seem to be more variable than in our previous observations of this species. The adult sex ratio of these mantled howler troops (total of 46 individuals) was 1:1.72 males to females. These values are similar to those reported by Estrada (1982) at Los Tuxtlas, Veracruz. The only other report of sympatric howler species was by Smith (1970) in Macuspana, Tabasco. The coexistence of *A. pigra* and *A. p. mexicana* at these sites in Campeche is extremely precarious. The groups we observed were living on private property and ranches that have been reducing their forest cover each year in order to increase cattle production.

Table 4. Sex–age composition and mean composition of troops of howler monkeys (*Alouatta pigra* and *Alouatta palliata*) and subgroups of spider monkeys (*Ateles geoffroyi yucatanensis*) sighted during surveys in Yucatan Peninsula, Mexico

Species and sex–age category	State	MDF	LDF	MANG	MSDF	MSEF	TEF	TSEF
Alouatta pigra								
Ad. male (mean ± SD)								
	Campeche	–	1	1	1.5 ± 0.7	1.6 ± 0.7	1.3 ± 0.5	1.4 ± 0.5
	Quintana Roo	–	–	–	1.2 ± 0.5	1.3 ± 0.5	–	1.6 ± 0.8
	Yucatan	–	–	–	–	1	–	–
Ad. female (mean ± SD)								
	Campeche	–	1	2	3	2.5 ± 0.9	2.1 ± 0.9	2.1 ± 0.9
	Quintana Roo	–	–	–	2.5 ± 1.3	2.2 ± 0.8	–	2.7 ± 0.7
	Yucatan	–	–	–	–	2	–	–
Juv. male (mean ± SD)								
	Campeche	–	2	–	0.5 ± 0.7	0.7 ± 0.7	0.6 ± 0.6	0.4 ± 0.5
	Quintana Roo	–	–	–	0.7 ± 0.5	0.8 ± 0.7	–	0.7 ± 0.4
	Yucatan	–	–	–	–	–	–	–
Juv. female (mean ± SD)								
	Campeche	–	–	0.5 ± 0.7	–	0.5 ± 0.6	0.3 ± 0.4	0.4 ± 0.5
	Quintana Roo	–	–	–	0.2 ± 0.5	0.6 ± 0.7	–	0.4 ± 0.5
	Yucatan	–	–	–	–	1	–	–
Infant (mean ± SD)								
	Campeche	–	–	1.5 ± 0.7	0.5 ± 0.7	0.7 ± 0.6	0.4 ± 0.6	0.4 ± 0.5
	Quintana Roo	–	–	–	1.2 ± 0.8	1.2 ± 0.8	–	1.1 ± 0.6
	Yucatan	–	–	–	–	2	–	–
Total (mean ± SD)								
	Campeche	–	4	5	5.5 ± 0.7	6.2 ± 1.7	4.7 ± 1.9	4.7 ± 1.4
	Quintana Roo	–	–	–	6.0 ± 1.8	6.3 ± 1.3	–	6.6 ± 1.7
	Yucatan	–	–	–	–	6	–	–

Ateles geoffroyi yucatanensis								
Ad. male (mean ± SD)								
	Campeche	1	4	2.2 ± 0.9	–	2.2 ± 2.1	3.0 ± 1.6	1.5 ± 0.7
	Quintana Roo	–	–	–	2.0 ± 1.4	3.7 ± 2.3	–	2.8 ± 1.5
	Yucatan	–	2.5 ± 0.7	–	2.5 ± 2.1	4.0 ± 2.8	–	–
Ad. female (mean ± SD)								
	Campeche	1	6	3.2 ± 1.2	–	3.3 ± 2.5	3.8 ± 1.2	2
	Quintana Roo	–	–	–	4.5 ± 2.1	5.6 ± 3.0	–	4.2 ± 1.8
	Yucatan	–	3.5 ± 0.7	–	4.5 ± 2.1	9.0 ± 5.6	–	–
Juv. male (mean ± SD)								
	Campeche	–	1	–	0.5 ± 0.7	1.0 ± 1.1	0.8 ± 0.6	0.5 ± 0.7
	Quintana Roo	–	–	–	–	1.2 ± 1.5	–	0.9 ± 0.9
	Yucatan	–	1	–	–	5.0 ± 2.8	–	–
Juv. female (mean ± SD)								
	Campeche	–	1	0.5 ± 0.5	–	1.0 ± 1.2	1.3 ± 0.6	0.5 ± 0.7
	Quintana Roo	–	–	–	–	1.3 ± 1.5	–	1.1 ± 0.9
	Yucatan	–	1	–	0.5 ± 0.7	7.0 ± 1.4	–	–
Infant (mean ± SD)								
	Campeche	1	–	0.7 ± 0.5	–	1.6 ± 1.0	1.6 ± 0.7	1.0 ± 1.4
	Quintana Roo	–	–	–	2.5 ± 0.7	2.1 ± 1.8	–	1.7 ± 1.3
	Yucatan	–	2.0 ± 1.4	–	1.5 ± 0.7	–	–	–
Total (mean ± SD)								
	Campeche	3	12	6.7 ± 2.5	–	9.1 ± 6.6	10.5 ± 3	5.5 ± 0.7
	Quintana Roo	–	–	–	9.5 ± 3.5	14.0 ± 9	–	11 ± 4.4
	Yucatan	–	10.0 ± 2.8	–	9.0 ± 5.6	25 ± 12.7	–	–
Alouatta palliata mexicana								
Ad. male (mean ± SD)								
	Campeche (*)	–	–	–	–	2.2 ± 0.8	–	–
	Quintana Roo (**)	–	–	–	–	3	–	–

(*Continued*)

Table 4. *(Continued)*

Species and sex–age category	State	MDF	LDF	MANG	MSDF	MSEF	TEF	TSEF
Ad. female (mean ± SD)								
	Campeche (*)	–	–	–	–	3.8 ± 0.8	–	–
	Quintana Roo (**)	–	–	–	–	2	–	–
Juvenile (mean ± SD)								
	Campeche (*)	–	–	–	–	1.0 ± 0.7	–	–
	Quintana Roo (**)	–	–	–	–	1	–	–
Infant (mean ± SD)								
	Campeche (*)	–	–	–	–	2.2 ± 0.4	–	–
	Quintana Roo (**)	–	–	–	–	1	–	–
Total (mean ± SD)								
	Campeche (*)	–	–	–	–	9.2 ± 1.9	–	–
	Quintana Roo (**)	–	–	–	–	7	–	–

Codes: Vegetation type: LDF = low, deciduous forest; MSDF = medium, semideciduous forest; MSEF = medium, semievergreen forest; TEF = tall, evergreen forest; TSEF = tall, semievergreen forest; MDF = medium, deciduous forest; and MANG = mangrove (Flores y Espejel, 1994). (*) Troops located in sympatric area; (**) Troops located in a five stars hotels out of their natural geographic distribution.

General Conservation Considerations Regarding the Yucatan Peninsula as a Priority Area for Primate Conservation in Mexico

Data presented in this chapter indicate that at present, the conservation status of Yucatan Peninsula's primate populations can be described as follows. There remain large tracks of forest as well as fragmented landscapes where primate populations continue to survive. However, there are areas that are being negatively impacted by the effects of habitat disturbance. For example, clearing areas for cattle ranching in Campeche, agriculture in Yucatan, and most damaging, the establishment of tourist areas in northern Quintana Roo have diminished forest cover and isolated several primate populations. This is the case for Puerto Morelos Botanical Garden (Location # 38), which is located near the extensive tourist infrastructure of Playa del Carmen and Cancun. Furthermore, the desire of hotels to attract more tourists and a lack of understanding by hotel administrators have led to errors of judgment that could have severe negative effects on the native primate populations. One example is the introduction of an *A. p. mexicana* troop (Location # 41) into the "Playacar" tourist complex, located in northern Quintana Roo. This is an area exclusively endemic to *A. pigra*. It is certainly possible that mantled howlers from this captive group could escape and contact and possibly join a nearby *A. pigra* group.

It is likely that expanding agriculture, timber harvesting, and cattle ranching in the near future will result in increased forest fragmentation, forest degradation, and habitat loss. This may result in the fragmentation of primate populations, population isolation, and may lead to demographic, social, and reproductive disruption. One example of this is our observation in southern Campeche that howler monkeys living in very small (<1 ha) forest fragments commonly walk, feed, and drink on the ground (Pozo-Montuy, 2003). Under such conditions, the howlers are extremely vulnerable to predation by carnivores such as coyotes (*Canis latrans*) (Pozo-Montuy, pers. obs.). A similar situation, and increased in time spent on the ground, was reported for *A. p. mexicana* in southeastern Veracruz (Serio-Silva and Rico-Gray, 2000a).

The future of the Yucatan Peninsula's primate populations remains uncertain. However, with informed conservation efforts howler and spider monkey populations can continue to persist. Because the Yucatan Peninsula still contains large tracks of forested habitats, this region must be considered among the highest priority conservation regions in Mesoamerica (Serio-Silva and Rico-Gray, 2000b). What is needed, are larger scale and long-term research programs staffed by biologists, primatologists, ecologists, and anthropologists. In this way, critical knowledge of the behavioral ecology and demography of Yucatan Peninsula's primate populations can be obtained, and this knowledge can serve as the basis for developing and evaluating effective conservation and management policies. Given the current status of Yucatan Peninsula's primate populations we recommend the following conservation guidelines for the region.

(1) Increase and support efforts to promote the habitat and population conservation status of three states on the Yucatan Peninsula.
(2) Develop and prioritize research projects focused on the basic ecology, behavior, management, and conservation of primate populations in their natural habitat, evaluate effects of habitat fragmentation, and promote student training in primatology in local universities.
(3) Establish links with the local governments to increase the number of protected forested areas and set up community-based conservation initiatives in specific localities.

SUMMARY

In this paper, we present the results of an area-wide survey (January 2000–April 2002) that provides information on current locations of *A. pigra*, *A. p. mexicana* and *A. g. yucatanensis* populations in the Yucatan Peninsula. Primates were encountered in 66 of 78 sites surveyed. Of these, 24 sites harbored *A. pigra*, 15 harbored *A. g. yucatanensis*, and both species were encountered in 24 sites. In total, we found 70 subgroups of *A. g. yucatanensis* of which 6 in Yucatan, 40 in Quintana Roo, and 24 in Campeche. A total of 149 *A. pigra* troops were encountered of which 1 in Yucatan, 39 in Quintana Roo, and 109 in Campeche. All four *A. p. mexicana* troops were found in Campeche. An important corollary is the new report of two sites in Campeche where *A. pigra* and *A. p. mexicana* coexist sympatrically.

In addition, surveys assessed the legal protection status of the habitats/sites in which primate populations were present. Using GIS, we identified 93,942.39 km^2 (63.9% of total) as potential habitat for the three primate species occurring in Mexico. In this sense, although there continues to remain a substantial area of habitat suitable to primates in the Yucatan Peninsula, of the total number of sites sampled, 34 (51.5%) were located in unprotected areas, with only 32 sites (48.5%) legally protected. We evaluated how vegetation types could be having clear effects on the possibility to find wild primates in the Yucatan Peninsula. We found that populations of the three primate species were more likely to be found in MSEF. Finally, even though the Yucatan Peninsula is considered one of the most important Mexican forested areas to promote effective conservation management (for primates), we found early evidences of negative impact on habitat disturbance as a consequence of tourism in some sites in the north of the Yucatan Peninsula.

We use this information to present a general assessment of the conservation status of primate populations and their habitat on the Yucatan Peninsula.

ACKNOWLEDGMENTS

We are grateful to the following institutions and researchers for participating directly and indirectly in the development and execution of this research: Primate Conservation, Inc., Lincoln Park Neotropical Fund, Conservation Committee of American Society of Primatologists, Pronatura Península de Yucatan AC (PPY), and Instituto de Ecología AC. Mrs. Joann Andrews (PPY), Dr. Arturo Gómez-Pompa (University of California), MC Antonio Sánchez Martínez (INIFAP-Escárcega). We are grateful to LeAndra Luecke (Washington University, St. Louis, MO) for assistance in developing the Map in Figure 1. We would also like to thank Drs. Alejandro Estrada (IB-UNAM), Paul Garber (University of Illinois), and Mary Pavelka (University of Calgary) for their extraordinary editorial aid to improve this manuscript.

REFERENCES

Barrueta, T., Estrada, A., Pozo, C., and Calmé, S. 2003, Reconocimiento demográfico de *Alouatta pigra* y *Ateles geoffroyi* en la Reserva El Tormento, Campeche, México. *Neotrop. Primates* 11(3):165–169.

Bolin, I. 1981, Male parental behavior in black howler monkeys (*Alouatta palliata pigra*) in Belice and Guatemala. *Primates* 22:349–360.

Bonilla-Moheno, M. 2002, *Evaluación de la incidencia parasitaria de primates silvestres en hábitat fragmentado y conservado en la península de Yucatan*. BSc Thesis, Facultad de Ciencias, UNAM, México, DF, p. 51.

Challenger, A. 1998, *Utilización y conservación de los ecosistemas terrestres de México: pasado, presente y futuro*. CONABIO, UNAM, Sierra Madre, México.

CONANP. 2004, *Comisión Nacional de Áreas Naturales Protegidas. Región XI Península de Yucatán.* http://regionxi.conanp.gob.mx/

Conservation International. 2000, *Vegetation of the Maya Forest. Geography of the Maya Tropical Forest.* http://www.conservation.org

Estrada, A. 1982, Survey and census of howler monkeys (*Alouatta palliata*) in the rain forest of "Los Tuxtlas," Veracruz, Mexico. *Am. J. Primatol.* 2(4):363–372.

Estrada, A. 2004, *Investigaciones con primates silvestres en el sureste de Mexico. Antecedentes generales: las selvas del trópico humedo.* http://www.primatesmx.com/monos.htm

Estrada, A., Castellanos, L., García, Y., Franco, B., Muñoz, D., Ibarra, A., Rivera, A., Fuentes, E., and Jiménez, C. 2002a, Survey of the black howler monkey, *Alouatta pigra* populations at the Mayan site of Palenque, Chiapas, Mexico. *Primates* 44: 51–58.

Estrada, A., Luecke, L., Van Belle, S., French, K., Muñoz, D., García, Y., Castellanos, L., and Mendoza, A. 2002b, The black howler monkey (*Alouatta pigra*) and the spider monkey (*Ateles geoffroyi*) in the Mayan site of Yaxchilan, Chiapas, Mexico: A preliminary survey. *Neotrop. Primates* 10(2):89–95.

Estrada, A., Luecke, L., Van Belle, S., Barrueta, E., and Meda, M. R. 2004, Survey of black howler (*Alouatta pigra*) and spider (*Ateles geoffroyi*) monkeys in the Mayan sites of Calakmul and Yaxchilan, Mexico and Tikal, Guatemala. *Primates* 45(1):33–39.

Flores, S. and Espejel, I. 1994, *Tipos de vegetación de la península de Yucatan. Etnoflora Yucatanense—Fascículo 3.* Universidad Autónoma de Yucatan, Mérida, Yucatan, México, p. 136.

Gonzalez-Kirchner, J. P. 1998, Group size and population density of the black howler monkey (*Alouatta pigra*) in Muchukux Forest, Quintana Roo, Mexico. *Folia Primatol.* 69(5):260–265.

Gonzalez-Kirchner, J. P. 1999, Habitat use, population density and subgrouping pattern of the Yucatan spider monkey (*Ateles geoffroyi yucatanensis*) in Quintana Roo, Mexico. *Folia Primatol.* 70(1):55–60.

Horwich, R. and Johnson, E. D. 1986, Geographical distribution of the black howler (*Alouatta pigra*) in Central America. *Primates* 27(1):53–62.

Lara, A. C. P. and Jorgenson, J. P. 1998, Notes on the distribution and conservation status of spider and howler monkeys in the state of Quintana Roo, Mexico. *Prim. Conserv.* 18:25–29.

National Research Council. 1992, *Techniques for the Study of Primate Population Ecology.* National Academic Press, Washington, D.C.

Navarro-Fernandez, E. 2000, *Distribución de primates (Cebidae) en Campeche, México: un análisis para su conservación.* Master's Thesis, El Colegio de la Frontera Sur, Chetumal, Quintana Roo, p. 48.

Ostro, L. E. T., Silver, S. C., Koontz, F. W., Young, T. P., and Horwich, R. H. 1999, Ranging behavior of translocated and established groups of black howler monkeys *Alouatta pigra* in Belize, Central America. *Biol. Conserv.* 87:181–190.

Pozo-Montuy, G. 2003, *Comportamiento de monos aulladores (Alouatta pigra) en hábitat fragmentado en la Ranchería Leona Vicario, Balancán, Tabasco.* BSc Thesis, Licenciatura en Biología, División Académica de Ciencias Biológicas, Universidad Juárez Autónoma de Tabasco, Villahermosa, p. 54.

Ramos-Fernández, G. and Ayala-Orozco, B. 2003, Population size and habitat use of spider monkeys at Punta Laguna, Mexico, in: L. K. Marsh, ed. *Primates in Fragments: Ecology and Conservation*, Kluwer Academic Plenum Publishers, New York, pp. 191–209.

Rangel-Negrín, A. 2003, *Niveles de cortisol en monos araña (Ateles geoffroyi yucatanensis) en diferentes condiciones de hábitat y cautiverio en la península de Yucatan.* Facultad de Ciencias, UNAM, México, DF, p. 66.

Serio-Silva, J. C. and Rico-Gray, V. 2000a, Use of a stream as water source by a troop of mexican howler monkeys (*Alouatta palliata mexicana*) during extreme environmental conditions. *Southwestern Nat.* 45(3):332–333.

Serio-Silva, J. C. and Rico-Gray, V. 2000b, Primates of the peninsula of Yucatan: current state and strategies for their conservation. *Am. Soc. Primatol. Bull.* 24(2):8–9.

Serio-Silva, J. C. and Rico-Gray, V. 2002, Interacting effects of forest fragmentation and howler monkey foraging on germination and dispersal of fig seeds. *Oryx* 36(3):266–271.

Shaker, J. K. 1999, *Pronatura—Nature Conservation in Mexico*, No. 7. El Eden. pp. 20–29.

Smith, J. D. 1970, The systematic status of the black howler monkey *Alouatta pigra* Lawrence. *J. Mammal.* 51(2):358–369.

Struhsaker, T. T. 1981, Census methods for estimating densities, in: *Techniques for the study of primate population ecology.* National Academy Press, Washington, pp. 36–80.

SYPR. 2000, *LCLUC-SYPR Project web site.* http://earth.clarku.edu/lcluc

Taube, K. 2003, Ancient and contemporary Mayan conceptions about field and forest, in: A. Gómez-Pompa, M. F. Allen, S. L. Fedick, and J. J. Jiménez-Osornio, eds., *The Lowland Mayan Area. Three millennia at the Human–Wildland Interface*, Food Products Press, Binghamton, New York, pp. 461–492.

Watts, E. S. and Rico-Gray, V. 1987, Los primates de la península de Yucatan, México: Estudio preliminar sobre su distribución actual y estado de conservación. *Biótica* 12:57–66.

Watts, E. S., Rico-Gray, V., and Chan, C. 1986, Monkeys of the Yucatan peninsula, Mexico: preliminary survey of their distribution and status. *Prim. Conserv.* 7:17–22.

CHAPTER TWENTY-ONE

A Metapopulation Approach to Conserving the Howler Monkey in a Highly Fragmented Landscape in Los Tuxtlas, Mexico

Salvador Mandujano, Luis A. Escobedo-Morales, Rodolfo Palacios-Silva, Víctor Arroyo-Rodríguez, and Erika M. Rodríguez-Toledo

INTRODUCTION

Habitat loss, forest fragmentation, and hunting are the critical forces that are driving primate populations to extinction (Cowlishaw and Dunbar, 2000). Deforestation affects primates in two basic ways. First, the fragmentation process

Salvador Mandujano • Departamento de Biodiversidad y Ecología Animal, Instituto de Ecologia A. C., km 2.5 Carretera Antigua a Coatepec No. 351, Congregacion del Haya, Xalapa 91070, Veracruz, Mexico. **Luis A. Escobedo-Morales, Rodolfo Palacios-Silva, Víctor Arroyo, and Erika M. Rodriguez-Toledo** • Division de Postgrado, Instituto de Ecologia A. C., km 2.5 Carretera Antigua a Coatepec No. 351, Congregacion del Haya, Xalapa 91070, Veracruz, Mexico.

New Perspectives in the Study of Mesoamerican Primates: Distribution, Ecology, Behavior, and Conservation, edited by Alejandro Estrada, Paul A. Garber, Mary S. M. Pavelka, and LeAndra Luecke. Springer, New York, 2005.

randomly distributes primates throughout forest fragments, the result being that only certain ones are inhabited by monkeys. Second, due to the inadequate size of forest fragments, particularly if the fragment is too small, primates may become extinct locally after disturbance has occurred (Marsh, 2003). As a result, conservation of many primate species depends on the capacity of fragmented forests to support these populations (Johns and Skorupa, 1987). However, primate plasticity in response to habitat loss and fragmentation varies depending on both species and ecological factors (Cowlishaw and Dunbar, 2000). For example, large primates that are mainly frugivores are the most vulnerable to altered habitats; and fragment occupation by different primate species is usually conditioned by fragment characteristics such as size (Johns and Skorupa, 1987). In general, the larger and better the quality of the fragment, the more individuals may inhabit it (Estrada and Coates-Estrada, 1996).

However, habitat fragmentation and isolation has the potential of significantly limiting an individual's capacity to move among habitat fragments (Swart and Lawes, 1996). In many cases, this reduction in dispersal ability forces primates to live in small fragments, which can cause changes in foraging and activity patterns, social organization, and physiological conditions (Bicca-Marques, 2003; Chiarello and de Melo, 2001; Clarke *et al.*, 2002; Estrada and Coates-Estrada, 1996; Ferrari and Diego, 1995; Gilbert, 2003; Gómez-Marin *et al.*, 2001; Juan *et al.*, 2000). In particular, fragment isolation reduces primates' dispersal movements, leading to inbreeding that in some cases can diminish genetic variability (Pope, 1992; Gonçalves *et al.*, 2003). Dispersal capacity in fragmented landscapes will depend on the specific characteristics of each species as well as the spatial configuration of the landscape (Clobert *et al.*, 2004). For species in which individual dispersal is an important aspect of metapopulation dynamics and habitat fragment isolation limits individual movement, the presence of corridors is fundamental for conservation (Swart and Lawes, 1996).

Metapopulation theory has been the focus of much discussion in population and conservation biology in fragmented habitats (Hanski and Gaggiotti, 2004). The essence of the metapopulation approach is that the presence of a given species in an area depends on the balance between the rates at which local populations become extinct and those at which new ones are established by migrants from other populations in the landscape. In consequence, metapopulations exist as various local populations within a fragmented system surrounded by a matrix. According to predictions based on metapopulation theory, if the landscape has been so severely transformed by deforestation that the number, size, quality, and connectivity of fragments are all quite low, the probability of persistence on a regional scale will decrease due to limited fragment occupation and reduced colonization of empty fragments (Ovaskainen and Hanski, 2004). This type of scenario clearly indicates the pressing need for a proposal of measures to mitigate and reverse fragmentation. Although there have been many studies on primates that inhabit fragments (see, for example, Marsh, 2003), few have addressed the problem of primate population conservation from a metapopulation perspective (see, Swart and Lawes, 1996; Cowlishaw and Dunbar, 2000; Chapman *et al.*, 2003).

The wide distribution of the genus *Alouatta* and the marked variability of its habitats are indicative of its great capacity to exploit different resources (Clarke and Zucker, 1994; Fedigan *et al.*, 1998; Clarke *et al.*, 2002), a feature that has permitted representatives of the genus to survive on very small habitat fragments where other species have been unable to do so (Crockett, 1998; Jones, 1999). In any case, this adaptability can be explained by a diet composed mainly of leaves, making them less dependent on seasonal fluctuations in the abundance of fruit; furthermore, they can feed on a large number of secondary species that are typical of clearings and fragment edges (Milton, 1980; Horwich, 1998; Bicca-Marques, 2003). *Alouatta* populations are divided into social groups that act as semi-closed reproductive units (Crockett and Eisenberg, 1987). Dispersal is a reproductive strategy that is often adopted by both juveniles and adults of both sexes (Glander, 1992; Jones, 1995), possibly to avoid inbreeding that may result in maintaining the genetic variability of populations (Pope, 1992). In a forest, 70% of howler monkeys have been observed to abandon their birth group (Glander, 1992). In particular, some data suggest that howler movements may follow a "stepping stone" pattern to move from one fragment to another (Glander, 1992). Stepping stones are defined as places where an organism only briefly interrupts its trajectory toward a habitat fragment (Söndgerath and Schröder, 2002). However, it is interesting to note that both the howlers' ability to exist on small fragments and their high degree of dispersal can cause stepping stone sites to become colonized habitats, a feature that may result in a metapopulation-type distribution.

In view of the extensive destruction, fragmentation, and conversion of primate habitats to anthropogenic vegetation in the Neotropics, the degree to which the primates living there can use a landscape consisting of forest fragments and agricultural habitats is key to understanding the ecological flexibility of the species involved; furthermore, such data are relevant to the design of conservation scenarios at landscape level (Estrada and Coates-Estrada, 1996). Los Tuxtlas, a region located in the Mexican state of Veracruz, is the northernmost area of lowland tropical rainforest in America. These forests are inhabited by *Alouatta palliata mexicana* and *Ateles geoffroyi vellerosus*, which are classified as low risk and vulnerable, respectively (Rylands *et al.*, 1995). In the region, a 74% and 84% decrease in numbers of *Ateles* and *Alouatta* has been estimated, respectively (Estrada and Coates-Estrada, 1996). This reduction in primate distribution and abundance in Los Tuxtlas are principally due to deforestation of tropical rainforest: 75% of native habitat has been lost, 20% has become isolated vegetation fragments, and only 5% consists of contiguous rain forest at high elevations (>800 m) (Estrada and Coates-Estrada, 1996). As a result, the remaining populations consist of groups inhabiting archipelagos of forest fragments that vary in size, isolation distance, and age, and that live in precarious demographic and ecological conditions. The creation of corridor systems and the adoption of a metapopulation approach to the problem of conserving isolated primate populations in fragmented landscapes in Los Tuxtlas, have been recommended as strategies to increase connectivity among isolated habitat patches occupied by primates (Estrada and Coates-Estrada, 1996; Juan *et al.*, 2000). Until now, there has, however, been no explicit proposal of either an analytic method or a specific landscape scenario for attempting to increase the persistence of monkeys by identifying priority fragments and specific routes that permit greater individual flow among fragments.

Objectives of this Chapter

In the present study, we use the theoretical and methodological foundations of landscape and metapopulation ecology to evaluate and propose conservation strategies for howler monkey populations inhabiting severely altered environments (less than 20% of the original landscape habitat, according to Andrén, 1994; Fahrig, 2003; With, 2004). To tackle this, we selected a landscape characterized by a high degree of forest loss and fragmentation in the southern part of Los Tuxtlas. In this paper, we provide a synthesis of the major results from this body of research in order to answer the following question: if our goal is to increase the viability of the howler monkey metapopulation in a highly altered landscape, what must we increase, the area of remaining habitat fragments, the connections among them, or both?

STUDY LANDSCAPE

General Characteristics

Los Tuxtlas, which lies in the state of Veracruz, Mexico, was decreed a Special Biosphere Reserve in 1998 (18°18′N and 94°45′W); it represents the northern limit of tropical forest distribution on the American continent (Dirzo and García, 1992). The Reserve covers 155,122 ha and has an elevation that ranges from 0 to 1780 m. Los Tuxtlas is naturally divided into two parts: the northern San Martin Tuxtlas Volcano and the southern San Martin Pajapan Volcano, separated by the Sierra Santa Marta. The climate is warm, with a mean annual temperature of 25°C; annual rainfall oscillates between 1850 and 4600 mm (Soto and Gama, 1997).

The study area was located at the base of the Santa Marta and San Martín Pajapan Volcanoes and includes eight pieces of communal land in the municipality of Tatahuicapan de Juárez; it is bordered by the Tecuanapa and Tilapa Rivers, the Mexican Gulf Coast, and the skirts of the Santa Marta Sierra at an altitude of 800 m (Figure 1a). The landscape, 4960 ha, is characterized by an irregular topography with slopes that often exceed 30°. Only in lowlands near the Gulf coast is there an area with very gentle slopes and permanent flooding. Since the establishment of the communal lands in the 1960s, a much accelerated deforestation process began, with an annual rate of original vegetation loss varying from 4% to 7% (Rodríguez-Luna *et al.*, 1987). Approximately 17,000 Zoque-Popoluca indigenous people inhabit the Sierra and neighboring areas (Silva-López and Portilla-Ochoa, 2002).

Methods

The habitat fragments were identified using a geographic information system (Rodríguez-Toledo *et al.*, 2003). A fragment was defined as a remnant of original tropical rain forest and secondary forest (>10 m height) higher than 0.5 ha size (see Estrada and Coates-Estrada, 1996). The landscape was digitalized using aerial photographs (1:20,000, INEGI, 1999), orthophotos (INEGI, 1996), digital data (INEGI, 1990), and field information, through ArcView 3.2 (ESRI©) software. Information on fragment size, isolation, and shape was

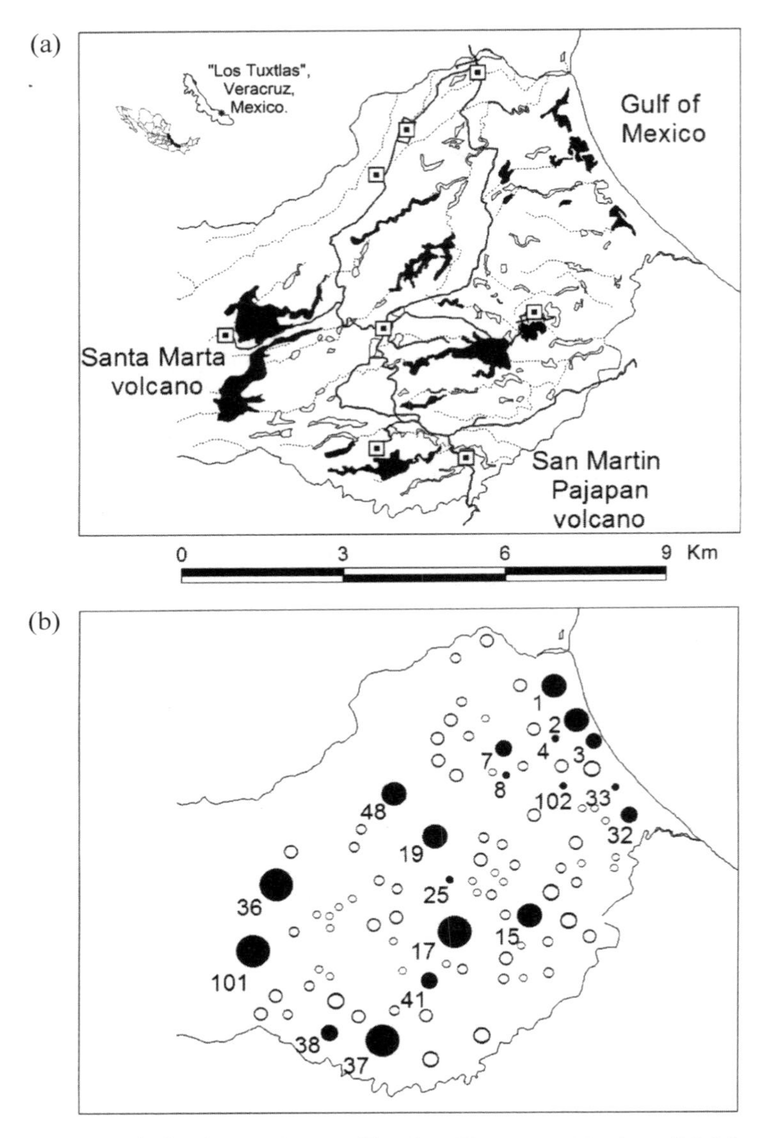

Figure 1. (a) Study landscape in Los Tuxtlas, Veracruz, Mexico. (b) Map of the patch network occupied by *Alouatta palliata* showing the relative patch sizes (indicated by the size of the circle) and their spatial locations in the study landscape area. Open circles represent empty fragments and black circles occupied fragments during the 2001–2003 surveys. Number represents the identity of howler monkeys troops as in Table 2.

estimated using the Patch Analyst 2.2 module. Isolation distances were measured to the following: nearest fragment, continuous forest, town, and river. To analyze the composition and plant structure of fragments, 15 were selected (Arroyo-Rodríguez and Mandujano, 2003). To ascertain the degree of vegetation disturbance, data were compared with those of Los Tuxtlas Biology Station, which covers 700 ha of well-preserved forest. We sampled vegetation in accordance with the Gentry (1982) protocol. On each fragment and in continuous forest, ten 50 m $\times$ 2 m transects were sampled. All tree, bush, vine, and palm species with DBH $\geq$2.5 cm were recorded. Species were categorized as follows, according to their germination light needs: primary (tolerant of shade), secondary (intolerant of shade), and non-secondary light demander species (NSLD). With the sum of density, frequency, and dominance, the importance value (IV) index was calculated for each species, both on fragments and in continuous forest.

Fragment Characteristics

The study landscape was severely altered, as of the 4960 ha of total area, only 547 ha (11%) represent tropical rain and secondary forests. The matrix surrounding forest fragments was made up principally of corn crops and pastures. In total, there were 92 fragments (Figure 1a), most (68%) located in the riparian zones of rivers or streams, although some (24%) were found on mountaintops and others (8%) ran along the ocean shore, establishing in permanently flooded areas. Eighty-one percent of the fragments were smaller than 5 ha, and only five (8%) exceeded 10 ha, the largest covering 76 ha. The mean distance from these forest fragments to continuous forest was 3625 $\pm$ 1587 m (SD) (range 144–6704 m); nevertheless, 15% of the fragments were less than 1000 m away. The mean distance from one fragment to the nearest was 111 $\pm$ 99 m (range 8–438 m), but 85% were less than 200 m away. The mean distance from a fragment to the nearest town, however, was 880 $\pm$ 656 m (range 0–2542 m). Fragment shape was irregular, as most ran along rivers or streams, making them quite long and narrow. As fragment shape became less regular, the relationship between surface area and perimeter decreased and the edge effect increased. If we consider a border to measure 100 m in width, 64 ha were estimated to comprehend the fragment interior. That is to say, of 547 ha of fragments, only 11.7% was estimated to correspond to forest less altered in composition and structure. In particular, of the 92 fragments in the study landscape, only six (F2, F15, F17, F36, F37, and F101) fragments had interior forest.

Floristic Composition and Vegetation Structure

On the fragments, the following families had the most individuals: Euphorbiaceae (8% of all plants recorded), Vochysiaceae (6%), Monimiaceae (6%), Violaceae (4%), and Asteraceae (3%); while in continuous forest Arecaceae (49%), Moraceae (13%), Meliaceae (5%), Fabaceae (3%), and Euphorbiaceae (3%). Considering the abundance, secondary species as *Siparuna andina*, *Croton schiedeanus*, and *Eupatorium galeotti*, as well as NSLD such as *Vochysia guatemalensis*, *Tapirira mexicana*, and *Rinorea guatemalensis*, represent 26% of all plants recorded on fragments (Table 1). In contrast, in continuous forest these last species were not recorded; while the palm *Astrocaryum mexicanum* represented 42% of the plants recorded, while on fragments it constituted only 6%.

Table 1. Abundance and basal area of the 10 species with the highest important value (IV) in the fragments and continuous forest. Species were categorized according to their germination light needs: primary (Pri), secondary (Sec), and non-secondary light demander species (NSLD).

Species	Family	Ecological group	Abundance		Basal area		IV[a]
			n	%	m^2	%	
Fragments							
Vochysia guatemalensis	Vochysiaceae	NSLD	227	5.6	6.6	6.7	15
Siparuna andina	Monimiaceae	Sec	260	6.4	1.5	1.5	14
Astrocaryum mexicanum	Arecaceae	Pri	233	5.8	0.6	0.6	12
Tapirira mexicana	Anacardiaceae	NSLD	163	4.0	6.4	6.5	12
Croton schiedeanus	Euphorbiaceae	Sec	150	3.7	1.3	1.4	10
Pseudolmedia oxyphyllaria	Moraceae	Pri	121	3.0	2.2	2.2	9
Rinorea guatemalensis	Violaceae	NSLD	143	3.5	0.6	0.6	8
Terminalia amazonia	Combretaceae	Pri	45	1.1	16	16	7
Dendropanax arboreus	Araliaceae	NSLD	65	1.6	2.7	2.8	6
Eupatorium galeotti	Araliaceae	Sec	117	2.9	0.3	0.3	6
Continuos forest							
Astrocaryum mexicanum	Arecaceae	Pri	90	42	0.2	1.7	53
Poulsenia armata	Moraceae	Pri	5	2.3	2.2	15	22
Brosimum alicastrum	Moraceae	Pri	3	1.4	2.1	14	17
Dussia mexicana	Fabaceae	Pri	1	0.5	2.1	15	16
Nectandra ambigens	Lauraceae	Pri	3	1.4	1.6	11	15
Pseudolmedia oxyphyllaria	Moraceae	Pri	13	6.0	0.4	2.8	15
Cordia megalantha	Boraginaceae	NSLD	2	0.9	1.4	9.9	13
Guarea glabra	Meliaceae	Pri	5	2.3	0.8	5.7	11
Chamaedorea tepejilote	Arecaceae	Pri	14	6.5	0.1	0.1	10
Cynometra retusa	Caesalpinaceae	Pri	7	3.3	0.2	1.5	10

[a] The importance value (IV) index was calculated for each species considering the sum of density, frequency, and dominance.

The average basal area of fragments was significantly lower (6.6 ± 2.4 m^2) than that recorded in continuous forest (15.0 m^2) ($t = 3.36$, $p = 0.005$). In terms of ecological groups, continuous forest had proportionally greater richness ($\chi^2 = 13.76$, $p = 0.001$) and abundance ($\chi^2 = 158.5$, $p = 0.001$) of primary species than that of fragments, on which secondary and NSLD species dominated. As to DBH range, there were more thin plants on fragments (<60 cm DBH) and less large trees (>60 cm DBH) than in continuous forest ($\chi^2 = 31.44$, $p = 0.01$). In particular, continuous forest had a greater proportion of large primary trees than that of fragments ($\chi^2 = 6.59$, $p = 0.05$) and a lower proportion of individuals of secondary ($\chi^2 = 122.99$, $p = 0.001$) and NSLD ($\chi^2 = 86.5$, $p = 0.001$) species with DBH below 60 cm.

Fragment size was negatively correlated to the number of secondary species ($R^2 = 0.48$, $p = 0.01$) and NSLD species abundance ($R^2 = 0.63$, $p = 0.01$); it was positively correlated to the basal area of primary species ($R^2 = 0.53$, $p = 0.01$). Furthermore, fragment size was negatively correlated to the frequency of plants with DBH between 10 and 20 cm ($R^2 = 0.37$, $p = 0.01$) and positively to the number of trees with DBH of over 60 cm ($R^2 = 0.36$, $p = 0.01$).

HOWLER MONKEY FRAGMENT OCCUPATION

Methods

Fieldwork was carried out from January of 2001 to July of 2003 (Rodríguez-Toledo *et al.*, 2003; Escobedo and Mandujano, in press). Two or three times a year, all fragments making up the study landscape were sampled. Due to the small size of fragments, it was possible to count all individuals inhabiting each fragment. Observations were made from 07:00 to 12:00 and 16:00 to 18:00 h. Individuals were categorized according to the characteristics that made their identification possible, such as face shape, scars, and coloring patterns on back, tail, and extremities.

Primate Distribution in Relation to Fragment Characteristics

Of the 92 fragments, *Alouatta* inhabited 18, representing 20% of the total (Figure 1b). Three howler groups lived on fragment F19, while one group used four (F5, F6, F7, and F8). The remaining fragments were inhabited by one group. On five fragments, only males were observed (Table 2). The mean

Table 2. Annual abundance and group composition of howler monkeys, and of forest fragments characteristics in the study area

		Isolation distances (m) to:			2001[a]	2002					2003				
Fragment	Size (ha)	Nearest fragment	Nearest town	Nearest continuous forest	Total	M	F	J	I	Total	M	F	J	I	Total
1	11	96	1438	6704	3	1	3	0	0	4	1	2	0	0	3
2	9.3	34	2125	6169	5	2	3	0	0	5	2	3	0	0	5
3	4.7	34	2542	5900	7	1	5	0	2	8	1	4	1	1	7
15	11.8	115	4	3675	10	3	7	3	2	15	4	5	3	2	14
17	57.2	18	307	3364	5	2	3	0	0	5	2	3	0	1	6
19[b]	29.9	196	562	3197	11	4	7	2	1	14	4	3	2	2	11
32	5.3	24	1988	4426	6	2	3	0	0	5	2	3	1	0	6
33	3.67	12	2186	4817	6	1	1	0	1	3	2	1	1	0	4
36	75.5	75	81	144	1	1	0	0	0	1	1	0	0	0	1
37	32.6	50	6	1164	1	2	0	0	0	2	1	0	0	0	1
38	5	23	192	1184	0	0	0	0	0	0	1	0	0	0	1
41	6.5	57	625	2850	5	1	4	0	0	5	1	4	0	0	5
48	13	15	557	2660	1	1	0	0	0	1	1	0	1	0	1
5, 6, 7, 8[c]	14.6	43	1941	5634	10	2	3	1	1	7	2	3		2	8
101[d]	71.0	75	438	206	?					?					2?

M = adult males, F = adult females, J = juveniles, I = infants.

[a] During 2001, specific data (sex, age) were not observed in detail in each group.

[b] Fragment 19 inhabited three groups.

[c] Fragments 5–8 used by one group, therefore the size is the sum of each fragment and isolation is the mean distance.

[d] Fragment 101 was sampled only once at the end of 2003, thus is lacking precise data.

isolation distance from one specific group to others was 2.7 ± 0.8 km; the mean distance from one group to the closest fragment was 0.3 ± 0.4 km, while the mean distance from howler groups to continuous forest was 6.2 ± 2.3 km. Specifically, 55% and 33% of individuals and groups, respectively, were located on fragments with an isolation distance greater than 200 m (Table 2). Groups F38, F17, F15, F41, F19, and F48 were the most isolated.

The incidence of primates on fragments was related to isolation distance and fragment size (Figure 2). As the latter increased (>20 ha), fragment occupation reached 100% (Figure 3a). In contrast, as fragment isolation increased (>200 m) fragment occupation dropped to zero (Figure 3b). In particular, of fragments less than 20 ha size, occupation increased if isolation distance between fragments was less than 150 m (Figure 2).

Fragment characteristics differed between those occupied by *Alouatta* and those that were not (Table 3). In general, occupied fragments were larger, less isolated, and contained a greater number of primary tree species represented by large individuals. In particular, variations related to species richness, number

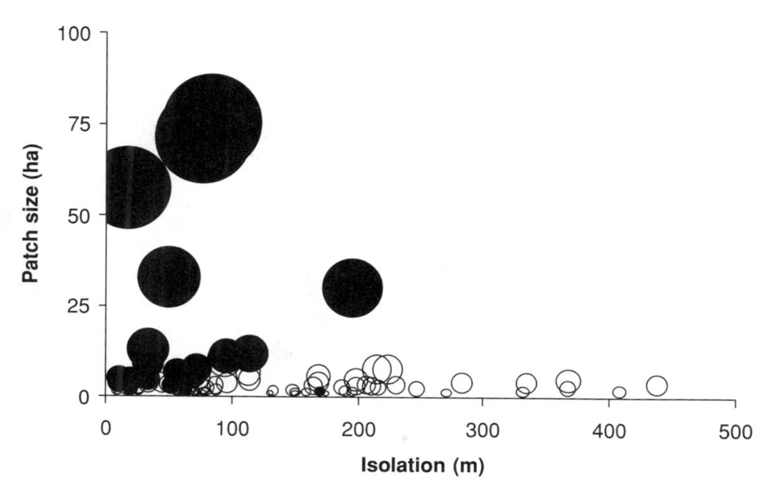

Figure 2. A plot of the fragment area (in ha) against the isolation (nearest fragment, in m). Open circles represent empty fragments and black circles occupied fragments during the 2001–2003 surveys. The size of the circles is relative to the fragment size.

of individuals, and total basal area for the 10 most important families in terms of *Alouatta* food sources. The basal area of these families was greater on occupied fragments ($t = 2.45$, $p = 0.03$).

LANDSCAPE CONNECTIVITY

Methods

Connectivity is an estimate of the functional relationship between the distance traveled by individuals and the spatial configuration of a habitat (With, 2004). The connectivity analysis estimates the ease with which organisms move from one fragment to another within a given landscape and is used to describe the effect of spatial habitat structure on movement (Taylor *et al.*, 1993). To estimate connectivity, nine scenarios resulting from the combination of two factors were simulated: the number of fragments and the distance monkeys traveled from one to another (Palacios and Mandujano, in press). Three levels were contemplated for the first factor: (1) all fragments in the landscape, (2) all those found on a 50 m fringe running along rivers, and (3) only those occupied by howler monkeys. For the second factor, three levels of monkey movement were considered: (1) traveling of $\leq$800 m between fragments, (2) traveling $\leq$200 m, and (3) traveling $\leq$200 m. For each scenario, the gamma and beta indexes proposed by Forman and Godron (1986) were calculated in order to evaluate connectivity. The gamma index (γ) provides the maximum value of 1 when

(a)

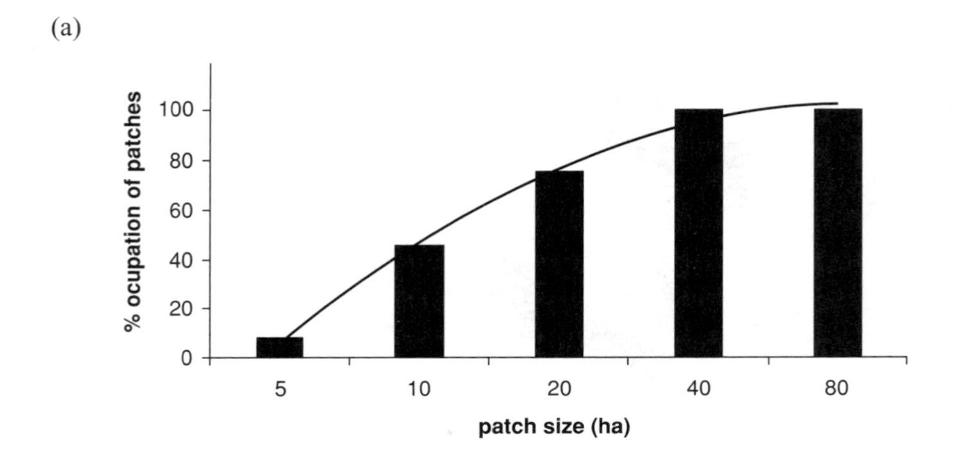

(b)

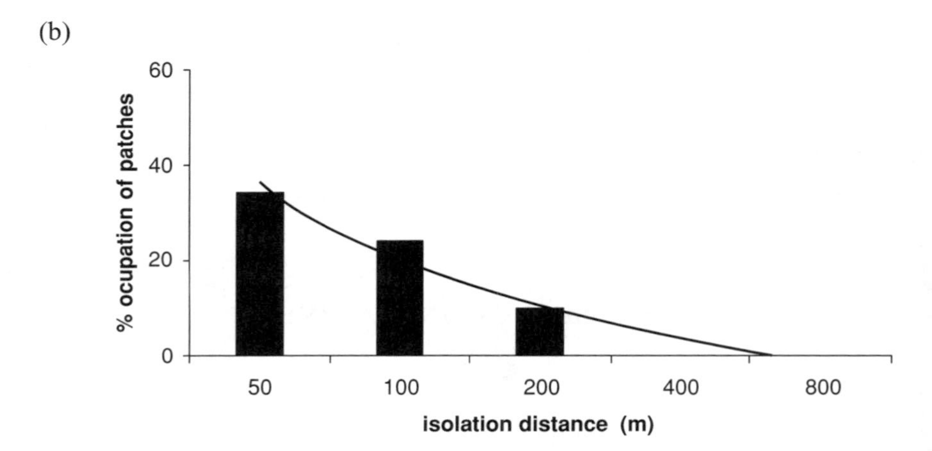

Figure 3. (a) Frequency occupation of patches in relation to patch size and (b) distance to nearest patch. The continuous line is illustrative to represent the probable incidence function.

landscape connectivity is defined as the maximum number of possible routes. The beta index (β) provides the maximum value of 1 when landscape connectivity is defined by the presence of the maximum number of routes with possible cycles.

Traveling Observations and Spatial Landscape Configuration

During the study, 10 cases of movement were detected (range 15–656 m). Of the movements recorded, 75% occurred at distances below 100 m. As the

Table 3. Comparison of fragment characteristics and tree structure between occupied and unoccupied by *Alouatta*

Variables	Fragments occupied by *Alouatta*	Fragments unoccupied
Number and percentage of fragments	18 (20%)	74 (80%)
Average size of fragments	20 ha	2.7 ha
Total surface of fragments	350 ha	197 ha
Interior surface of fragments[a]	64 ha	0 ha
Isolation to neighbor fragment	65 m	122 m
Isolation to continuous forest	6000 m	6200 m
Isolation to neighbor town	1000 m	800 m
Richness of tree species/1000 m^2	63 ± 7	68 ± 10
Number of tree individuals/1000 m^2	243 ± 35	299 ±51
Percentage of primary species	50 ± 6	44 ± 11
Percentage of secondary species	18 ± 6	23 ± 8
Basal area (m^2) of primary species	4.6 ± 2.1	2.4± 1.9
Basal area (m^2) of secondary species	0.6 ± 0.5	0.9 ± 0.5
Basal area (m^2) of principal families in howler diet[b]		
Moraceae	6.9	3.3
Fabaceae	3.0	2.2
Lauraceae	1.1	1.2
Sapotaceae	2.3	2.3
Boraginaceae	1.2	1.2
Cecropiaceae	1.3	2.0
Burseraceae	5.7	4.1
Annonaceae	8.5	5.5
Euphorbiaceae	2.5	1.2
Anacardiaceae	4.3	2.7

[a] Considering a border 100 m around the fragments.
[b] According to data of Estrada (1984), Silva-López *et al.* (1993) and Juan et al. (2000).

isolation distance between fragments increased, less movements were observed. Travel data fit the negative exponential ($R^2 = 0.86$, $F = 24.1$, $p = 0.008$).

Landscape Connectivity from the Perspective of *Alouatta*

Only in three scenarios was a connectivity percentage of between 80% and 100% obtained; in the other six, the landscape structure was segmented, causing connectivity to be under 50% (Figure 4). That is to say, maximum landscape connectivity could only be achieved if monkeys walked along the ground 400–800 m in order to reach a group inhabiting another fragment. Nevertheless, as field data show that the frequency of monkeys traveling such long distances is extremely low, the actual landscape connectivity is at present under 30% ($\gamma = 0.3$ and $\beta = 0.1$). In particular, six fragments (F1, F2, F3, F4, and F32)

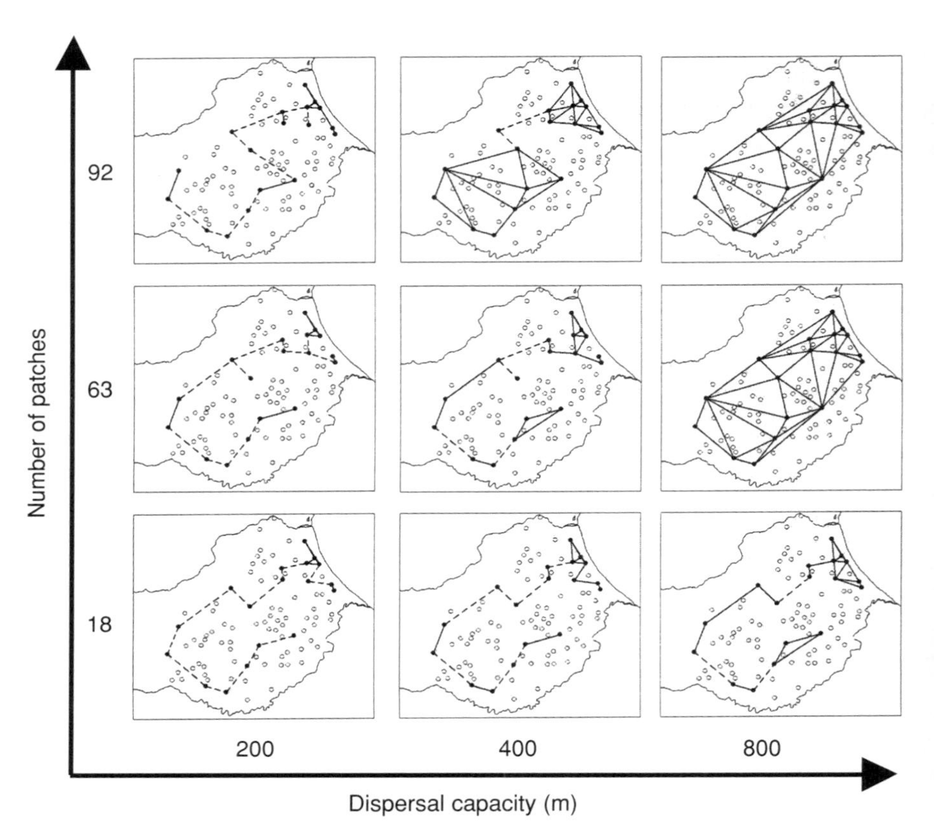

Figure 4. Network connectivity in the nine simulated landscapes scenarios according to hypothetical travel capacity of howlers and number of fragments in the landscapes. The black lines represent edges connecting groups inhabiting different fragments. The edges were estimated using the minimum spanning tree.

seem to have greater relative importance to landscape connectivity. This is true because the fragments serve as stepping stones or gathering sites on route to other fragments. Unfortunately, these fragments are found only in the lowest parts of the landscape and do not maintain the largest primate populations (see Table 2).

METAPOPULATION VIABILITY

Methods

The RAMAS Metapop Program (Akçakaya, 2002) was used to simulate deterministic and stochastic factors affecting howler monkeys group dynamics. As a

Table 4. Parameters and values used in the simulations of the viability analysis of the *Alouatta* metapopulation in the study area

Parameters	Value used in the simulations
Duration	30 years
Repetitions	5000
Sexes used in the simulation	Males and females
Type of system of mating	Poligamic
Dense-dependence type	Scramble competition
Age states	Infants, juvenile and adult
Rate of fecundity	0.2, single mature females reproduce
Proportion of sexes when being born	0.5
Survival rate	Infants 0.85, juvenile 0.5, and adults 0.8
Initial population	14 populations, 73 individuals
Carrying capacity	1.5 individuals/ha
Stochastic demographic	Correlation of the rates of fecundity and survival with the carrying capacity
Catastrophes	Hurricanes: annual probability of 0.1, reduction of 30% in the carrying capacity, regional effect. Illnesses: annual probability of 0.01, reduction of 60% in the abundance, local effect and transmitted by dispersores
Dispersal	Juvenile of both sexes they are dispersed, and dense-dependent
Correlation of the vital rates among populations	Populations present a high correlation
Depression for endogamia and Allee effect	Not considered

metapopulation can be defined as a set of local populations that interact through individual migration from one population to another (Hanski and Gaggiotti, 2004), this study considered a "local population" to be a group inhabiting the same fragment, and "metapopulation" as this set of groups inhabiting different fragments (Escobedo and Mandujano, in press). A summary of the parameters used in the simulations appears in Table 4. Simulations were done in order to evaluate the possible effects of various connectivity levels among fragments as well as how changes in habitat surface influenced metapopulation dynamics. The connectivity levels considered were (1) total isolation among populations, (2) present landscape with poor fragment connectivity (<30%), and (3) potential landscape with increased connections among fragments. In terms of changes in habitat surface, the three trends were as follows: (1) constant habitat loss at an annual rate of 4% of fragment surface, (2) arrested deforestation, and (3) fragment regeneration or reforestation at an annual rate of 1%. Response variables were the total number of individuals expected in the next 30 years, the likelihood of local extinction, and identification of the most important fragments in terms of metapopulation maintenance.

Abundance and Demographic Rates

It was estimated that 71–75 howler monkeys inhabit the study landscape. The ecological density (individuals/fragment size) was 1.1 ± 1.9 (SD) individuals/ha (range from 0.02 to 7.9). The structure in terms of ages and sexes appears in Table 2. In general, no significant differences in demographic parameters were found from 2002 to 2003. Fecundity was recorded as 0.20 and infant survival as 0.85. Mortality was particularly difficult to determine; the only confirmed deaths were of three adult females on fragment F19. This means that the finite population growth rates calculated for the 2001–2002 and 2002–2003 periods were 1.05 and 0.97, respectively.

Effects of Fragment Size and Connectivity

The simulations of the nine scenarios gave four possible monkey abundances over the next 30 years (Figure 5). First, in the three scenarios with 4% annual habitat loss, the expected abundance was 0–10 individuals. Second, for

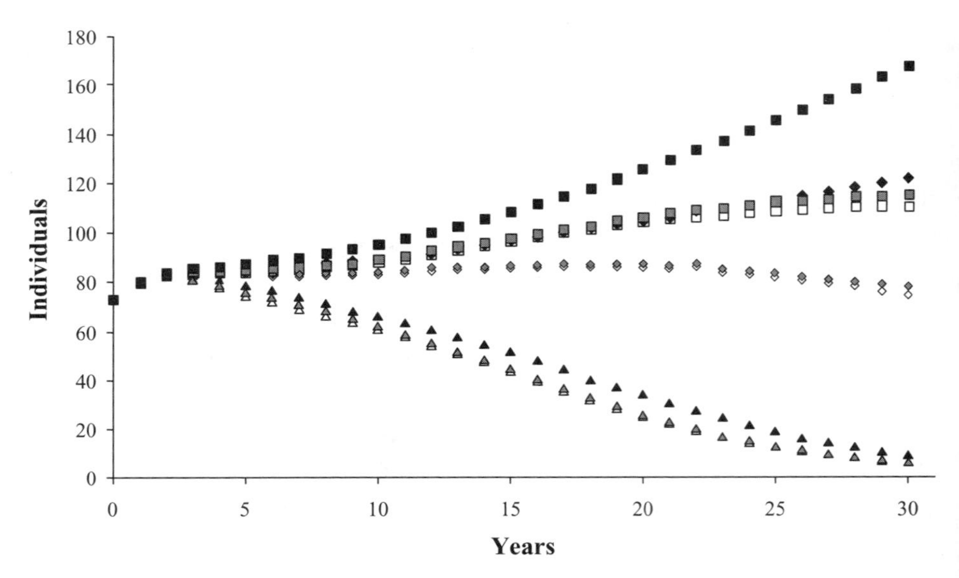

Figure 5. Expected metapopulation tendencies under the nine simulated scenarios: fragment area was increased (squares), maintained (rhombi), and decreased (triangles); potential connectivity (black), present connectivity (gray), and total isolation (white).

two scenarios in which the habitat was maintained but the possibility of individual dispersal among fragments was minimal, population abundance was expected to remain at 73–78 individuals. Third, an increase in abundance of 110–121 individuals was expected when habitat was maintained but dispersal among fragments was high, and when habitat increased by 1% annually but the possibility of individual dispersal among fragments was minimal. Fourth, the final maximum abundance expected was 166 individuals in a scenario of 1% annual habitat recovery and where dispersal possibility was high due to the use of corridors. In particular, at the three scenarios with 4% annual habitat loss, the likelihood of extinction was estimated at 35%. In contrast, the expected extinction probability was 0.1% for scenarios in which the habitat was maintained or increased at an annual rate of 1%. That is, both the amount of habitat ($F = 12.4$, $p = 0.001$) and landscape connectivity ($F = 3.88$, $p = 0.02$) significantly affected the risk of extinction and variation in metapopulation abundance.

DISCUSSION

If our goal is to increase the viability of the howler monkey metapopulation in a highly altered landscape, what must we increase, the area of remaining habitat fragments, connections that link them, or both? According to our results, fragment occupation by howler monkeys is associated with fragment size, isolation, and quality. Specifically, the bigger the fragment, the shorter the isolation distance among fragments, and the higher the number of trees with DBH >60 cm, the more the fragments were occupied by howlers. Our data show that the likelihood of extinction at a metapopulation level will be high (35%) in this landscape if deforestation continues to reduce fragment surface area and to increase isolation. Moreover, if during selective felling, the tallest and most important trees for these primates are removed, foraging and resting sites will become scarce. The results also indicate that the spatial configuration of habitat is vital to metapopulation persistence, as not only fragment size but location with respect to other fragments determines landscape connectivity. This aspect is crucial because the observed movements of howler monkeys suggest that mean traveling distance between fragments is less than 200 m. Thus, a conservation strategy must deal with the protection of large, occupied fragments, avoiding further loss of area and, to the greatest extent possible, supporting programs that favor gradual habitat recovery and increased connectivity. The simulated expected extinction probability of the howler monkey metapopulation is only

0.1% for scenarios in which habitat size and connectivity are increased at a reforestation rate of 1% and connectivity level of 30%, respectively. Thus, our results suggest two strategies for conserving howler monkeys in the study landscape. The first is to protect priority fragments—that is, those that maintain the largest groups of howlers and that have the greatest surface area, connectivity, and habitat quality. The second is to develop reforestation strategies in order to recover and restore habitat by increasing the area of certain occupied fragments and by establishing corridors and/or stepping stones.

Our results suggest the necessity of conserving fragments larger than 20 ha. Large fragments can sustain a greater number of individuals due to an increase in carrying capacity (Estrada and Coates-Estrada, 1996), decreasing the likelihood of local extinction (Escobedo and Mandujano, in press). Several studies have illustrated that an increase in the area and number of protected sites is vital to conserving endangered populations. For example, Gilbert (2003) suggested that areas larger than 100 ha are necessary for the conservation of six primate populations, *Alouatta seniculus* among them, in the Amazon. Chapman *et al.* (2003) determined that *Colobus guereza* populations, despite their ability to colonize and survive in altered habitats, might disappear due to the rapid habitat loss. Furthermore, Lawes *et al.* (2000) found that *Cercopithecus mitis* populations are more vulnerable to extinction due to habitat loss and their limited dispersal ability, as they form a metapopulation in an imbalanced state. Harcourt (1995) indicated that the main threat to mountain gorilla populations was habitat loss. For *Brachyteles arachnoides*, Strier (1993) also found that the persistence probability after 100 years was maximized by increasing available habitat and allowing the population to expand naturally. As a consequence, in our study landscape, fragments that would merit priority protection would be sites F1–F8, F15, F17, F19, F32, F33, and F41 (Figure 1).

The second strategy reflects the need to restore monkey habitat. For this, two measures are necessary: increasing the size of small (<20 ha) fragments that are presently occupied and increasing the connectivity of fragments inhabited by monkeys. For example, according to Gilbert (2003), secondary vegetation corridors are important for primate populations in Central Amazonia. Swart and Lawes (1996) evaluated different management strategies designed to creating corridors connecting patches in a *C. mitis* metapopulation and concluded that in the long run (>200 years), corridors would increase metapopulation persistence. In terms of the study region, the value of stepping stones has been discussed for non flying mammals (Estrada *et al.*, 1994), birds (Estrada *et al.*, 2000), and bats (Estrada and Coates-Estrada, 2001). In particular, some data suggest that howler movements may follow a stepping stone pattern when they travel from one fragment to another (Glander, 1992). For primates, a stepping stone can be a group of isolated trees, live fences, riparian zones, or remnants of arboreal vegetation and/or habitat patches that are substantially smaller than an animal's home range.

Based on reforestation options and considering the results of network analysis in which the ideal net is defined as the possibility of maximum individual exchange among fragments (see Figure 4), an estimate was made of the number of hectares that must be reforested in order to create a scenario in which the probability of extinction at the metapopulation level is under 1%. To increase minimum size to 20 ha on small fragments that are presently occupied, 107 ha would need to be reforested (Table 5). As to the possibility of creating corridors to connect fragments, this can be done with live trees or arboreal strips. Rodríguez-Toledo (2002) proposed the use of a vegetation strip of 15 m in width, being as wide as an occupied fragment (e.g., fragment F48 in Figure 1); given the location and topography of fragments inhabited by monkeys, river edges could be used to create riparian corridors. Thus, if live fences (5 m wide) and strip corridors (15–50 m wide) were created, this would allow reforestation scenarios of 5, 17, or 64 ha, depending on the width of the vegetation strip (Table 5). The creation of stepping stone patches at 200 m intervals could be

Table 5. Management strategies and area (in ha) of reforestation habitat needed to increase the persistence probability of the howler monkey metapopulation in the study landscape. See a graphic representation in Figure 6

Management strategy	Reforestation habitat surface (ha)
I. Increasing actually occupied fragment less than 20 ha	107
II. Increasing connectivity	
Corridors	
5 m strip	5
15 m strip	17
50 m strip	64
Stepping stones over 200 m	
25 m ratio (0.2 ha)	6
50 m ratio (0.8 ha)	25
75 m ratio (1.7 ha)	56
III. Increasing total surface	
I + II a i	112
I + II a ii	123
I + II a iii	170
I + II b i	113
I + II b ii	132
I + II b iii	163

an alternative management measure that would increase connectivity allowing movements among primate groups. Once again, the area of these patches may vary but it must be large enough for monkeys to identify it from a distance but not so large that they choose to occupy it permanently. We estimated that if the stepping stones created are assumed to be round and to have a radius of 25, 50, or 75 m, an area of 6, 25, or 56 ha would be reforested, respectively (Table 5).

Consequently, the total reforestation area, taking into account increased both fragment size and connectivity, could vary from 112 to 170 ha (Table 5). A reforestation scenario that employs corridors or stepping stones is presented schematically in Figures 6a and 6b, respectively. In both hypothetical scenarios, both fragment size and connectivity permit the establishment of a metapopulation with >99% of persitence probability. Species that are particularly useful for increasing size and reconnecting fragments are those trees that are vital to howler monkeys as sources of food and shelter; they should also serve some use to the area's human inhabitants. For example, Rodríguez-Toledo (2002) proposed the following species for restoration of the study landscape: *Ficus yoponensis* Desv., *Cecropia obtusifolia* Bertol., *Brosimum alicastrum* Sw., *Pseudolmedia oxyphyllaria* Donn Smith, *Dialium guianense* (Aubl.) Sandwith, *Poulsenia armata* (Miq.) Standl., *Spondias mombin* L., *Nectandra ambigens* (Blake) Allen, and *Bursera simaruba* (L.) Sarg. The majority of these tree species are also used by local inhabitants for various purposes (e.g. construction, medicinal, ornmental, forage, etc.).

In conclusion, our results stress the urgency of implementing conservation approach in which increase of area of habitat island and connectivity are of fundamental importance for persistence of primates in the landscape. In this scenario it would be important to involve the local communities. This would require to emphasize the ecological services that conservation and expansion of area of forest fragments would have on capturing water, decreasing erosion, and enhancing available medicinal, alimentary, ornamental and woody species reservoirs, among other benefits (Silva-López and Portilla-Ochoa, 2002). Parallel environmental education program would serve to raise awareness among local people regarding the importance of preserving the primates in their forests.

SUMMARY

As a result of tropical forest loss, fragmentation, and modification, the distribution and abundance of howler monkeys (*A. palliata*) have diminished in Los Tuxtlas, Mexico. The remaining population consists of groups inhabiting

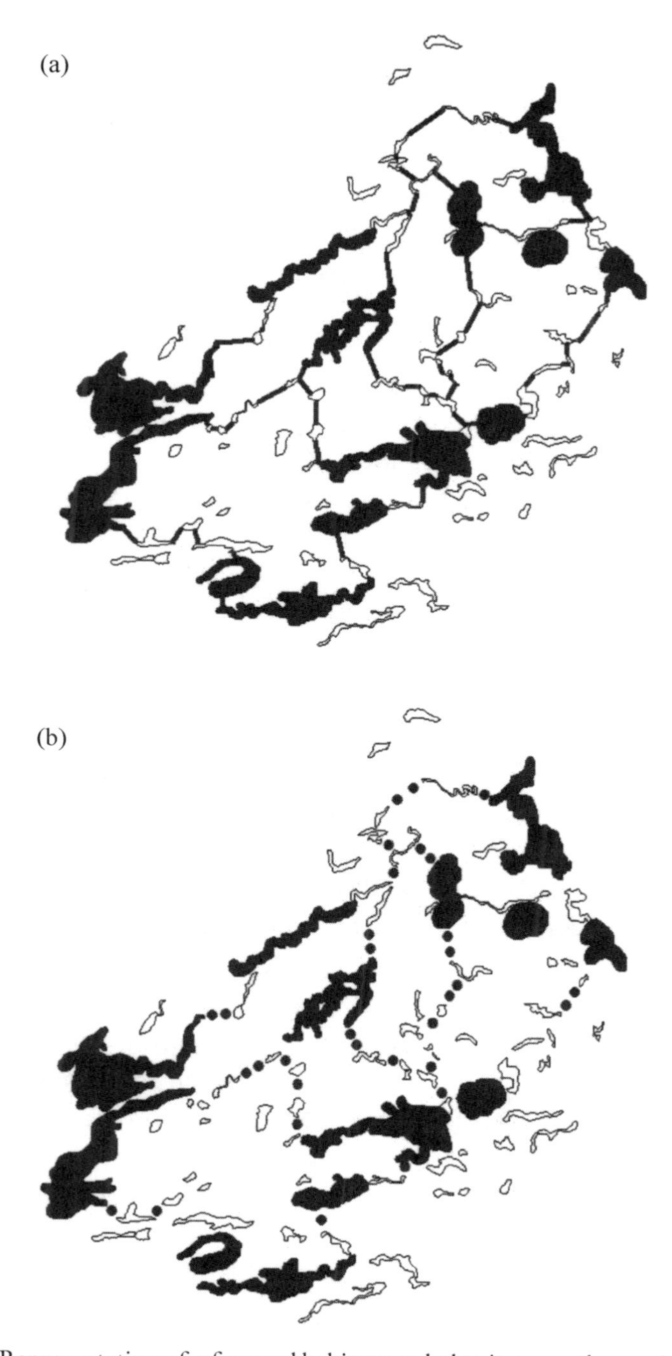

Figure 6. Representation of reforested habitat needed to increase the persistence probability of the howler monkey metapopulation in the study landscape, using corridors (a) or stepping stones (b). Black fragments represent those occupied by monkeys and white represent unoccupied fragments. See Figure 1a to compare the present landscape with the potential scenarios presented herein. The total habitat recovery in scenarios A and B represents only 170 and 163 ha, respectively.

archipelagos of forest patches. Metapopulation theory predicts a high persistence as more habitat patches are occupied and dispersal among populations becomes possible. We studied howler monkeys inhabiting a landscape of 4960 ha, where only 11% of the original habitat existed and was fragmented in 92 patches (size range: 0.5–76 ha). From 2001 to 2003, we surveyed groups and estimated the metapopulation persistence probability using the population viability program RAMAS Metapop and ecological network models derived from landscape ecology. In particular, we evaluated the importance of habitat area changes and connectivity levels. A total of 75 individuals inhabited 18% of all patches. The average isolation distance among primate groups was 2.8 and 5.8 km to continuous forest. No corridors connecting fragments existed. Hence, monkeys traveled a mean distance of less than 200 m along the ground or low brushes to move from one fragment to another. At landscape level, the actual connectivity of the metapopulation was therefore low (<30%). When we simulated the total number of individuals expected over the next 30 years, we found that habitat area change had a higher impact on metapopulation viability than on connectivity level. An extinction probability of 35% was estimated if the present rate of deforestation (4% annually) continues over the next 30 years in this landscape. A priority strategy must therefore address the protection of occupied fragments that are large (>20 ha size), less isolated (<200 m), and have better quality habitat (trees size DBH >60 cm), preventing further loss of area, and must support programs that favor habitat recovery and increased connectivity. Habitat restoration measures should include increasing the size of occupied fragments and creating corridors and stepping stones to reconnect fragments. An estimated 112–170 ha of recovered forest is needed in order to reduce the extinction probability to 1%. The species used in such programs must be native trees that are important to both primates and people. In conclusion, the integration of metapopulation ecology and landscape ecology is promising from the perspective of regional conservation. The results presented herein shed light on this new approach.

ACKNOWLEDGMENTS

The authors would like to thank the Mateo-Gutiérrez family for their hospitality. We are also grateful to F. García-Orduña and D. Canales of the University of Veracruz' Institute of Neuroethology. The comments and suggestions of the following people have helped to improve various versions on which this chapter was based: G. Silva-López, J. C. Serio-Silva, R. Manson, G. Castillo, A. Cuarón, and M. Equihua. We are particularly grateful to A. Estrada and P. Garber for inviting to write this paper and for constructive editorial help. The National Council on Science and Technology (CONACYT) granted research scholarships for the co-authors, and the Institute of Ecology's Department of Ecology and Animal Behavior provided the support necessary for the completion of this research. A small grant from the American Society of Primatology (ASP) and Primate Conservation Inc. provided partial support for this study.

REFERENCES

Akçakaya, H. R. 2002, *RAMAS Metapop: Viability Analysis for Stage-Structured Metapopulations (Version 4.0)*. Applied Biomathematics, Setauket, New York.

Andrén, H. 1994, Effects of habitat fragmentation on birds and mammals in landscapes with different proportion of suitable habitat: A review. *Oikos* 71:340–346.

Arroyo-Rodríguez, V. and Mandujano, S. 2003, Comparación de la estructura vegetal entre fragmentos desocupados y ocupados por *Alouatta palliata mexicana* en el sureste de México. *Neotrop. Primates* 11:168–171.

Bicca-Marques, J. C. 2003, How do howler monkeys cope with habitat fragmentation? in: L. K. Marsh, ed., *Primates in Fragments: Ecology and Conservation*, Kluwer Academic/Plenum Publishers, New York, NY, pp. 283–303.

Chapman, C. A., Lawes, M. J., Naughton-Treves, L., and Gillespie, T. 2003, Primate survival in community-owned forest fragments: Are metapopulation models useful amidst intensive use? in: L. K. Marsh, ed., *Primates in Fragments: Ecology and Conservation*, Kluwer Academic/Plenum Publishers, New York, NY, pp. 63–78.

Chiarello, A. G. and de Melo, F. R. 2001, Primate population densities and sizes in Atlantic forest remnants of northern Espírito Santo, Brazil. *Int. J. Primatol.* 22:379–396.

Clarke, M. R., Collins, A. D., and Zucker, E. L. 2002, Responses to deforestation in a group of mantled howler (*Alouatta palliata*) in Costa Rica. *Int. J. Primatol.* 23:365–381.

Clarke, M. R. and Zucker, E. L. 1994, Survey of the howling monkey population at La Pacifica: A seven-year follow-up. *Int. J. Primatol.* 15:61–73.

Clobert, R., Ims, R. A., and Rousset, F. 2004, Causes, mechanisms and consequences of dispersal, in: I. Hanski and O. E. Gaggiotti, eds., *Ecology, Genetics, and Evolution of Metapopulations*, Elsevier Academic Press, Burlington, MA, pp. 307–335.

Cowlishaw, G. and Dunbar, R. 2000, *Primate Conservation Biology*. The University of Chicago Press, Chicago, IL.

Crockett, C. M. 1998, Conservation biology of the genus *Alouatta*. *Int. J. Primatol.* 19:549–578.

Crockett, C. M. and Eisenberg, J. F. 1987, Howlers: Variation in group size and demography, in B. B. Smuts, D. L. Cheney, R. M. Seyfarth, R. W. Wrangham, and T. T. Strunhsaker, eds., *Primate Societies*, University of Chicago Press, Chicago, pp. 54–68.

Dirzo, R. and García, M. C. 1992, Rates of deforestation in Los Tuxtlas, a neotropical area in Veracruz, Mexico. *Conserv. Biol.* 6:84–90.

Escobedo, L. A. and Mandujano, S. Probabilidad de extinción del mono aullador en un paisaje altamente fragmentado de México, in: J. Saénz and C. Harvey, eds., *Evaluación y Conservación de la Biodiversidad en Paisajes Fragmentados en Mesoamérica*, Universidad Nacional, Costa Rica, in press.

Estrada, A. 1984, Resource use by howler monkeys (*Alouatta palliata*) in the rain forest of Los Tuxtlas, Veracruz, Mexico. *Int. J. Primatol.* 5:105–131.

Estrada, A., Cammarano, P., and Coates-Estrada, R. 2000, Bird species richness in vegetation fences and in strips of residual rain forest vegetation at Los Tuxtlas, Mexico. *Biodiver. Conserv.* 9:1399–1416.

Estrada, A. and Coates-Estrada, R. 1996, Tropical rain forest fragmentation and wild populations of primates at Los Tuxtlas. *Int. J. Primatol.* 5:759–783.

Estrada, A. and Coates-Estrada, R. 2001, Bat species richness in live fences and in corridors of residual rain forest vegetation at Los Tuxtlas, Mexico. *Ecography* 24:94–102.

Estrada, A., Coates-Estrada, R., and Meritt, D. Jr. 1994, Non flying mammals and landscape changes in the tropical rain forest region of Los Tuxtlas, Mexico. *Ecography* 17:229–241.

Fahrig, L. 2003, Effects of habitat fragmentation on biodiversity. *Annu. Rev. Ecol. Evol. Syst.* 34:487–515.

Fedigan, L. M., Rose, L. M., and Morera-Avila, R. 1998, Growth of mantled howler groups in a regenerating Costa Rican dry forest. *Int. J. Primatol.* 19:405–432.

Ferrari, S. F. and Diego, V. H. 1995, Habitat fragmentation and primate conservation in the Atlantic Forest of Eastern Minas Gerais, Brazil. *Oryx* 29:192–196.

Forman, R. T. T. and Godron, M. 1986, *Landscape Ecology*. John Wiley & Sons, New York.

Gentry, A. H. 1982, Patterns of neotropical plant species diversity. *Evol. Biol.* 15:1–85.

Gilbert, K. A. 2003, Primates and fragmentation of Amazon forest, in: L. K. Marsh, ed., *Primates in Fragments: Ecology and Conservation*, Kluwer Academic/Plenum Publishers, New York, pp. 145–157.

Glander, K. E. 1992, Dispersal patterns in Costa Rica mantled howling monkeys. *Int. J. Primatol.* 13:415–436.

Gómez-Marin, F., Vea, J. J., Rodríguez-Luna, E., García-Orduña, F., Canales-Espinosa, D., Escobar, M., and Asensio, N. 2001, Food resources and the survival of a group of howler monkeys (*Alouatta palliata mexicana*) in disturbed and restricted habitat at Los Tuxtlas, Veracruz, Mexico. *Neotrop. Primates* 9:60–67.

Gonçalves, E. C., Ferrari, S. F., Silva, A., Coutinho, P. E. G., Menezes, E. V., and Schneider, M. P. C. 2003, Effects of habitat fragmentation on the genetic variability of silvery marmosets, *Mico argentatus*, in: L. K. Marsh, ed., *Primates in Fragments: Ecology and Conservation*, Kluwer Academic/Plenum Publishers, New York, pp. 17–28.

Hanski, I. and Gaggiotti, O. E. 2004, *Ecology, Genetics, and Evolution of Metapopulations.* Elsevier Academic Press, Burlington, MA.

Harcourt, A. H. 1995, Population viability estimates: Theory and practice for a wild gorilla population. *Conserv. Biol.* 9:134–142.

Horwich, R. H. 1998, Effective solutions for howler conservation. *Int. J. Primatol.* 19:579–598.

Instituto Nacional de Estadística Geográfica e Informática (INEGI) 1990, Conjuntos de datos vectoriales de la carta San Juan Volador E15A74, escala 1:75000. Xalapa.

Instituto Nacional de Estadística Geográfica e Informática (INEGI) 1996, Ortofotos digitales E15A74b, E15A74c, E15A74e y E15A74f. de la carta San Juan Volador E15A74, escala 1:75000. Xalapa.

Instituto Nacional de Estadística Geográfica e Informática (INEGI) 1999, Fotografías aereas escala 1:20000. Vuelo especial Sierra de Los Tuxtlas. Xalapa.

Johns, A. D. and Skorupa, J. P. 1987, Responses of rain-forest primates to habitat disturbance: A review. *Int. J. Primatol.* 8:157–191.

Jones, C. B. 1995, Dispersal in mantled howler monkeys: A threshold model. *Mastozool. Neotrop.* 2:207–211.

Jones, C. B. 1999, Why both sexes leave: Effects of habitat fragmentation on dispersal behavior. *Endang. Species Updat.* 16:70–73.

Juan, S., Estrada, A., and Coates-Estrada, R. 2000, Contrastes y similitudes en el uso de recursos y patrón diarios de actividades en tropas de monos aulladores (*Alouatta palliata*) en fragmentos de selva en Los Tuxtlas, México. *Neotrop. Primates* 8:131–135.

Lawes, M. J., Mealin, P. E., and Piper, S. E. 2000, Patch occupancy and potential metapopulation dynamics of three forest mammals in fragmented afromontane forest in South Africa. *Conserv. Biol.* 14:1088–1098.

Marsh, L. K., ed., 2003, *Primates in Fragments: Ecology and Conservation.* Kluwer Academic/Plenum Publishers, New York.

Milton, K. 1980, *The Foraging Strategy of Howler Monkeys: A Study in Primate Economics.* Columbia University Press, New York.

Ovaskainen, O. and Hanski, I. 2004, Metapopulation dynamics in highly fragmented landscapes, in: I. Hanski and O. E. Gaggiotti, eds., *Ecology, Genetics, and Evolution of Metapopulations*, Elsevier Academic Press, Burlington, MA, pp. 73–103.

Palacios, R. and Mandujano, S. Conectividad de parches de hábitat para los primates en un paisaje altamente fragmentado en el sureste de México, in: J. Saénz and C. Harvey, eds., *Evaluación y Conservación de la Biodiversidad en Paisajes Fragmentados en Mesoamérica*, Universidad Nacional, Costa Rica, in press.

Pope, T. R. 1992, The influences of dispersal patterns and mating system on genetic differentiation within and between populations of the red howler monkey (*Alouatta seniculus*). *Evolution* 46:1112–1128.

Rodríguez-Luna, E., García-Orduña, F., Silva-López, G., and Canales-Espinosa, D. 1987, Primate conservation in Mexico. *Primate Conserv.* 8:114–118.

Rodríguez-Toledo, E. M. 2002, Propuesta sobre la conexión de fragmentos como alternativa para la conservación del mono aullador (*Alouatta palliata mexicana*) en un paisaje alterado en Los Tuxtlas, Veracruz. MSc Thesis, Instituto de Ecologia A. C., Xalapa, Mexico.

Rodríguez-Toledo, E. M., Mandujano, S., and García-Orduña, F. 2003, Relationships between characteristics of forest fragments and howler monkeys (*Alouatta palliata mexicana*) in southern Veracruz, México, in: L. K. Marsh, ed., *Primates in Fragments: Ecology and Conservation*, Kluwer Academic/Plenum Press, New York, pp. 79–97.

Rylands, A. B., Mittermeier, R. A., and Rodríguez-Luna, E. 1995, A species list for the New World primates (Platyrrhini): Distribution by country, endemism, and conservation status according to the Mace-Lande System. *Neotrop. Primates* 3:113–160.

Silva-López, G., Jimenez-Huerta, J., and Benitez-Rodríguez, J. 1993, Availability of resources to primates and human in a forest fragment of Sierra de Santa Marta, Mexico. *Neotrop. Primates* 1:3–5.

Silva-López, G. and Portilla-Ochoa, E. 2002, Primates, lots and forest fragments: Ecological planning and conservation in the Sierra de Santa Marta, México. *Neotrop. Primates* 10:9–11.

Söndgerath, D. and Schröder, B. 2002, Population dynamics and habitat connectivity affecting the spatial spread of populations—a similation study. *Landsc. Ecol.* 17:57–70.

Soto, A. and Gama, L. 1997, Climas, in: E. González-Soriano, R. Dirzo, and R. C. Vogt, eds., *Historia Natural de Los Tuxtlas*, UNAM, México, DF, pp. 7–23.

Strier, K. B. 1993, Viability analyses of an isolated population of Muriqui monkeys (*Brachyteles arachnoides*): Implications for primate conservation and demography. *Primate Conserv.* 14/15:43–52.

Swart, J. and Lawes, M. J. 1996, The effect of habitat patch connectivity on samango monkey (*Cercopithecus mitis*) metapopulation persistence. *Ecol. Model.* 93:57–74.

Taylor, P. D., Fahring, K. H., and Merriam, G. 1993, Connectivity is vital element of landscape structure. *Oikos* 68:571–573.

With, K. A. 2004, Metapopulation dynamics in highly fragmented landscapes, in: I. Hanski and O. E. Gaggiotti, eds., *Ecology, Genetics, and Evolution of Metapopulations*, Elsevier Academic Press, Burlington, MA, pp. 23–44.

CHAPTER TWENTY-TWO

Quantifying Fragmentation of Black Howler (*Alouatta pigra*) Habitat after Hurricane Iris (2001), Southern Belize

Shelley M. Alexander, Mary S. M. Pavelka, and Nicola H. Bywater

INTRODUCTION

Hurricane Iris made landfall on October 8, 2001, in the Monkey River area in the southern coast of Belize, catastrophically disturbing the surrounding vegetation. The goals of this chapter are to quantify changes to habitat after Hurricane Iris and to infer possible threats to a regional black howler (*Alouatta pigra*)

Shelley M. Alexander • Department of Geography, University of Calgary, Alberta, Canada.
Mary S. M. Pavelka • Department of Anthropology, University of Calgary, Alberta, Canada.
Nicola H. Bywater • Department of Geography, University of Calgary, Alberta, Canada.

New Perspectives in the Study of Mesoamerican Primates: Distribution, Ecology, Behavior, and Conservation, edited by Alejandro Estrada, Paul A. Garber, Mary S. M. Pavelka, and LeAndra Luecke. Springer, New York, 2005.

population. We used remote sensing (RS) and geographic information system (GIS) technologies to quantify the fragmentation of habitat. Here, we use the term habitat to refer to the assemblage of environmental features (e.g., forest cover, water) within the study area of a known population of black howlers (Pavelka *et al.*, 2003).

Habitat fragmentation can reduce the total amount of forest and isolate remaining forest patches (Meffe and Carroll, 1997). The effects of fragmentation are quantified by measuring changes to structure, function, and composition of habitat, using metrics defined by the discipline of landscape ecology (Forman and Godron, 1986). For example, a common measure of landscape fragmentation is the patch (habitat) to matrix (disturbance) ratio (Forman and Godron, 1986), which can quantify the extent of structural change of habitat; the lower the patch–matrix ratio, the more fragmented the habitat and the more likely sensitive wildlife species will fail to thrive (Hobbs, 1993; Debinski and Holt, 2000). The biological or ecological consequences of habitat fragmentation (i.e., functional and compositional change) are more challenging to quantify because they require knowledge of a focal species behavioral responses to fragmentation, and how changes in those responses might relate to system stability. Generally, conservationists must extrapolate life history characteristics and behavioral responses from captive animals and infer possible responses to fragmentation. It is difficult to set up a rigorous fragmentation experiment in nature, especially with sensitive species, because such research requires significant intervention (i.e., defaunation or habitat alteration) and true replicates and controls, which generally are not possible in nature. However, where experimental research has been possible, it has confirmed that a loss of habitat area and patch isolation can affect community structure by altering species composition, competitive interactions, and predator–prey dynamics (Fahrig and Merriam, 1985; Hobbs, 1993; Palomares *et al.*, 1996; Debinski and Holt, 2000). These changes may be sufficient to drive vulnerable species to local or global extinction, depending on the scale of disturbance (Burkey, 1995; Weaver *et al.*, 1996). The severity of any disturbance also is related to the spatial arrangement and composition of remaining habitat patches: their size, shape, the amount of edge habitat, isolation from other habitat patches, and the biotic qualities of the intervening space (Saunders *et al.*, 1991; Collinge, 1996). Moreover, species experience fragmentation effects differently depending upon their perceptual scale, which incidentally tends to be related to their body and home range size (Holling, 1992).

Our current understanding of this link between fragmentation and wildlife persistence is inadequate, because it is rare to have detailed population data for any species, and the tools to link population dynamics to spatial change are not well developed, or plagued with limitations (i.e., deterministic rules). Yet, habitat fragmentation is occurring at a global scale and is considered one of the principal causes of our current conservation crisis (Wilcox and Murphy, 1985). The tropics are no exception; they are the most species-rich regions of the world, contain the greatest levels of biodiversity, and are experiencing more intense development than other regions, while being highly susceptible to natural disasters (Saunders *et al.*, 1991; Collinge, 1996; Turner, 1996; Scariot, 1999). Fortunately, a growing body of research examines the effects of forest fragmentation in tropical regions. One key finding is that species distribution in tropical forests are naturally patchy (likely across trophic levels), which may increase the likelihood of extinction (Turner, 1996), and this may be more critical in tropical forests because these regions tend to be plagued by regular catastrophic disturbance, such as hurricanes.

Tropical hurricanes can fragment forests substantially by reducing canopy cover, particularly the upper canopy, which then changes the vertical distribution of foliage and increases low canopy area (Brokaw and Grear, 1991; Burslem and Whitmore, 1999; Boutet and Weishampel, 2003). In tandem, light penetration tends to increase after hurricane disturbance, which promotes growth and regeneration (Brokaw and Grear, 1991) but also tends to desiccate forests. Other research indicates that hurricanes transition canopy and ecosystems toward randomness, increase local diversity over the long term, and in fact may be necessary to promote regeneration and turnover (Brokaw and Grear, 1991; Burslem and Whitmore, 1999; Boutet and Weishampel, 2003). However, when large-scale natural disturbance occurs in concert with human development (e.g., citrus plantations, shrimp farms), landscapes may be altered more rapidly than succession processes can respond, at which point biodiversity and the landscape may decline or cease to be resilient (Weaver *et al.*, 1996).

We combined two important research tools, GIS and RS, to measure changes to the structure, composition, and function of habitat in the Monkey River area that resulted from Hurricane Iris. We interpreted this habitat change relative to needs of a focal species, the Belizean black howler. The focal approach uses a single species, usually one that is sensitive to disturbance, and assumes that the disturbance effects may be extrapolated from that species to the broader community or ecosystem. Our research is unique, as few studies integrate GIS to examine habitat fragmentation effects on primates, and no known research has quantified habitat fragmentation locally or measured the effects of hurricanes on black howler monkey habitat in Belize.

RS, GIS, and Fragmentation Metrics

Remote sensing data provide researchers with up-to-date information, which when used with a GIS serves as a platform to quantify habitat fragmentation (Johnsson, 1995; Lunetta and Elvidge, 1998). RS and GIS have been used successfully to test relationships, make predictions, and describe change in floristic heterogeneity and multi-scalar ecological processes (Hargis *et al.*, 1998; Schumaker, 1996). Integrating these tools enables scientists and government agencies to view disturbance at multiple scales; locally, and in the context of the entire landscape (Lugo, 1995). Moreover, GIS provides a more quantitative basis for decision-making, habitat management, and species conservation (Collinge, 1996). It has grown more popular because: (1) landscape data are becoming easier to obtain as the cost of commercial satellite imagery continues to decline (Tischendorf, 2001); (2) GIS provides a straightforward platform to model landscape patterns at different scales (Luoto *et al.*, 2001); (3) computer processing capabilities are increasing, making it easier and faster to handle large volumes of data (Luoto *et al.*, 2001); and (4) specialized softwares, such as Fragstats 3.3 (defined subsequently), have been coupled with GIS and provide a means to conduct advanced fragmentation analysis.

However, deriving useful information from multi-spectral satellite images of tropical forests is not without its challenges. Importantly, the use of RS imagery in spatial analysis remains less developed in the tropics, relative to the northern hemisphere, and there is much work necessary to identify and mediate inherent problems. In general, satellite imagery is hampered by seasonal variability in cloud cover, moisture, and spectral characteristics of vegetation; understandably, this is a more significant issue in moist tropical regions (Foody and Hill, 1996; Mayaux and Lambin, 1997; Mas, 1999). Tropical forests receive a high amount of annual rainfall (>2500 mm), resulting in few days that are cloud-free, which makes obtaining high-quality multi-spectral images almost impossible (Foody and Hill, 1996; Hill, 1999). This seasonal (even daily) variability in moisture influences spectral signatures, which makes time series comparison of images collected over weeks, months, or years less reliable.

Consequently, dry season images often are preferred because they facilitate more accurate classifications (Foody and Hill, 1996; Mayaux and Lambin, 1997; Mas, 1999). However, this constrains analysis to one season that may not fully reflect resource availability or disturbance effects, and perhaps more critically does not ensure that cloud-free images will be available. For instance, we were unable to attain a single cloud-free Landsat image for our study site up to 2 years after Hurricane Iris. Problems also may be encountered when collecting training site data (i.e., data that define the vegetation type of a specific location on the ground, and that are used to classify spectral reflectance in the Landsat imagery). For example, dense vegetation and remoteness bias data collection to sites near roads, waterways, or any such area that have more open canopy (Shelley Alexander, pers. obs., 2003). This is not an exclusive problem to the tropics, but can result in disproportionate sampling of vegetation types. In addition, tropical residential areas tend to be nested within vegetation and may use palm thatch for materials, which further complicates spectral separation of forested, nonforested, and residential sites. To improve RS capability and minimize the previous biases, we concur that at a minimum *a priori* ground-based experience be acquired before interpreting results (Mas, 1999).

Black Howlers and Habitat Fragmentation

The geographic range of *Alouatta pigra* is limited to regions of Mexico, Belize, and Guatemala (see Rylands *et al.*, this volume); it is presumed small because black howlers may have more narrow habitat requirements than other howler species, such as a heavy dependence on riverine and seasonally flooded areas (Horwich and Johnson, 1986; Estrada *et al.*, 2002). Black howlers recently have been upgraded to endangered status on the IUCN Red List, as a result of documented population decline that has resulted primarily from human-induced habitat loss (IUCN, 2003). Information on this species is scant relative to other howler species (Chapman and Balcomb, 1998; Estrada *et al.*, 2002; Pavelka *et al.*, 2003), but disease, habitat destruction (both natural- and human-induced), and hunting are known to contribute to black howler population decline and in some cases local extinction (James *et al.*, 1997). In addition, mountain ranges, major rivers, and possibly *Alouatta palliata* (mantled howler monkey) limit expansion of black howlers to other parts of Central and South America (see Ford, this volume; Horwich and Johnson, 1986).

Estrada *et al.* (2002) studied the effects of habitat fragmentation on black howlers in Chiapas, Mexico, and found that these primates did not inhabit forest fragments that were isolated from other forest by distances of 0.61–2.6 km (mean = 1.6 km). They found no relationship between troop size and fragment size, but troops in fragments had lower numbers of adult males. Estrada *et al.* (1999) noted that isolation and loss of habitat area will reduce black howler populations, but they also observed the same howlers occupying new types of habitat, such as plantations that offered shading from large trees (see also Estrada *et al.*, this volume). Presently, it is unknown whether black howlers in habitat fragments are under dietary stress due to lack of resources, or whether dietary stress predisposes the species to greater parasitic infestation or results in higher infant mortality rates. Moreover, we lack detailed information on how reduced forest connectivity affects dispersal, influences reproductive potential, or affects viability (Estrada *et al.*, 2002). Ostro *et al.* (2000) studied translocated black howlers and found that it is important to maintain lowland forest corridors for dispersal, recolonization, and to sustain existing populations. In addition, knowledge how human disturbance and catastrophic events disrupt howler habitat size, quality, and connectivity is essential to the conservation of this species. Identification, protection, and enhancement of important habitat and movement corridors will help preserve the species over the long term, but must begin immediately (Estrada *et al.*, 2002). RS and GIS are essential tools in this effort; combined, they can improve the database of ecological information required to enhance conservation of black howlers (Estrada *et al.*, 2002).

Study Area

Research was conducted in the Stann Creek District of Southern Belize (Figure 1). Our focal study site included the Monkey, Bladen, and Swayze River Basins (Figure 2), all characterized by semi-evergreen and evergreen rain forests (Horwich and Johnson, 1986; Pavelka *et al.*, 2003). The climate of Belize consists of a distinct wet and dry season, and the transition between the two is abrupt. The rainy period lasts from June to November and the dry season from December to May (Belize Government, 2003). The mean annual rainfall is 1524 mm in the north and 4064 mm in the south (Belize Government, 2003). Tropical storm peak activity usually occurs in September and October, and their numbers and intensity vary from year to year.

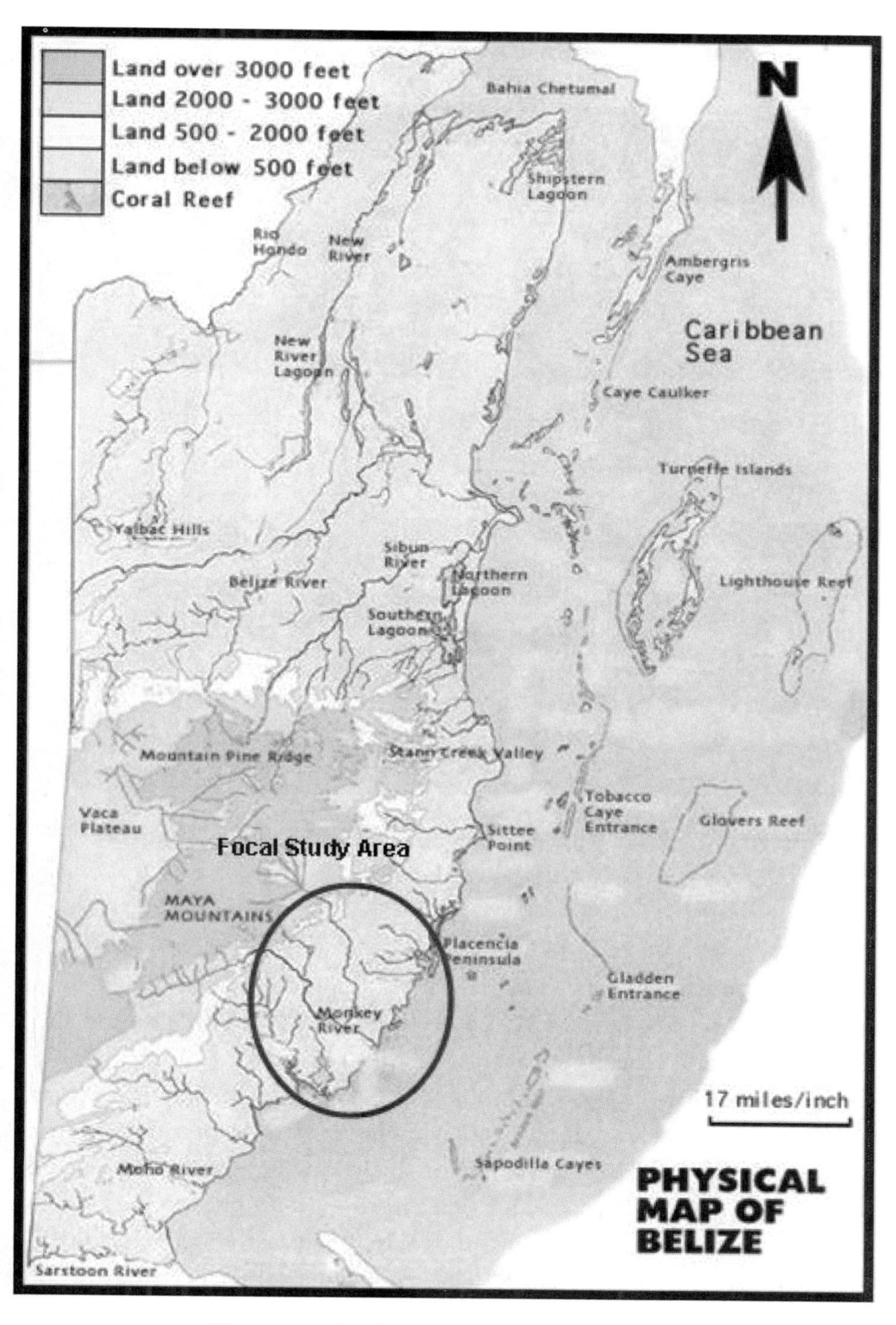

Figure 1. Study area, Monkey River, Belize.

Figure 2. Landsat image pre-hurricane, showing agriculture, forests, savannah, and settlements.

METHODS

We used a Landsat 5 image (March 28, 2001) captured prior to the hurricane and Landsat 7 image (January 30, 2003) recorded 14 months after the hurricane. Both satellite images were acquired for approximately 100 km × 100 km, but were reduced to the above-mentioned focal site for this fragmentation analysis (Figure 2). We know of no major modifications to habitat that occurred between the two Landsat images, and for analytical purposes, we assumed no significant change. We acknowledge that anthropogenic zones are steadily, and sometimes rapidly, expanding, but we have no imagery to document this and based our assumption on local and personal ground experience. We provided a close range image of the pre- and post-hurricane area for examination by the reader (Figure 3).

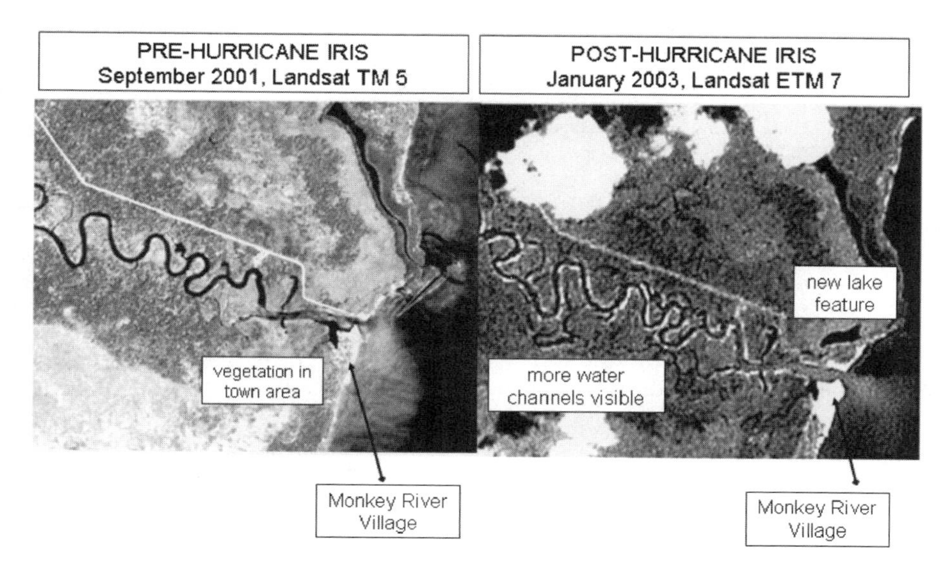

Figure 3. Pre- and post-hurricane Landsat imagery, focal area, Southern Belize.

Landsat data were radiometrically corrected, processed in PCI Geomatica 8.2.1. We used a supervised classification for pre- and post-hurricane images, using 123 ground data points collected (January 2003). Texture measures were incorporated to improve identification of vegetation features and patterns. The resulting spatial data layers (UTM-16 projection, WGS84 datum, 25 m resolution, and ±50 m error) were imported and analyzed in ArcGIS 8.3 and Fragstats 3.3 (hereafter, ArcGIS and Fragstats). For more information on the use of GIS and RS, assumptions of various models, and strengths and limitations of various approaches, we refer the reader to Scott *et al.* (2002) and Jensen (1996).

Fragstats links with ArcGIS and uses spatial metrics to quantify fragmentation pattern at patch, class, and landscape scales (McGarigal and Marks, 1995). Here, we conducted a landscape level analysis. Three key metrics were used to measure spatial character, including size, shape, and core area (McGarigal *et al.*, 2002). Measures of spatial placement included isolation and proximity, contagion and interspersion, and connectivity (McGarigal *et al.*, 2002). Spatial composition measures the variety and abundance of patch types, and metrics we used included richness, evenness, and diversity (McGarigal *et al.*, 2002). We calculated metrics at the landscape and class level, but report only the landscape here because we believe more ground data are necessary for accurate and reliable vegetation class calculations. The principal metrics employed are summarized in Table 1, and defined in the subsequent section in more detail.

Table 1. Summary of metrics

Category	Metric	Description
Spatial character	Size	Mean patch area
	Shape	Length and density of edge
	Core area	Secure habitat after 200 m edge removed
Spatial placement	Isolation/Proximity	Density of patches per area
	Contagion	Connections to other similar habitat
	Interspersion	Variability of habitat types
	Connectance	Functional connections at 1.0, 1.6, and 2.0 km
Spatial composition	Richness/Evenness/ Diversity	Number of different habitat patch types, distribution relative to each other

Area, Edge, and Density

Landscape (or patch) area, density, and edge measure the degree of fragmentation in a landscape (McGarigal *et al.*, 2002): Higher area, density, and lower edge signify less fragmented habitat. Habitat patch density reflects landscape heterogeneity and isolation of patches, while edge density measures the perimeter relative to area of patches (McGarigal *et al.*, 2002).

Core Area

Core area is "the area that is unaffected by the edges of the patch" (McGarigal *et al.*, 2002), where edge distance is a user-specified buffer within which edge-effects penetrate a patch. We found no literature about black howler response to edge, and thus based our buffer size on the assumption that "edge-effects" generally penetrate 200 m into a patch (McGarigal *et al.*, 2002). We computed the total core area, core area percentage of landscape, and the number of core areas (or disjunct core habitat) for pre- and post-hurricane images.

Contagion and Interspersion

Texture measures that were calculated included contagion (i.e., measured the aggregation of cells of similar class) and interspersion (i.e., measured the intermixing of patches of different types) (Hargis *et al.*, 1998). Contagion summarizes the overall "clumpiness" of vegetation classes, where −1 indicates disaggregation, 0 is randomly dispersed, and +1 signifies perfect aggregation.

This represents the homogeneity and proximity of like vegetation types. The aggregation index was computed "as a percentage based on the ratio of the observed number of similar adjacencies relative to the maximum number of the same" (McGarigal *et al.*, 2002). Higher interspersion values result when patches are well mixed, or have a salt and pepper appearance.

Connectivity

Connectivity examines the degree to which a landscape facilitates or impedes ecological flows or functionality (Keitt *et al.*, 1997; McGarigal *et al.*, 2002). These flows may include the exchange of mass or energy across the landscape, such as movement of terrestrial mammals (or other animal species) and transport of water or nutrients. A system that facilitates natural flows is considered to be ecologically functional. In Fragstats, the user defines connectivity based on the life history characteristics of a focal species, such as home range size, mean dispersal distance, habitat requisites for movement, among many others (Weaver *et al.*, 1996).

We measured patch cohesion to quantify the connection of habitat perceived by a species dispersing, which was calculated using patch area and perimeter (McGarigal *et al.*, 2002). In addition, we determined connectance indices, which calculate the number of functional connections, based on a black howlers perception of connectivity as a percentage of maximum possible connections (McGarigal *et al.*, 2002). We defined howler connectivity at three tolerance thresholds: 1, 1.6, and 2 km (based on Estrada *et al.*, 2002). The connectivity metrics may have the greatest importance to black howler persistence, because safe access to other habitat may be necessary after a catastrophic disturbance depletes resources in established territories. Alternatively, if adequate resources remain, the functional joins may facilitate immigration of new howlers, which can help repopulate and facilitate persistence in the short and long term.

Diversity

Diversity metrics measure a wide array of ecological effects and quantify the richness and evenness of forest types and resource distribution (McGarigal *et al.*, 2002). We calculated three standard measures of diversity including patch richness density, shannon's diversity index, and shannon's evenness index.

Statistical Comparison of Pre–Post Hurricane Fragmentation Metrics

We used a paired *t*-test to evaluate the statistical significance of all fragmentation metrics before and after Hurricane Iris (SPSS Inc., 2003). Paired evaluations were used to compare pre- and post-hurricane results using all metrics, and then for sets of metrics within the noted categories (Table 1): spatial character, spatial placement, and spatial composition. Significance levels are reported as *p*-values within each relevant section.

RESULTS

We aggregated our supervised classification into broad vegetation types of coniferous, broadleaf, savannah, other vegetation, water, anthropogenic disturbance, and no data (i.e., cloud-covered area). Fragmentation metrics were calculated at the landscape scale, which means that patches were calculated for all classes of vegetation simultaneously. The results of pre- and post-hurricane image analysis were compared descriptively and statistically to determine the nature and significance of observed change in habitat structure, function, and composition.

Area, Edge, and Density

Our area, edge, and density calculations provided estimates of the general character of the landscape (Table 2). We observed a significant difference ($p = 0.014$) between pre- and post-hurricane images. In fact, there was 3.6 times more patches and approximately 2 times (1.81 times) more edge post-hurricane ($N = 18{,}950$, edge = 6469 km) than pre-hurricane ($N = 5260$, edge = 3560 km). Edge density also increased almost two-fold after the Hurricane Iris. In addition, patch density increased 3.5-fold post-hurricane indicating more patches

Table 2. Area, edge, density metrics at the landscape level

Metric	Pre-hurricane	Post-hurricane	Ratio post–pre
Number of patches	5260	18,950	3.60
Patch density	8.45	29.89	3.54
Largest patch index (%)	15.23	5.67	0.37
Total edge (km)	3560	6469	1.81
Edge density (m per ha)	57.19	102.05	1.78

Table 3. Core area metrics at the landscape level

Metric	Pre-hurricane	Post-hurricane	Difference: post-pre
Total core area (m^2)	7609	2106	−5503
Number disjunct core areas	280	261	−19

(and more patch types) relative to area (see also Bywater, 2003). Lastly, the largest patch index was about one-third times smaller after the hurricane (i.e., ratio of patch index, which indicates percentage of land covered by that patch) (Table 2). Revisiting Figure 3, the increase in patch number is evident visually by comparing the pre- and post-hurricane images and focusing primarily on the extent of observable open water channels. The reader also is directed to the new body of water exposed just inland from the coast, on the north side of the river.

Core Area

Core area metrics measure the total area within a patch and beyond the user-specified edge depth of 200 m, and offer different insight than those above because they integrate both patch area and shape into their calculations. We observed that the landscape lost a significant ($p = 0.00$) amount of total core area habitat (approximately 5500 m^2) since the hurricane (Table 3). The disjunct core area metric indicated that 19 core areas were lost as a result of the hurricane.

Contagion and Interspersion Metrics

The contagion index is expressed as a percentage, where a value close to 0 indicates that patch types are maximally disaggregated, and a value approaching 100 reveals that the landscape is maximally aggregated (McGarigal *et al.*, 2002). Contagion and interspersion were significantly different after the hurricane ($p = 0.002$). The pre-hurricane image had a lower contagion index value than the post-hurricane image (Table 4), which means the habitat was more aggregated after the event. The percentage of like adjacencies index also measures the degree of aggregation, where a value approaching 0 indicates that there are no like adjacencies (McGarigal *et al.*, 2002) and our calculations indicated greater aggregation post-hurricane.

Table 4. Contagion and interspersion metrics at the landscape level

Metric	Pre-hurricane	Post-hurricane
Contagion index (%)	29.39	42.94
Percentage of like adjacencies	53.38	84.86
Interspersion & juxtaposition index (%)	79.22	81.70

Interspersion and juxtaposition metrics are alternative measures of patch adjacency, where a value close to 0 indicates uneven adjacencies, and a value near 100 reveals that all patch types are equally adjacent to other patch types (McGarigal *et al.*, 2002). We found that both landscapes are well interspersed (i.e., salt and pepper like appearance), with slightly greater interspersion after the event (Table 4).

Connectivity

Prior to the hurricane, the landscape was not well connected functionally for black howlers at any distance threshold (i.e., 1.0, 1.6, 2.0 km) (Table 5). Importantly, we found the habitat was significantly less well connected after the hurricane ($p = 0.002$), particularly at the distances of 1.0 and 1.6 km. In terms of primate habitat, this suggests fewer direct canopy connections, which would reduce opportunity for individual animals to move under the protection of canopy, or to travel without dropping to the ground. The reduced canopy connectivity may also be observed visually in the post-hurricane Landsat image, which shows many more open waterways and a new water body on the north side of Monkey river (Figure 3).

Diversity

The patch richness density was standardized to per unit area (number of patch types per 100 ha or 1 km^2) for ease of comparison amongst landscapes, and

Table 5. Connectivity metrics at the landscape level

Metric	Pre-hurricane	Post-hurricane
Connectance index (%) (1 km)	0.93	0.72
Connectance index (%) (1.6 km)	1.81	1.46
Connectance index (%) (2.0 km)	2.60	2.09

Table 6. Diversity metrics at the landscape level

Metric	Pre-hurricane	Post-hurricane
Patch richness density	0.019	0.021
Shannon's diversity index	2.162	2.304
Shannon's evenness index	0.869	0.898

required the integration of the broad vegetation categories listed at the start of "Results" section. We showed that habitat patch richness was low and significantly different before and after the hurricane ($p = 0.009$) (Table 6). Post-hurricane diversity was slightly higher (0.021), which indicated that this transitioned landscape is somewhat richer or more varied. The Shannon's diversity index reported that the diversity for both landscapes was similar, although post-hurricane it was higher (2.16 pre versus 2.3 post). Likewise, we found the images had similar indices of evenness, or an even distribution of area among patch types. Again, a visual inspection of Figure 3 shows changes that suggest new cover types. For example, on the south shore of Monkey River at its confluence with the Caribbean sea, a large white swatch shows the remains of a once vegetated town. A ground visit would confirm that not only vegetation, but also houses were removed from this area.

DISCUSSION

We undertook a spatial analysis using GIS and RS to quantify the changes in habitat structure, function, and composition that resulted around Monkey River from Hurricane Iris (2001). We quantified these changes using fragmentation metrics including patch number, area, edge, density, core area, contagion, interspersion, connectivity, and diversity. Although these fragmentation metrics are quantitative, landscape ecology has been labeled (unfortunately) a qualitative science and significantly different from the mathematical, empirically tested science of metapopulations (Minta and Kareiva, 1994). Yet, trends in these fragmentation metrics have been explored and linked significantly with mammalian species persistence in many other studies (see "Introduction" section, this paper). Moreover, we evaluated pre- and post-hurricane metrics using statistical techniques. Based on the findings of research on other mammalian species and the observed sensitivity of howlers to fragmentation (Estrada *et al.*, 2002), we argue that it is reasonable to extrapolate other species-specific fragmentation effects and predictions of black howlers (within reason). We acknowledge that there will be species-specific differences, but we assume general trends will apply, which include those founded largely on tests of the Equilibrium Theory of Island Biogeography (MacArthur and Wilson, 1967): (1) A reduction in patch size (area), core habitat area, habitat connectivity, and diversity likely reduces persistence (Palomares *et al.*, 1996; Weaver *et al.*, 1996; Estrada *et al.*, 2002); (2) An increase in patch number and edge may forebode population declines and community disruption (Fahrig and Merriam, 1985; Debinski and Holt, 2000); and (3) Diversity may underscore ecosystem stability.

We found that Hurricane Iris significantly altered the study area for all fragmentation metrics ($p < 0.014$). As expected in most post-disturbance landscapes, we observed an increase in the total number of patches and a decrease in patch size available to resident wildlife populations. This has implications for black howlers if the mean patch size falls below the habitat requirements of local howler populations, because then we may see a decline in howler numbers. For instance, black howlers establish home range, which range in size from 1 ha (0.01 km^2) (Kitchen, 2004) to 24 ha (approximately 0.24 km^2). Thus, a loss of even small habitat areas is likely to result in population decline.

In addition, we observed approximately a two-fold increase in edge habitat (length and density); black howlers tend not to occur in deforested or transition habitat, which may be consistent with the type of edge habitat observed after Hurricane Iris. Increases in edge habitat will intensify light and sound penetration into habitat patches (Saunders *et al.*, 1991; Hobbs, 1993; Debinski and Holt, 2000) altering plant community structure. Coincidentally, increased edge habitat may compromise immigration and emigration rates (Forman and Alexander, 1998; Stamps *et al.*, 1987). Although black howlers have shown daily ranges of between 40 and 700 m (Crockett and Eisenberg, 1987), they rarely came to the ground in Monkey River before the hurricane (Pavelka *et al.*, 2003) and some individuals may remain reluctant to do so, especially if the hurricane has altered the composition of vegetation between patches, excluded canopy cover that links patches, or has facilitated the colonization by edge-tolerant competitors or predators. Consequently, we expect that the quality of habitat with respect to accessibility and use has been substantially degraded for black howlers in this study area.

We calculated core area for remnant patches after excluding 200 m of edge from all patches (i.e., reducing the patches by 200 m on all sides to simulate the effects of edge), and observed approximately a 3.5-fold decrease in total core habitat and greater variability of patch size after the hurricane. Higher edge and smaller habitat cores increase the edge-to-area ratio, and we can expect that non-resilient (edge-sensitive) species such as the black howler may not persist or may move out and be replaced by edge-tolerant species (Noss, 1983; Weaver *et al.*, 1996). The marked reduction in core habitat and reduction of patch size may require howlers to reconfigure home ranges or immigrate in order to survive. As such, we again would predict a reduction in total howler numbers over time and/or reconfiguration of home ranges, to increase home range size and include more resources. If core habitat size has been reduced below thresholds required by black howlers, we expect extinction risk will increase (especially, if disjunct populations elsewhere cannot repopulate the area over time).

The effects of changes to diversity, contagion, and interspersion require careful consideration. These metrics were more difficult to interpret because they are species and process dependent. For instance, a habitat specialist that needs highly secure core areas may be negatively affected by an increase in habitat diversity, if that results in less of the preferred habitat, or if the diversity increases because of habitat transition to disturbed or edge type habitats. Alternatively, edge-tolerant habitat generalists may thrive with greater diversity even if it includes high proportions of disturbed habitat.

The post-hurricane landscape was comprised of more patches of smaller size than prior to the event. Therefore, even though the contagion indexes reported higher aggregation for this image, this does not necessarily indicate superior habitat quality for black howlers. For instance, the higher aggregation values may be the result of contiguous but highly varied suitable habitat transitioning to more aggregated areas of disturbed habitat. One could argue that high aggregation of habitat would be good, except in the identified case where it is dominated by edge habitat that is unsuitable to the focal species. The interspersion indexes indicated that patch types are well interspersed and connectivity metrics indicated that the physical connectivity was good post-hurricane. However, the functional connectivity was consistently low before and worse after the hurricane for all distance thresholds: 1.0, 1.6, and 2.0 km. This reduction in functional connectivity may decrease the likelihood of individuals or species moving across the matrix to alternative patches. We did not have data on average dispersal distances, but do know that black howlers are reluctant to move to the ground under natural conditions. Consequently, we suspect that dispersal or immigration opportunities have been reduced. We suspect also that this may impair access to alternative food sources, impede opportunities to reproduce and recruit young into the population, and may subject the population to inbreeding as well as decline (Weaver *et al.*, 1996; Debinski and Holt, 2000).

Diversity metrics are useful for quantifying the overall composition of a landscape. We found that the pre- and post-hurricane landscapes had similar diversity measurements and that both landscapes have high richness. The patch richness index, Shannon's diversity, and evenness index indicated greater richness and diversity after Hurricane Iris. These results were consistent with findings of Boutet and Weishampel (2003), who suggested hurricane disturbance may be a necessary component of tropical forest dynamics. Our findings likely represent the addition of new patch types to the post-hurricane landscape (Bywater, 2003), but given the extent and expansion of human development in the region, there may not be enough resilience (in terms of space or time) for this natural succession process to occur and be a functional and positive change for the forest ecosystem. The addition of another vegetation class (disturbance) coupled with the loss of patch size and increased isolation may be detrimental to feeding opportunities for black howlers in the area. Silver *et al.* (1998) observed that one-third of feeding time was spent eating fig tree products (i.e., fruit, and young and mature leaves). We observed substantive loss of large trees, especially figs in the focal study area, and although Silver *et al.* (1998) suggest that dietary composition may switch to a less frugivorous diet, such behavioral adaptability has limits. Consequently, there may have emerged a significant conservation problem in our study area: if the behavioral adaptability and dietary tolerance of black howlers to fewer fruit bearing trees are insufficient relative to vegetation regeneration, co-incident with rapid human encroachment, and limited potential for new individuals to emigrate, then immigration of this population may not persist.

SUMMARY

Fragstats proved effective for measuring hurricane disturbance and possible implications for black howler habitat, and we suggest continued application and development in this area of research. We postulated possible biological, ecological, and demographic effects to the species as a result of the observed fragmentation. The exploratory nature of the present research has contributed valuable information and methodological tools to the study of tropical forest and primate conservation. Understanding the effects to persistence will require further investigation and the integration of demographic data with PVA.

We examined the fragmentation effects of Hurricane Iris (2001) on landscape structure, composition, and function. We showed that black howler habitat in Monkey River and the surrounding area (Bladen and Swayse Rivers) has experienced marked changes for the worse. Patch size has declined, while the number and isolation of patches have increased. Combined, these effects may be expected to reduce survival or at least increase emigration to areas with sufficient resources. Emigration may be difficult because edge habitat has increased. Estimates show that diversity has increased in tandem with edge, suggesting that this edge habitat may be comprised significantly of disturbed regenerating forest. The latter may require black howlers to move to the ground for travel, and expose them to increased disease and predation risk. With confidence, we can say that the previously small population of black howlers ($N < 50$) is now under significant stress, given the comprehensively negative fragmentation effects of the hurricane, and especially with the continued and rapid incursion of human development.

In future analyses, we will investigate changes to home range size, shape and composition, canopy class transitions, and black howler persistence scenarios, combining population data with our fragmentation results within a spatially explicit population viability analysis. Such work will facilitate approximation of persistence under catastrophic disturbance and examine potential management or recovery scenarios. Canopy class transitions will highlight which vegetation classes may have been most severely affected by the hurricane.

The integration of spatial technologies in primate conservation has not been exploited fully. The current state of all wildlife species' conservation needs will ensure RS and GIS applications continue to evolve and expand. We urge researches to explore and enhance capability of spatial technologies for use in tropical environments and at scales relative to small-scale habitat specialists. The ability to conduct such analyses with confidence will provide managers and researchers with information necessary to make more informed decisions regarding the long-term survival of this and other primate species.

REFERENCES

Belize Government 2003, The Climate of Belize. Available: http://www.hydromet.gov.bz/climate. Accessed on April 3, 2003.

Boutet, J. C. Jr. and Weishampel, J. F. 2003, Spatial pattern analysis of pre- and post-hurricane forest canopy structure in North Carolina, USA. *Landscape Ecol.* 18:553–559.

Brokaw, N. V. L. and Grear, J. S. 1991, Forest structure before and after hurricane Hugo at three elevations in the Luquillo Mountains, Puerto Rico. *Biotropica* 23(4a):386–392.

Burkey, T. V. 1995, Extinction rates in archipelagos: Implications for populations in fragmented habitats. *Conserv. Biol.* 9(3):527–541.

Burslem, D. F. R. P. and Whitmore, T. C. 1999, Species diversity, susceptibility to disturbance and tree population dynamics in tropical rain forest. *J. Veg. Sci.* 10:767–776.

Bywater, N. H. 2003, The effects of Hurricane Iris (2001) on howler monkey habitat fragmentation effects in Monkey River, Belize. MGIS Thesis, University of Calgary, Alberta.

Chapman, C. A. and Balcomb, S. R. 1998, Population characteristics of howlers: Ecological conditions or group history. *Int. J. Primatol.* 19(3):385–403.

Collinge, S. K. 1996, Ecological consequences of habitat fragmentation: Implications for landscape architecture and planning. *Landscape Urban Plan.* 36(1):59–77.

Crockett. C. M. and Eisenberg, J. F. 1987, Howlers: Variations in group size and demography, in: B. B. Smuts, D. L. Cheney, R. M. Seyfarth, R. W. Wrangha, and T. T. Struhsaker, eds., *Primate Societies*, University of Chicago Press, Chicago, pp. 54–68.

Debinski, D. M. and Holt, R. D. 2000, A survey and overview of habitat fragmentation experiments. *Conserv. Biol.* 14(2):342–355.

Estrada, A., Anzures, A., and Coates-Estrada, R. 1999, Tropical rain forest fragmentation, howler monkeys (*Alouatta palliata*), and dung beetles at Los Tuxtlas, Mexico. *Am. J. Primatol.* 48:253–262.

Estrada, A., Mendoza, A., Castellanos, L., Pacheco, R., Van Belle, S., Garcia, Y., and Munoz, D. 2002, Population of the black howler monkey (*Alouatta pigra*) in a fragmented landscape in Palenque, Chiapas, Mexico. *Am. J. Primatol.* 58:45–55.

Fahrig, L. and Merriam, G. 1985, Habitat patch connectivity and population survival. *Ecology* 66(6):1762–1768.

Foody, G. M. and Hill, R. A. 1996, Classification of tropical forest classes from Landsat TM data. *Int. J. Remote Sens.* 17(12):2353–2367.

Forman, R. T. T. and Alexander, L. E. 1998, Roads and their major ecological effects. *Annu. Rev. Ecol. Syst.* 29:207–231.

Forman, R. T. T. and Godron, M. 1986, *Landscape Ecology*. John Wiley & Sons, New York.

Hargis, C. D., Bissonette, J. A., and David, J. L. 1998, The behaviour of landscape metrics commonly used in the study of habitat fragmentation. *Landscape Ecol.* 13:167–186.

Hill, R. A. 1999, Image segmentation for humid tropical forest classification in Landsat TM data. *Int. J. Remote Sens.* 20(5):1039–1044.

Hobbs, R. J. 1993, Effects of landscape fragmentation on ecosystem processes in the western Australian wheatbelt. *Biol. Conserv.* 64:193–201.

Holling, C. S. 1992, Cross-scale morphology, geometry and dynamics of ecosystems. *Ecol. Monogr.* 62:447–502.

Horwich, R. H. and Johnson, E. D. 1986, Geographical distribution of the black howler (*Alouatta pigra*) in Central America. *Primates* 27(1):53–62.

IUCN Red List of Threatened Species. 2003, http://www.redlist.org/. Accessed on February 23, 2004.

James, R. A., Leberg, P. L., Quattro, J. M, and R. C. Vrijenhoek. 1997, Genetic diversity in black howler monkeys (*Alouatta pigra*) from Belize. *Am. J. Phys. Anthropol.* 102:329–336.

Jensen, J. R. 1996, *Introductory Digital Image Processing: A Remote Sensing Perspective,* 2nd Edn. Prentice-Hall, Inc., Upper Saddle River, New Jersey.

Johnsson, K. 1995, Fragmentation index as a region based GIS operator. *Int. J. Geogr. Inf. Syst.* 9(2):211–220.

Keitt, T. H., Urban, D. L., and Milne, B. T. 1997, Detecting critical scales in fragmented landscapes. *Conserv. Ecol.* 1(1):1–27.

Kitchen, D. M. 2004, Alpha male black howler monkey responses to loud calls: Effect of numeric odds, male companion behaviours and reproductive investment. *Anim. Behav.* 67:125–139.

Lugo, A. R. 1995, Management of tropical biodiversity. *Ecol. Appl.* 5(4):956–961.

Lunetta R. S. and Elvidge, C. D. 1998, *Survey of Multispectral Methods for Land Cover Change Analysis.* Ann Arbor Press, New York.

Luoto, M., Rekolainen, S., Salt, C. A., and Hansen, H. S. 2001, Managing radioactively contaminated land: Implications for habitat diversity. *Environ. Manage.* 27(4):595–608.

MacArthur, R. H. and Wilson, E. O. 1967, *The Theory of Island Biogeography.* Princeton University Press, Princeton, NJ.

Mas, J. F. 1999, Monitoring land-cover changes: A comparison of change detection techniques. *Int. J. Remote Sens.* 20(1):139–152.

Mayaux, P. and Lambin, E. F. 1997, Tropical forest area measured from global land-cover classifications: Inverse calibration models based on spatial textures. *Remote Sens. Environ.* 59(1):29–43.

McGarigal, K., Cushman S. A., Neel, M. C., and Ene, E. 2002, FRAGSTATS: Spatial pattern analysis program for categorical maps. Produced by the authors at the University of Mass., Amherst. www.umass.edu/landeco/research/fragstats/fragstats.html

McGarigal, K. and Marks, B. J. 1995, FRAGSTATS: Spatial pattern analysis program for quantifying landscape structure. General Technical Report, US Department of Agriculture, Forest Service, 1–122.

Meffe, G. K. and Carroll, R. C. eds. 1997, *Principles of Conservation Biology*, 2nd Edn. Sinauer Associates, Sunderland, MA.

Minta, S. C. and Kareiva, P. M. 1994, A conservation science perspective: Conceptual and experimental improvements, in: T. W. Clark, R. P. Reading, and A. L. Clarke, eds., *Endangered Species Recovery—Finding the Lessons, Improving the Process*, Wiley, New York, pp. 275–304.

Noss, R. 1983, A regional landscape approach to maintain diversity. *BioScience* 33(11):700–706.

Ostro, L. E. T, Silver, S. C., Koontz, F. W., and Young, T. P. 2000, Habitat selection by translocated black howler monkeys in Belize. *Anim. Conserv.* 3:175–181.

Palomares, F., Ferreras, P., Fedraini, J. M., and Delibes, M. 1996, Spatial relationships between Iberian lynx and other carnivores in an area of south-western Spain. *J. Appl. Ecol.* 33:5–13.

Pavelka, M. S. M, Brusselers, O. T., Nowak, D., and Behie, A. M. 2003, Population reduction and social disorganization in *Alouatta pigra* following a hurricane. *Int. J. Primatol.* 24(5):1037–1055.

Saunders, D. A., Hobbs, R. J., and Margules, C. R. 1991, Biological consequences of ecosystem fragmentation: A review. *Conserv. Biol.* 5(1):18–32.

Scariot, A. 1999, Forest fragmentation effects on palm diversity in central Amazonia. *J. Ecol.* 87:66–76.

Schumaker, N. H. 1996, Using landscape indices to predict habitat connectivity. *Ecology* 77(4):1210–1225.

Scott, J. M., Heglund, P. J., Morrison, M. L., Haufler, J. B., Raphael, M. G., Wall, W. A., and Samson, F. B. eds. 2002, *Predicting Species Occurrences: Issues of Accuracy and Scale.* Island Press, Washington, Covelo, London.

Silver, S. C., Ostro, L. E. T., Yeager, C. P., and Horwich, R. 1998, Feeding ecology of the black howler monkey (*Alouatta pigra*) in northern Belize. *Am. J. Primatol.* 45(3):263–279.

SPSS INC, 2003. 233 S. Wacker Drive, 11th floor, Chicago, Illinois, 60606. http://www.spss.com/

Stamps, J. A., Buechner, M., and Krishnan, V. V. 1987, The effects of edge permeability and habitat geometry on emigration from patches of habitat. *Am. Nat.* 129(4):533–552.

Tischendorf, L. 2001, Can landscape indices predict ecological processes consistently? *Landscape Ecol.* 16(3):235–254.

Turner, I. M. 1996, Species loss in fragments of tropical rain forest: A review of the evidence. *J. Appl. Ecol.* 33:200–209.

Weaver, J. L., Paquet, P. C., and Ruggiero, L. F. 1996, Resilience and conservation of large carnivores in the Rocky Mountains. *Conserv. Biol.* 10(4):964–976.

Wilcox B. A. and Murphy, D. D. 1985, Conservation strategy: The effects of fragmentation on extinction. *Am. Nat.* 125:879–887.

PART FIVE

Synopsis and Perspectives

CHAPTER TWENTY-THREE

New Perspectives in the Study of Mesoamerican Primates: Concluding Comments and Conservation Priorities

Paul A. Garber, Alejandro Estrada, and Mary S. M. Pavelka

INTRODUCTION

The monkeys of Mesoamerica represent an ecologically diverse and successful radiation of non-human primates that inhabit the tropical and subtropical forests from Mexico south and east into Guatemala, Belize, Honduras, Nicaragua, Costa Rica, and the border between Panama and Colombia. This includes a maximum of 9 species and as many as 21 subspecies (Rylands *et al.*,

Paul A. Garber • Department of Anthropology, University of Illinois, Urbana, IL. **Alejandro Estrada** • Estación de Biología Los Tuxtlas, Instituto de Biología, Universidad Nacional Autonoma de México. **Mary S. M. Pavelka** • Department of Anthropology, University of Calgary, Calgary, Alberta, Canada.

New Perspectives in the Study of Mesoamerican Primates: Distribution, Ecology, Behavior, and Conservation, edited by Alejandro Estrada, Paul A. Garber, Mary S. M. Pavelka, and LeAndra Luecke. Springer, New York, 2005.

this volume). Although howlers, capuchins, spider monkeys, night monkeys, squirrel monkeys, and tamarins represent only a small proportion of the enormous biodiversity of the Mesoamerican region, they are an important component of the community of arboreal mammals in tropical forests, accounting for approximately 61% of the 55 recognized taxa (15 rodent taxa, 13 marsupial taxa, 4 carnivore taxa, and 2 sloth taxa) (Reid, 1997; Emmons, 1990). Further, populations of these species are likely to play a critical role in the recycling of matter, nutrients and energy in the ecosystem (Estrada and Coates-Estrada, 1993). Primates also serve as pollinators and seed dispersers of tropical plants, and in forest regeneration (Lambert and Garber, 1998). Moreover, as outlined by Ford (this volume), the biogeography and dispersal of primates into Mesoamerica occurred as part of a complex set of geological, climatic, ecological, and evolutionary events that have shaped the history of Mesoamerica over the past million years. Different primate lineages entered Mesoamerica at different times, and therefore represent distinct colonization and speciation events.

The study of Mesoamerican primates also holds special significance for the discipline of primatology. From 1931 to 1933, Clarence Raymond Carpenter conducted the first long-term field study of a primate in the wild. Carpenter (1934: 6) studied the "behavior, social relations, and ecology of howling monkeys" (*Alouatta palliata*) on Barro Colorado Island, Panama. Writing the Foreword to Carpenter's (1934) monograph, Robert Yerkes predicted, "Looking forward, it is certain that Doctor Carpenter's contribution may be counted on to command the attention and stir the enthusiasm of other investigators" (Carpenter, 1934: 4). This has certainly been the case. Over the past 70 years, the efforts of many dedicated researchers have contributed importantly to our understanding of the distribution, ecology, behavior, and conservation of Mesoamerican primates.

STUDIES OF MESOAMERICAN PRIMATES

The primary goal of this volume was to integrate and synthesize current information on primates in the Mesoamerican region and examine how anthropogenic factors (such as deforestation, agriculture, and habitat change) as well as natural disturbances (such as hurricanes) affect the current distribution, demography, behavioral ecology, and conservation status of individual taxa. During much of the 20th century, several areas of Mesoamerica have been characterized by political instability, civil war, poverty, and devastating natural and human-induced habitat destruction. For countries such as El Salvador, Guatemala,

Honduras, and Nicaragua, only now are we beginning to collect surveys and basic scientific information on their remaining primate populations. El Salvador maintains the highest human population density of any country in Central America, poverty is extreme, and much of the original forests has been cut (see Estrada *et al.*, Chapter 1). The country's entire populations of howler monkeys and capuchins may now be extinct. Spider monkeys continue to exist in El Salvador, but their density, habits, and viability are poorly known. However, recent efforts by local biologists are beginning to provide information about the current distribution of spider monkey populations in that country (Morales, 2002).

The countries of Mexico, Belize, Costa Rica, and Panama, are the sites of the five long-term primate field study sites in Mesoamerica: Los Tuxtlas, Mexico (*A. palliata* and *Ateles geoffroyi*), Baboon Sanctuary, Belize (*Alouatta pigra*), Santa Rosa, Costa Rica (*Cebus capucinus*, *A. geoffroyi*, and *A. palliata* at the site), Hacienda La Pacifica, Costa Rica (*A. palliata*), and Barro Colorado Island, Panama (*A. geoffroyi*, *C. capucinus*, *Saguinus geoffroyi*, and *A.palliata*). While these long-term study sites have provided detailed behavioral, ecological, and demographic data for primate populations that span decades, they have focused principally on three primate species: mantled howlers, black howlers, and white-faced capuchin monkeys. These taxa, however, are among the least endangered Mesoamerican primates, and given the number of studies that have focused on them (Estrada *et al.*, Chapter 1), it is not surprising that this volume is heavily weighted to studies of these three species.

Despite many decades of research throughout the Mesoamerican region, several questions concerning primate behavior, ecology, and conservation remain unanswered. In the case of *A. palliata*, *A. pigra*, and *Cebus*, long-term studies have been conducted in only a small number of localities (long-term studies of *C. capucinus* have been concentrated in the dry tropical forests of Costa Rica) and therefore, we have limited information on variability in ecology and behavior across a spatial scale encompassing the range of habitats exploited by these species. Moreover, only two of these long-term study sites (Los Tuxtlas, Mexico and Santa Rosa National Park, Costa Rica) represent large, continuous forested areas. In contrast, the Community Baboon Sanctuary in Belize and Hacienda La Pacifica in Costa Rica consist of fragmented forests and/or linear strips of vegetation along rivers. Barro Colorado, Panama is a relatively small island (1600 ha) and lacks the normal range of predators found on the mainland. Clearly, there is a need to study Mesoamerican primates in areas of continuous forests.

For species such as *Saimiri oerstedii* (Boinski, 1987a,b; 1994), *S. geoffroyi* (Dawson, 1975; Garber, 1980, 1984a,b), and *A. geoffroyi* (Coelho *et al.*, 1976; Cant, 1986; Chapman, 1987, 1988a,b; Milton, 1981a,b) we have data for only a single group over the course of 1 year, or studies of a few groups over shorter periods. In the case of *Aotus zonalis*, there exist virtually no published studies and we continue to lack even the most basic natural history information (Moynihan, 1964; Thorington *et al.*, 1976). Even for those species for which we have detailed information, we know very little about dietary and habitat flexibility, and how these species respond to changing environmental conditions associated with human disturbance.

Recent studies of fragmented landscapes (Murcia, 1995; Restrepo *et al.*, 1999) indicate that edges represent dynamic components of an ecosystem, and that edge effects change over time. For example, Restrepo *et al.* (1999) have described changes in fruit abundance, leaf area, water availability, soil fertility, temperature, and humidity not only between edges and interior habitats, but also between older and younger edges. In other cases, however, edges and interior forest zones may contain similar plant species. Williams-Linera (1990) reports that in a lowland rainforest in Panama, there were no significant differences in tree or seedling species composition in the forest edges and the forest interiors. Given that edge effects are neither uniform nor standard, from site-to-site (Murcia, 1995), it is difficult to predict exactly how a primate species will respond to the specific conditions of a forest fragment (edge plus interior), and the effect that factors such as the distribution of resting or refuse sites, gaps in the forest canopy, suitable routes for arboreal travel, and exposure to predators have on the ability of individual primate species to survive in human-disturbed forest landscapes (Chapman and Peres, 2001)

Aotus, Saimiri, and Saguinus

The Mesoamerican primates that we know least about, *Aotus*, *Saimiri*, and *Saguinus*, are small-bodied platyrrhines that differ significantly in behavioral ecology, social organization, mating systems, and life history strategies. *Aotus* is reported to live in pair-bonded nuclear family social groups and is the only species of higher primates to adopt a nocturnal lifestyle (Wright, 1981). For their body mass (approximately 900–1000 g), night monkeys are characterized by an extremely short gestation period (133 days), females give birth once per year to a single infant, and adult males (presumably fathers) help care for the young

(Garber and Leigh, 1997). Moreover, *Aotus* is reported to occupy small home ranges of 4–10 ha (Wright, 1989). Given the limited quantitative data available, patterns of habitat preference and diet in Mesoamerican night monkeys are poorly understood. Hladik *et al.* (1971) examined the stomach contents of *Aotus* from Panama. These authors report that fruits (65%), foliage (30%), and insects (5%) account for the majority of the night monkeys' diet. There are no published studies concerning the ability of Mesoamerican night monkeys to exploit disturbed forests. However, Wright (1981: 214) cites Cassidy (pers. comm.) indicating that *Aotus* was found to inhabit "shady trees adjacent to coffee plantations in Colombia."

Saguinus geoffroyi (adult body mass 450–500 g) lives in small multimale–multifemale groups (6–10 individuals) that are characterized by cooperative infant care, polyandrous and polygynous matings, the production of twin offsprings, female reproductive suppression, and the potential for a group's sovereign breeding female to give birth twice per year (Garber and Leigh, 1997). However, on a mainland site (Agua Clara, Panama) located approximately 5 km from Barro Colorado Island, Garber (pers. comm.) found that females tended to give birth only once per year, and that it was not uncommon for a female to lose one of her twin infants during the first few months of life.

Panamanian tamarins have home ranges of 10–30 ha (Dawson, 1975; Garber, 1984a) and, in addition to consuming ripe fruits in the canopy, and resources such as plant exudates and small vertebrates found on the trunks of large trees, spend approximately 70% of their foraging time searching for insects (large-bodied orthopterans) (Garber, 1984a,b). Garber (1984a) reports, that in a dry tropical forest in Panama, tamarins restricted much of their insect foraging behavior to areas of disturbed secondary vegetation such as forest edges and tree fall gaps. In this regard, Panamanian tamarins may be able to survive in forest fragments characterized by a high ratio of forest edge to forest interior.

Saimiri oerstedii (adult body mass 650–900 g) lives in the largest social groups of any Mesoamerican primate (40–65 individuals) and utilizes home ranges of 76–110 ha (Boinski, 1987c; Janson and Boinski, 1992). These groups may contain as many as 14 reproductively adult females. Squirrel monkeys are characterized by an extremely short breeding season in which males attain a "fatted"state immediately prior to mating (associated with hormonally mediated water retention), birth synchrony within groups, and an elongated period of juvenile development (Boinski, 1987a; Janson and Boinski, 1992). Infant *Saimiri* weigh approximately 20% of their mother's weight at birth (the largest of any anthropoid primate), achieve 95% of adult brain size at 3 months of age, and have an interbirth interval of up to 2 years (Garber and Leigh, 1997). Whereas male night monkeys and tamarins reach full adult reproductive maturity by 2–2.5 years of age, male *Saimiri* may not reach full adult body mass until 4–5 years of age (Boinski, 1992). Costa Rican squirrel monkeys devote approximately 90% of their foraging time to the pursuit and capture of insect prey (Boinski, 1986). Given the energetic requirements of large group size, large home range size, and a slow life history, forest fragmentation is likely to impact populations of *Saimiri* more severely than *Saguinus* or possibly *Aotus*. Recent studies of *Saimiri* in fragmented landscapes in Costa Rica indicate, however, that populations of squirrel monkeys can persist in small forest patches or across linear strips of vegetation (J. Saenz, pers. comm.). How this affects diet, group size, group cohesion, and reproductive success is unclear.

In short, field research on *Aotus*, *Saimiri*, and *Saguinus* is sorely needed in order to examine the natural history, ecology, and behavior of these taxa. Moreover, deforestation in Costa Rica and Panama, where populations of these primates are found, has reduced the original forest cover to about 40% and 39%, respectively (Estrada *et al.*, Chapter 1). Although these two countries still contain large extensions of forest cover in proportion to their territory, current deforestation rates in Costa Rica are estimated at −0.77% per year, and in Panama these are −1.65% per year (see Estrada *et al.*, Chapter 1). Thus, lack of information about the current distribution, basic ecology, and behavior of populations of *Aotus, Saimiri, and Saguinus*, coupled with gradual and rapid modifications of landscapes in Costa Rica and Panama by human activity, make the task of identifying specific conservation recommendations exceedingly difficult.

Ateles

Ateles geoffroyi is the largest bodied Mesoamerican primate (adult body mass is 7.5–8.2 kg; Ford and Davis, 1992). Although it is widely distributed from the Yucatan peninsula into Panama, 60% of all published reports come from sites in Mexico and Costa Rica (Estrada *et al.*, Chapter 1). Given their rapid speed of travel (tail-assisted suspensory locomotion), fission–fusion social system, and large home range (several hundred hectares), long-term field studies of *A. geoffroyi* have yet to be conducted. Geoffroy's spider monkey is highly frugivorous (ripe fruits account for >70% of feeding time; Chapman, 1987).

Van Roosmalen and Klein (1988) report that *A. geoffroyi* may be the most "flexible" spider monkey species and is found to exploit a wide range of habitat types including mangrove, primary forest, evergreen forest, semi-deciduous forest, and deciduous forest. *A. geoffroyi* also is found to range from sea level to 2500 m above sea level (Rowe, 1996). Along with *Cebus*, *Ateles* has a "more highly developed and the most fissurated brain" of any Mesoamerican primate (Hershkovitz, 1977: 359) and a slow life history. A female does not produce her first infant until approximately age 7–9, has a long period of gestation (226–232 days), and an interbirth interval of approximately 3 years (Fedigan and Rose, 1995). Due to its large size, palatable meat to humans, and the attractiveness of its infants as pets, *Ateles* is extremely vulnerable to human hunting and capture (Kinzey, 1997; Duarte and Estrada, 2003).

Given its slow life history and reproductive pattern, one might expect *Ateles* to be among the first Mesoamerican primate species to become locally extinct in areas of forest fragmentation or adjacent to human habitation. Although this may often be the case, populations of *Ateles* are found to exist in fragmented landscapes in El Salvador, Honduras, and Mexico. Investigations focusing on the ecological, local, and historical conditions that allow *Ateles* to survive in these areas are a high priority. In a recently published study of a population of spider monkeys inhabiting a highly fragmented landscape in Los Tuxtlas, Mexico, it was shown that larger forest fragments contained larger populations of spider monkeys and that the presence, relative abundance and basal area of emergent fruiting trees appeared to explain the persistence of *Ateles* in such environments (González-Zamora and Mandujano, 2003).

Milton and Hopkins (this volume) provide detailed information on the reintroduction of *A. geoffroyi* into protected forested areas on BCI, Panama. This began more than 40 years ago with the introduction of several juveniles. However, only one male and four females survived the initial reintroduction and have served as the genetic founders of the current population. Their data indicate that for more than 30 years, spider monkey population growth either increased extremely slowly or not at all. This highlights the difficulties that species with a slow life history may have in colonizing new environments. Since the late 1990s the population on BCI has increased to 28 animals. Milton and Hopkins conclude that even in the absence of other large-bodied frugivores, mammalian predators, and being introduced into a productive and protected environment, spider monkey populations require extremely long periods of time to increase to a point where persistence is likely. Conservation policies governing primate reintroductions need to consider a species' life history traits and the minimum population size to maximize the likelihood of success.

Cebus and Alouatta

Despite exhibiting very different adaptive patterns, life history strategies, diets, and foraging behavior, *Cebus* and *Alouatta* represent the most geographically widespread genera of New World monkeys. *Cebus* and *Alouatta* also are among the best-studied primates. Capuchins are highly encephalized, are the only group of platyrrhines that can move their digits independently, possess a pseudo-opposable thumb, are reported to hunt for vertebrates in a coordinated manner, exploit embedded or hidden foods, and may on very rare occasions use a tool to solve a foraging problem (Janson and Boinski, 1992; Garber and Brown, 2004). In the case of *C. capucinus* (adult body mass 2.2–3.2 kg), Fedigan and Rose (1995) and Fragaszy *et al.* (2004) report that females have their first offspring at 6–7 years of age, and are characterized by an interbirth interval of 26.4 months. In contrast, sympatric mantled howlers are considerably larger (4.7–7.5 kg; Glander, this volume), have an earlier age at first reproduction (3.5 years of age), and a shorter interbirth interval (19.9 months). Moreover, adult male white-faced capuchins reach full adult body mass by age 10 (Fragaszy *et al.*, 2004), whereas adult male mantled howlers reach full adult body mass at age 5 (Glander, this volume).

Capuchins and howlers also differ significantly in dietary profile. Mantled howlers are characterized by a slow rate of food passage, and exploit a diet principally of fruits, mature and immature leaves, and flowers (Milton, 1980, 1984; Estrada, 1984). Milton (1980) has referred to mantled howlers as behavioral folivores. Black howlers, studied in Belize and more recently in Mexico, are reported to have a similar diet (Silver *et al.*, 1988; Barrueta, 2003; Pavelka and Knopff, 2004; Rivera and Calme, this volume). In contrast, white-faced capuchins exploit a broader-based diet composed of soft fruits, hard fruits, palm nuts, shoots, flowers, leaves, vertebrates, invertebrates, eggs, and insect larvae (Fragaszy *et al.*, 2004). Hard fruits and seeds are often opened or broken by being pounded against a hard substrate (Panger, 1998). The exploitation of embedded resources, enhanced manipulative abilities, and large brain size has been associated with the evolution of complex cognitive skills in capuchins (Fragaszy *et al.*, 2004).

Given these distinctions in diet and life history traits, *Cebus* and *Alouatta* appear to have evolved very different adaptive solutions to ecological problems associated with exploiting a wide range of forest types. These may offer advantages in persisting in highly fragmented landscapes. *Alouatta* may survive in small forest patches by adopting an energy-minimizing foraging pattern (small day range, small home range, and extended periods of rest) associated with the consumption and fermentation of leaves during fruit-limited periods of the year (Milton, 1980), and by expanding the spectrum of plant species used as sources of food (Estrada *et al.*, 1999; García del Valle *et al.*, 2001; Gonzáles-Picazo *et al.*, 2001; Bicca-Marques, 2003; Fuentes *et al.*, 2003). *Cebus* may survive in highly fragmented landscapes by exploiting an extremely broad-based diet including hard-to-locate resources, and by traveling on the ground between forest patches or by using man-made landscapes as stepping-stones or corridors (DeGamma-Blanchet and Fedigan, this volume; Estrada *et al.*, this volume). However, when in proximity to human settlements, capuchins are sometimes considered pests because they raid agricultural crops and gardens (Estrada *et al.*, this volume). Under similar circumstances, howler monkeys may be able to co-exist in a more commensurate relationship with humans (Estrada *et al.*, this volume).

USE OF NEW TECHNOLOGIES AND THEORETICAL MODELS TO STUDY MESOAMERICAN PRIMATES

A second goal of papers in this volume was to use traditional and new technologies and analytical tools to investigate current problems in primate behavioral ecology and conservation. A major question in conservation management is the relationship between the health status and persistence of primate populations living in fragmented landscapes. One measure of persistence and sustainability is censusing the size, composition, age structure, and density of primate groups in forests that vary in size, ratio of edge to interior, and degree and type of disturbances. Several papers in the volume present these important data. In addition, levels of stress hormones such as cortisol obtained from fecal samples and information on parasite loads represent equally important indicators of population health and viability (Stoner and Gonzlez Di Pierro, this volume). Primates can maintain heavy parasite loads that compromise their immune system and reproductive fitness in the absence of outwardly visible indicators of poor health (Gillespie *et al.*, 2004). Information on parasite inventories and measures of parasite incidence or load in primate populations in continuous forests can provide the needed baseline information against which we can compare populations of the same species existing in fragmented landscapes. The study by Stoner and González Di Pierro on *A. pigra* in this volume is a good case in point, stressing the need for more studies on the parasite ecology of Mesoamerican primates.

Recent advances in reproductive endocrinology allow field researchers to non-invasively obtain information on ovarian function, mating during non-fertile periods, reproductive suppression, and female mate choice in primates by measuring steroid hormone levels in feces (Carnegie *et al.*, this volume). Previously, such studies were restricted to captive primates housed in controlled settings. Non-conceptive matings have been reported in several species of capuchins and tamarins (Manson *et al.*, 1997; Carnegie *et al.*, this volume; Garber, 1997). In the case of capuchins, non-conceptive matings have been suggested to reflect a reproductive strategy used by females to discourage infanticide (Fedigan, 2003). In tamarins, non-conceptive mating has been suggested to reinforce a socio-sexual bond between group males and the breeding female to insure male care-giving behavior (Garber, 1997).

The use of non-invasive techniques to monitor steroid hormonal levels in wild primates will play an increasingly important role in assessing fertility and reproductive seasonality in primates inhabiting environments differentially altered by human modification. For example, Van Belle and Estrada (this volume) report that the mean population density of *A. pigra* was significantly higher in forest fragments than in extensive forests, suggesting crowding and possibly populations living above sustainability thresholds in these forest fragments. However, what remains to be determined is whether population density is a reliable measure of population health and habitat quality (as often assumed), or whether forest fragments contain "refuge" populations in which individuals are characterized by lower fertility, greater parasite loads, and experience higher levels of stress.

The ongoing work at Monkey River, Belize, integrates the investigation of fecal parasites and stress hormones in determining primate densities (see Pavelka and Chapman, this volume). In the future, the implantation of biotelemetry devices in wild primates will also contribute to this endeavor. For example, Susan Williams (pers. comm.) is conducting pioneering field research using biotelemetry to determine the range of mechanical demands placed on the masticatory system of mantled howler monkeys when naturally exploiting resources in a Costa Rican forest. Collaborative studies designed to collect complementary data on behavior, diet, habitat utilization, demography, reproductive endocrinology, population persistence, and population health are needed for effective management of primates and the ecosystems they inhabit.

The use of non-invasive techniques to extract nuclear and mitochondrial DNA from the roots of shed or collected hair, and epithelial cells expelled in feces now allow field primatologists, in conjunction with geneticists, to conduct studies of genetic variation, gene flow, and paternity in wild primates. DNA profiles derived from microsatellite markers facilitate the identification of individuals, their contribution to the gene pool of the population, an assessment of mate choice (see Jack and Fedigan, this volume, for an example with *Cebus*), dispersal patterns and kinship, and genetic variation and genetic distances in primate populations (see García del Valle, 2004 for an example with populations of *A. pigra*). In adopting these analytical techniques, Mesoamerican primate research is beginning to achieve an ever increasing level of precision in addressing questions linking observed behavioral patterns and social networks with individual reproductive success, as well as a greater understanding of how the demographic and genetic features of individual populations vary in response to alternative ecological conditions (e.g. continuous versus fragmented forests).

Mesoamerican primate research also has taken a leading role in using experimental field approaches to examine questions concerning cognition, decision-making, and sensory adaptations in non-human primates (Garber and Brown, this volume; Garber, 2000; Garber and Brown, 2004). Experimental field studies build on the strengths of laboratory and traditional field investigations by presenting wild primates with social and ecological problems analogous to those they naturally encounter, but under systematic and controlled conditions. By varying temporal, spatial, and quantity information available to a forager, the researcher can test hypotheses concerning the degree to which certain cues are more salient than others in the decision-making process, as well as evidence of age- or sex-based differences in cognitive ability (Bicca-Marques and Garber, 2005).

As indicated earlier, Mesoamerican primates are characterized by significant differences in developmental trajectories. Papers by Bezanson (this volume) and MacKinnon (this volume) offer critical frameworks for examining the ontogeny of diet, foraging, and locomotor behavior in howlers and capuchins. There is evidence that, in some primate species, neuromuscular development associated with locomotor skills may have become dissociated from neuromuscular development required for fine motor control, extractive foraging, prey manipulation, or object manipulation. In the case of tamarin monkeys, individuals reach locomotor independence by 3 months of age, but are still provisioned with insects by adult caretakers at 9 months of age (Garber and Leigh, 1997). In other primate species, locomotor skills and fine manipulative skills appear to develop early and at approximately the same stage of development (i.e. *S. oerstedii*; Boinski and Fragaszy, 1989). In this regard, studies of primate locomotion, diet, and cognition should be placed within the context of primate life history strategies and evaluated in terms of patterns of somatic and neural growth and development, age-related survivorship, and the requirements of efficiently exploiting particular resources and forest habitats (Garber, in press).

Finally, the expanded use of Geographic Information Systems (GIS) and remotely sensed (RS) satellite data represent critical conservation and research tools for identifying vegetation types and for documenting changes in vegetation cover at various landscape scales over time. Such landscape changes can be caused by natural events (e.g. hurricanes and fires) or by human activity (e.g. mining, oil exploration, timber extraction, among others). In our volume, the chapter by Alexander *et al.* on the impact of Hurricane Iris on the habitat of a population of black howler monkeys in coastal Belize serves as a case study of how GIS can be applied to evaluate habitat change and its effect on primate populations.

Continued work linking satellite imagery, forest cover, vegetation types, habitat fragmentation, climate, topography, human land-use patterns, and relationships between areas of human population centers, ecotourism, and primate survivorship is needed. Progress in this direction is being made through the unfolding of the Mesoamerican Biological Corridor Project (see next section of this chapter), where remote sensing is being used to map the current system of natural protected areas in each Mesoamerican country and to project, in intermediate areas, the vegetation corridors that could be protected and/or established to enhance long-term species viability. Moreover, such information coupled with sorely needed surveys of primate population in many localities can update our "maps" regarding the current distribution of species and their populations in Mesoamerica (see Serio-Silva *et al.*, this volume). Thus the combination of layered data sets containing information on primate population distribution, land-use patterns, human settlements, geological and climatological features, and vegetation types can provide the diagnostics required to identify "hot spots" of conservation or risk for individual primate populations, or species in particular countries, or geographic localities in the region.

KEY ISSUES IN MESOAMERICAN PRIMATE CONSERVATION

Chapters in this volume have identified major issues and priorities for the conservation of Mesoamerican primates. We summarize these below.

Negative Impact of Land-Use Patterns Upon the Persistence of Tropical Rain Forest Vegetation and Primate Populations in the Region

Currently, only 30% of the original forest cover remains in Mesoamerica. Deforestation continues to fragment forested landscapes and this constitutes an important pressure upon extant primate populations. Countries such as Belize, Honduras, Nicaragua, and Guatemala contain the largest extensions of forest vegetation in their territories, but are also countries with the highest deforestation rates (Estrada *et al.*, Chapter 1). We continue to lack systematic and updated information regarding the current distribution of primate species and populations for these countries. Belize, Honduras, Nicaragua, and Guatemala represent a priority in primate conservation research.

Positive Impact of Some Agricultural Practices Upon the Persistence of Primate Populations

There is a general perception that agricultural activities are the principal threat to biodiversity in the tropics and a major cause of local extinctions, including primates. Such a binary view perceives conservation as a conflict between agriculture and tropical rain forests. The investigations of primates in agro-ecosystems reported in this book for landscapes in Mexico, Guatemala, and Costa Rica and by others elsewhere (Estrada and Coates-Estrada, 1996; McCann *et al.*, 2003; Harvey *et al.*, 2004), suggest that there is an alternative view that needs to be considered. Using a landscape perspective allows one to focus on the interactions among forests, agro-ecosystems, and the needs of the human population as important components in the conservation equation. The concurrent cultivation in some Mesoamerican localities of shaded and unshaded arboreal crops has resulted in fragmented landscapes that, in some cases, seem to contribute to the persistence of primate populations. These situations merit further investigation, as they open the possibility of enhancing the conservation of primates in human-modified landscapes. The landscape view also requires that attention be placed on investigating ways in which local subsistence economies can be diversified, involving the participation of multidisciplinary research teams. It also stresses the need to document the economic and ecological benefits for people of maintaining land-use patterns in which heterogeneous landscapes containing arboreal crops may play an important role in the persistence of primate populations and species.

Expanding Human Population

Environmental pressures on native vegetation in the region also come from an expanding human population. Mesoamerica is characterized by a high growth rate of 3% per year, and an expected doubling of current population from 45 to 84 million people in the next 25–35 years. This, combined with extreme poverty in the majority of the population, exerts direct pressure on land-use and the quality of life of the human inhabitants. Primate conservation research must consider the needs of rural people and indigenous populations in developing viable and individual conservation plans for the various regions of Mesoamerica. This needs to be integrated into educational outreach programs to stress the fact that the primate fauna is an integral part of the cultural and natural patrimony of the people of this region.

Impact of Natural Events on Primate Distribution and Density

Specific localities of Mesoamerica are regularly or occasionally affected by hurricanes, volcanic activity, earthquakes, torrential rains, flooding, and other natural events which are likely to have an important ecological impact on primate habitats, primate population dynamics (including human primates), and population viability. In conjunction with anthropogenic disturbance, these events may contribute significantly to habitat loss and fragmentation, and forest degradation, with direct impacts on primate population health and survivorship. Understanding the impact of these natural events remains an important issue in Mesoamerican primate conservation.

Economic Incentives for the Conservation of Primate Populations

Economic incentives can play an important role in primate conservation. These incentives relate to specific patterns of land-use that currently exist in several regions and landscapes across Mesoamerica. In Mexico, Belize, Guatemala,

Honduras, and El Salvador the government protects forested areas that harbor Maya archeological remains. There also exist ecological reserves dedicated to research and/or ecotourism in every Mesoamerican country. Excellent examples of these are the Los Tuxtlas field station in Mexico, Barro Colorado Island in Panama, and La Selva field station in Costa Rica. Recently, there have been several attempts to preserve forest habitats and generate revenue by developing educational field courses for university students and biological field stations to promote primate research in northeastern Costa Rica (e.g. La Suerte Biological Field Station), Isla de Ometepe, Nicaragua (Ometepe Biological Field Station), and Bocas del Toro, Panama (ITEC).

Conservation Initiatives by Mesoamerican Countries

In spite of poverty, overcrowding, and underdevelopment, the countries of Mesoamerica have expressed great concern over the need to conserve their biodiversity. Between 1993 and 1994, all Mesoamerican countries ratified the international convention on biological diversity, which led to the consolidation of existing protected areas and the creation of new, naturally protected areas in each country. There currently exists a total of 420 protected areas, encompassing about 15 million ha, or about 20% of the area of Mesoamerica. Mesoamerican countries have gone one step further with the interest of protecting their biodiversity, while at the same time improving the quality of life for their rural populations. The result of such action is the Mesoamerican Biological Corridor (MBC) project, a unique program in Latin America linking conservation efforts by several governments. Each nation has proposed a system of corridors that will connect the existing system of naturally protected areas. This will serve to avoid habitat fragmentation and isolation, enhance the viability of species and populations, and promote sustainable use of the land and forest remnants in intermediate areas. The MBC project completed a 5-year-long diagnostic phase in 2004, and will proceed to a phase of consolidating agreements (paralleled by field projects) among Mesoamerican countries, with the general goal of "improving the connectivity of ecosystems, the sustainable use of the land and the services generated for the region's development" (CBM, 2004). The MBC project has the potential to enhance the persistence of primate species and populations and their habitats throughout the region. However, mapping the location and state of conservation of such species and populations within this framework is still a task to be accomplished.

Contribution by Primatologists to Conservation

Primatologists have been investigating primate species and populations in the region since the 1930s and thus have a critical role to play in Mesoamerica. Their contribution over the last 70 years has focused mainly on providing documentation of the natural history, ecology, behavior, and evolutionary history of Mesoamerican primates. This has resulted in a large body of scientific and technical literature on the primates of the region. In spite of these efforts, however, we still lack sufficient and current information on several species. For example, the absence of individual chapters in this volume dedicated to species such as *A. zonalis*, *S. geoffroyi*, and *S. oerstedii* are a clear indication that much work needs to be done. Studies of these primates are a research priority, as no systematic and detailed field studies have been published on any of these species since the 1980s.

While primatologists have contributed detailed longitudinal information on primate life-history traits that further our understanding of primate biology, ecology, and behavior, and on the plasticity of responses primates show to various environmental conditions, success in translating these efforts and information into conservation initiatives has been more limited. In this regard, we call upon primate researchers in Mesoamerica to focus on the empirical, conceptual, and theoretical tools needed to develop explicit conservation recommendations for individual primate populations, individual primate species, and threatened habitats.

Finally, it is important to point out that despite seven decades of field research in Mesoamerica by primatologists, many of the countries in the region continue to lack trained primate specialists. Mexico is the only Mesoamerican country that maintains a small contingent of professionally trained primatologists, a possible reason for the rapid increase in field data and publications on primate species and populations in that country over the last two decades (Estrada and Mandujano, 2003). Most Mesoamerican countries lack primate scientists native to the region, and therefore experience great difficulty in sustaining long-term conservation initiatives and research. Countries such as Nicaragua, Honduras, and El Salvador are still in the earliest stages of collecting basic information on the presence, location, and viability of their primate populations. And, as forests continue to be cut and non-human primates continue to be captured as pets or hunted for food, we face greater and greater challenges in developing an effective plan of conservation for the Mesoamerican region. However, political stability, revenues generated through ecotourism, and a generation of young Mesoamerican scientists offer hope that effective changes to conservation policy and increases in financial resources devoted to conservation efforts become national priorities.

It is the hope of the editors and the contributors to this volume that we have identified critical, new, and important issues in primate research and conservation, allowing the reader to achieve a greater level of understanding, and integration of Mesoamerican primate taxonomy, biogeographic history, behavior, ecology, and conservation. We emphasize that the human population of Mesoamerica has for several thousand years, and continues to this day, to be an important component of the tropical ecosystems and must be considered when developing conservation strategies to insure the persistence of primate species and populations. The impact of humans on the native ecosystems is likely to be more pervasive and harmful today than in the distant past. Therefore, primatologists must pay special attention to the social, economic, and political forces at play in the region. This includes consideration of the need for sustainable land-use patterns and equity for the human population. Human and non-human primates have coexisted in Mesoamerica for thousands of years. We are hopeful that the Mesoamerican landscape can sustain the needs of human primates and the needs of non-human primates in an equitable way. In our view, it is imperative that we perceive primate research, not only as a way to enhance scientific knowledge, but also as means to insure the conservation of the natural and cultural patrimony of the nations in this biologically important region of the world.

ACKNOWLEDGMENTS

We are grateful to each contributing author for their efforts and enthusiasm in translating ideas, data, theories, and perspectives into a single volume. We are also grateful to the American Society of Primatologists. The ASP played an important role in supporting the symposium from which the book project originated, at their annual meetings at the University of Wisconsin–Madison in June of 2004. We also acknowledge the scholarship of participants at the ASP symposium on Mesoamerican primates, some of whom, although not contributors to the volume, nonetheless provided important intellectual insights. We recognize the efforts of several graduate students who participated as coauthors in some of the chapters in this book. The quality of their contributions attests to their high level of scientific investigation, and to their enthusiasm in conducting field research in the spirit of early Mesoamerican primatologists such as C. R. Carpenter and Ch. Southwick. As always, PAG thanks Sara and Jenni for being Sara and Jenni. My hope is that their children will live in a world that continues to contain large and viable non-human primate populations and sustainable tropical forests in Mesoamerica, South America, Africa, and Asia. AE is grateful to Dr. Karen Strier, from the Department of Anthropology, University of Wisconsin-Madison, for supporting his sabbatical year in her laboratory, a period during which the book project was developed. Further acknowldgement by AE goes to Dr. Joseph W. Kemnitz, Director of the Wisconsin Regional Primate Center, for supporting the library work needed during the sabbatical, and in the book project. Lastly, AE recognizes the economic support of the National Autonomous University of Mexico, and of the Scott Neotropic Fund in the development of the ASP symposium and the book project.

REFERENCES

Barrueta, T. 2003, Reconocimiento demográfico y dieta de *Alouatta pigra* en un fragmento de selva en El Tormento, Campeche, México. MSc Thesis, El Colegio de la Frontera Sur, México, p. 49.

Bicca-Marques, J. C. 2003, How do howler monkeys cope with habitat fragmentation? in: L. K. Marsh, ed., *Primates in Fragments: Ecology and Conservation*, Kluwer Academic/Plenum Publishers, New York, pp. 283–303.

Bicca-Marques, J. C. and Garber, P. A. 2005, Use of social and ecological information in tamarin foraging decisions. *Int. J.Primatol.* 26:

Boinski, S. 1986, *The Ecology of Squirrel Monkeys in Costa Rica*. PhD Thesis, University of Texas, Austin.

Boinski, S. 1987a, Birth synchrony in squirrel monkeys (*Saimiri oerstedii*). *Behav. Ecol.Sociobio.* 21:283–310.

Boinski, S. 1987b, Mating patterns in squirrel monkeys (*Saimiri oerstedii*). *Behav. Ecol.Sociobiol.* 21:13–21.

Boinski, S. 1987c, Habitat use by squirrel monkeys (Saimiri oerstedi) in Costa Rica. *Folia Primatol.* 49:151–167.

Boinski, S. 1992, Olfactory communication among Costa Rican squirrel monkeys: A field study. *Folia Primatol.* 59:127–136.

Boinski, S. 1994, Affiliation patterns among male Costa Rican squirrel monkeys. *Behaviour* 130:191–209.

Boinski, S. and Fragaszy, D. M. 1989, The ontogeny of foraging in squirrel monkeys, *Saimiri oerstedii. Anim.Behav.* 37:415–428.

Cant, J. H. G. 1986, Locomotion and feeding postures of spider and howling monkeys: Field study and evolutionary interpretation. *Folia Primatol.* 46:1–14.

Carpenter, C. R. 1934, A field study of the behavior and social relations of howling monkeys. *Comp.Psychol.Monogr.* 10, 48:1–168.

CBM. 2004, Resumen corredor biologico Mesoamericano-CBM programa estrategico regional. Draft. CBM, CCAD and SICA, 6 October, 2004, Managua, Nicaragua.

Chapman, C. A. 1987, Flexibility in diets of three species of Costa Rican primates. *Folia Primatol.* 49:90–105.

Chapman, C. A. 1988a, Patterns of foraging and ranging use by three species of neotropical primates. *Primates* 29:177–194.

Chapman, C. A. 1988b, Patterns of foraging and range use by three species of Neotropical primates. *Primates* 29:177–194.

Chapman, C. A. and Peres, C. A. 2001, Primate conservation in the new millennium: The role of scientists. *Evol.Anthropol.* 10:16–33.

Coelho, A. M. Jr., Coelho, L., Bramblett, C., Bramblett, S., and Quick, L. 1976, Ecology, population characteristics, and sympatric associations in primates: A socioenergenetic analysis of howler and spider monkeys in Tikal, Guatemala. *Yearbk.Phys.Anthropol.* 20:96–135.

Dawson, G. A. 1975, Behavioral ecology of the Panamanian tamarins, *Saguinus oedipus* (Callitrichidae, Primates). PhD Thesis, East Lansing, Michigan State University, Michigan.

Duarte, A. and Estrada, A. 2003, Primates as pets in Mexico City: An assessment of species involved, source of origin and general aspects of treatment. *Am.J.Primatol.* 61:53–60.

Emmons, L. 1990, *Neotropical Rainforest Mammals.A Field Guide.* The University of Chicago Press, Chicago.

Estrada, A. 1984, Resource use by howler monkeys (*Alouatta palliata*) in the rainforest of Los Tuxtlas, Veracruz, Mexico. *Int.J.Primatol.* 5:105–131.

Estrada, A. and Coates-Estrada, R. 1993, Aspects of ecological impact on howling monkeys (*Alouatta palliata*) on their habitat: A review, in: A. Estrada, E. Rodriquez Luna, R. Lopez-Wilchis, and R. Coates-Estrada, eds., Avances en: Estudios Primatologicos en Mexico I. Asociacion Mexicana de Primatologia, A.C. y Patronatto Pro-Universidad Veracruzana, A.C. Xalapa, Veracruz, Mexico, pp. 87–117.

Estrada, A. and Coates-Estrada, R. 1996, Tropical rain forest fragmentation and wild populations of primates at Los Tuxtlas. *Int.J.Primatol.* 15:759–783.

Estrada, A., Juan Solano, S., Ortíz Martínez, T., and Coates-Estrada, R. 1999, Feeding and general activity patterns of a howler monkey (*Alouatta palliata*) troop living in a forest fragment at Los Tuxtlas, Mexico. *Am. J. Primatol.* 48:167–183.

Estrada, A. and Mandujano, S. 2003, Investigaciones con *Alouatta* y *Ateles* en Mexico. *Neotrop.Primates* 11(3):147–156.

Fedigan, L. 2003, Impact of male takeovers on infant deaths, births and conceptions in *Cebus capucinus* at Santa Rosa, Costa Rica. *Int.J.Primatol.* 24:723–741.

Fedigan, L. M. and Rose, L. M. 1995, Interbirth interval variation in three sympatric species of neotropical monkeys. *Am.J.Primatol.* 37:9–24.

Ford, S. and Davis, L. 1992, Systematics and body size: Implications for feeding adaptations in New World monkeys. *Am.J.Phys.Anthropol.* 88:415–468.

Fragaszy, D. M., Visalberghi, E., and Fedigan, L. M. 2004, *The Complete Capuchin: The Biology of the Genus Cebus.* Cambridge University Press, Cambridge.

Fuentes, E., Estrada, A., Franco, B., Magaña, M., Decena, Y., Muñoz, D., and García, Y. 2003, Reporte preliminar sobre el uso de recursos alimenticios por una tropa de monos aulladores, *Alouatta palliata*, en El Parque La Venta, Tabasco, México. *Neotrop.Primates* 11:24–29.

Garber, P. A. 1980, Locomotor behavior and feeding ecology of the Panamanian tamarin (*Saguinus oedipus geoffroyi*, Callitrichidae, Primates). *Int.J.Primatol.* 1:185–201.

Garber, P. A. 1984a, Use of habitat and positional behavior in a neotropical primate, *Saguinus Oedipus*, in: P. S. Rodman and J. G. H. Cant, eds., *Adaptations for Foraging in Nonhuman Primates*, Columbia University Press, New York, pp. 112–133.

Garber, P. A. 1984b, The proposed nutritional importance of plant exudates in the diet of the Panamanian tamarin, *Saguinus oedipus geoffroyi*. *Int.J.Primatol.* 5:1–15.

Garber, P. A. 1997, One for all and breeding for one: Cooperation and competition as a tamarin reproductive strategy. *Evol.Anthropol.* 5:135–147.

Garber, P. A. 2000, Evidence for use of spatial, temporal, and social information by primate foragers, in: S. Boinski and P. A. Garber, eds., *On the Move: How and Why Animals Travel in Groups*, University of Chicago Press, Chicago, IL, pp. 261–298.

Garber, P. A. Primate locomotor behavior and ecology, in: S. Bearder, C. J. Campbell, A. Fuentes, K. C. MacKinnon, and M. Panger, eds.,*Primates in Perspective*, Oxford University Press, Oxford, in press.

Garber, P. A. and Brown, E. 2004, Wild capuchins (*Cebus capucinus*) fail to use tools in experimental field study. *Am.J.Primatol.* 62:165–170.

Garber, P. A. and Leigh, S. R. 1997, Ontogenetic variation in small-bodied New World primates: Implications for patterns of reproduction and infant care. *Folia Primatol.* 68:1–22.

García del Valle, Y. 2004, Efecto de la fragmentación en la variación genética de poblaciones de monos aulladores *Alouatta pigra* en la selva Lacandona, Chiapas. México. Master's Thesis, Colegio de la Frontera Sur, San Cristóbal de las Casas, Chiapas.

García del Valle, Y., Muñoz, D., Estrada, A., Franco, B., and Magaña, M. 2001, Uso de plantas como alimento por monos aulladores, *Alouatta palliata*, en el parque Yumká, Tabasco, México. *Neotrop.Primates* 9:112–118.

Gillespie, T. R., Greiner, E. C., and Chapman, C. A. 2004, Gastrointestinal parasites of the guenons of western Uganda. *J.Parasitol.* 90: 1356–1360.

González-Picazo, H., Estrada, A., Coates-Estrada, R., and Ortíz-Martínez, T. 2001, Consistencias y variaciones en el uso de recursos alimentarios utilizados por una tropa de monos aulladores (*Alouatta palliata*) y deterioro del habitat, en Los Tuxtlas, Veracruz, Mexico. *Universidad y Ciencia* 17:27–36.

González-Zamora, A. and Mandujano, S. 2003, Uso de fragmentos por *Ateles geoffroyi* en el sureste de México. *Neotrop.Primates* 11:174–177.

Harvey, C., Tucker, N., and Estrada, A. 2004, Can live fences, isolated trees and windbreaks help conserve biodiversity within fragmented tropical landscapes? in: G. Schroth, G. Fonseca, C. Gascon, H. Vasconcelos, A. M. Izac, and C. Harvey, eds., *Agroforestry and Conservation of Biodiversity in Tropical Landscapes*, Island Press Inc., New York, pp. 261–289.

Hershkovitz, P. 1977, *Living New World Monkeys (Platyrrhini)*, Vol. 1. University of Chicago Press, Chicago.

Hladik, C. M., Hladik, A., Bousset, J., Valdebouze, P., Viroben, G., and Delort-Laval, J. 1971, La regime alimentaire des primates de l'ile de Barro-Colorodo (Panama). Resultats des analyses quantitatives. *Folia Primatol.* 16:85–122.

Janson, C. H. and Boinski, S. 1992, Morphological and behavioral adaptations for foraging in generalist primates: The case of the cebines. *Am.J.Phys.Anthropol.* 88:483–498.

Kinzey, W. G. 1997, *New World Primates. Ecology,Evolution and Behavior.* Aldine Press, New York.

Lambert, J. E. and Garber, P. A. 1998, Ecological and evolutionary implications of primate seed dispersal. *Am.J.Primatol.* 45:9–28.

Manson, J. H., Perry, S., and Parish, A. R. 1997, Nonconceptive sexual behavior in bonobos and capuchins. *Int.J.Primatol.* 18:767–786.

McCann, C., William-Guillen, K., Koontz, F., Roque, A., Martinez, J. C., and Koontz, C. H. 2003, Shade coffee plantations as wildlife refuges for mantled howler monkeys (*Alouatta palliata*) in Nicaragua, in: L. K. Marsh, ed., *Primates in Fragments: Ecology and Conservation*, Kluwer Academic/Plenum Publishers, New York, pp. 321–340.

Milton, K. 1980, *The Foraging Strategy of Howler Monkeys: A Study in Primate Economics.* Columbia University Press, New York.

Milton, K. 1981a, Diversity of plant foods in tropical forests as a stimulus to mental development in primates. *Am.Anthropol.* 83:534–548.

Milton, K. 1981b, Food choice and digestive strategies of two sympatric primate species. *Am.Nat.* 117:476–495.

Milton, K. 1984, The role of food-processing factors in primate food choice, in: P. S. Rodman and J. G. H. Cant, eds., *Adaptations for Foraging in Nonhuman Primates*, Columbia University Press, New York, pp. 249–279.

Morales H. K. 2002, Wild populations of spider monkeys (*Ateles geoffroyi*) in El Salvador, Central America. *Neotrop.Primates* 10:153–154.

Moynihan, M. 1964, Some behavior patterns of platyrrhine monkeys 1. The night monkey (*Aotus trivirgatus*). *Smithson.Misc.Coll.* 146:1–84.

Murcia, C. 1995, Edge effects in fragmented forests: Implications for conservation. *Trends Ecol.Evol.* 10:58–62.

Panger, M. 1998, Object-use in free-ranging white-faced capuchins (*Cebus capucinus*) in Costa Rica. *Am.J.Phys.Anthropol.* 106:311–321.

Pavelka, M. S. M. and Knopff, K. 2004, Diet and activity in *A. pigra* in Southern Belize: Does degree of frugivory influence activity level? *Primates* 45:105–112.

Reid, F. 1997, *A Field Guide to the Mammals of Central America and Southeast Mexico.* Oxford University Press, London.

Restrepo, C., Gomez, N., and Heredia, S. 1999, Anthropogenic edges, treefall gaps, and fruit–frugivore interactions in a neotropical montane forest.*Ecology* 80:668–685.

Rowe, N. 1996, *A Pictorial Guide to the Living Primates.* Pogonias Press, Charlestown, RI.

Silver, S. C., Ostro, L. E. T., Yeager C. P., and Horwich R. 1988, Feeding ecology of the black howler monkey (*Alouatta pigra*) in northern Belize. *Am.J.Primatol.* 45:263–279.

Thorington, R. W. Jr., Muchenhirn, N. A., and Montgomery, G. G. 1976, Movements of a wild night monkey, *Aotus trivirgatus*, in: R. W. Thorington Jr. and P. G. Heltne, eds., *Neotropical Primates*, National Academy of Sciences, Washignton, D.C., pp. 32–34.

Van Roosmalen, M. G. M. and Klein, L. L. 1988, The spider monkeys, genus *Ateles*, in: R. A. Mittermeier, A. B. Rylands, A. F. Coimbra-Filho, and G. A. B. da Fonseca, eds.,*Ecology and Behavior of Neotropical Primates*, Vol. 2, World Wildlife Fund, Washington, D.C., pp. 455–537.

Williams-Linera, G. 1990, Vegetation structure and environmental conditions of forest edges in Panama. *J.Ecol.* 78:356–373.

Wright, P. C. 1981, The night monkeys, genus *Aotus*, in: A. F. Coimbra-Filho and R. A. Mittermeier, eds., *Ecology and Behavior of Neotropical Primates*, Vol. 1, Avademia Brasileira de Ciencias, Rio De Janeiro, pp. 211–240.

Wright, P. C. 1989, The nocturnal primate niche in the New World. *J.Human Evol.* 18:635–658.

SPECIES INDEX

Page numbers in *italics* = in figure; page numbers in **bold** = in table

Primates

Other Animals

Plants

SUBJECT INDEX

Page numbers in **bold** = in table; page numbers in *italic* = in figure.